JOHN DEERE

SHOP MANUAL JD-201

Series ■ 40 ■ 320 ■ 330 ■ 420 ■ 430 ■ 440

Models ■ 435D ■ 440ID

Series ■ 820 (3 Cyl.) ■ 830 (3 Cyl.)

Models ■ 720 Diesel ■ 730 Diesel

Models ■ 80 ■ 820 ■ 830 (2 Cyl. Diesel Models)

I&T SHOP MANUALS

Information and Instructions

This shop manual contains several sections each covering a specific group of wheel type tractors. The Tab Index on the preceding page can be used to locate the section pertaining to each group of tractors. Each section contains the necessary specifications and the brief but terse procedural data needed by a mechanic when repairing a tractor on which he has had no previous actual experience.

Within each section, the material is arranged in a systematic order beginning with an index which is followed immediately by a Table of Condensed Service Specifications. These specifications include dimensions, fits, clearances and timing instructions. Next in order of arrangement is the procedures paragraphs.

In the procedures paragraphs, the order of presentation starts with the front axle system and steering and proceeding toward the rear axle. The last paragraphs are devoted to the power take-off and power lift sys-

tems. Interspersed where needed are additional tabular specifications pertaining to wear limits, torquing, etc.

HOW TO USE THE INDEX

Suppose you want to know the procedure for R&R (remove and reinstall) of the engine camshaft. Your first step is to look in the index under the main heading of ENGINE until you find the entry "Camshaft." Now read to the right where under the column covering the tractor you are repairing, you will find a number which indicates the beginning paragraph pertaining to the camshaft. To locate this wanted paragraph in the manual, turn the pages until the running index appearing on the top outside corner of each page contains the number you are seeking. In this paragraph you will find the information concerning the removal of the camshaft.

More information available at Clymer.com
Phone: 805-498-6703

Haynes Publishing Group
Sparkford Nr Yeovil
Somerset BA22 7JJ England

Haynes North America, Inc
859 Lawrence Drive
Newbury Park
California 91320 USA

ISBN-10: 0-87288-359-0
ISBN-13: 978-0-87288-359-8

JOHN DEERE
(PREVIOUSLY JD-15)

Series ■ 40 ■ 320 ■ 330 ■ 420 ■ 430 ■ 440

SHOP MANUAL
JOHN DEERE

SERIES 40-320-330-420-430-440
MODEL IDENTIFICATION SUFFIX

H—Hi-Crop	T—Tricycle
I—Special Utility (420 only)	U—Utility
I—Industrial (440 only)	V—Special
S—Standard	W—Row Crop Utility

Tractor serial number is stamped on left side of center frame.

INDEX (By Starting Paragraph)

CONDENSED SERVICE DATA

GENERAL	Series 40	Series 320-330	Series 420-430	Series 440 Prior 448001	Series 440 After 448000
Engine Make	Own	Own	Own	Own	Own
Cylinders	2	2	2	2	2
Bore—Inches	4	4	4¼	4¼	4¼
Stroke—Inches	4	4	4	4	4
Displacement—Cubic Inches	100.5	100.5	113.3	113.3	113.3
Compression Ratio	6.5:1	(G)6.5:1(AF)4 7:1	(G)7.0:1(AF)5.15:1	(G)7.0:1(AF)5.15:1	(G) 7.5:1
Pistons Removed From?	Above	Above	Above	Above	Above
Main Bearings, Number of	2	2	2	2	2
Main Bearings, Adjustable?	No	No	No	No	No
Rod Bearings, Adjustable?	No	No	No	No	No
Cylinder Sleeved?	No	No	No	No	No
Forward Speeds	4	4	4 or 5	4 or 5	4 or 5
Carburetor Make	M-S	M-S	M-S	M-S	M-S
Carburetor Model (gas)	TSX530	TSX245	TSX641	TSX756	TSX777
Carburetor Model (all-fuel)	TSX562	TSX245	TSX678	TSX768	......
Carburetor Model (LP-Gas)	......	......	TSG80S	......	......
Convertor Model (LP-Gas)	......	......	UT725	......	......
Distributor Model (Delco-Remy)	1111709	1111709	1112571	1112571	1112571
Generator Model (Delco-Remy)	1101859	1101859	1101859	1101859	1101859
Starter Model (Delco-Remy)	1107127	1107127	1107127**	1107165	1107165
Regulator Model (Delco-Remy)	1118308	1118780	1118780	1118780	1118780
Electrical System—Voltage	6	6	6	6	6
Battery Ground Polarity	Positive	Positive	Positive	Positive	Positive
TUNE-UP					
Compression—Gage Pounds	(G)125	(G)110(AF)90	(G)135(AF)100	(G)135(AF)100	(G)145
Tappet Gap—Inlet and Exhaust	0.012C	0.012C	0.015C	0.015C	0.015C
Inlet Valve Face Angle	30°	30°	29°	29°	29°
Exhaust Valve Face Angle	45°	45°	44°	44°	44°
Inlet Valve Seat Angle	30°	30°	30°	30°	30°
Exhaust Valve Seat Angle	45°	45°	45°	45°	45°
Timing Mark Location	Flywheel	Flywheel	Flywheel	Flywheel	Flywheel
Advance Timing Degrees	25	31	28	28	28
Breaker Point Gap	0.022	0.022	0.022	0.022	0.022
Spark Plug Size	14 mm	14 mm	14 mm	14 mm	14 mm
Electrode Gap	0.025	0.025	0.025	0.025	0.025
Engine High Idle—RPM	2025	1825	2025	2025	2200
Engine Loaded—RPM	1850	1650	1850	1850	2000
Belt Pulley Loaded—RPM	1270	1246	1270	1270	1373
Belt Pulley High Idle—RPM	1390	1376	1390	1390	1510
Power Shaft High Idle—RPM	612	608	612	612	665
Power Shaft Loaded—RPM	560	550	560	560	605
SIZES—CAPACITIES—CLEARANCES					
Crankshaft Journal Diameter	2.3975	2.3975	2.3975	2.3975	2.3975
Crankpin Diameter	2.4985	2.4985*	2.4985	2.4985	2.4985
Camshaft Journal Diameter	1.810	1.810	1.810	1.810	1.810
Piston Pin Diameter	1.1878	1.1878	1.1878	1.1878	1.1878
Valve Stem Diameter—Intake	0.3720	0.3720	0.3720	0.3720	0.3720
Valve Stem Diameter—Exhaust	0.3720	0.3720	0.3720	0.3720	0.3720
Compression Ring—Width	3/32-inch	3/32-inch	3/32-inch	3/32-inch	3/32-inch
Oil Ring—Width	1/4-inch	1/4-inch	1/4-inch	1/4-inch	1/4-inch
Main Bearing Running Clearance	0.001-0.0035	0.001-0.0035	0.001-0.0035	0.001-0.0035	0.001-0.0035
Rod Bearing Running Clearance	0.001-0.0035	0.001-0.0035	0.001-0.0035	0.001-0.0035	0.001-0.0035
Camshaft Bearings Running Clearance	0.0015-0.0035	0.0015-0.0035	0.0015-0.0035	0.0015-0.0035	0.0015-0.0035
Crankshaft End Play	0.003-0.007	0.003-0.007	0.003-0.007	0.003-0.007	0.003-0.007
Camshaft End Play	0.003-0.007	0.003-0.007	0.003-0.007	0.003-0.007	0.003-0.007
Piston Skirt Clearance	0.004-0.006	0.004-0.006	0.003-0.005	0.003-0.005	0.003-0.005
Cooling System	3½ Gallons	3½ Gallons	2¾ Gallons	2¾ Gallons	2¾ Gallons
Crankcase Oil	5 Quarts	5 Quarts	5 Quarts	5 Quarts	5 Quarts
Fuel Tank	10.5 Gallons	10.5 Gallons	10.5 Gallons	10.5 Gallons	10.5 Gallons
Starting Tank (all-fuel)	0.9 Gallons	0.9 Gallons	0.9 Gallons	0.9 Gallons	0.9 Gallons
Transmission and Differential	6½ Quarts***	6½ Quarts***	8 Quarts***	9½ Quarts	9½ Quarts
Final Drives—Each	See Note	See Note	See Note	3½ Quarts	3½ Quarts
Touch-O-Matic:					
Single	4½ Quarts	4½ Quarts	4½ Quarts	......	......
Dual	5½ Quarts	5½ Quarts	5½ Quarts	5½ Quarts	5½ Quarts
Belt Pulley	½ Pint	½ Pint	½ Pint	½ Pint	½ Pint
TIGHTENING TORQUE—FT.-LBS.					
Cylinder Head Bolts	100-110	100-110	85-95	85-95	85-95
Main Bearing Bolts	140	140	140	140	140
Rod Bearing Bolts	55-60	55-60	55-60	55-60	55-60
Rocker Arm Assembly Nuts	25-30	25-30	25-30	25-30	25-30
Flywheel Cap Screws	75-80	75-80	75-80	75-80	75-80

*Series 320 prior to serial number 320,885 is 2.2495.
**Series 420 after serial number 100,000, use model 1107165.
***40T, 7½ quarts; 330S and 330U, 8 quarts; 420I, 420T, 420W, 430T and 430W, 9½ quarts.

NOTE: 40H, 1½ pints; 40U, 320U, 330U, 420U and 430U, 2½ pints; 40S, 40W, 320S, 330S, 420S and 430S, 3½ pints; 40V, 420V and 430V, 1¾ quarts; 40T, 420H, 420T, 430H and 430T, 2 quarts; 420I, 420W and 430W, 3½ quarts.

3

FRONT SYSTEM

LOWER SPINDLE OR FORK
Models 40T-420T-430T

1. **R&R AND OVERHAUL.** The procedure for removing either the fork and wheel assembly on the single wheel version or the lower spindle and wheels assembly on the dual wheel version is evident after an examination of the unit and reference to Figs. JD800 and 800A. It will be noted, however, that the fork or spindle are retained by four heat treated Allen head cap screws (27). Ordinary cap screws should never be used at this point.

When installing the dual front wheels, tighten the bearing adjusting nut to a torque of 35-40 Ft.-Lbs., then back the nut off one castellation and insert the cotter pin.

To adjust the bearings on the fork mounted single front wheel, install more than enough shims (40—Fig. JD800A) and measure the end play of axle shaft (37). Then, remove a total thickness of shims equal to 0.000-0.003 more than the measured end play. For example, if the measured end play was 0.005, then remove 0.005-0.008 thickness of shims. This pro-

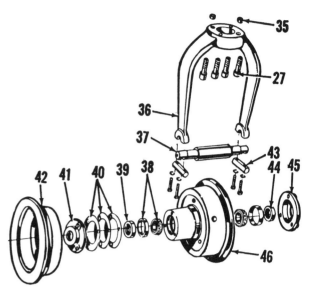

Fig. JD800A — Lower assembly of front end unit for Deere 40T, 420T and 430T tractors having a single front wheel. Wheel fork (36) is bolted to upper spindle in the same manner as the lower spindle (26) assembly shown in figure JD800.

27. Screws (heat-treated)
35. Dowels
36. Wheel fork
37. Wheel axle
38. Outer bearing
39. Grease seal
40. Adjusting shims
41. Bearing retainer
42. Rim half
43. Axle clamp
44. Oil seal
45. Bearing retainer
46. Hub & rim

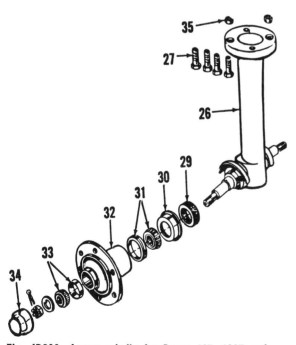

Fig. JD800—Lower spindle for Deere 40T, 420T and 430T tractors having dual front wheels. Item (26) is bolted to upper spindle by means of heat-treated screws (27) in the same manner as the tricycle fork (36) shown in figure JD800A.

26. Lower spindle
27. Screws (heat-treated)
29. Seal
31. Inner bearing
32. Wheel hub
33. Outer bearing
34. Hub cap
35. Dowels

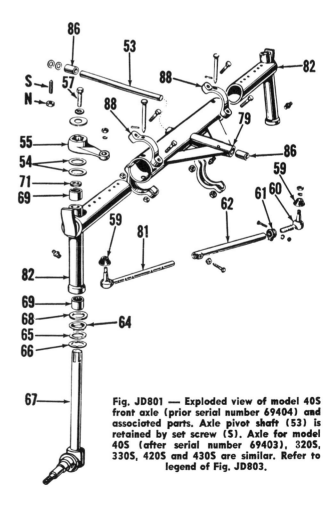

Fig. JD801 — Exploded view of model 40S front axle (prior serial number 69404) and associated parts. Axle pivot shaft (53) is retained by set screw (S). Axle for model 40S (after serial number 69403), 320S, 330S, 420S and 430S are similar. Refer to legend of Fig. JD803.

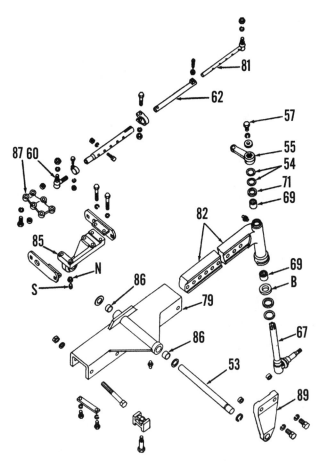

Fig. JD802–Exploded view of Model 40, 420 and 430 I-beam type adjustable front axle. The tubular type is shown in Fig. JD803. The 40W axle is similar. Refer to legent of Fig. JD803.

Notches in flange of knuckle. Install thrust washer (68) so that tangs of same fit into notches of vertical spindle tube. On I-beam type axles, install thrust bearing (B–Fig. JD802) with numbered side down. On all models, when installing the steering arms (55), vary the number of 0.010 and 0.027 thick shims (54) to give the spindle an end play of 0.000-0.030 and tighten cap screw (57) to a torque of 60-65 Ft.-Lbs.

When installing front wheels, tighten the bearing adjusting nuts to a torque of 35-40 Ft.-Lbs., then back the nut off one castellation and insert the cotter pin.

TIE RODS & TOE IN

All Models So Equipped

3. The procedure for removing the tie-rods and/or tie ends is evident. When re-assembling, vary the length of each tie-rod to obtain the recommended toe-in of $\frac{1}{8}$-$\frac{1}{2}$ inch.

NOTE: Steering gear must be in its mid, or high point position and front wheels must be poining straight ahead when making the toe-in adjustment. Be certain to adjust each tie-rod an equal amount.

AXLE PIVOT SHAFT AND BUSHINGS

All Models So Equipped

4. To renew the axle pivot shaft and bushings, first remove the axle assembly from tractor as follows:

cedure will give the taper roller bearings the desired pre-load of 0.000-0.003. Shims are available in thicknesses of 0.002, 0.005 and 0.010.

STEERING KNUCKLES

All Models So Equipped

2. **R&R AND OVERHAUL.** To remove the knuckles (67), shown in the accompanying illustrations, first support front of tractor and remove wheel and hub assemblies. Remove cap screws (57) and bump steering arms (55) from knuckles. Knuckles can be withdrawn from below and needle bearings (69) can be driven from the axle extensions.

New needle bearings should be pressed or driven into position using a special driver such as John Deere tool number AM-457T. Apply the driver against the end of the bearing which has the manufacturers number stamped on it. When reinstalling knuckles in tubular type axles on 40S prior to serial number 69404, make certain that tangs on thrust washer (64–Fig. JD801) fit through slots in seal retainer washer (66) and into

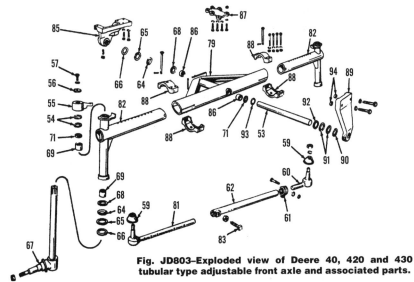

Fig. JD803–Exploded view of Deere 40, 420 and 430 tubular type adjustable front axle and associated parts.

53. Pivot shaft	65. Seal	85. Pivot shaft rear support
54. Shims	66. Seal retaining washer	86. Bushing
55. Steering arm	67. Spindle and knuckle	87. Center steering arm
56. Washer	68. Thrust washer	88. Clamp
57. Cap screw	69. Needle bearing	89. Pivot shaft front support
59. Dust cover	71. Felt seal	90. Snap ring
60. Tie rod end	79. Axle center member	91. Spring washer
61. Clamp	81. Adjustable tie rod end	92. Washer
62. Tie rod tube	82. Axle extension	93. Retainer
64. Thrust washer	83. Set screw	94. Dowels

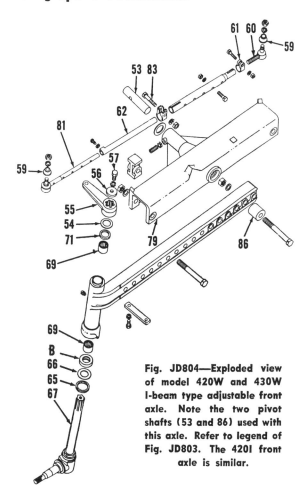

Fig. JD804—Exploded view of model 420W and 430W I-beam type adjustable front axle. Note the two pivot shafts (53 and 86) used with this axle. Refer to legend of Fig. JD803. The 420I front axle is similar.

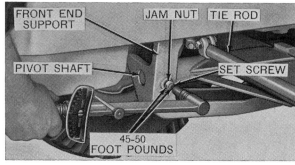

Fig. JD805A—View of the set screw which retains the 440I front axle pivot shaft. Set screw and jam nut are torqued to 45-50 Ft.-Lbs.

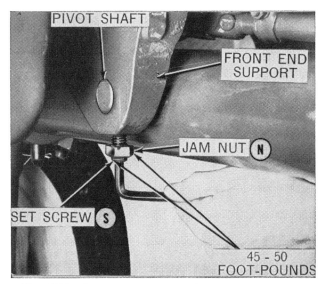

Fig. 805B—On some models, the front axle pivot shaft is retained in the front end support by set screw (S). The set screw and jam nut (N) should be tightened to a torque of 45-50 Ft.-Lbs.

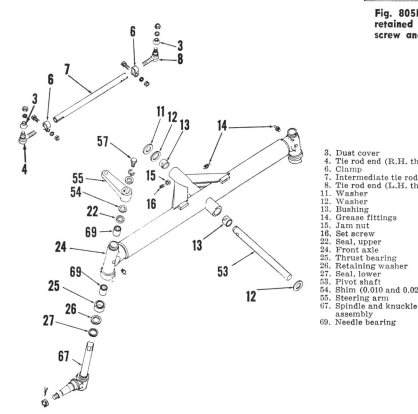

Fig. JD805 — Exploded view of the tubular, non-adjustable front axle used on 440I models.

3. Dust cover
4. Tie rod end (R.H. thread)
6. Clamp
7. Intermediate tie rod
8. Tie rod end (L.H. thread)
11. Washer
12. Washer
13. Bushing
14. Grease fittings
15. Jam nut
16. Set screw
22. Seal, upper
24. Front axle
25. Thrust bearing
26. Retaining washer
27. Seal, lower
53. Pivot shaft
54. Shim (0.010 and 0.027)
55. Steering arm
67. Spindle and knuckle assembly
69. Needle bearing

On models 40, 420, and 430, disconnect tie-rod ends from center steering arm (87–Figs. JD802 or 803), unbolt the pivot brackets (85 and 89) and roll the axle assembly away from tractor. On 420I, 420W and 430W, disconnect tie-rod ends from center steering arm and remove one axle extension and wheel assembly. Remove the front pivot shaft (86–Fig. JD804). Loosen lock nut and set screw and bump rear pivot shaft (53) forward until shaft clears support. Withdraw the axle assembly from front support.

On 440I, raise front wheels slightly and remove the wheels and hubs. Disconnect the tie-rod ends from steering arms, them remove jam nut and set screw from front support as shown in Fig. JD805A. Drive pivot shaft forward and out of front support, then slide front axle assembly out of front support.

On all other models, disconnect tie-rod ends, loosen jam nut (N) and using an Allen wrench, unscrew set screw (S) as shown in Fig. JD805.

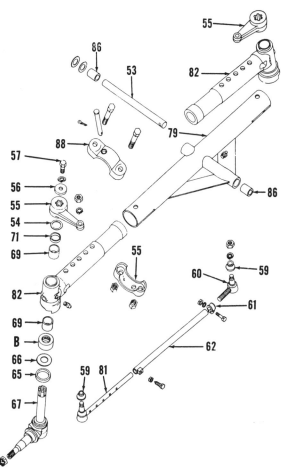

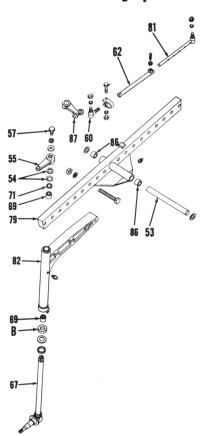

Fig. JD807 — Exploded view of model 420V and 430V I-beam type adjustable front axle. Front axles for 40V, 40H, 420H and 430H are similar. Refer to legend of Fig. JD803.

Fig. JD806—Exploded view of 40U, 330U, 420U and 430U front axle and component parts. Pivot shaft (53) is retained by set screw (S) shown in Fig. JD805B. Refer to legend of Fig. JD803.

Fig. JD808 — View showing grille of 440I prior to removal. Notice dowels being removed using a hollow pipe, cap screw and washer.

Bump the pivot shaft out and withdraw the axle assembly.

On all models, check the pivot shaft bushings inside diameter against the values which follow:

```
40S .................... 1.004-1.006
420I-420W-430W (front) 2.002-2.004
               (rear)  1.377-1.379
All Others ............ 1.377-1.379
```

If the bushings are excessively worn, install new bushings, using a closely piloted arbor, and ream them to the desired inside diameter.

MANUAL STEERING SYSTEM
R&R FRONT END UNIT
All Models Except 440I

5. To remove the complete front end unit including front support, wheels and axle or lower spindle or fork, proceed as follows: Remove hood and grille, drain cooling system and dis-

connect upper and lower radiator hoses. On early models, remove the Allen head cap screw from front of steering shaft and slide the steering shaft rearward and out of way. Note: On models serial number 125001 and up, it will be necessary to drive roll pin from coupling at rear of steering shaft before shaft can be moved rearward. On 40T, 420T and 430T, unbolt steering gear support from front end support and lift the gear unit from tractor. On all other models, unbolt steering gear housing from gear support and turn the gear housing until the stub (worm) shaft will clear the radiator; also unbolt fan shroud from radiator.

On all models, remove the radiator mounting bracket bolts and lift radiator from tractor. After radiator is removed, support front half of tractor, unbolt front support from engine and roll front end unit away from tractor. When reassembling, reverse the disassembly procedure and tighten the front support retaining cap screws to a torque of 85-90 Ft.-Lbs.

Model 440I

5A. Remove the hood and the two grille supports. Remove the screen from front of grille and place a chain or strap sling around top of grille as shown in Fig. JD808. Remove the two cap screws which pilot the grille to the front support, pull the hollow dowels and lift off grille. Note: If dowels are stuck they can be pulled by

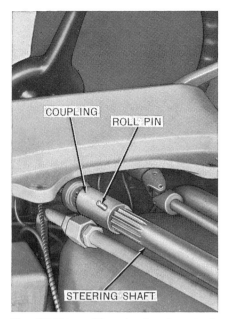

Fig. JD808A — View showing the roll pin which must be removed before the 440I steering shaft can be moved rearward.

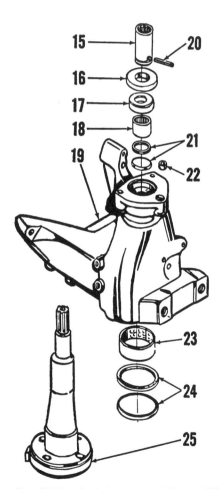

Fig. JD809—Exploded view of Deere 40T, 420T and 430T front support.

15. Coupling	21. Seal
16. Thrust washer	22. Dowel
17. Thrust bearing	23. Bushing
18. Needle bearing	24. Seal
20. Groov pin	25. Upper spindle

using a piece of hollow pipe with a cap screw and washers as shown. Drive out roll pin from coupling at rear of steering shaft as shown in Fig. JD808A, then, disconnect steering shaft from steering gear worm shaft and slide steering shaft rearward. Loosen cap screws attaching steering gear housing to steering gear support and rotate steering gear housing clockwise (viewed from top) until steering gear worm shaft will clear radiator.

Drain cooling system and remove left hand side rail, fan shroud and radiator. Support front of tractor, then remove the cap screws that attach front support to engine and roll front support, steering gear support and axle assembly away from tractor.

FRONT SUPPORT

Models 40T-420T-430T

6. **OVERHAUL.** To overhaul the front support (19—Fig. JD809), first remove hood and grille. On early models, remove the Allen head cap screw from front of steering shaft and slide the steering shaft rearward.

NOTE: On tractors serial number 125,001 and up, it will be necessary to drive out the roll pin from the coupling at rear of steering shaft before shaft can be moved rearward. Remove the cap screws joining the steering gear support to the front end support (19) and lift the steering gear housing and support, as a unit, from tractor. Raise front of tractor and remove the lower spindle, wheel fork or axle and center steering arm from tractor. Remove Groove pin (20) and withdraw spindle (25) from below. Inspect seal and retainer (24), bushing (23), seal (21), needle bearing (18) and thrust bearing (17).

To install needle bearing (18), use a suitable driver against end of bearing which has numbers stamped on it. Bushing (23) can be removed by collapsing. The bushing is presized, and if carefully installed, using a closely piloted arbor, will require no final sizing.

Pack the needle bearing (18) with bearing grease prior to reassembly. Note: The retainers for upper seal (21) and lower seal (24) should be installed with cupped side down.

Models 420I-420W-430W

6A. **OVERHAUL.** To overhaul the front support (32—Fig. JD810), first remove hood and grille. On early 420I and 420W models, remove the Allen head cap screw from front of steering

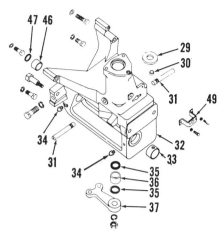

Fig. JD810 — Exploded view of the 420I, 420W and 430W front support and associated parts.

29. Thrust washer	35. Oil seal
30. Hollow dowel	36. Bushing
31. Radiator support	37. Steering arm
32. Front support	46. Rear bushing
33. Front bushing	47. Plug
34. Grease fitting	49. Bracket

shaft and slide the steering shaft rearward. Note: On tractors serial number 125,001 and up, it will be necessary to drive out the roll pin from coupling at rear of steering shaft before shaft can be moved rearward.

Disconnect tie-rods from center steering arms, then remove the center steering arm (37) from vertical steering shaft. Remove the cap screws joining the steering gear support to the front support and lift the steering gear and support as a unit from tractor.

Remove the three axle clamp bolts. Note that the center bolt also positions the front pivot shaft. Loosen the two spacer bars on bottom side of main member, then pull axle extensions from axle main (center) member. Remove the front pivot shaft. Loosen the jam nut and set screw which retains the rear pivot shaft and drive, or pull, the rear pivot shaft forward until it clears the front support, then remove the axle main (center) member from front support. The rear pivot shaft can now be removed from the axle main member.

Inspect oil seals (35) bushing (36) and renew as necessary. Inspect front and rear pivot shaft bushings (33 and 46) and renew as necessary. Bushings are pre-sized and will require no final sizing if carefully installed using a piloted driver. Be sure oil holes in bushings align with oil holes in front support.

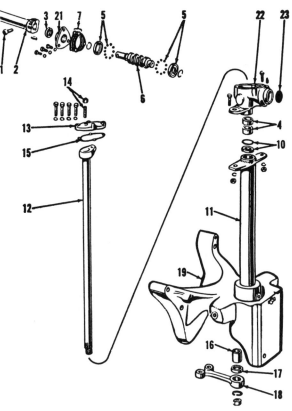

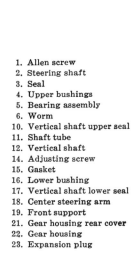

1. Allen screw
2. Steering shaft
3. Seal
4. Upper bushings
5. Bearing assembly
6. Worm
10. Vertical shaft upper seal
11. Shaft tube
12. Vertical shaft
14. Adjusting screw
15. Gasket
16. Lower bushing
17. Vertical shaft lower seal
18. Center steering arm
19. Front support
21. Gear housing rear cover
22. Gear housing
23. Expansion plug

Fig. JD811—Exploded view of Deere 40S prior to serial number 69404 and 40U prior to serial number 63120 steering gear, front support and associated parts. Shims (7) control worm shaft end play.

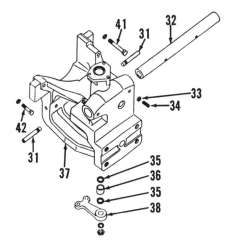

Fig. JD813 — Exploded view of series 440I front support. Notice bar (32) used to support grille.

31. Radiator support	36. Bushing
32. Grille support	37. Front support
33. Jam nut	38. Steering arm
34. Set screw	41. Special cap screw
35. Oil seal	42. Special cap screw

Suport tractor and loosen the pivot pin retaining jam nut and set screw as shown in Fig. JD805B. Drive pivot shaft from front support and remove front axle assembly.

Inspect seals (17—Fig. JD811 or 812) and bushing (16) in lower part of front support and renew as necessary.

Model 440I

6C. **OVERHAUL.** To overhaul the front support (37—Fig. JD813), remove hood and the two grille supports. Remove the screen from front of grille and place a chain or strap hoist around top of grille as shown in Fig. JD808. Remove the two cap screws which pilot the grille to the front support, pull the hollow dowels, and lift off grille. Note: If dowels are stuck they can be pulled by using a piece of hollow pipe with a cap screw and washers as shown. Drive out roll pin from coupling at rear of steering shaft as shown in Fig. JD808A, then disconnect steering shaft from steering gear worm shaft and slide steering shaft rearward.

Disconnect tie-rods from center steering arm and knuckle steering arms; then remove center steering arm from vertical shaft. Remove the cap screws joining steering gear support to

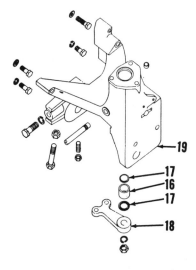

Fig. JD812—Exploded view of front support and associated parts used on models 40H, 40S (after serial number 69403), 40U (after serial number 63119), 40V, 40W, 320S, 320U, 330S, 330U, 420H, 420S, 420U, 420V, 430H, 430S, 430U and 430V.

16. Bushing	18. Center steering arm
17. Oil seal	19. Front support

Models 40H-40S-40U-40V-40W-320S-320U-330S-330U-420H-420S-420U-420V-430H-430S-430U-430V

6B. **OVERHAUL.** To overhaul the front support (19—Fig. JD811 or 812), first remove hood and grille. On early models, remove the Allen head cap screw from front of steering shaft and slide the steering shaft rearward. Note: On tractors serial number 125,001 and up, it will be necessary to drive out the roll pin from coupling at rear of steering shaft before shaft can be moved rearward.

Disconnect tie-rods from center steering arm, then remove the center steering arm from vertical steering shaft. Remove the cap screws joining the steering gear (or support) to the front support and lift the steering gear and support as a unit from tractor.

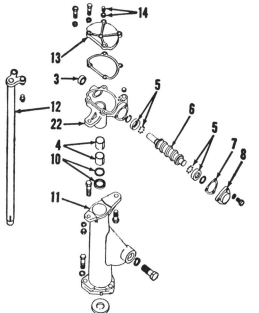

3. Seal
4. Bushings
5. Bearing assembly
6. Worm
7. Shims
8. Front cover
10. Vertical shaft seal
11. Gear support
12. Vertical shaft
13. Top cover
14. Adjusting screw

Fig. JD814—Exploded view of manual steering gear and support typical of all models except early 40S and 40U.

Slide the steering shaft rearward until free from worm shaft. Note: On tractors serial number 125,001 and up, it is necessary to drive out roll pin from coupling at rear of steering shaft before steering shaft can be moved rearward. Remove the bolts attaching the gear housing to the shaft tube or support (11—Fig. JD811 or 814). On axle type tractors disconnect the center steering arm from the vertical shaft. Lift steering gear unit from tractor.

The procedure for disassembling and reassembling the removed unit is evident after referring to the accompanying exploded views. Bushings (4) should be reamed after installation to an inside diameter of 0.999-1.000 for models 40S prior to serial number 69,404, 40U prior to serial number 63,-120 and 1.124-1.125 for all other models.

front support and lift steering gear and support from tractor as a unit. Remove left hand side rail, drain cooling system and remove fan shroud and radiator.

Raise front wheels slightly and remove the wheel and hub units. Remove jam nut and set screws from front support, drive pivot shaft forward out of front support, then slide front axle out of front support.

Inspect seals (35—Fig. JD813) and bushing (36) and renew as necessary. When renewing seals install both seal lips down and position upper seal so top surface of same is 1/32-inch below top of counterbore. Grille support bar can be removed after loosening the two set screws.

STEERING GEAR

For the purpose of this manual, the steering gear assembly will include the gear housing, gear support and all parts contained therein.

It should be noted that model 40S at serial number 69404 and 40U at serial number 63120, changed from a one stud to a two stud lever shaft.

All Models

7. **ADJUSTMENT.** The cam and lever type steering gear units are provided with two adjustments; one for the worm shaft bearings, the other for mesh or backlash.

8. WORM SHAFT BEARINGS. The worm shaft bearings should be adjusted to provide a slight amount of rotational drag. To adjust the bearings on models 40S prior to serial number 69404 and 40U prior to serial number 63120, vary the number of shims (7—Fig. JD811) which are located between the gear housing rear cover (21) and the housing. On all other models, adjust the bearings by varying the number of shims (7—Fig. JD814) which are located between the gear housing front cover (8) and the housing. Shims are available in thicknesses of 0.002, 0.003 and 0.010.

9. BACKLASH. With the worm shaft bearings adjusted as outlined in the preceding paragraph 8, support front of tractor to remove any unnecessary load from the gear unit and place the steering gear in its mid-position by turning the steering wheel half way between full right and full left turn. Loosen the lock nut and turn the adjusting screw (14—Figs. JD811 or 814) **down** until a barely perceptible drag is experienced when turning the steering wheel through the mid or high point position. When properly adjusted, the steering gear will have a slight amount of backlash at all other positions.

10. **R&R AND OVERHAUL.** To remove the steering gear unit from tractor, first remove hood, grille and on early models, remove the Allen head cap screw from front of steering shaft.

POWER STEERING SYSTEM

Various combinations of power steering equipment have been used on the Dubuque built tractors beginning with tractor serial number 80,001. On tractors with serial numbers 80,001 through 131,817, the pressurized working fluid was supplied by the "Touch-O-Matic" system pump and reservoir and was regulated, and directed to the power cylinder, by a flow divider valve (K—Fig. JD817). Effective at tractor serial number 131,-818 and up, the flow divider valve was discontinued and a separate, belt driven pump with a reservoir was installed as shown in Fig. JD816.

On tractors serial numbers 80,001 through 125,000, a power cylinder was incorporated in the steering linkage as shown at (H) in Fig. JD817. Effective serial number 125,001 and up, the power cylinder was removed from the steering linkage and an integral power cylinder, rack and pinion assembly was incorporated into the steering gear support as shown in Fig. JD821A.

The control valve, except for a change from Woodruff keys to pins in the ends of the shafts (at serial number 133665), has remained the same.

Summing up the above, the following power steering combinations have existed:

Tractor Serial Numbers 80,001 to 125,001
  Control valve
  Flow divider valve
  External power cylinder
  Working fluid from "Touch-O-Matic"
Tractor Serial Numbers 125,001 to 131,818
  Control valve
  Flow divider valve
  Integral power cylinder, rack and pinion
  Working fluid from "Touch-O-Matic"
Tractor Serial Numbers 131,818 and Up
  Control valve
  Separate pump
  Integral power cylinder, rack and pinion

LUBRICATION
Models Prior to Ser. No. 125,001

11. Grease fittings are provided at outer (tie-rod) ends for all models and at inner (steering knuckle) end of model 420U.

Models Ser. No. 125,001 & Up

11A. Fill rack and pinion reservoir to filler hole level with S.A.E. 80 gear oil. Maintain at this level.

BLEEDING
Models Prior to Ser. No. 131,818

11B. All air must be bled from system to prevent noisy and/or erratic steering. To bleed system, start engine and turn steering wheel full left and full right several times. Then stop engine and recheck the "Touch-O-Matic" reservoir fluid level.

Models Ser. No. 131,818 & Up

11C. All air must be bled from system to prevent noisy and/or erratic steering action. To bleed system, fill pump reservoir to "full" mark with Type "A" automatic transmission fluid. Start engine and while observing fluid in reservoir, turn steering wheel full left and full right several times. When air bubbles and turbulence cease in

Fig. JD815—Pressure gage installation used to check operating pressure of power steering system prior to tractor serial number 131,818. Pressure should be 950-1050 psi with optimum pressure 1000 psi.

reservoir, stop engine and refill reservoir to "full" mark if necessary. Install reservoir cover and be sure slots in washer (2—Fig. JD816A) face cover as they act as a breather for the system.

TROUBLE SHOOTING
All Models

11D. The accompanying table lists troubles which may be encountered in operating the power steering system. The procedure for correcting most of the troubles is evident; however, for those not readily remedied, refer to the appropriate subsequent paragraphs.

OPERATING PRESSURE AND RELIEF VALVE
Models Prior to Ser. No. 131,818

12. A pressure check of the power steering system relief valve can be

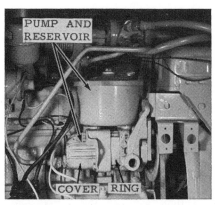

Fig. JD816 — View of the separate belt driven power steering pump which was installed in production at tractor serial number 131,818.

made in the following manner: Connect a high pressure test gage in series between the upper cylinder hose and steering valve as shown in Fig. JD815. Start engine and allow to run until hydraulic system oil reaches normal operating temperature. With steering wheel held hard against stop at either extreme left or extreme right hand position and with engine running at high idle speed, the gage pressure should be 950-1050 psi. If pressure is not as specified, remove relief valve plug and vary the number of shims under relief spring. Adding shims will increase pressure.

If operating pressure cannot be restored with shims, it will be necessary to renew or recondition the hydraulic pump as outlined in paragraph 120 or 121.

Models Ser. No. 131,818 & Up

12A. A pressure check of the power steering system relief valve can be made in the following manner: Con-

TROUBLE-SHOOTING CHART

	Loss of Power Assistance	Power Assistance in One Direction Only	Unequal Turning Radius	Erratic Steering Control	Steering Cylinder Bottoming
Binding, worn or bent mechanical linkage	★		★	★	
Insufficient fluid in reservoir	★				
Low pump pressure	★				
Binding or sticking steering valve	★	★		★	
Damaged or restricted hose or tubing	★	★			
Improperly adjusted tie-rods or drag links			★		★
Steering arms not positioned properly			★		★
Air in system	★			★	
Faulty cylinder	★				
Faulty flow divider valve	★			★	
Sticking flow divider relief valve				★	
Plugged divider valve piston	★			★	
Faulty or loose pump drive belt	★			★	
Timing marks on rack and pinion not aligned			★	★	
Faulty flow control valve in pump				★	
Plugged filter element		★		★	

nect a high pressure test gage (at least 1500 psi) in series with the pump outlet line. Start engine and allow to run until working fluid is warmed to operating temperature. With steering wheel held hard against stop at either extreme left or extreme right and the engine running at high idle speed, the gage pressure should be 950-1050 psi (1000 psi preferred).

If the pressure is more than 1050 psi, the relief valve spring is too strong or the relief valve is stuck in the closed position. If the pressure is less than 950 psi, the relief valve spring is too weak or the pump requires overhauling. In any event, the first step in eliminating trouble is to remove and thoroughly clean (or renew) the flow control and relief valve (15—Fig. JD816A). Components of the flow control and relief valve are not available separately.

To remove the flow control and relief valve unit, proceed as follows: With hood off, remove pump reservoir cover and withdraw as much oil as possible with a suction gun. Remove the filter screen assembly from reser-

voir; then, remove reservoir, manifold and gaskets from pump. Refer to Fig. JD816. Disconnect the pump pressure line, then remove the cap screws which join the pump cover to pump body and separate cover from pump ring, but be careful not to allow the flow control valve assembly to drop from pump cover. Remove the valve unit and renew or clean same as required. If cleaning and/or renewal of the flow control and relief valve does not restore the pressure, overhaul the pump as outlined in paragraph 13B.

PUMP
Models Prior to Ser. No. 131,818

13. Power steering system operating pressure is supplied by the "Touch-O-Matic" system pump. For information pertaining to same, refer to paragraph 120 or 121.

Models Ser. No. 131,818 & Up

13A. **REMOVE AND REINSTALL.** Remove hood from tractor and the cover from pump reservoir. Refer to Fig. JD816. Use a suction gun and remove as much oil as possible from

pump reservoir; then, disconnect the pressure and return lines. Loosen generator drive belt adjusting cap screw and remove drive belt. Remove lower bolt from headlight bracket and tilt bracket upward. Remove pump drive pulley retaining nut, then remove pulley. Remove cap screws retaining pump and pull pump from tractor. NOTE: In some cases, the removal of the pump pulley is prevented by the "Touch-O-Matic" inlet pipe. When this occurs, remove pump by pulling pump rearward while unscrewing the attaching cap screws.

Reverse above procedure to reinstall, then fill and bleed system as outlined in paragraph 11C.

13B. **OVERHAUL.** With pump removed as outlined in paragraph 13A, refer to Fig. JD816A and proceed as follows: Remove filter, filter retainer and spring; then unbolt and lift manifold (8), reservoir (9) and gaskets (10) from pump body. Place pump body in a soft faced vise with pump cover upward. Remove the four pump body bolts and lift pump cover (11) from pump. Be sure flow control and relief valve (15) does not drop out. Lift the pressure plate (16) from the locating dowels (21). Note the position of the arrow that is embossed on the pump ring (18) and lift pump ring from dowels. Remove the vanes (20) from rotor (19); then remove the small snap ring from drive shaft, lift rotor from shaft and remove shaft from pump body. Dowels and oil seal can be removed from pump body if necessary.

Clean all parts in a suitable solvent and inspect same for nicks, burrs and scoring.

Check contour of pump ring. If chatter marks, gouges or other wear can be detected with the finger, renew both ring and vanes. Place vanes in rotor slots and check fit. Vanes should fit snugly yet be free to slide. If vanes or rotor is worn enough to allow vanes to tilt, renew rotor and/or vanes. Check pump body and pressure plate faces for signs of scoring. If faces are only lightly scored they can be refinished by lapping same on a flat surface; however, if heavy scoring is present, renew parts. Check pump drive shaft for worn splines or wear on oil seal surface. If any wear is found, renew shaft. If oil seal in pump body shows any sign of leakage, renew same. Prior to installing seal, pack the cavity between lips of seal with Lubriplate or equivalent and install seal with lettering toward front of pump using John Deere tool No.

Fig. JD816A—Exploded view of the separate belt driven power steering pump used on tractors from serial number 131,818 and up.

2. Special washer	11. Pump cover	16. Pressure plate	21. Dowel
5. Spring	14. Flow control	18. Pump ring	23. Pump body
6. Retainer	valve spring	19. Rotor	24. Seal
7. Filter	15. Flow control and	20. Rotor vane	25. Snap ring
8. Manifold	relief valve	(10 used)	26. Pump shaft

M614T or equivalent. Check pump cover, reservoir and manifold for any damage. If filter screen is plugged or damaged, renew same. While the flow control and relief valve assembly can be disassembled and cleaned, the individual parts are not catalogued separately. If any damage is found, renew the complete assembly.

13C. Reassembly is the reverse of disassembly; however, the following points should be kept in mind.

When installing the rotor to the pump drive shaft, be sure the chamfered end of the splines face the pump body. The rounded edges of the vanes face outward and the pump ring is installed so the embossed arrow on outer edge points in the direction of pump rotation. The flow control valve assembly is installed in the pump cover so the hex head end is toward the aft end of pump.

Tighten the pump body bolts alternately and to a torque of 25 Ft.-Lbs. Tighten the manifold and reservoir attaching cap screws to a torque of 8-10 Ft.-Lbs.

Reinstall pump on tractor, then fill and bleed system as outlined in paragraph 11C. Be sure slots in washer (2) face cover as they act as a breather for the system.

Fig. JD817—View showing the position of the component parts which comprise the power steering system on models prior to 125,001.

A. Steering shaft
B. Steering valve
E. Pressure line
F. Return line
G. "Touch-O-Matic" pump
H. Steering cylinder
I. Cylinder hoses
K. Flow divider valve

STEERING VALVE
All Models

The steering valve is located under the hood and is interposed between steering shaft and steering gear housing. See Fig. JD817 which shows the installation of the power steering components on early models. The steering valve directs the hydraulic oil from the flow divider valve, or separate pump, to either the right turn, or left turn port of the steering cylinder, or to the return line of the hydraulic pump.

14. REMOVE AND REINSTALL. Disconnect power cylinder lines, pressure line and return line from steering valve. Remove socket head screws and/or roll pin from steering shaft and steering shaft extension, then slide steering shaft rearward and remove Woodruff key. On models prior to 131,-818, remove bolts retaining steering gear housing to support, rotate housing clockwise until steering valve will clear governor control bracket, then slide steering valve from steering shaft extension and remove Woodruff key.

On models serial number 131,818 and up, slide steering shaft rearward; then, remove the steering valve support bracket attaching cap screws and lift off steering valve and bracket.

Reinstall valve unit by reversing the removal procedure and bleed the system as outlined in paragraph 11B or 11C.

Note: End of steering valve having port marked "IN" must be toward front of tractor.

14A. OVERHAUL. With the steering valve off tractor, remove the three cap screws from each end cap and remove (tap stub shaft if necessary) end caps. Note: On models serial number 133,665 and up, it will be necessary to

remove pins prior to removing end caps. Slide spool and sleeve from housing as a unit. Refer to Fig. JD818 and remove pin (J) which retains front stub shaft to spool (H) and sleeve (I). Make provisions for restraining or catching spring loaded detent balls (F) before separating sleeve and spool.

Thoroughly clean all parts in a suitable solvent.

Check and renew all parts found to be unserviceable paying particular at-

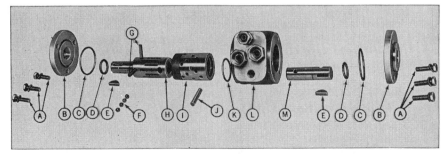

Fig. JD818—Exploded view of steering valve. Pin (J) retains sleeve (I), spool (H) and shaft (M) together as a unit. Effective at tractor serial number 133,665, the Woodruff keys (E) were deleted and pins were installed.

A. Cap screws
B. End cap
C. "O" ring
D. "O" ring
E. Woodruff key
F. Detent balls
G. Detent springs
H. Spool
I. Sleeve
J. Pin
K. "O" ring
L. Housing
M. Stub shaft

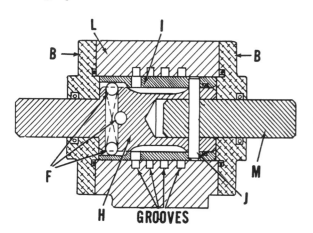

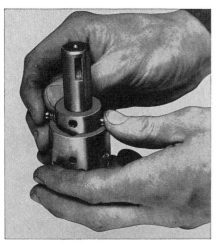

Fig. JD818B—View showing method of inserting detent springs and balls during assembly of steering valve sleeve and spool. Note that holes in spool are offset, thereby determining position of spool in sleeve. Early type unit is shown.

tention to machined surfaces for nicks and burrs. Make sure pin (J) is straight and not excessively worn.

When reassembling, renew all "O" ring seals and refer to Fig. JD818 and 818A. Slide spool (H) into detent end of sleeve (I) and before entering all the way, insert detent springs and balls as shown in Fig. JD818B. Be sure "O" ring is in place inside of sleeve, then insert stub shaft (M—Fig. JD 818A) in spool and insert pin (J). Reinstall spool and sleeve unit in housing (L), install end caps (B) and secure with cap screws.

FLOW DIVIDER VALVE
Models Prior to Ser. No. 131,818

15. **REMOVE AND REINSTALL.** Remove ignition coil. Refer to Fig. JD 817 and disconnect oil lines from flow divider valve (K). If tractor is equipped with headlights, disconnect wires at base of right hand headlight and remove same. Remove the two hex mounting studs and remove flow divider valve.

Fig. JD818A — Cut-away view of steering valve showing relative position of parts after assembly.

Reinstall the valve unit by reversing removal procedure, and tighten all oil line fittings before completely tightening the flow divider valve mounting studs. Bleed the system as in paragraph 11B.

15A. **OVERHAUL.** With valve unit removed from tractor, refer to Figs. JD819 and 819A and proceed as follows: Remove large hex plug, divider valve piston and spring. Remove small hex plug, spring and the nylon relief valve being careful not to lose the pressure adjusting shims.

Thoroughly clean all parts and examine housing for cracks or other damage. Examine nylon relief valve for excessive wear and piston for nicks and burrs. Make sure that valve

Fig. JD819—Power steering system flow divider and component parts in their relative positions. Note shims which control relief valve pressure. This unit was discontinued at tractor serial number 131,817.

Fig. JD819A — Cut-away view of flow divider valve showing the four metering orifices, divider valve piston and relief valve. This unit was discontinued at tractor serial number 131,817.

A. Inlet port
B. Piston
C. Spring
D. Metering orifices (4)
E. Steering valve port
F. "Touch-O-Matic" port
G. Relief valve
H. Spring
I. Relief valve by-pass port

piston slides freely in its bore. Examine springs for rust or damage. Renew any parts found to be unserviceable and reassemble.

STEERING CYLINDER (EXTERNAL)
Models Prior to Ser. No. 125,001

16. **REMOVE AND REINSTALL.** Disconnect oil lines from steering cylinder and outer end of cylinder from steering arm. On 420U, remove screw plug, spring and outer ball seat from cylinder inner pivot. On 420I, remove pin retaining inner end of cylinder to axle main member.

To reinstall the cylinder, proceed as follows: On 420U, screw ball joint end on inner end of cylinder until there is 1/16-inch of threaded surface exposed as shown in Fig. JD820. Mount cylinder on ball end of axle pivot shaft, install outer seat, spring and screw plug.

On 420I, attach cylinder to front axle main member with attaching pin and cotter pin.

On both models, be sure that top hole in steering cylinder faces rearward.

Adjust tie-rod end of cylinder so that cylinder will not bottom, either in the extended or retracted position, when front wheels are turned to extreme right and left positions. After correct length has been determined, attach tie-rod end to right hand steering arm. Be sure hoses pass through

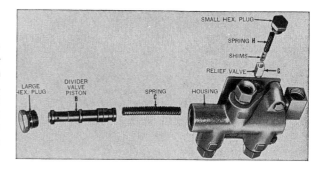

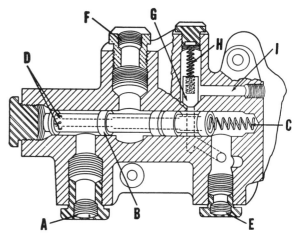

Fig. JD820—Model 420U (prior to serial number 125,001) steering knuckle (ball joint end) is screwed on inner end of cylinder until only 1/16-inch of threaded area is left exposed as shown.

loop of flow divider return line and that hose from top hole of steering valve attaches to top hole of steering cylinder. Bleed the system as in paragraph 11B.

16A. **OVERHAUL.** Refer to Fig. JD 820A and remove tie-rod end (T). Straighten detent locks, unscrew retaining nut (S) and pull rod (L) and piston (J) from cylinder (H). Remove outer seal retainer washer (R), seal (Q), inner seal retainer washer (P), spring (O), rubber gasket (N) and end cap (M). Remove nut (I) from rod (L), and withdraw piston (J).

Thoroughly clean all parts and inspect cylinder for damage, wear or scoring. Be sure that piston rod is straight and not worn excessively. Inspect piston rings for wear or damage.

Pay particular attention to seals and their retaining washers. Renew any questionable parts.

It is recommended that a ring compressor be used when installing piston and rod assembly; however, if a ring compressor is not available, a screw driver can be used to compress the rings if caution is exercised.

POWER CYLINDER, RACK AND PINION ASSEMBLY
Models After Ser. No. 125,000

Effective at tractor serial number 125,001 and up, the linkage booster type of power steering was discontinued and a power cylinder which operates a rack and pinion assembly was made an integral part of the steering gear support. This rack and piston assembly is shown installed in Fig. JD821.

17. **REMOVE AND REINSTALL.** To remove the power cylinder, rack, pinion shaft and pinion housing as an assembly, proceed as follows: Remove hood and grille, turn the front wheels to the straight ahead position and on axle type tractors, remove the center steering arm.

Disconnect one hose from power cylinder and the other hose from the steering valve. Now, connect the loose end of the hose which is still connected to the power cylinder, to the open fitting on the power cylinder and the loose end of the hose which is still connected to the steering valve to the open fitting on the valve. With hoses connected in this manner, loss of oil from the system as well as entry of foreign material into the system will be prevented.

Refer to Fig. JD821 and proceed as follows: On models with a vertically mounted steering wheel, remove socket head cap screws from the steering extension shaft, slide steering shaft and steering valve assembly rearward and remove extension shaft. On models with slanted steering wheel, a

similar procedure can be used for removing the extension shaft, except it will be necessary to remove roll pin from coupling at rear end of steering shaft before steering shaft can be moved rearward. Also, on tractors serial number 131,818 and up, it will be necessary to remove the steering valve support bracket cap screws.

The steering gear and/or steering gear support can be removed at this time, or if desired, can be removed after the power cylinder, rack and pinion and housing assembly has been removed from front support. In either case, remove the cap screws retaining rack and pinion housing to front support and remove same.

17A. When reinstalling the assembly, both the steering gear and the rack and pinion will need to be timed to insure that front wheels will turn an equal amount in both directions.

Rotate the pinion shaft (item 5 in Fig. JD821A) through its complete range of travel and mark the extreme positions; then, turn pinion shaft to a point midway between the two marked positions. Rotate steering gear worm (stub) shaft through its full range of travel and count the total number of turns. Turn the shaft back one-half of the total number of turns and hold in this position. With the hollow dowel (2), coupling (3) and gasket (7) in place, and the front

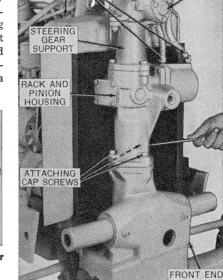

Fig. JD821 — View of the integral power cylinder, rack and pinion assembly which is used on tractors serial number 125,001 and up.

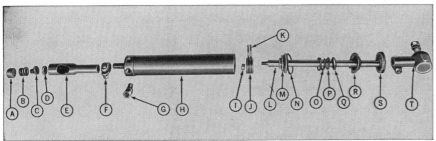

Fig. JD820A — Exploded view of 420U (prior serial number 125,001) steering cylinder showing component parts and their relative positions. Model 420I is similar.

A. Screw plug	H. Cylinder	N. Rubber gasket
B. Spring	I. Nut	O. Spring
C. Outer ball seat	J. Piston	P. Inner retainer washer
D. Inner ball seat	K. Piston rings	Q. Seal
E. Steering knuckle	L. Piston rod	R. Outer retainer washer
F. Clamp	M. End cap	S. Retaining nut
G. Fitting		T. Tie rod end

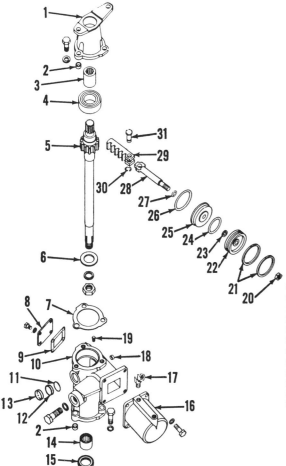

1. Steering gear support
2. Hollow dowel
3. Coupling
4. Ball bearing
5. Pinion shaft
6. Thrust washer
7. Gasket
8. Cover
9. Gasket
10. Rack and pinion housing
11. Shims (0.002, 0.005 and 0.010)
12. Thrust pad
13. Cupped plug
14. Needle bearing
15. Oil seal
16. Cylinder housing
17. Union (45°)
18. Cupped plug
19. Pipe plug
20. Self-locking nut
21. Piston rings
22. Piston
23. Washer
24. "O" ring
25. Cylinder head
26. "O" ring
27. "O" ring
28. Piston rod
29. Rack
30. Snap ring
31. Pin

Fig. JD821A—Exploded view of the integral power cylinder, rack and pinion assembly used on all tractors serial number 125,001 and up.

shims (11) are free at this time and care must be exercised not to lose same. Tap pinion shaft (5) and bearing (4) out top end of housing and be careful not to damage thrust washer (6).

Pull rack, cylinder head, piston rod and piston from cylinder (16) then remove locknut (20) and pull piston and cylinder head from piston rod as shown in Fig. JD821B. Remove rings (21) from piston. Remove snap ring from pin which joins rack and piston rod and drive out pin (31). Discard all "O" rings.

Clean all parts in a suitable solvent and inspect. Check cylinder housing for cracks and/or scored walls and renew as necessary.

Examine piston and piston rings. If ring lands in piston, or piston rings are worn or cracked, renew as necessary. Be sure "O" ring bore and lands of cylinder head are in good condition and that cylinder head is not warped or distorted. Examine rack to see that teeth are not chipped or that rear (thrust) side is not worn or scored. Piston rod must be free of scoring or scratches.

Examine housing (10) for cracks or other damage which would render it unsatisfactory. Inspect pinion shaft (5) for chipped teeth, damaged splines or any other damage which would make it unserviceable. When install-

wheels in the straight ahead position, install the rack and pinion housing assembly (10), steering gear and support on the front support and secure. Be careful not to move the previously positioned shafts.

Reconnect the steering shaft, steering valve and hoses. On all tractors except tricycle, install the center steering arm (parallel to tractor centerline) on bottom of vertical steering shaft and before tightening, rotate steering wheel full left and full right to be sure steering mechanism will operate through its entire range.

Refill rack and pinion housing to top of filler hole with specified lubricant.

18. **OVERHAUL.** With the rack and pinion assembly removed as outlined in paragraph 17, refer to Fig. JD821A and proceed as follows: Place assembly on a bench and separate steering gear and its support (1) from rack and pinion housing (10). Remove cap screws and separate power cylinder and rack assembly from pinion housing (10). Note: Thrust pad (12) and

Fig. JD821B—View showing piston and cylinder head removed from piston rod. Note position of "O" rings. Refer to legend of Fig. JD821A.

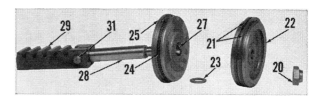

Fig. JD821C—View showing method of checking clearance prior to establishing the pre-load between rack and pinion. Refer to text.

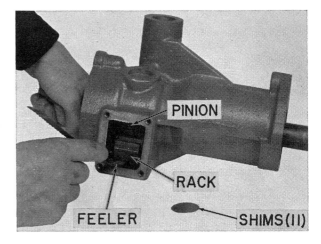

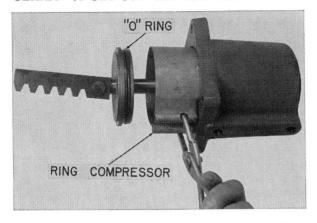

Fig. JD821D — A ring compresser, such as that shown, can be used to install the piston and rings assembly in the cylinder. Note "O" ring and the position of the cylinder head.

on face of cylinder and use heavy grease to hold same in position.

Rotate the pinion shaft until the marked tooth, shown in Fig. JD821E, is visible; then, mate this marked tooth on pinion with the first tooth space of the rack. Rotate pinion shaft to pull rack inward; then secure cylinder to rack and pinion housing with hose ports on top side. Use a new gasket and install inspection cover.

18A. Refer to paragraph 17A to reinstall the complete unit to tractor. Fill rack and pinion reservoir to filler hole level with specified lubricant. Fill and bleed power steering system as outlined in paragraphs 11B or 11C.

ing ball bearing (4—Fig. JD821A) on pinion shaft, use a sleeve type driver which will contact only inner race of bearing. Renew thrust washer (6) if same is worn. Inspect the needle bearing (14) in bottom of housing and if renewal is necessary, press needle bearing into housing until top of same is $\frac{1}{16}$-inch below thrust washer contact surface. Install new seal (15) with lips facing upward (toward bearing).

When reassembling, be sure to renew all "O" rings, place thrust washer (6) on pinion shaft and install shaft and bearing unit into housing. Use caution not to damage needle bearing (14) and/or oil seal (15).

NOTE: When mating rack to pinion shaft, it is necessary to vary the thickness of shim pack (11) under thrust pad (12) to obtain desired 0.000-0.001 pre-load against rack when rack is in mid-position. To make this adjustment, proceed as follows: Place thrust pad (12) in its bore with NO shims. Insert rack and move same to its mid-position; then, using a feeler gage as shown in Fig. JD821C, measure the distance between thrust pad and thrust side of rack. Now, remove

rack and thrust pad and insert the necessary amount of shims (11) to provide the 0.000-0.001 pre-load. Shims are available in thicknesses of 0.002, 0.005 and 0.010.

Join rack to piston rod with pin (31—Fig. JD821A) and install snap ring (30). Place "O" ring (27) in bore of cylinder head and slide same on piston rod with largest diameter of cylinder head toward rack. Place the flat washer (23) on piston rod, then install piston, flat side toward rack and secure with the self-locking nut (20). Use a ring compressor such as that shown in Fig. JD821D and install piston and ring assembly in cylinder. Be sure "O" ring is on outside diameter of cylinder head, position cylinder head in cylinder and be sure that outer face of cylinder head is flush with machined surface of cylinder. Place the large "O" ring in the groove

STEERING GEAR
All Models

19. Steering gear units used on tractors prior to serial number 125,001, which are equipped with power steering, are the same as those used on models without power steering. Removal, reinstallation and servicing procedures remain the same as those for manual steering which are covered in paragraphs 7 through 10.

19A. Steering gear units used on tractors serial number 125,001 and up, which are equipped with power steering are basically the same as the earlier units, with the exception of the shorter lever shaft. Removal, reinstallation and servicing are obvious after reference to paragraphs 7 through 10 and Fig. JD821F.

Fig. JD821E—View showing marked tooth of pinion shaft. When installing rack, mate the first tooth space of rack with marked tooth of pinion shaft.

3. Oil seal
4. Bushings
5. Bearing assembly
6. Steering worm
7. Shims (0.002, 0.003 and 0.010)
8. Front cover
10. Oil seal
11. Steering gear support
12. Lever shaft
13. Top cover
14. Adjusting screw

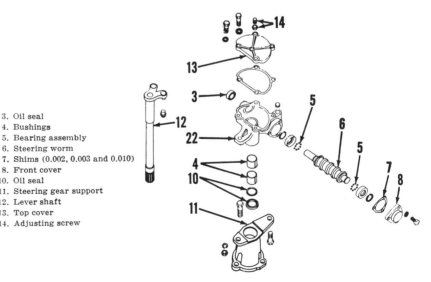

Fig. JD821F—Exploded view of the steering gear unit and support used on models 125,001 and up, when equipped with power steering. Except for the shorter lever shaft (12), this unit is similar to earlier types.

ENGINE AND COMPONENTS

R&R ENGINE WITH CLUTCH
All Models

20. To remove engine and clutch as a unit, first drain cooling system and if engine is to be disassembled, drain oil pan; then, remove the complete front end unit as outlined in paragraph 5 or 5A. Disconnect wires from generator, coil, lights and hour meter on 40 series, or "Tachometer" cable from drive coupling on models so equipped. On late model tractors, disconnect wire from the oil pressure sending unit. Unclip wiring harness from rocker arm cover and lay harness rearward and out of way. Disconnect the "Touch-O-Matic" lines from pump and top of instrument panel and remove the lines. Unhook governor spring from control rod and disconnect governor connecting rod from control rod; then, remove the control rod bracket attaching cap screws. Disconnect air cleaner hose and fuel line from carburetor, disconnect carburetor from manifold and choke wire from carburetor. Disconnect heat indicator sending unit from engine and remove dust shield from lower front end of center frame. Sup-

port engine in a hoist and tractor under center frame, unbolt engine from center frame and move engine forward and away from tractor.

CYLINDER HEAD
All Models

21. To remove cylinder head from tractor, first drain cooling system and remove hood. Unhook governor spring from control rod and disconnect governor connecting rod from control rod; then, remove the control rod bracket attaching cap screws. Disconnect air cleaner hose and fuel line from carburetor and remove manifold. Remove wiring from rocker arm cover and remove cover, rocker arms assembly and push rods. On 420, 430 and 440 series, disconnect heat indicator bulb and water by-pass line from water outlet manifold. Disconnect hose from water outlet manifold and remove the water manifold. Disconnect oil line from rear of cylinder head. Disconnect spark plug cables, remove the cylinder head retaining cap screws and lift head from tractor.

When installing cylinder head, tighten the cap screws progressively from center outward and to a torque

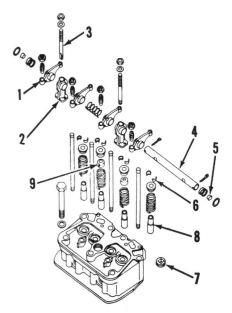

Fig. JD823—Deere 420, 430 and 440 series tappet levers and shaft assembly. Rear stud (3) provides oil passage for tappet lever lubrication. Note oil deflector (9) for intake valve stems.

1. Tappet lever	6. Split cone lock
2. Shaft support	7. Pipe plug
3. Oil passage stud	8. Valve guide
4. Tappet lever shaft	9. Oil deflector
5. Cork stopper	

of 100-110 Ft.-Lbs. for 40, 320 and 330 series; 85-95 Ft.-Lbs. for 420, 430 and 440 series. If the tappet lever shaft support studs were removed, make certain that rear stud (3—Fig. JD822 and 823) is installed in the correct location. This stud has a flat on one side and provides an oil passage for lubrication of the tappet levers.

VALVES AND VALVE SEATS
All Models

22. Intake and exhaust tappet gap should be set cold to 0.012 for the 40, 320 and 330 series; 0.015 for the 420, 430 and 440 series. Inlet and exhaust valves, which seat directly in the cylinder head have a stem diameter of 0.3715-0.3725, and are not interchangeable. Valve stems should have a clearance of 0.002-0.0045 in the guides.

Valve face angle, valve seat angle and desired seat width are as follows:

Inlet Valve

40-320-330 face angle	30°
420-430-440 face angle	29°
40-320-330-420-430-440 seat angle	30°
40-320-330 seat width	⅛ inch
420-430-440 seat width	3/32 inch

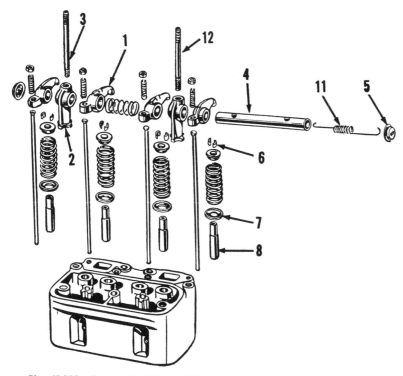

Fig. JD822—Deere 40, 320 and 330 series tappet levers and shaft assembly. Rear stud (3) provides oil passage for tappet lever lubrication.

1. Tappet lever	5. End cap
2. Shaft support	6. Split cone lock
3. Oil passage stud	8. Valve guide
4. Tappet lever shaft	11. Retaining spring

Exhaust Valve

 40-320-330 face angle..........45°
 420-430-440 face angle.........44°
 40-320-330-420-430-440
  seat angle ....45°
 40-320-330-420-430-440
  seat width ....5/64 inch
 Seats can be narrowed, using 20 and 70 degree stones.

VALVE GUIDES
All Models

23. Inlet and exhaust valve service guides (8—Fig. JD822 and 823) are interchangeable and should be installed with smaller O. D. of guide up using a piloted drift. On the 40, 320 and 330 series, press guides in until port end measures 2 3/32 inches from gasket surface of cylinder head. On the 420, 430 and 440 series, the installation dimension is 1 7/16 inches from top of combustion chamber as shown in Fig. JD824. After installation, ream the guides to an inside diameter of 0.3745-0.3760.

VALVE SPRINGS
All Models

24. Inlet and exhaust valve springs are interchangeable. E a c h spring should require not less than 39 pounds pressure to compress it to a height of 2 inches. Springs which are rusted, distorted, or do not meet the foregoing pressure specifications should be renewed.

VALVE TAPPET LEVERS
(ROCKER ARMS)
All Models

25. To remove valve tappet levers and shaft assembly, remove hood and tappet cover (rocker arm cover). Remove nuts retaining tappet lever shaft

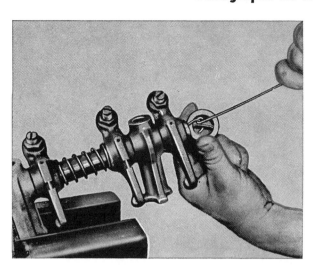

Fig. JD825—Using hooked tool to disassemble the 40, 320 and 330 series tappet levers assembly. Tool can be made from welding rod.

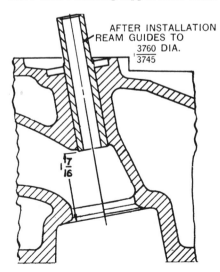

AFTER INSTALLATION
REAM GUIDES TO
3760 DIA.
3745

7
16

Fig. JD824—When installing guides on 420, 430 and 440 series engines, press guide in until port end measures 1 7/16 inches from top of combusion chamber as shown.

supports to cylinder head and lift assembly from head. To disassemble the 40, 320 and 330 series, pull one of the end caps out from the shaft, then engage spring (11—Fig. JD822) with a wire hook and lift off the end cap (5) as in Fig. JD825. Tappet levers, supports and springs can then be removed from shaft.

To disassemble the 420, 430 and 440 series, remove cotter pin from one end of shaft and slide parts from shaft. Refer to Fig. JD823.

Valve contacting surface of tappet lever can be refaced but original arc must be maintained by performing this operation on a fixture especially designed for refacing tappet levers. Diameter of tappet lever shaft is 0.810-0.8105. Tappet lever bore (inside diameter) is 0.812-0.813. Clearance in excess of 0.006 between unbushed bore of lever and shaft is corrected by renewing tappet lever and/or shaft.

Inlet and exhaust tappet levers are not interchangeable on 40, 320 and 330 series engines. On 420, 430 and 440 series engines, tappet levers are interchangeable for the individual cylinders but are not interchangeable between cylinders. If shaft support studs have been removed, be sure they are reinstalled in their correct location; rear stud (3) is machined to provide an oil passage for lubrication of tappet levers.

CAM FOLLOWERS
All Models

26. The mushroom type cam followers (6—Fig. JD828) operate directly in machined bores of the cylinder block. Cam followers are supplied only in the standard size and should

have an I&T suggested clearance of 0.0005-0.0025 in block bores. Any tappet can be removed after removing the camshaft as outlined in paragraph 30.

VALVE TIMING
All Models

27. On 40, 320 and 330 series engines, inlet valve opens at top dead center and exhaust valve closes 5 degrees after top dead center. On 420, 430 and 440 series engines, inlet valve opens 1 degree after top dead center and exhaust valve closes 4 degrees after top dead center. Valves are correctly timed when mark on camshaft gear is meshed with an identical mark on crankshaft gear as in Fig. JD826.

To check valve timing when engine is assembled, adjust tappets to 0.012 on the 40, 320 and 330 series, 0.015 on the 420, 430 and 440 series. Turn engine until front cylinder inlet valve is just opening. At this time the D/C

Fig. JD826 — 40, 320, 330, 420, 430 and 440 series valve timing marks on camshaft and crankshaft gears.

mark on flywheel, shown in Fig. JD 827, should be within ¼ inch of index line on side of timing hole. If marks do not register within ¼ inch, remesh camshaft gear with crankshaft gear.

TIMING GEAR COVER
All Models

28. To remove the timing gear cover, first remove the complete front end unit as outlined in paragraph 5 or 5A. Remove fan assembly on 40, 320 and 330 series and on 420, 430 and 440 series, remove water pump. Remove generator and hydraulic pump. Disconnect controls and remove governor. Remove pulley from crankshaft and unbolt cover from cylinder block. When reinstalling cover, refer to note in paragraph 29.

CRANKSHAFT FRONT OIL SEAL
All Models

29. To renew the crankshaft front oil seal, first remove the complete front end unit as per paragraph 5 or 5A. Remove fan assembly on 40, 320 and 330 series and on 420, 430 and 440 series, remove water pump. Remove pulley from crankshaft. Remove front oil seal plate (14—Fig. JD830). Oil seal (15) can be pressed out of plate and renewed.

Note: Soak new seal in oil and install in plate with lip of seal facing toward rear of tractor. Slide seal and plate assembly on crankshaft using a piece of shim stock as a sleeve to prevent damaging seal.

CAMSHAFT AND BEARINGS
All Models

30. To remove camshaft, first remove the timing gear cover and tappet levers (rocker arms) assembly. Re-

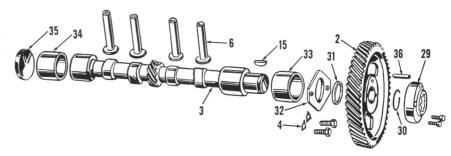

Fig. JD828 — 40, 320 and 330 series camshaft, bushings and cam followers. Hydraulic pump drive coupling (29) is bolted to camshaft gear. 420, 430 and 440 series camshaft and related parts are similar.

2. Camshaft gear	6. Cam follower	31. Spacer washer	34. Rear bushing
3. Camshaft	15. Gear key	32. Thrust plate	35. Expansion plug
4. Tab washers	30. Snap ring	33. Front bushing	36. Dowel pin

move oil pan, oil pump and ignition distributor. Unlock thrust plate cap screws and remove screws from block as shown in Fig. JD829. Hold cam followers up and carefully withdraw camshaft. To remove gear from shaft, remove hydraulic pump drive coupling (29—Fig. JD828) and snap ring (30); then press shaft out of gear.

Camshaft journals rotate in two renewable presized bushings. Shaft journal size is 1.8095-1.8105. Inside diameter of bushings is 1.812-1.813. The maximum permissible clearance between camshaft journals and the bushings is 0.005 and when it exceeds this amount, it will be necessary to renew the bushings.

To renew camshaft bushings, remove engine and flywheel to gain access to expansion plug (35) behind rear bushing. Bushings can be driven out or pressed out of block and new ones pressed or driven in. Bushings are presized and, if carefully installed, using John Deere driver tool number AM-458-T or equivalent, require no final sizing. Be sure oil hole in each bushing registers with oil slot in cylinder block.

Position expansion plug (35) with cupped side of plug facing rearward. Press the plug into the block until front surface of plug is 13 7/32 inches from front face of block at front camshaft bearing bore. The expansion plug must be installed in this position to assure an adequate oil supply to the tappet lever (rocker arm) oil passage.

Install thrust plate spacer washer (31) with chamfered edge of spacer bore facing front camshaft journal. Install thrust plate (32) and Woodruff key (15) and press gear on shaft until gear seats firmly against spacer washer. Length of spacer washer (31) and thickness of thrust plate control

camshaft end play. For correct end play, the measured thickness of washer (31) should be 0.003 to 0.007 greater than the thickness of thrust plate (32).

Mesh single mark on camshaft gear with single mark on crankshaft gear as shown in Fig. JD826. Install oil pump as outlined in paragraph 39 and ignition distributor as in paragraph 51.

TIMING GEARS
All Models

31. **CAMSHAFT GEAR.** Gear is keyed to shaft and the press fit is usually such as to require removal of the camshaft if the gear is to be renewed.

CRANKSHAFT GEAR. To R&R the crankshaft gear it is first necessary to remove the crankshaft which in turn necessitates the removal of the engine from the tractor. Refer to paragraph 20 for engine removal. To facilitate installation of gear on crankshaft, heat gear in boiling water.

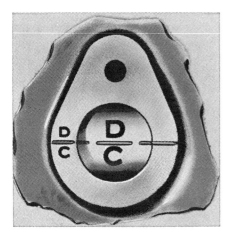

Fig. JD827 — 40, 320, 330, 420, 430 and 440 series top center position is when flywheel mark DC registers with index on right side of engine.

Fig. JD829—Removing 40, 320, 330, 420, 430 and 440 series cap screws which retain camshaft thrust plate to cylinder block. Cap screws have tab type lock washers.

GOVERNOR GEAR. Removal requires governor removal; refer to GOVERNOR section.

ROD AND PISTON UNITS
All Models

32. Piston and connecting rod assemblies are removed from above after removing cylinder head and oil pan.

Remove carbon accumulation and ridge from unworn portion of cylinders to prevent damaging ring lands when pistons are withdrawn. Connecting rods are not numbered and should be identified in some manner for reassembly. Install connecting rods and caps so that raised (pip) marks face camshaft side of engine and tighten the connecting rod cap screws to 55-60 Ft.-Lbs. torque.

PISTONS AND RINGS
Series 40-320-330

33. Aluminum pistons are supplied in standard, 0.020 and 0.040 oversize. Diameter of the standard piston skirt is 3.995-3.996 inches. Diameter of the standard cylinder bore is 4.000-4.001. Rebore cylinder if out-of-round more than 0.003. Clearance of piston skirt in cylinder should be 0.004-0.006. Raised (pip) mark on piston heads should face toward front of engine.

There are three compression rings and one ventilated oil ring per piston. All rings are stamped either "TOP" or with a small "pip" mark to indicate the top portion of the ring. Recommended end gap is 0.010-0.020 for compression rings, 0.010-0.018 for the oil control rings. Recommended side clearance for the top compression ring is 0.0015-0.003; for the other rings, 0.001-0.0025. Rings are supplied in standard as well as 0.020 and 0.040 oversizes.

Series 420-430-440

33A. Aluminum pistons are supplied in standard as well as 0.020 and 0.040 oversizes. Diameter of the standard piston skirt is 4.246-4.247 inches. Diameter of the standard cylinder bore is 4.250-4.251. Rebore cylinder of out-of-round more than 0.003. Clearance of piston skirt in cylinder should be 0.003-0.005. Raised (pip) mark on piston heads should face toward front of engine.

There are three compression rings and one ventilated oil ring per piston. All rings are stamped either "TOP" or with a small "pip" mark to indicate the top portion of the ring. Recommended end gap is 0.013-0.023 for all

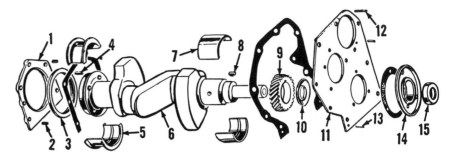

Fig. JD830—40 series crankshaft and main bearings. Renewal of rear oil seal (3) requires detaching engine from center frame and removal of flywheel. 320, 330, 420, 430 and 440 series are similar.

1. Oil seal retainer	6. Crankshaft	9. Crankshaft gear	12. Dowel pin (upper)
2. & 4. Dowel pin	7. Front bearing	10. Oil slinger	13. Dowel pin (lower)
3. Oil seal	8. Gear key	11. Gear cover	14. Oil seal plate
5. Rear bearing			15. Oil seal

rings. Recommended side clearance for top compression ring is 0.0015-0.003; for the other rings 0.001-0.0025. Rings are supplied in standard as well as 0.020 and 0.040 oversizes.

PISTON PINS AND BUSHINGS
All Models

34. The 1.1877-1.1879 inch diameter floating type piston pins which are retained in the piston bosses by snap rings, are available only in the standard size. Fit pins to a thumb press fit in piston. Clearance in connecting rod bushing is 0.0006-0.0009 for 40 series; 0.0004-0.0009 for 320, 330, 420, 430 and 440 series.

CONNECTING RODS AND BEARINGS
All Models

35. Connecting rod bearings are steel-backed, babbitt-lined, non-adjustable, precision type and can be renewed without removing connecting rod and piston assembly. Replacement rods are not marked and should be installed with the small raised (pip) mark on rod and cap facing the camshaft side of engine. Bearings are available in standard size and undersizes of 0.003, 0.030 and 0.033.

Crankpin diameter:
Series 320 prior 320855 . . .2.249-2.250
Series 320 after 320854 . . .2.498-2.499
Series 40-330-420-430-
   440 . . . . .2.498-2.499
Running clearance 0.001-0.0035
Cap screw torque. . . . . . .55-60 Ft. Lbs.

CRANKSHAFT AND BEARINGS
All Models

36. Crankshaft is supported on two steel-backed, babbitt-lined, non-adjustable, precision type bearings, renewable from below after removing the oil pan. Renew worn bearings; do

not file caps. Remove main bearing caps and turn bearing upper halves out of cylinder block with a reworked rivet inserted in oil hole of crankshaft journal. Rear bearing controls crankshaft end play of 0.003-0.007. Bearings are available in standard size and undersizes of 0.003, 0.005, 0.030 and 0.033.

The quickest method of removing the crankshaft is to remove the engine from the tractor as outlined in paragraph 20.

Check crankshaft journals for wear, scoring and out-of-round condition against the values which follow:
Journal diameter 2.397-2.398
Running clearance 0.001-0.0035
Renew shaft if
   out-of-round more than 0.003
Cap screw torque . . . . .140 ft. lbs.

CRANKSHAFT REAR OIL SEAL
All Models

37. To renew the treated leather oil seal (3-Fig. JD830), which is mounted in a one-piece retainer (1), split tractor (detach engine from center frame), as outlined in paragraph 69 and remove flywheel. Install new oil seal with the lip of same facing towards front of engine. Soak seal assembly in oil before installing. Slide seal and plate assembly on crankshaft using a piece of shim stock as a sleeve to prevent damaging seal.

Seal contacting surface of crankshaft must be smooth and true. Surface of crankshaft and bore of seal plate must be concentric within 0.005.

FLYWHEEL
All Models

38. To remove flywheel, split tractor (detach engine from center frame) as outlined in paragraph 69 and remove clutch. Remove cap screws retaining

flywheel to crankshaft and pry flywheel off dowels.

To install a new ring gear, heat same to 500 degrees F. and install gear with beveled edge of teeth facing rear of tractor.

OIL PUMP
All Models

39. **REMOVE AND REINSTALL.** To remove oil pump, remove flywheel dust shield and oil pan. Remove lock nut from oil pump attaching screw located on right side of upper crankcase, then the Allen screw, as shown in Fig. JD832 and withdraw pump from bottom of crankcase.

When reinstalling pump, rotate engine crankshaft until number one (front) cylinder is on compression stroke and mark "DC" on flywheel indexes with mark at timing hole. Install pump and mesh gears (pump to camshaft) so that distributor driving groove in pump gear is parallel to the crankshaft and narrow part of groove

wall is toward outside of tractor. Recheck ignition timing.

40. **OVERHAUL.** End clearance between pump body gears and lower cover (11—Fig. JD833) should be 0.002-0.006. The recommended radial clearance between the gears and pump body is 0.0005-0.002. Check pump drive shaft and its mating surface for wear. New shaft diameter is 0.4825-0.4830 with a pump body bore diameter of 0.4835-0.4845. No gasket is used between lower cover and pump body; be sure surfaces are clean, true and not distorted.

OIL RELIEF VALVE
All Models

41. The spool type relief valve (3—Fig. JD833) is located in oil pump body and can be adjusted with shims (5) inserted between the spring and relief valve retaining plug to maintain a pressure of 30 psi. Oil pressure adjustment requires removal of the oil pan.

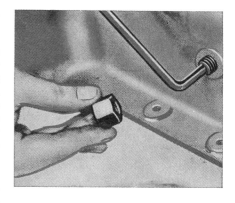

Fig. JD832 — 40, 320, 330, 420, 430 and 440 series oil pump is retained in position by an Allen head screw located on right side of upper crankcase.

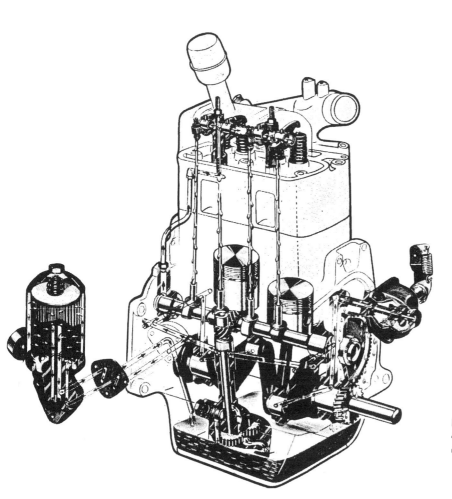

Fig. JD831—Phantom view of 40 series lubrication system. Rear tappet lever shaft support stud provides an oil passage to lubricate tappet levers assembly. Pressure relief valve (shim adjustment) is located in pump body. 320, 330, 420, 430 and 440 series are similar.

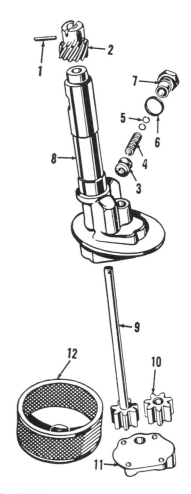

Fig. JD833 — 40, 320, 330, 420, 430 and 440 series oil pump. Pressure relief valve adjustment is controlled by shim washers (5) at relief valve spring.

1. Gear pin	7. Plug
2. Drive gear	8. Pump body
3. Relief valve	9. Shaft & pump gear
4. Valve spring	10. Idler gear
5. Shims	11. Pump cover
6. Gasket	12. Pump screen

CARBURETOR (EXCEPT LP-GAS)

42. Marvel-Schebler carburetors are used and their applications are as follows:

40 Series
    Gasoline TSX530
    All-fuel TSX562
320 Series
    Gasoline TSX245
    All-fuel TSX245
330 Series
    Gasoline TSX245
420-430 Series
    Gasoline TSX641
    All-fuel TSX678
440 Series
    Gasoline (Prior 448001)....TSX756
    Gasoline (After 448000)....TSX777
    All-fuel (Prior 448001).....TSX768

Carburetor must be removed to adjust the fuel level. Float should be set to ¼ inch when measured from bowl cover gasket surface to nearest edge of float. The idle mixture screw controls the flow of air in the idle system and turning the screw toward its seat richens the mixture. Power range is controlled by a power jet needle which reduces fuel flow and leans the mixture when turned toward its seat. Refer to table on page 72 for calibration data.

LP-GAS SYSTEM

Total fuel tank capacity is 24 gallons on series 420 and 430; but tank should NEVER be filled more than 85 per cent full of fuel (21 gallons). This allows room for expansion of the fuel due to a possible rise in temperature.

CAUTION: LP-Gas expands readily with any decided increase in temperature. If tractor must be taken into a warm shop to be worked on during extremely cold weather, make certain that fuel tank is as near empty as possible. LP-Gas tractors should never be stored or worked on in an unventilated space.

TROUBLE SHOOTING
Series 420-430

42A. The following trouble shooting paragraphs list troubles that can be attributed directly to the fuel system; however, many of the troubles can be caused by derangement of other parts such as valves, battery, spark plugs, distributor, coil, etc.

The procedure for remedying many of the causes of trouble is evident.

The following paragraphs will list the most likely causes of trouble, but only the remedies which are not evident.

42B. **HARD STARTING.** Hard starting could be caused by:

a. Improperly blended fuel.

b. Excess-flow valve in withdrawal valve closed. Close withdrawal valve to reset excess-flow valve, then open withdrawal valve slowly.

c. Incorrect starting procedure.

d. Restricted fuel strainer.

e. Liquid fuel in lines.

f. Automatic fuel shut-off on strainer not operating properly. A "click" should be heard when ignition switch is turned on. If no click is heard, check wiring and check solenoid on strainer.

g. Plugged vent on back of convertor. The vent is a ¼-inch tapped hole.

h. Defective low pressure diaphragm in convertor.

i. Stuck high pressure valve or broken high pressure spring in convertor.

j. Restricted fuel lines.

42C. **ENGINE SHOWS NOTICEABLE LOSS OF POWER.**

a. Throttle not opened sufficiently due to maladjusted governor or carburetor linkage.

b. Plugged vent on back of convertor. The vent is a ¼-inch tapped hole.

c. Clogged fuel strainer (if strainer shows frost, it is probably clogged).

d. Plugged fuel lines or restrictions in withdrawal valves (indicated by frost). With engine cold, both withdrawal valves closed and lines and filter empty of gas, remove plug at bottom of strainer, open the liquid withdrawal valve slightly and check for fuel flow.

e. Closed excess flow valves in vapor or liquid withdrawal valves (Indicated by frosted withdrawal valve). Close frosted valve to seat excess flow valve, then re-open slowly.

f. Lean mixture caused by restricted or altered fuel lines or hoses.

g. Sticking high pressure valve in convertor.

h. Restricted low pressure valve in convertor.

i. Defective convertor diaphragms.

j. Engine not up to operating temperature. Check thermostat.

k. Improperly adjusted carburetor.

l. Faulty adjustment of throttle linkage.

m. Faulty converter adjustment.

n. Faulty gasket between carburetor and manifold.

o. Leaking fuel hose between convertor and carburetor.

p. Air entering between carburetor throttle body and air horn.

q. Clogged air filter.

42D. **POOR FUEL ECONOMY.** Could be caused by any of the conditions listed in paragraph 42C, plus:

a. Improperly filled fuel tank.

b. Faulty fuel.

42E. **ROUGH IDLING.** Could be caused by faulty ignition system plus:

a. Faulty adjustment of carburetor.

b. Faulty adjustment of throttle linkage.

c. Faulty carburetor to manifold gasket.

d. Leaking hose between carburetor and convertor.

42F. **POOR ACCELERATION.**

a. Faulty idle speed adjustment.

b. Faulty load adjustment.

c. Faulty low pressure diaphragm in convertor.

d. Restricted convertor to carburetor hose.

42G. **ENGINE STOPS WHEN THROTTLE IS BROUGHT TO SLOW IDLE POSITION.**

a. Faulty slow idle speed adjustment.

b. Faulty convertor to carburetor hose.

c. Faulty carburetor gaskets.

d. Faulty gasket between carburetor and air horn.

e. Leaking convertor back cover gasket (Indicated by fuel bubbles in radiator).

42H. **OVERHEATING.** Could be caused by defective cooling system plus:

a. Lean mixture due to faulty adjustment of linkage.

42J. **CONVERTOR FREEZES UP WHEN ENGINE IS COLD.**

a. Running on liquid fuel before engine is warm.

b. Leaking convertor high pressure valve. With ignition switch turned on, this can be detected by odor of gas or hissing sound.

42K. **CONVERTOR FREEZES UP DURING NORMAL OPERATION.** Could be caused by a defective cooling system plus:

a. Water circulating backwards through convertor.

b. Restrictions in water piping or convertor.

c. Running on liquid fuel before engine is warmed up.

42L. FROST ON WITHDRAWAL VALVE.

a. Closed excess flow valve. Close withdrawal valve to reset the excess flow valve; then, open withdrawal valve slowly.

b. Water in fuel tank will sometimes cause ice to form in liquid withdrawal valve. Empty all fuel from filler hose, pour one pint of alcohol in hose, attach hose to fuel tank and inject alcohol into fuel tank. Alcohol will act as an antifreeze and water will be dissipated through the engine.

42M. LACK OF FUEL AT CARBURETOR.

a. Empty fuel tank or withdrawal valve closed.

b. Excess flow valve in withdrawal valve closed. Close the withdrawal valve to reset the excess flow valve; then, open the withdrawal valve slowly.

c. Restriction in withdrawal valve.

d. Restricted fuel strainer.

e. Faulty wiring to strainer shut off valve or faulty valve.

f. Faulty convertor high pressure valve.

g. Restricted fuel lines.

42N. FUEL IN COOLING SYSTEM.

This trouble is usually caused by a ruptured convertor back cover gasket.

CARBURETOR
Models 420-430

42P. **ADJUSTMENT.** Before making any carburetor adjustments, be sure that the air cleaner is connected to the carburetor and is filled to the proper level with specified oil (same as crankcase oil).

42R. **LOAD ADJUSTMENT.** With engine at operating temperature and running at high idle speed, loosen jam

Fig. JD833A—External view of carburetor used on models equipped with LP-gas equipment. Refer to text for load adjustment.

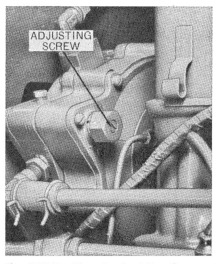

Fig. JD833B—View showing idle adjustment point on convertor. Refer to text for procedure.

nut and turn load adjusting needle in (clockwise) until the engine rpm starts to drop. Now turn needle out (counter-clockwise) until the engine rpm picks up and engine operation smooths out; then, open the load needle one additional turn and tighten jam nut. Refer to Fig. JD833A.

42S. **IDLE ADJUSTMENT.** With the load adjustment set as outlined in paragraph 42R, proceed as follows: Set throttle to low idle position and allow engine to stabilize. Note: This may take as long as two minutes as the engine low idle speed is controlled by the governor action and engine may hunt (surge) for several minutes before stabilizing. Carburetor throttle lever must NOT be against the throttle stop screw at this time.

Adjust the vertical governor control rod to provide a slow idle speed of 650-750 (700 preferred) rpm. Then, after engine rpm has again stabilized, adjust the idle fuel mixture to obtain the best low idle speed by means of the idle adjusting screw on the regulator as shown in Fig. JD833B. This adjustment is approximately twelve turns from the full rich (seated) position. Clockwise rotation of the idle adjusting screw richens the mixture. Readjust the governor vertical control rod, if necessary, to get the recommended 700 rpm low idle speed. Note: if an appreciable change in engine rpm has occurred, repeat the adjustment of the idle adjusting screw on the regulator.

After governor linkage is properly adjusted, manually hold the throttle lever of the carburetor against the throttle stop screw and adjust stop screw to give an engine speed of 600

rpm. Release the lever and recheck the 700 rpm setting by moving throttle lever from high idle position to low idle position several times and allowing engine to stabilize at slow idle. Do not be concerned if engine hunts (surges) several minutes before stabilizing.

FUEL STRAINER AND SHUT-OFF VALVE
Series 420-430

42T. All of the fuel must pass through the strainer before reaching the convertor. The strainer contains a filter element which consists of a felt pad and a chamois disc backed by a brass screen on each side. The purpose of the filter is to remove all solids from the fuel before the fuel reaches the convertor valves. A solenoid operated, automatic fuel shut-off is located on top of the strainer. Whenever the ignition switch is turned on, the solenoid opens the valve with an audible "click."

If the strainer shows frost, it is probably clogged and needs cleaning.

42U. **CLEANING.** To clean the strainer first make certain that both fuel tank withdrawal valves are closed, engine is cold and lines and filter are empty of gas. Note: Lines and filter will be empty if engine was properly stopped.

Remove plug (14—Fig. JD833C)

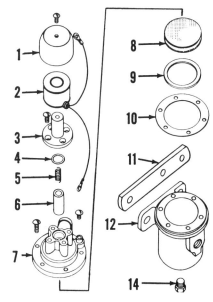

Fig. JD833C—Exploded view of the LP-gas fuel strainer and automatic shut-off assembly.

1. Fuelock case	8. Filter pack
2. Fuelock coil	9. Retaining ring
3. Plunger housing	10. Gasket
4. "O" ring	11. Bracket
5. Plunger spring	12. Strainer body
6. Plunger	14. Drain plug
7. Strainer housing	

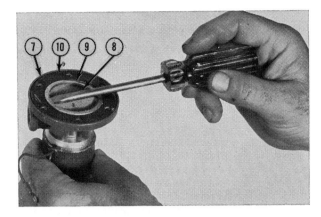

Fig. JD833D — Removing filter pack retaining ring from LP-gas fuel strainer cover.

7. Strainer cover
8. Filter pack
9. Retainer ring
10. Gasket

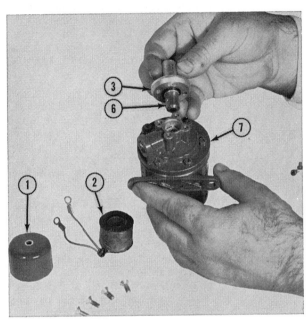

Fig. JD833E — Removing the fuel shut-off plunger and plunger housing from LP-gas fuel strainer cover.

1. Case
2. Fuelock coil
3. Plunger housing
6. Plunger
7. Housing

from bottom of strainer and open the liquid withdrawal valve slightly; thus allowing pressure from the fuel tank to blow out any accumulation of dirt.

42V. R&R AND OVERHAUL. To remove the strainer, close both fuel tank withdrawal valves, disconnect fuel lines and remove strainer.

Remove cover from strainer body and remove the filter pack by prying out the retainer ring with a screw driver as shown in Fig. JD833D. Filter pack can be cleaned in a suitable solvent. Reinstall filter pack with chamois disc toward top.

Disconnect wire and remove case from shut-off valve. Lift off coil (2—Fig. JD833E), remove plunger housing (3) and lift out plunger and spring. Inspect and renew any damaged parts.

Assemble the solenoid coil, plunger and spring on plunger housing.

To test solenoid, connect it to a battery to see if plunger compresses spring in plunger housing.

When reassembling, fasten one of the coil wires to screw which retains case (1). After unit is installed on tractor, test for leaks by using soapy water around all connections.

CONVERTER (REGULATOR)
Models 420-430

42W. HOW IT OPERATES. With the ignition switch turned on, the fuel shut-off valve solenoid is energized which opens the shut-off valve and permits fuel to flow through the filter to the primary chamber (A—Fig. JD-833F) of the convertor through primary valve (1) which is held open by spring (2). As the fuel enters the primary chamber, it passes through a series of passages (B) which accelerate vaporization. As the pressure builds up in the primary chamber, the primary diaphragm (3) is forced outward and compresses spring (2). The primary valve (1) is returned to its seat by spring (5) which forces primary valve lever (4) to follow the outward movement of the primary diaphragm. This action of the diaphragm and primary valve maintains approximately 2½-5 psi pressure in the primary chamber.

Vacuum created by the engine is applied to the secondary chamber (E) which in turn reduces the pressure in the secondary chamber. This causes the secondary diaphram (11) to move inward, putting pressure on the secondary lever (10) which opens the secondary valve (6). Since the passage uncovered by the secondary valve communicates with primary chamber (A), fuel will flow from the primary chamber to the secondary chamber in an amount equal to the demand of the engine.

Fig. JD833F — Sectional views of convertor. Refer to text for operation.

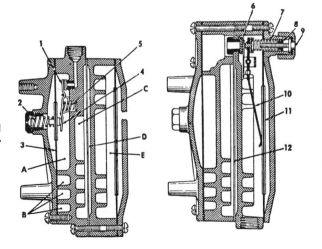

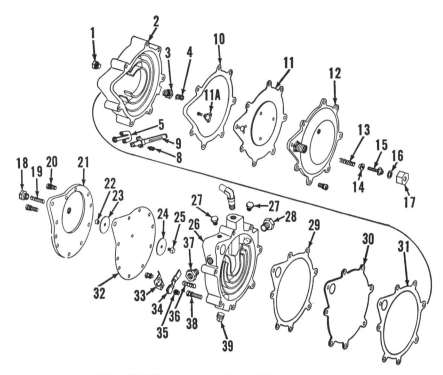

Fig. JD833G—Exploded view of LP-gas convertor.

1. Secondary valve seat	15. Idle adjusting screw
2. Secondary body	16. Rubber gasket
3. Bushing	17. Cap
4. Nipple	18. Cap
5. Secondary valve	19. Primary spring
8. Screw	20. Screw
9. Secondary valve lever	21. Primary cover
10. Secondary gasket	22. Nut
11. Secondary diaphragm	23. Washer
11A. Sealing plate	24. Washer
12. Secondary cover	25. Button
13. Secondary spring	26. Primary body
14. Retainer	

27. Plug
28. Bushing
29. Gasket
30. Baffle plate
31. Gasket
32. Primary diaphragm
33. Primary lever bracket
34. Primary lever
35. Primary valve seat
36. Spring
37. Bushing
38. Screw
39. Plug

42X. R&R AND OVERHAUL. Close both vapor and liquid withdrawal valves, start engine and allow to run until all fuel is exhausted from convertor, fuel lines and carburetor. Shut off ignition. Drain the tractor cooling system and the convertor. Remove tractor hood, then disconnect the water hoses and fuel lines from convertor. Remove mounting bolts and lift convertor from its mounting bracket.

Remove cap (18—Fig. JD833G) and the primary spring (19). Remove the cap screws which retain primary cover (21) and lightly tap the primary cover until it begins to separate from the primary body (26). Now carefully free the primary diaphragm (32) from primary cover and primary body, then remove primary cover and diaphragm.

Remove the cap screws which retain the primary body (26) to secondary body (2), separate the two parts and remove the baffle plate (30) and gaskets (29 & 31). Gaskets (29 & 31) will probably adhere to parts and be

broken, so it will be necessary to scrape off the broken pieces.

Remove cap (17) and gasket (16), then remove the idle adjusting screw (15), retainer (14) and spring (13).

Remove screws retaining secondary cover (12) to secondary body (2). Tap secondary cover gently and when cover starts to separate from body, carefully free diaphragm from secondary body. Remove screws from sealing plate 11A, then remove sealing plate and secondary diaphragm from secondary cover.

Remove the two screws from secondary valve lever (9) and secondary valve leaf lever (5) and remove same from body (2).

Remove the two screws and free the primary lever bracket (33) and remove bracket, primary lever (34), spring (36) and seat (35).

42Y. Wash all parts EXCEPT diaphragms and valves in a suitable solvent and dry with compressed air. Diaphragms are normally clean; however, if cleaning is necessary, wipe same with a cloth moistened with

cleaning solvent. Clean valves by wiping with a clean, dry cloth.

Check all parts as follows: Inspect diaphragms for signs of chafing, tears, thin spots or excessive wrinkling. If condition of diaphragms is questionable, renew same. Check all valves and levers for wear and misalignment. Be sure to renew valves if the faces are scratched, scored or grooved. Check the knife edge of the primary lever bracket (33) and if this edge is worn, renew the bracket. Check all springs for signs of wear or distortion and renew any springs that are the least questionable. It is recommended that the primary lever spring and the primary diaphragm spring be renewed each time the convertor is disassembled, unless the convertor has had very little service.

42Z. Install secondary valve seat (1) in secondary body (2) and tighten securely. Place the gaskets (29 & 31) in position on baffle plate (30), then place baffle plate and gaskets in position on secondary body (2) and install retaining screws (38). Tighten screws progressively in diametrically opposed pairs.

Install the primary valve seat (35) in the primary body (26) and tighten seat with a thin walled socket. Use a plastic or some other similar sealant on threads. Do not use shellac or similar compounds. Place the primary lever spring (36) in recess of primary body, then engage the primary lever assembly (34) in the primary lever bracket (33) and attach the primary bracket to the primary body. Make sure that spring (36) fits over the projection on the primary lever (34). Actuate the lever by hand and observe its action. Valve should operate smoothly with no binding and the valve cap should come to rest squarely on the valve seat.

Place diaphragm assembly (32) (includes items 22, 23, 24 & 25) on primary body (26). Place the lockwashers on screws and insert same through the holes in the primary cover (21). Position cover on primary body and apply enough pressure to flatten diaphragm, then start all screws. Now press primary cover down and tighten screws in diametrically opposed pairs. CAUTION: When tightening cover screws, be sure that diaphragm does not wrinkle. Place the primary diaphragm spring (19) into the recess of retainer cap (18), then install the spring and cap in the primary cover and tighten cap securely.

Position the secondary leaf valve (5) in the secondary body (2); install

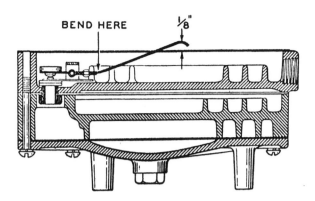

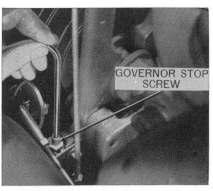

Fig. JD833H—View showing adjustment of secondary valve lever. If adjustment is necessary, bend lever at point indicated. Refer to text.

Fig. JD835—Deere 40, 320 and 330 series governed speed is controlled by a stop screw which is located in the left side of the fan bracket. The stop screw can be adjusted as shown. 420, 430 and 440 series are similar except that stop screw is located in water pump housing.

the secondary valve lever (9) on top of the secondary leaf valve and secure with screws. Check to be sure that lever assembly is not binding and that the leaf lever has not been warped or distorted by the tightening of the retaining screws. The valve on the leaf lever should seat squarely on the seat (1). Adjust the secondary valve lever (9) to extend ⅛-inch, plus or minus 1/64-inch, above the face of the secondary body as shown in Fig. JD833H. If necessary to adjust, bend lever at the location shown.

Place the secondary diaphragm assembly (11—Fig. JD833G) in position against secondary cover (12) and align holes. Place the diaphragm sealing plate (11A) in position on diaphragm (11) and secure with the three screws. Place lock washer on screws and insert screws through secondary cover (12) and diaphragm (11), then place secondary gasket (10) over screws. Position the secondary cover, diaphragm and gasket on secondary body (2) and while making certain diaphragm lies flat and smooth, start screws but do not completely tighten same until after idle screw has been installed.

Install valve spring (13) in opening of cover (12) and make sure spring engages its seat on secondary valve lever (9). Start the retainer (14) on the idle adjusting screw (15) and rotate retainer until it is approximately half-way up on the adjusting screw. Insert the adjusting screw and retainer into the hexagonal guide in the cover (12) and be sure retainer slides freely in guide. Lubricate the rubber gasket (16) and press same on head of adjusting screw (15). Place cap (17) over rubber gasket and head of adjusting screw and while holding adjusting screw with a screw driver, tighten cap. Note: Tighten cap (17) until sufficient pressure is put on rubber gasket to prevent idle adjusting screw from turning due to vibration, yet allow it to be free enough to turn if adjustment is necessary. Now tighten screws

holding secondary cover (12) to secondary body (2). Be sure to tighten screws progressively and in diametrically opposed pairs.

Set idle adjusting screw twelve turns open (counter-clockwise) for an initial setting and reinstall any plugs and/or reducers removed during disassembly. Reinstall convertor assembly to tractor by reversing the removal procedure and adjust as outlined in paragraphs 42R and 42S.

GOVERNOR

All Models

43. **SPEED ADJUSTMENT.** Governed speed is controlled on 40, 320 and 330 series tractors by a stop screw which is located in the left side of the fan mounting bracket. On 420, 430 and 440 series tractors, stop screw is located at lower left hand side of water pump. Adjust the screw as shown in Fig. JD835 to obtain the recommended speeds which follow:

Series 40-420-430-440 (Prior 448001)
Crankshaft rpm (Load)........1850
  (No Load) 2025
Belt Pulley rpm (Load).......1270
  (No Load) 1390
Power Take-Off rpm (Load)... 560
  (No Load) 612

Series 440 (After 448000)
Crankshaft rpm (Load)......2000
  (No Load) 2200
Belt Pulley rpm (Load).......1373
  (No Load) 1510
Power Take-Off rpm (Load).. 605
  (No Load) 665

Series 320-330
Crankshaft rpm (Load).......1650
  (No Load) 1825
Belt Pulley rpm (Load).......1246
  (No Load) 1376
Power Take-Off rpm (Load)... 550
  (No Load) 608

44. **LINKAGE ADJUSTMENT.** Free-up and align all linkage to remove any binding. Adjust or renew any parts causing lost motion. Back out stop screw (Fig. JD835) located in engine

fan bracket or water pump housing. With the hand control in the lowest speed position and governor to carburetor rod disconnected, push down on governor lever (C—Fig. JD836). Now adjust the rod connecting the hand lever control shaft to governor lever so that hole in rod yoke and hole in governor lever are in register; then lengthen the rod two turns. Place hand control in full speed position and with carburetor throttle valve in wide open position, adjust length of carburetor throttle rod (E) so that the rod just enters hole in carburetor throttle shaft lever. Then

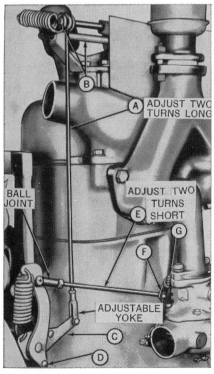

Fig. JD836—Deere 40, 320 and 330 series governor and carburetor linkage, showing control rods and linkage adjustments. 420, 430 and 440 series are similar and adjusted the same.

shorten the rod two turns. Adjust engine speed with the governor stop screw as outlined in preceding paragraph.

45. **R & R GOVERNOR.** Remove hood and grille. Remove air cleaner and disconnect governor linkage. Loosen the fan adjusting nut on 40, 320 and 330 series tractors, or pulley flange on 420, 430 and 440 series tractors and remove the fan belt. Remove cap screws retaining governor housing to timing gear cover and remove housing. Withdraw governor shaft assembly being careful not to drop thrust washer (4—Fig. JD837). If the washer is dropped, it will be necessary to remove either the oil pan or timing gear cover to retrieve the washer.

46. **DISASSEMBLE AND OVERHAUL GOVERNOR.** Remove spring. Remove upper expansion plug (14—Fig. JD837). Bump pin (10) out of fork (9) and operating shaft. Remove expansion plug (11) from side of case and bump operating shaft out of case. Remove seal retainer (16) from side of case and renew oil seal. Weight pins (6) can be removed from carrier after grinding off the staked portion at end of pins. Shaft front bushing (12) can be renewed after removing lower front expansion plug (13). The bushing (12) is of the blind type and when installing the bushing, press the bush-

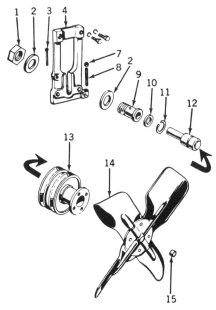

Fig. JD838—Deere 40, 320 and 330 series fan assembly and associated parts. John Deere recommends renewing the complete fan assembly as a unit.

1. Adjusting nut	9. Sleeve
2. Washer	10. Rubber washer
3. Cotter pin	11. Snap ring
4. Mounting bracket	12. Bearing assembly
7. Jam nut	13. Pulley and hub
8. Speed adjusting	14. Fan blades
screw	15. Cap

ing in until blind end is 2.275-2.280 from the machined gasket surface of the governor case. To remove shaft rear bushing (3) from cylinder block, tap bushing for a ½ inch SAE thread

and pull bushing with a cap screw. The bushing should be installed so that front end of bushing is 0.010-0.020 rearward from the front face of the bushing mounting boss in cylinder block. Bushings are presized and should be installed using a piloted drift.

COOLING SYSTEM

FAN

Series 40-320-330

48. An exploded view of the fan assembly and associated parts is shown in Fig. JD838. The procedure for removing the fan is evident after removing the radiator as outlined in paragraph 48A. The John Deere factory recommends renewing the complete fan assembly as a unit.

RADIATOR

All Models Without Power Steering Except 440I

48A. To remove the radiator, remove the hood and grille, drain cooling system and disconnect upper and lower radiator hoses. Remove the Allen head screw retaining steering shaft to the steering gear worm shaft and slide steering shaft rearward. On models with slanted steering wheel it will be necessary to drive roll pin from the coupling at rear of steering shaft before shaft can be moved rearward. On tricycle models, unbolt steering gear support from front end support and lift unit from tractor. On all other models, unbolt steering gear housing from steering gear support and turn the gear housing until worm shaft will clear the radiator. Unbolt fan shroud from radiator, then remove the radiator mounting bracket bolts and lift radiator from tractor. See Fig. JD838A.

All Models With Power Steering Except 440I

49. On models with power steering, remove hood and grille, drain cooling system and disconnect upper and lower radiator hoses. On tractors with vertical steering wheel, disconnect front steering shaft from steering gear worm shaft and move steering shaft and control valve rearward enough to clear worm shaft. On tractors with slanted steering wheel, connect the two valve to cylinder lines in series as follows: Disconnect one line from power cylinder and the remaining line from control valve. Now, connect the open end of the line which is still attached to the power cylinder, to the open connection on the power cylinder. Follow the same procedure with the

1. Dowel pin (long)	11. Plug (right)
2. Dowel pin (short)	12. Front bushing
3. Rear bushing	13. Plug (lower)
(in block)	14. Plug (upper)
4. Thrust washer	15. Housing
5. Shaft & gear	16. Seal retainer
6. Weight pin	17. Seal
7. Weight	18. Lever
8. Thrust bearing	19. Operating lever
9. Fork	& shaft
10. Fork pin	20. Spring

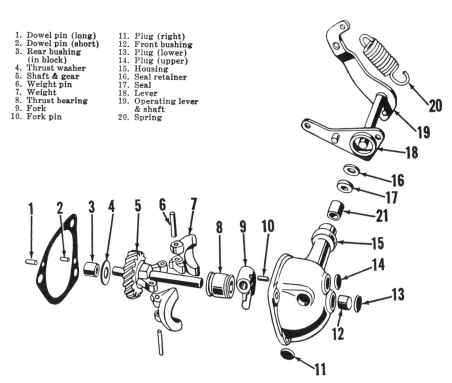

Fig. JD837—Exploded view of Deere 40, 420, 430 and 440 series governor. Bushings 3, 12 and 21 are pre-sized. 320 and 330 series are similar.

Fig. JD838A—View showing model 420T with steering gear support and radiator removed. Model 430T is similar.

Fig. JD839A—View showing 420T with steering gear support, radiator and water pump removed. Model 430T is similar.

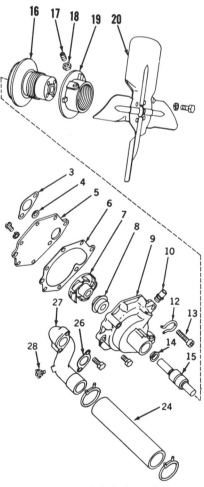

Fig. JD839 — Exploded view of the water pump and component parts used on 420, 430 and 440 series tractors. Bearing and shaft assembly (15) is purchased as a unit. See text for dimension when installing slinger (14).

4. Shim	14. Slinger
5. Cover	15. Bearing and shaft
7. Impeller	assembly
8. Seal	27. Inlet elbow
9. Housing	28. Drain cock
12. Retainer	

line which is still attached to the control valve. Disconnect front steering shaft from steering gear worm shaft, drive roll pin from coupling at rear of steering shaft and move steering shaft and control valve rearward enough to clear steering gear worm shaft. NOTE: On tractors serial number 133,665 and up, it will also be necessary to remove the cap screws which retain the control valve bracket.

On tricycle models, unbolt steering gear support from front end support and lift unit from tractor. On all other models, unbolt steering gear housing from the gear support and turn steering gear housing until worm shaft will clear radiator. Unbolt fan shroud from radiator, then remove the radiator mounting bracket bolts and lift radiator from tractor.

Model 440I

49A. On 440I models, the work outlined in paragraph 48A or 49 will have to be done; in addition, the left hand side rail will have to be removed in order to gain access to the radiator mounting bolts.

WATER PUMP
Series 420-430-440

49B. **REMOVE AND REINSTALL.** To remove water pump, first remove radiator as in paragraph 48A, 49 or 49A. Then remove cap screws retaining fan to fan hub and remove fan. Loosen the adjustable flange lock nut (18—Fig. JD839) and set screw (17) then remove the adjustable flange (19) and fan belt from the fan hub. Remove cap screw from generator adjusting bracket and move bracket

upward out of the way. Disconnect by-pass hose from nipple (10) and the hydraulic pressure line from hydraulic pump. Remove water inlet elbow (27). Support water pump, remove attaching socket head screws and withdraw water pump assembly from tractor, being careful not to lose the shim washer (4) located back of the upper right hand attaching lug. See Fig. JD839A.

Using heavy grease to hold the inlet elbow gasket (26—Fig. JD839) and shim washer (4) in position, rein-

Fig. JD839B—View showing method of removing fan hub from water pump shaft. Refer to text and Fig. JD839H for installation dimension.

stall the pump by reversing the removal procedure.

49C. **OVERHAUL.** With pump removed as in paragraph 49B, use suitable puller and remove fan hub from pump shaft as shown in Fig. JD839B. Remove rear cover plate and gasket. Remove bearing retainer snap ring as shown in Fig. JD839C. Use a drift smaller than the diameter of water pump shaft and press shaft and bearing assembly from impeller and pump housing. See Fig. JD839D.

Inspect all parts for damage or wear and renew those found unserviceable.

Leakage at drain hole in bottom of housing indicates a leaking seal (8—Fig. JD839). To install the seal, drive only on metal portion. Lubricate lip of seal with Lubriplate or its equivalent. If a new slinger is installed, press same on shaft until the measured distance between the shoulder of slinger and end of shaft is 1½ inches as shown in Fig. JD839E. Shaft and bearing is available only as an assembly.

To reassemble, press the pump shaft and bearing unit into housing until groove in bearing aligns with snap ring slot. Then install snap ring. With pump shaft supported at front end, press the impeller on shaft (vanes toward front of pump) until there is 0.037-0.057 clearance between impeller vanes and pump housing as shown in Fig. JD839F. Using a drift smaller than the diameter of the pump shaft, support pump shaft at impeller end as shown in Fig. JD839G and press fan hub on shaft until measured distance between forward end of shaft and forward end of fan hub is 1.095-1.115 inches as shown in Fig. JD839H.

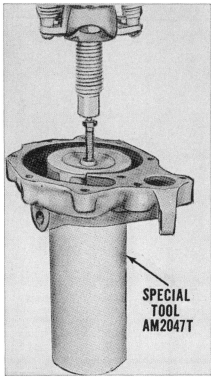

Fig. JD839D — View showing method of pressing water pump shaft and bearing assembly from impeller and housing, using John Deere special tool AM 2047T.

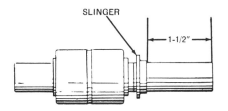

Fig. JD839E — When reinstalling slinger, press it on shaft until shoulder measures 1½ inches from impeller end of shaft as shown.

NOTE: Use caution not to disturb position of impeller while installing the fan hub. To do so will change the clearance already established between impeller vanes and pump housing. Use new gasket and install rear cover plate.

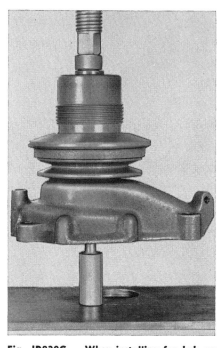

Fig. JD839G — When installing fan hub on water pump shaft, use a drift smaller than diameter of pump shaft and support shaft as shown. See text and Fig. JD839H for installation dimension.

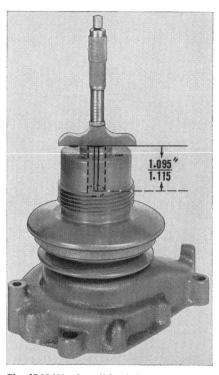

Fig. JD839H—Install fan hub on water pump shaft until distance from end of shaft and shoulder of fan hub is 1.095-1.115 as shown.

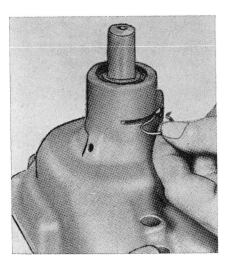

Fig. JD839C—Remove retainer (snap ring) as shown prior to pressing shaft and bearing assembly from impeller and housing.

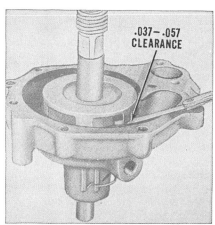

Fig. JD839F—Press impeller on water pump shaft until clearance between impeller vanes and pump housing is 0.037-0.057 as shown.

IGNITION & ELECTRICAL SYSTEM

(See Wiring Diagram on Page 69)

50. John Deere 40, 320 and 330 series tractors are equipped with a model 1111709 Delco-Remy distributor and gasoline models use Champion H10, AC 45L or Auto-Lite A7 spark plugs. Recommended spark plug for all-fuel engine is Champion J14. Electrode gap setting is 0.025.

50A. John Deere 420, 430 and 440 series tractors are equipped with a model 1112571 Delco-Remy distributor and gasoline models use Champion H10, AC 45L or Auto-Lite A7 spark plugs. Recommended spark plug for all-fuel engine is Champion H12. Electrode gap setting is 0.025.

IGNITION TIMING
All Models

51. Distributor rotation is clockwise when viewed from the drive end. Set breaker contacts to a 0.022 gap. To time the ignition distributor, first crank engine until No. 1 piston (front) is on compression stroke and the flywheel mark "SPARK" (advanced timing mark) aligns with timing port index as shown in Fig. JD840. At this time, the distributor drive groove in oil pump drive gear should be parallel to crankshaft and with narrow part of groove wall towards outside of tractor.

Fig. JD840—Deere 40, 320, 330, 420, 430, and 440 series flywheel mark "SPARK" is the fully advanced running timing mark of number one cylinder. Refer to ignition timing for correct timing procedure.

Install distributor with retaining cap screws only finger tight. Rotate distributor cam as far as it will go in direction of normal rotation. While holding cam in fully advanced position, rotate distributor body in the opposite direction until points just begin to open. Tighten distributor retaining cap screws and install number

one (front) cylinder spark plug wire in the terminal over the rotor contact arm.

As indicated above, the distributor is timed manually with the breaker cam in the fully advanced position. The unit can also be timed in the fully advanced position by operating the engine at fast idle rpm and using a timing light. To check action of the distributor governor, operate the engine at 600 rpm or less at which time the spark should occur 5° ATC for series 40, 1° BTC for series 320 and 330 and 2° BTC for series 420, 430 and 440.

ELECTRICAL UNITS

52. Delco-Remy electrical units are used and their applications are as follows:

Generator
    All Models 1101859
Regulator
    Series 40 1118308
    Series 320-330 1118780
    Series 420-430-440 1118780
Starting Motor
    Series 40-320-330 1107127
    Series 420 prior
      serial 100,001 1107127
    Series 420 after
      serial 100,000-430-440 1107165

CLUTCH

All Models (Single Plate)

60. The 420 and 430 series without continuous running power take-off and all 40, 320, 330 and 440 series are equipped with a single plate, dry type clutch shown in Figs. JD840A and JD 841. Component replacement parts of the clutch cover assembly are not available separately.

61. **ADJUSTMENT.** Clutch pedal free travel for series 40 tractors is 1½ inches. For series 320, 330, 420 (prior to ser. No. 131,809) and 430 tractors with no direction reverser, clutch pedal free travel is 1-inch. Adjust free travel by loosening lock nut and turning the adjusting stud on front lower left side of center frame as shown in Fig. JD842. Turning stud clockwise increases pedal free travel.

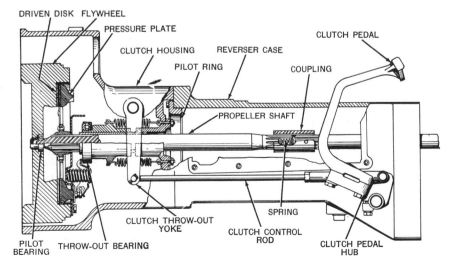

Fig. JD840A—View of the clutch, center frame (2-piece) and clutch operating linkage.

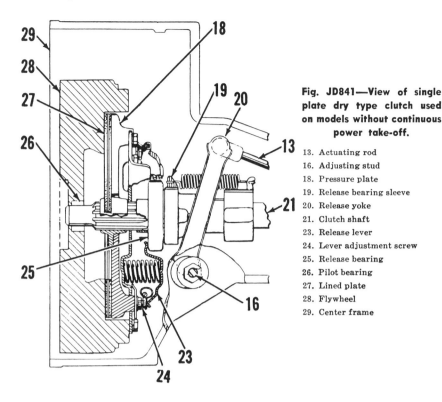

Fig. JD841—View of single plate dry type clutch used on models without continuous power take-off.

13. Actuating rod
16. Adjusting stud
18. Pressure plate
19. Release bearing sleeve
20. Release yoke
21. Clutch shaft
23. Release lever
24. Lever adjustment screw
25. Release bearing
26. Pilot bearing
27. Lined plate
28. Flywheel
29. Center frame

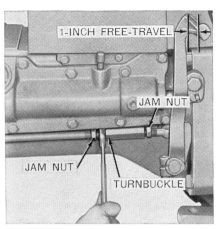

Fig. JD841B — On tractors serial number 131,809 and up, which are equipped with direction reverser, clutch pedal free travel of 1-inch is obtained by loosening jam nuts and rotating turnbuckle.

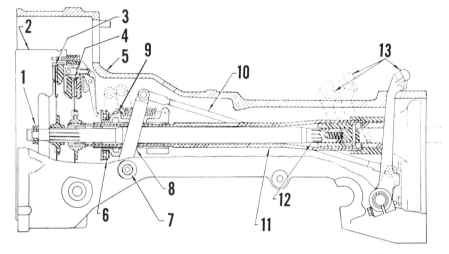

Fig. JD841A—View of dual plate clutch used when tractors are equipped with continuous-running power take-off.

1. Pilot bearing
2. Flywheel
3. 11" driven disc
4. 9" driven disc
5. Center frame
6. Throw-out bearing
7. Adjusting stud
8. Throw-out yoke
9. Release bearing carrier
10. Control rod
11. Outer clutch shaft
12. Inner clutch shaft
13. Clutch pedal

renewed if linings are worn. If any part of the cover assembly is worn or damaged, it will be necessary to renew the complete unit as component parts are not available separately.

Before reinstalling the clutch, pack the cavity in front of pilot bearing one-half full of high temperature fibrous grease.

The long hub of lined plate should be installed away from flywheel. Align driven plate with a clutch aligning tool or a spare propeller (clutch) shaft when reinstalling clutch unit.

Series 420-430 (Dual Plate)

63. The 420 and 430 series equipped with a continuous running power take-off are equipped with a dual plate clutch as shown in Fig. JD841A. Depressing the clutch pedal half way stops the motion of the tractor but does not interrupt the power flow to the pto shaft. Completely depressing the clutch pedal stops the motion of the tractor as well as the pto shaft.

64. **ADJUSTMENT.** Clutch pedal should have 1-inch free travel. To adjust the free travel, loosen the lock-nut and turn the adjusting stud on left side of center frame as shown in Fig. JD842. Turning the stud clockwise increases the pedal free travel.

65. **R & R AND OVERHAUL.** To remove the clutch, it is first necessary to detach (split) engine from center frame as outlined in paragraph 69. The clutch release bearing can be renewed at this time and the procedure for doing so is evident. Remove the cap screws retaining the clutch cover assembly to flywheel and withdraw the cover assembly and the 11-inch lined disc. Examine the clutch

On series 420 (ser. No. 131,809 and up) 430 and 440 tractors which are equipped with the latest (differential) type direction reverser, the clutch pedal free travel of 1-inch is obtained by loosening both jam nuts and rotating turnbuckle as shown in Fig. JD841B.

62. **R & R AND OVERHAUL.** To remove the clutch, it is first necessary to detach (split) engine from center frame as outlined in paragraph 69.

The clutch release bearing can be renewed at this time and the procedure for doing so is evident. Remove the cap screws retaining the clutch cover assembly to flywheel and remove the cover assembly and lined plate. Examine the clutch shaft pilot bearing in flywheel and renew the bearing if same is damaged or worn.

On some early models, the driven plate linings are available for field installation, while on later models, the complete driven plate must be

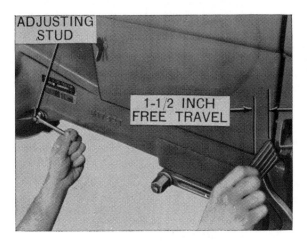

Fig. JD842 — Deere 40 series clutch pedal free travel of 1½ inches is obtained by turning the adjusting stud as shown. Turning the stud clockwise increases the pedal free travel. Adjust 320, 330 and the models of 420 and 430 series without the differential type direction reverser to 1-inch free travel in same manner.

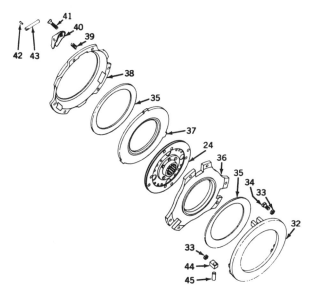

Fig. JD842A — Exploded view of dual clutch and related parts. Dual clutch is used with five speed transmissions to provide contiunous-running power take-off.

24. 9" driven disc
32. 11" pressure plate
33. Jam nuts
34. Cap screw (special)
35. Belleville washer
36. Flywheel plate
37. 9" pressure plate
38. Bracket (cover)
39. Spring
40. Release lever
41. Adjusting screw
42. Roll pin
43. Pivot pin
44. Rod end
45. Pin

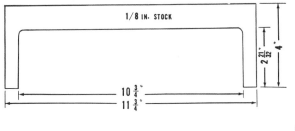

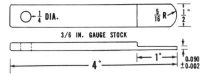

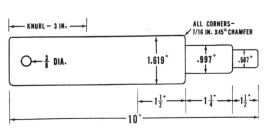

Fig. JD842B—Special tools for installing and adjusting the dual clutch used on models with continuous - running power take-off. Tools can be made, using the dimensions shown.

shaft pilot bearing in flywheel and renew the bearing if same is damaged or worn.

Renew the 11-inch driven disc if linings are worn or if disc is damaged. Separate linings are not available for field installation.

To disassemble the clutch cover assembly, proceed as follows: Carefully place punch marks on all of the components so they can be reassembled in the same relative position. Place the assembly in a press with the cover assembly up. Place a bar across the cover and compress the assembly until the release levers are just free. Loosen the lock nuts and completely unscrew the finger adjusting screws (41—Fig. JD842A). Release the compressing pressure and disassemble the remaining parts.

NOTE: In lieu of a suitable press, the finger adjusting screws can be removed after bolting the cover assembly in its normal position on the flywheel.

Inspect all parts and renew those showing damage or wear. New facings can be riveted to the 9-inch disc if the disc is in otherwise good condition.

To reassemble the unit, refer to Fig. JD842A and proceed as follows: Place the 11-inch pressure plate (32) on work bench with friction face down. Install the red marked Belleville washer (35) on pressure plate with convex side up and outer edge in counterbore of pressure plate. Align the previously affixed punch marks and install flywheel plate (36). Position the 9-inch driven disc (24) on flywheel plate with hub of same upward; then, again aligning punch marks, install the 9-inch pressure plate (37). Install second (yellow marked) Belleville washer (35) on pressure plate (37) with concave side up. (NOTE: If second Belleville washer is a replacement part the identification mark will be red). Align punch marks and place clutch bracket (cover) (38) in position, using caution to see that outer edge of Belleville washer is in counterbore of bracket (cover) (38). To facilitate starting of clutch finger adjusting screws, drop three capscrews through bracket (38) and flywheel plate (36) equidistant around outer bolt circle. Start clutch finger adjusting screws (41) and tighten only enough to hold clutch parts in position.

Install a NEW 11-inch driven disc in counterbore of flywheel, with hub pointing away from engine. Position clutch assembly on flywheel and start the retaining capscrews. Install aligning tool (pilot) and tighten capscrews. See Fig. JD842C.

NOTE: A new 11-inch driven disc must be installed while making adjustments.

If original disc is serviceable it can be installed after making the following adjustments:

To adjust clutch, refer to Fig. JD-842B for tools required and proceed as follows: With clutch finger height gage in position as shown in Fig. JD-842D and while holding down slightly on finger to take out the slack, turn finger adjusting screw until the 2 21/32-inch dimension is obtained and tighten locknut. Repeat same procedure on other two fingers. It is IMPORTANT that clutch fingers be adjusted to within 0.015 of each other. Use a box end wrench, or similar tool, and actuate each finger through its normal range of travel several times and recheck adjustment.

With finger adjustment made, refer to Fig. JD842E and set the clearance between the 9-inch pressure plate and adjusting capscrews. Loosen locknuts and turn adjusting screw until the 0.090 thickness gage can just be inserted between head of adjusting screw and 9-inch pressure plate. Tighten locknuts.

CLUTCH SHAFT

Series 40 Without Direction Reverser—320-330-420 Prior 100,001 Without Direction Reverser

66. To remove the clutch shaft, first detach (split) transmission from center frame as in paragraph 70. Then withdraw the clutch shaft from center frame.

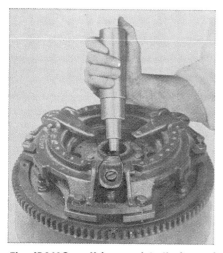

Fig. JD842C — Using special aligning tool when installing the 420 and 430 series dual clutch to flywheel.

Series 420 After 100,000—430 With No Direction Reverser and No Continuous-Running Power Take-Off

67. To remove the clutch shaft, detach (split) transmission from center frame as in paragraph 70 and withdraw the clutch shaft. When reassembling, be sure spring (2—Fig. JD843) is properly positioned in rear bore of shaft.

Series 420 After 100,000—430 With Continuous-Running Power Take-Off

67A. To remove the clutch shafts (outer and inner) detach (split) transmission from center frame as in paragraph 70 and withdraw both clutch shafts. When reassembling, be sure spring (4—Fig. JD843A) is properly positioned in rear bore of inner clutch shaft.

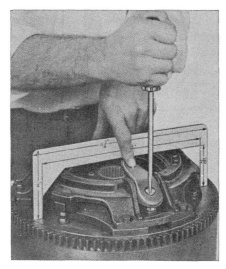

Fig. JD842D—Using special tool to adjust 420 and 430 series dual clutch release lever height. Adjustment must be made with a NEW 11-inch driven disc installed.

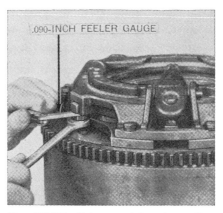

Fig. JD842E—Using special tool to adjust clearance between the 9-inch pressure plate and the releasing cap screws. Clearance is 0.090 as shown.

Series 40-420 Prior to 131,309 With Direction Reverser

68. To remove the clutch shaft, it is necessary to disassemble the direction reversing mechanism as outlined in paragraph 71.

Series 420 After 131,309 430-440 With Direction Reverser

68A. To remove the clutch shaft, it is necessary to disassemble the direction reversing mechanism as outlined in paragraphs 73 and 74.

CLUTCH RELEASE BEARING AND YOKE

All Models

68B. On models with one piece center frame (prior to ser. No. 131,-809), any service or repair is obvious after separating engine from clutch housing.

On models with two piece center frame (ser. no. 131,809 and up), throw-out bearing and carrier can be removed (after engine has been removed), by unhooking return springs. When pressing new bearing on carrier, be sure to align the index mark on bearing with notch in carrier. Yoke can be removed by driving Groov-pin from yoke shaft on right side, then driving out yoke shaft. If throw-out bearing sleeve is to be renewed it will be necessary to separate clutch housing from center frame (reverser housing) before sleeve can be unbolted and removed.

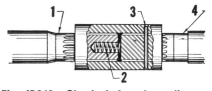

Fig. JD843—Clutch shaft and coupling arrangement on 420 series tractors serial number 100,001 and up and 430 series, when not equipped with direction reverser or continuous-running power take-off. Note that clutch shaft is held in position by spring (2).

1. Clutch shaft
2. Spring
3. Roll pin
4. Input shaft

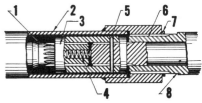

Fig. JD843A—Clutch shafts (inner and outer) and coupling arrangement on 420 series tractors serial number 100,001 and up and 430 series tractors when equipped with continuous-running power take-off. Note that inner clutch shaft is held in position by spring (4).

1. Inner clutch shaft
2. Outer clutch shaft
3. Coupling
4. Spring
5. Roll pin
6. Input shaft
7. Snap ring
8. Pto drive gear

TRACTOR-SPLIT
(Engine From Center Frame)

All Models

69. To detach engine from center frame, proceed as follows: Remove hood and grille. Detach steering shaft from steering worm shaft. Disconnect wires from generator, coil, lights and hour meter or cable from tachometer drive. Unclip wiring harness from rocker arm cover and lay harness rearward and out of way. Disconnect the "Touch-O-Matic" lines from front of instrument panel. Unhook governor spring from control rod and disconnect governor connecting rod from control rod; then, remove the control rod bracket attaching cap screws. Disconnect air cleaner hose and fuel line from carburetor, disconnect carburetor from manifold and choke wire from carburetor. On late model tractors, disconnect wire from the oil pressure sending unit. Disconnect heat indicator bulb from cylinder block or water outlet manifold and remove dust shield from lower front end of center frame. On 440 series remove side rails. Support engine in a hoist and tractor under center frame, unbolt engine from center frame, and move engine forward and away from tractor.

TRACTOR SPLIT
(Center Frame From Transmission)

All Models

70. To detach transmission and final drives assembly from center frame (clutch housing), proceed as follows: Disconnect battery ground strap and tail light wires. Drain the "Touch-O-Matic" system and disconnect the hydraulic lines at base of instrument panel. Support both halves of tractor separately, remove nuts and cap screws retaining transmission case to center frame and proceed as follows:

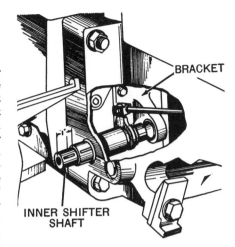

Fig. JD843C—Cutaway view showing inner shifter shaft and bracket of 420 series (serial number 100,001 through 131,308) when equipped with early type direction reverser.

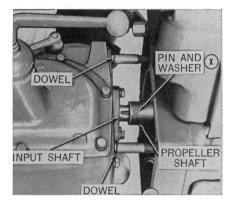

Fig. JD843D—Separating transmission housing from center frame to obtain access to input and output shaft bearing quills on Deere 40 series, 320 series and 420 series prior serial number 100,001.

On series 420 after 100,000, 430 and 440 without direction reverser, carefully separate the tractor halves.

Fig. JD844—Front view of transmission housing showing the installation of the direction reversing mechanism used prior to tractor serial number 131,309.

1. Clutch (propeller) shaft	50. Inner shifter shaft bracket
31. Direction reverser housing	54. Control rod

On series 40 without direction reverser, series 320, 330 and 420 prior to 100,001 without direction reverser, slide front half of tractor forward approximately two inches, remove coupling pin and washer (X—Fig. JD-843D), then separate the tractor halves.

On series 40 with direction reverser and series 420 (ser. no. 100,001 through 131,308) with direction reverser, remove the direction reverser control handle and Woodruff key. Pull the inner shifter shaft outward about 1-inch as shown in Fig. JD843C, then move the front half of tractor forward about two inches. Disconnect the control rod from the control shaft and separate the tractor halves.

On series 420 (ser. no. 131,309 and up), 430 and 440 with direction reverser, carefully separate tractor halves.

DIRECTION REVERSER

Two types of direction reverser units have been used. On tractors prior to serial number 131,309, the type shown in Fig. JD844C was used. For service information on this type refer to paragraphs 71 through 71B.

On the 420 series from serial number 131,309 and up, as well as the 430 and 440 series tractors, a differential type direction reverser such as that shown in Fig. JD846 has been used. For service information on this type, refer to paragraphs 72 through 74G.

Series 40-420 (Prior 131,309)

71. **R & R AND OVERHAUL.** To remove the direction reverser, first detach (split) transmission from center frame as outlined in paragraph 70.

Unbolt and withdraw the direction reverser housing (31—Fig. JD844) and component parts from transmission being careful not to damage or lose shims (15—Fig. JD844A) on transmission input shaft.

Visually examine the idler gear needle bearings (16) in the reverser mounting plate. If the needle bearings and mounting plate are in satisfactory condition, it will not be necessary to remove the plate. If needle bearings require renewal, unbolt and remove mounting plate (17) and drift out the worn bearings. Install new bearings so that rear face of bearings are flush with rear face of mounting plate. If mounting plate (17—Fig. JD-

844C) requires renewal, drift out bearing cup (19) and be careful not to damage or lose shims (18) which control the end play of the transmission input shaft. When reassembling, use the original number of shims (18) and using a dial indicator as shown in Fig. JD844B, check the transmission input shaft end play which should be 0.002-0.004. If the end play is not as specified, it will be necessary to remove mounting plate and bearing cup and vary the number of shims (18) which are available in thicknesses of 0.002, 0.005 and 0.010.

71A. Support direction reverser housing with propeller shaft pointed downward, refer to Fig. JD844C and remove internal parts as follows: Remove thrust plate (40) and the two needle thrust bearings (38) from top of idler gears (39 and 42). Remove the idler gears, lower thrust bearings

(38) and direct drive gear (14). Remove bronze thrust washer (13) and reverse drive gear (10). Lift out upper thrust bearing (7), spacer washer (8), lower thrust bearing (7) and direct drive member (6). Remove small thrust bearing (5) and thrust washer (4) from inner end of propeller shaft.

Support shift coupling (9) on John Deere tool No. AM2047T or equivalent and press or drift propeller shaft from ball bearing in front of housing.

Remove oil filler plug and groov-pin (26). Withdraw shifter rail (27), then remove shifter fork (25) and coupling (9) from housing. Usually it is not necessary to remove the detent spring cup plug (44); however, removal of same will facilitate reinstallation of ball (45) and detent spring (46). Remove front oil seals (28 and 29) and ball bearing (2).

Thoroughly check all parts and renew any which are damaged, worn

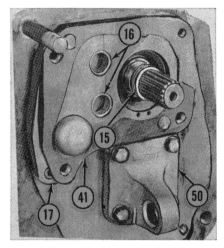

Fig. JD844A—On all models with direction reverser prior to tractor serial number 131,309, shims (15) control end play of the clutch (propeller) shaft. The procedure for adjusting the end play (0.010-0.030) is outlined in text.

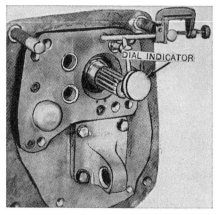

Fig. JD844B—On all models with direction reverser prior to tractor serial number 131,309, check the end play of the transmission input shaft as shown. Recommended end play is 0.002-0.004.

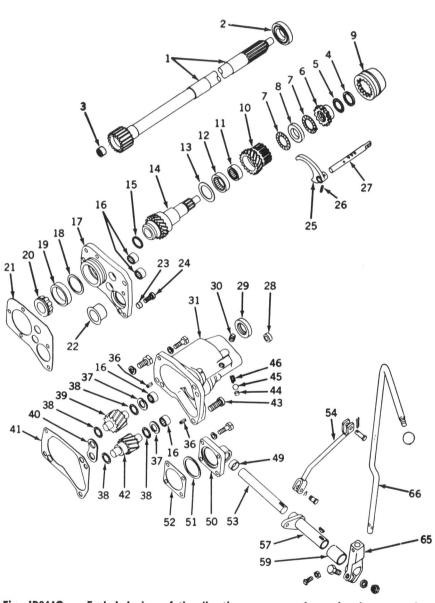

Fig. JD844C — Exploded view of the direction reverser and associated parts used on models prior to serial number 131,309. Shaft (53), bracket (50) and bushing (49) are used on models after serial number 100,000.

1. Clutch shaft	14. Direct drive gear	27. Shifter shaft	44. Cupped plug
2. Ball bearing	15. Shims	28. Oil seal	45. Ball
3. Needle bearing	16. Needle bearings	29. Oil seal	46. Spring
4. Thrust washer	17. Mounting plate	30. Pipe plug	49. Bushing
5. Thrust bearing	18. Shims	31. Housing	50. Bracket
6. Drive member	19. Bearing cup	36. Roll pin	51. Shims
7. Thrust bearing	20. Bearing cone	37. Washer	52. Shims
8. Washer	21. Gasket	38. Thrust bearing	53. Inner shaft
9. Coupling	22. Sealing cup	39. Idler gear	54. Control rod
10. Reverse drive gear	23. Dowel	40. Thrust plate	57. Outer shaft
11. Needle bearing	24. Socket head screw	41. Gasket	59. Spacer
12. Needle bearing	25. Shifter fork	42. Idler gear	65. Control pivot
13. Thrust washer	26. Groov-pin	43. Socket head screw	66. Control lever

or heat discolored. Small burrs can sometimes be removed with a fine stone or Crocus cloth. NOTE: When renewing needle bearings, always press on the numbered side and refer to the following:

Press needle bearing (3) in end of propeller shaft until bearing is 1/32-inch below face of counterbore. Press idler gear needle bearings in housing until bearings are 1/64-inch below face of counterbores as shown in Fig. JD844D. When renewing needle bearings in the reverse drive gear (10), press smaller bearing (11) in until bottom of bearing is 5/32-inch from bottom of counterbore and press larger bearing (12) in until top of bearing is $\frac{1}{32}$-inch below top of counterbore as shown in Fig. JD844E.

When new roll pins (36—Fig. JD-844C) are installed in housing, be sure they do not extend beyond their respective bores more than 7/64-inch as shown in Fig. JD844D.

Oil seals (28 and 29—Fig. JD844C) must bottom in counterbores and lips must face inward.

71B. Before assembling the direction reverser, an interference check and run-out check must be made on the direct drive gear as follows:

Place a thin layer of grease on front end of transmission input shaft and install the direct drive gear. Push the direct drive gear on the transmission input shaft as far as possible; then, remove the gear and examine grease on end of transmission shaft for evidence of the direct drive gear bottoming. If the drive gear is bottoming, grind off excess material on end of transmission input shaft until some clearance is obtained.

To check the drive gear run-out, proceed as follows: Position the direct drive gear on transmission input shaft, and with transmission in neutral, rotate the drive gear and check the total amount of drive gear run-out as shown in Fig. JD845. Remove the drive gear, reposition same on the transmission shaft splines and again

check the runout. Continue this procedure until the position is found which yields the least amount of runout. Now, determine the total plus runout and the total minus runout and zero the indicator dial midway between the two extreme positions. Rotate the direct drive gear to one of the extreme positions; at which time, it should be possible to manually push the end of the direct drive gear back to the zero runout position, or beyond. If the drive gear cannot be pushed back to where the indicator dial reads at least zero, renew the direct drive gear and recheck. After the desired position of the direct drive gear with respect to the transmission input shaft has been determined, mark the meshing splines so the parts can be reassembled in the same position.

Refer to Figs. JD844C and 845A and install ball bearing (2) in housing using suitable driver. Assemble shifter fork (25) into groove of coupling (9) and install coupling and shifter fork into housing with tapered edge of coupling and hub of shifter fork facing toward front of housing. Install detent spring and ball (45 and 46) into housing and while depressing ball and spring, start shifter rail (27) into position. Align shifter rail with hub of shifter fork and tap rail into fork until groov-pin holes align. Install groov-pin by working through filler plug hole.

Install propeller shaft (1) through coupling and ball bearing and align splines of coupling and propeller shaft. Support ball bearing on INNER race and drift propeller shaft in position. Install oil seals (28 and 29) with lips facing inward. NOTE: Before assembling the drive mechanism parts, move shifter fork to the reverse position (rail inward).

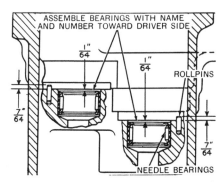

Fig. JD844D—Install needle bearings in direction reverser housing to the dimensions shown above. Press on numbered side of bearings only. Note installation dimensions of roll pins.

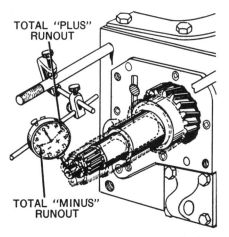

Fig. JD845—To check run-out of direct drive gear, mount dial indicator as shown and refer to text for procedure. Note that dial indicator is mounted on pilot end of direct drive gear.

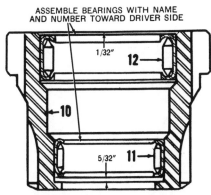

Fig. JD844E—Reverse drive gear has two needle bearings in its inner bore. Install bearings to dimensions shown and with numbered side toward driver.

Fig. JD845A — Sectional view of direction reverser used prior to tractor serial number 131,309. Notice the 1/4-inch distance between direct drive gear teeth and mounting surface of housing. Refer to legend for Fig. JD844C.

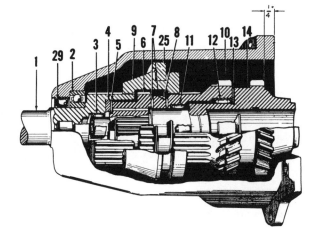

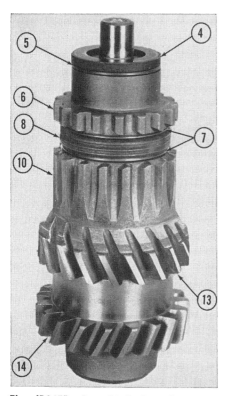

Fig. JD845B—Assembled view of reverser unit drive mechanism, showing relative position of parts. Refer to legend for Fig. JD844C.

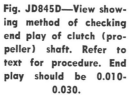

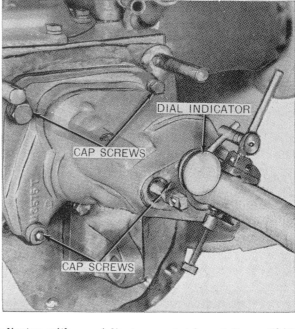

Fig. JD845D—View showing method of checking end play of clutch (propeller) shaft. Refer to text for procedure. End play should be 0.010-0.030.

Refer to Figs. JD845A and 845B and assemble the drive mechanism by reversing the disassembly procedure. It is IMPORTANT that the distance from outer face of direct drive gear to mounting surface of housing be approximately ¼-inch as shown in Figs. JD845A and 845C. If distance is less than ¼-inch, it indicates that components are improperly assembled or not properly seated.

With no shim pack on the transmission input shaft, install direction reverser, making sure that the previously identified splines of direct drive gear and transmission input shaft are mated. Tighten attaching screws se-

curely and mount dial indicator with button end contacting housing as shown in Fig. JD845D. Move propeller shaft back and forth (tap if necessary) and determine the amount of end play. Remove the direction reverser and install sufficient shims (15—Fig. JD-844A) to remove all but 0.010-0.030 end play.

Reinstall direction reverser and tighten securely, shift reverser into forward position (shift rail forward) and reconnect transmission to center frame by reversing removal procedure.

Series 420 (131,309 and Up) 430-440

72. ADJUST DIRECTION REVERSER CLUTCHES. Before making adjustments, lubricate all linkage points and check same for freedom of movement. Accurate clutch adjustments are important and it is recommended by John Deere that a spring scale be used to measure the 25-35 pound effort required to actuate the lever as shown in Fig. JD847.

Adjust rear clutch as follows: Shift reverse control lever to the middle position so that both the front and rear clutches are released. Remove the barrel plug from right rear side of reverser housing and while depressing lock pin, as shown in Fig. JD-847A, rotate adjusting ring clockwise one notch at a time until the specified 25-35 pounds pull is required to snap control lever into its rearward position. Make several checks to insure setting is accurate. Reinstall barrel plug.

Adjust front clutch as follows: Shift reverser control to the rearward position, then remove the barrel plug from the left forward end of shift cover. Shift tractor transmission into neutral, disengage tractor main clutch and working through access hole, rotate direction reverser differential assembly until the spring loaded lock pin is accessible. Re-engage the tractor main clutch and place the reverser control lever in the middle (neutral) position. Now adjust front clutch by

Fig. JD845C—With all parts installed and properly seated, teeth of direct drive gear should measure ¼-inch from mounting surface as shown.

Fig. JD846—Cut-away view of the differential type direction reverser. This unit is available on 420 (serial number 131,309 and up), 430 and 440 series.

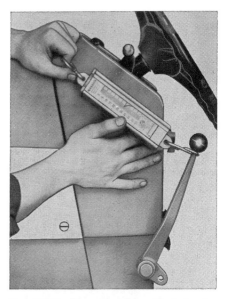

Fig. JD847—View showing method of using a spring scale to adjust direction reverser clutches. Refer to text.

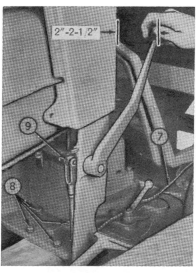

Fig. JD847B — Adjust vertical control rod clevis (9) until control lever is positioned as shown when unit is in reverse.

depressing the spring loaded lock pin and rotating the adjusting ring clockwise one notch at a time, until the 25-35 pounds pull required to snap the control lever into its forward position is obtained. Recheck several times to insure setting is accurate then reinstall barrel plug.

72A. ADJUST CONTROL LEVER. With control lever in reverse (rear) position, the distance between control lever knob and instrument panel should be 2-2½ inches as shown in Fig. JD847B. If control lever requires adjustment to obtain this measurement, disconnect clevis at top of vertical control rod and rotate same either way as required.

73. REMOVE AND REINSTALL. When work is needed only on the direction reverser, split tractor in the following manner: Remove hood and attach hoist to engine. NOTE: If tractor is a four wheel model, wedge front axle with wooden blocks. If tractor is a tricycle model, use a chain under clutch housing or engine to prevent front section of tractor from tipping. Place rolling floor jack under tractor transmission in such a manner as to allow rear section of tractor to be rolled rearward. Shift direction reverser unit into reverse operating position (control handle rearward) and remove battery and battery tray. Disconnect the hydraulic system pressure and return lines at bottom of instrument panel; then disconnect wiring from rear light and transmission top cover. Disconnect the direction reverser shifting linkage, then remove pivot bolt and linkage. Remove the cap screws which retain the instrument panel to the direction re-

verser housing, raise instrument panel slightly and retain in this position by inserting a wooden block between starter and fuel tank as shown in Fig. JD848.

Remove the cap screws retaining direction reverser housing to clutch housing. Split tractor by moving the rear section of the tractor rearward away from clutch housing and at the same time, unscrew turnbuckle from clutch rod. NOTE: Use caution during this operation to avoid releasing the tension of engine clutch. To do so will allow the clutch disc to shift and difficulty will be encountered when joining tractor halves. Move rear section of tractor rearward enough to allow a sling to be attached to direction reverser housing. Be sure sling unit is attached so that direction reverser unit is balanced; then, unbolt and remove direction reverser from transmission and place on work bench.

When reinstalling, be sure the hollow dowels are in place before joining direction reverser housing to transmission and keep unit level in order to avoid damage to input shaft and oil seals. Partially engage propeller (clutch) shaft in engine clutch, then start turnbuckle on clutch rod. Take-up turnbuckle as rear section of tractor is moved forward. Balance of reassembly is the reverse of disassembly. Adjust vertical control rod as in paragraph 72A, until control lever is 2-2½ inches rearward from instrument panel when lever is in reverse position. Adjust direction reverser clutches as outlined in paragraph 72 and engine clutch as outlined in paragraph 61.

74. OVERHAUL CLUTCHES AND DIFFERENTIAL. An exploded view of the complete direction reverser unit is shown in Fig. JD849. With direction reverser removed from tractor as outlined in paragraph 73, proceed

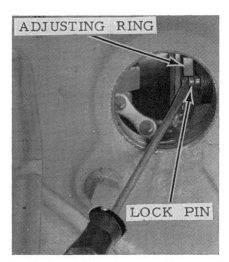

Fig. JD847A—View showing method of adjusting rear clutch. Rotating adjusting ring clockwise (downward) tightens the clutch.

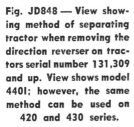

Fig. JD848 — View showing method of separating tractor when removing the direction reverser on tractors serial number 131,309 and up. View shows model 440I; however, the same method can be used on 420 and 430 series.

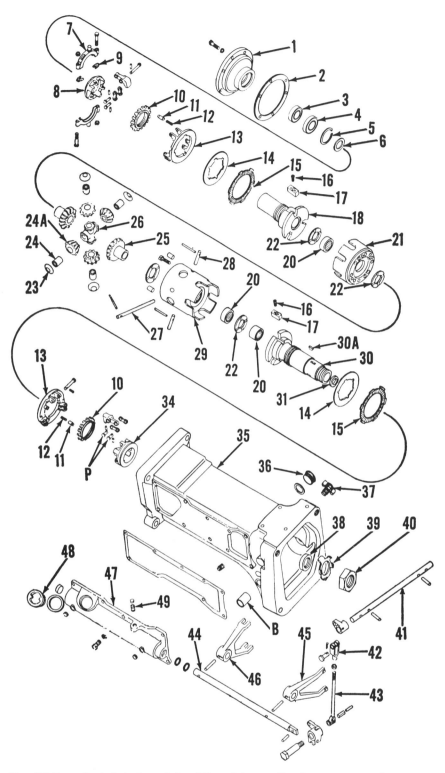

Fig. JD849 — Exploded view of the differential type direction reverser used on tractor models serial number 131,309 and up. Item (37) was discontinued at serial number 141,215.

P. Pins
B. Bushing
1. Quill
2. Gasket
3. Oil seal
4. Ball bearing
5. Snap ring
6. Shims (0.010 and 0.020)
7. Collar
8. Sliding sleeve
9. Shims
10. Adjusting ring (2 used)

11. Lock pin (2 used)
12. Spring (2 used)
13. Floating plate (2 used)
14. Steel disc (8 used)
15. Sintered disc (10 used)
16. Phillips screw
17. Drive key
18. Front clutch hub
20. Needle bearing
21. Housing cover
22. Thrust washer
23. Thrust washer
24. Needle bearing

24A. Pinion gear
25. Side gear
26. Spider
27. Pinion shaft (long)
28. Pinion shaft (short)
29. Housing
30. Rear clutch hub
30A. Woodruff keys (3 used)
31. Oil seal
34. Sliding sleeve
35. Reverser case
36. Plug

37. Test cock*
38. Oil seal
39. Lock plate
40. Nut
41. Control shaft
42. Clevis
43. Control rod (vertical)
44. Shifter rail
45. Rear fork
46. Front fork
47. Cover
48. Plug
49. Breather

*Test cock (37) no longer used.

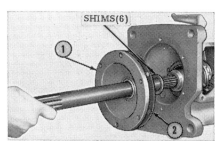

Fig. JD850 — Propeller (clutch) shaft and quill can be removed after removing cap screws which retain quill. Note shims (6) at rear of bearing.

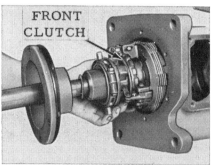

Fig. JD850A — Remove front clutch from housing as shown.

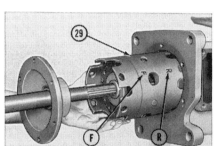

Fig. JD850B—Removal of differential unit is facilitated by using the previously removed propeller shaft as shown. Note "F" and "R" marks on differential housing.

as follows: Drain oil from housing and remove clutch pedal. NOTE: On models with the two piece clutch pedal it is necessary only to remove the clutch pedal from pivot. Be sure unit is in reverse operating position (shift rail forward) and remove the cap screws from the shifter cover and fork assembly. When removing the shifter cover and fork assembly, the rear end must be moved outward slightly in advance of the front end in order to gain the necessary clearance. Remove cap screws from propeller shaft bearing quill (1) and remove propeller shaft and quill as shown in Fig. JD850. Do not lose or damage the shims (6). Remove front clutch and hub assembly as shown in Fig. JD850A. Insert rear end of the previously removed propeller shaft into forward end of differential unit and pull same from housing as shown in Fig. JD850B. Note the front (F)

Fig. JD850C — Large nut can be removed after straightening tabs on lock plate. Torque nut to 175 Ft.-Lbs. when reassembling.

Fig. JD850E — Adjusting ring (33) can be removed by depressing the spring loaded lock pin (11).

justing ring (33) as shown in Fig. JD850E. Remove the floating plate (32—Fig. JD850F) and the clutch discs (5 sintered metal and 4 steel) from clutch hub.

Inspect clutch hub threads, oil seal, needle bearing and drive keys. The three drive keys (17—Fig. JD850G) at the flanged end are retained by counter-sunk Phillips-head screws (16) which are staked in position. When renewing keys, tighten screws securely and stake. If there is any doubt as to the condition of the oil seal (31—Fig. JD849) at rear inside diameter of clutch hub, renew same using John Deere tool M602T, or equivalent. When renewing needle bearing (20) in hub (30), drive old bearing out flanged end of hub. When installing new bearing, align oil hole in bear-

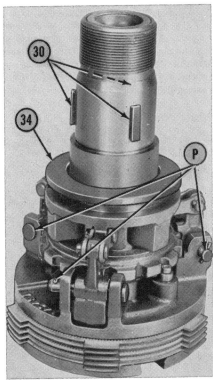

Fig. JD850D — View showing the removed rear clutch and hub assembly. Front clutch and hub is very similar.

P. Pins
30. Woodruff keys
34. Sliding collar

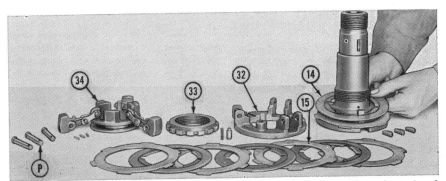

Fig. JD850F—View showing rear clutch and hub disassembled. Notice how sintered and steel clutch discs are alternated.

14. Steel disc　　　　　32. Floating plate　　　　　33. Adjusting ring
15. Sintered disc　　　　　　　　　　　　　　　　　　　34. Sliding collar

note the position of the spring loaded pin which locks the rear clutch adjusting ring and the oil hole at the top of the clutch hub. These must be in the same positions when reassembling.

NOTE: With the exception of the clutch hubs and the sliding sleeves, the front and rear clutch assemblies are identical. These differences are obvious upon examination and service procedure for one will suffice for the other. Hence, the service procedure given below for the rear clutch assembly can also be applied to the front clutch assembly.

74A. To overhaul the rear clutch and hub assembly, proceed as follows: Refer to Fig. JD850D and remove the Woodruff keys (30) from hub. Drive out the roll pins from ends of pins (P) which attach levers to sliding sleeve and remove pins. Note: Driving of roll pins can be made easier by engaging clutch to hold pins in proper position. Lift off the sliding sleeve (34) and levers assembly. Depress the spring loaded adjusting ring lock pin (11) and unscrew the ad-

and rear (R) marks stamped on the reverser housing. Straighten tabs of the lock plate which locks the large nut (40—Fig. JD850C) at rear of the reverser housing; then, either partially loosen nut and drive rear clutch forward or, completely remove nut and press rear clutch from housing. NOTE: Before completely removing the rear clutch and hub assembly,

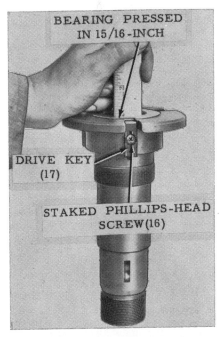

Fig. JD850G — Drive keys are retained to hub by staked Phillips-head screws. Needle bearing is properly installed when edge is 15/16-inch from machined surface as shown. Refer also to Fig. JD850H.

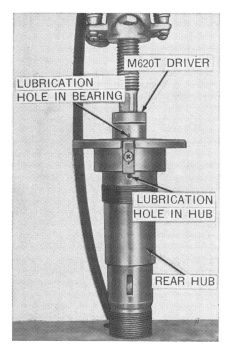

Fig. JD850H — Prior to installing needle bearing in rear hub, be sure oil holes in bearing and hub are aligned and press bearing in until oil holes are in register.

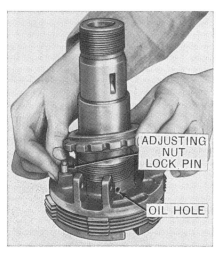

Fig. JD850K—When installing the floating plate, the lock pin must be positioned so the pin and oil hole in hub are in the position shown in order that the lock pin will register with access hole.

ing with the oil hole in clutch hub as shown in Fig. JD850H and press bearing into bore until oil holes are in register; at which time, the bearing will be approximately $\frac{15}{16}$-inch from flanged end of hub as shown in Fig. JD850G. When installing the bearing, press only on numbered end.

Inspect clutch plates (14 & 15—Fig. JD850F) for undue wear, scoring or distortion. Renew as necessary. Inspect floating plate (32), sliding sleeve (34) and linkage for undue wear or damage. If it is necessary to

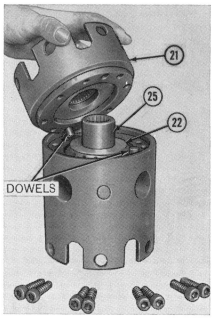

Fig. JD850L—After the eight socket head cap screws are removed, the differential housing cover can be removed. Notice the locating dowels.

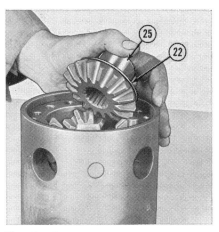

Fig. JD850M — With differential housing cover removed, the front drive (side) gear and thrust washer can be removed.

disassemble any of the linkage, use new roll pins when reassembling.

74B. When reassembling, place a sintered clutch disc (15) on clutch hub first; then alternate steel discs (14) and sintered discs as shown. Install the floating plate and the spring loaded lock pin. Be sure the lock pin and the oil hole are positioned as shown in Fig. JD850K. NOTE: Lock pin must be located in this position in order that clutch can be adjusted through barrel plug opening. Depress lock pin and install adjusting ring. Install sleeve and linkage, install pins and retain same with new roll pins. Adjust clutch with adjusting ring until it can just be engaged by hand pressure.

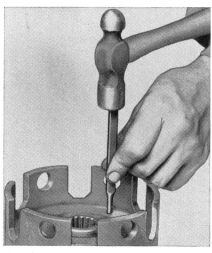

Fig. JD850N—The groov-pins which retain the pinion shafts can be removed by using a punch and driving from rear end of differential.

With clutch disengaged, place assembly in end of differential housing and align clutch discs. After discs are aligned, engage clutch so discs will be retained in this position to facilitate reassembly.

74C. To disassemble and overhaul the differential proceed as follows: Remove the eight socket head screws which retain housing cover (21—Fig. JD850L) to differential housing and remove cover. These screws have been staked and will be hard to remove. It may also be necessary to tap cover gently to dislodge it from the locating dowels. Remove drive (side) gear (25) and thrust washer (22) as shown in Fig. JD850M. Invert differential unit and drift out the three groov-pins which retain the differential pinion shafts as shown in Fig. JD850N. Again invert differential housing and remove pinion shafts (27 and 28—Fig. JD849), spider (26), pinion gears (24A) and thrust washers (23). Remove rear drive (side) gear (25) and thrust washer (22) from differential housing.

Inspect needle bearings (20 and 24) in cover, differential housing and pinion gears. When renewing, always press on numbered end of bearing. When installing bearings in cover and differential housing, press bearing in until end of bearing is $\frac{3}{32}$-inch below edge of machined surface as shown in Fig. JD850P. Press needle bearings in pinion gears until end of bearing is 1/64-inch from flat face of gear as shown in Fig. JD850R. Renew any thrust washers which show any undue wear and/or damage. Install pinion shafts as shown in Fig. JD850S.

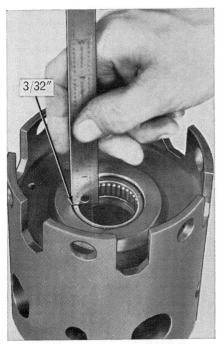

Fig. JD850P — When installing the needle bearings in the differential housing and differential housing cover, press bearings in until the bearing end is 3/32-inch below machined surface as shown.

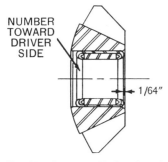

Fig. JD850R—Press needle bearings in pinion gears until bearing end is 1/64-inch below the flat side of gear as shown.

Assembly is the reverse of disassembly. Torque socket head screws to 25-28 ft.-lb. and stake in position.

74D. If any components of the front bearing quill assembly need servicing, proceed as follows: Remove snap ring (5—Fig. JD849) at rear of bearing and press shaft and bearing from quill (1). Bearing (4) can be pressed from shaft and a new one installed; however, press only on inner race of bearing. Seal (3) can be driven in from front side of quill using John Deere tool M603T or equivalent.

74E. Oil seal (38) at rear of direction reserver housing can be renewed using John Deere tool M605T or equivalent. Clutch cross-shaft bushing (B) can be renewed after removing

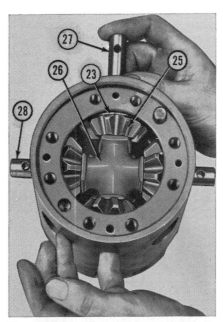

Fig. JD850S — View showing pinion shafts being installed. Hold one pinion gear in position with finger while installing the long pinion shaft.

Fig. JD850T—Adjust end play of propeller shaft by means of shims (6).

clutch shaft by using John Deere tool M620T or equivalent. Bushing is presized and if carefully installed will need no final sizing.

74F. Install the reverser mechanism in the direction reverser housing as follows: With the clutch discs aligned as previously outlined and the clutch engaged, place rear clutch and hub assembly into housing with oil hole on top side and the spring loaded lock pin at approximately the 2 o'clock

position when viewed from rear. Drive or press the assembly into position, install locking plate and large nut and torque nut to 175 ft.-lb. Lock nut by bending down tabs of lock plate.

Differential housing is stamped with an "F" and "R" (See Fig. JD-850B) indicating front and rear. Be sure the bronze thrust washer is over hub of rear drive (side) gear. Install differential assembly "R" end first into housing and position same so slots in differential housing fit over lugs of clutch discs. The propeller shaft can be used to install differential assembly into housing if so desired. With clutch discs of front clutch and hub assembly aligned and the clutch engaged, install the bronze thrust washer on the hub of the front drive (side) gear; then, position the clutch disc lugs in the slots of the differential housing and slide clutch into position. Place gasket (2—Fig. JD850T) on bearing quill (1) and the original shim pack (6) on propeller shaft and install propeller shaft. NOTE: It may be necessary to disengage the front clutch before propeller shaft will enter the front drive (side) gear. Tighten quill retaining cap screws to 15-18 ft.-lb.

At this point the end play of the reverser must be checked and adjusted, if necessary. Mount a dial indicator with the contact button resting on the front of the differential housing as shown in Fig. JD850U. Disengage both clutches, move differential housing back and forth and check the end play which should be 0.010-0.020. If end play is not within these limits remove front quill and add or subtract shims (6—Fig. JD-850T). Shims are available in thicknesses of 0.010 and 0.020.

When installing shifter cover, engage rear clutch and disengage front clutch. Start front fork into housing about ¼-inch ahead of rear fork. Engage shifter forks and secure cover with cap screws.

Fig. JD850U — When checking end play of differential unit, mount a dial indicator as shown. End play is controlled by shims at the rear of bearing in front quill.

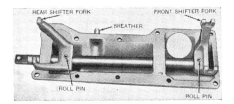

Fig. JD850Y — View showing shifter cover and forks. The forks can be removed from shift rail after removing plugs from shifter cover and driving out roll pins.

Fill unit to within ½-¾-inch below the access hole on right side of housing with Type "A" automatic transmission oil. Capacity is 3 U. S. quarts.

Install direction reverser in tractor as outlined in paragraph 73 and adjust both clutches as outlined in paragraph 72.

74G. R&R AND OVERHAUL SHIFTER COVER. The removal and overhaul of the shifter cover is obvious after an examination of the unit and reference to Fig. JD850Y. Roll pins can be driven from shifter forks after plugs in cover have been removed. Be sure breather is open and clean.

TRANSMISSION AND CONNECTIONS

The 420, 430 and 440 series tractors are available with either a four or five speed transmission, whereas the 40, 320 and 330 series are available with a four speed transmission only. Unless otherwise noted, the following overhaul procedures apply to all transmissions.

TRANSMISSION CASE REAR COVER
Models 40S-320S-330S-420S-430S (Single Touch-O-Matic)

75. R & R AND OVERHAUL. Drain transmission, disconnect upper lift link assembly from anchor yoke and remove the power shaft guard. Remove lock and adjusting nuts from the load control spring stud. Remove the control arm retaining cap screws (Fig. JD851) and tap the control arm free from the hollow dowels. Drive out pivot pin retaining roll pins and using a piece of pipe and a 7/16 inch cap screw, pull pivot pins from anchor yoke and rockshaft housing as shown. Remove the cover retaining cap screws and withdraw the transmission case rear cover.

The procedure for disassembling the removed cover is evident. Ream bushing (13—Fig. JD852), after installation, to an inside diameter of 1.378-1.379. Install seal (14) with lip of same facing front of tractor. Ream bushings (5) to an inside diameter of 0.862-0.863.

Note: Before installing the transmission case rear cover, the adjustment of the stop for the load and depth control lock-out screw should be checked and adjusted if necessary. To make the adjustment, proceed as follows: With set screw (10) removed, place cover on flat surface and measure the distance from front face (gasket face) of cover to rear end of lockout screw as shown in Fig. JD853. Turn the screw in or out of the cover until the measured distance is 2 15/16

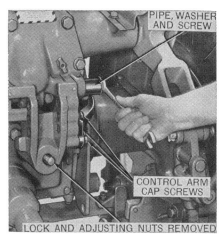

Fig. JD851—Pulling the anchor yoke pivot pin on Deere model 40S, 320S, 330S, 420S and 430S when equipped with single cylinder "Touch-O-Matic". Notice that the load control spring lock and adjusting nuts have been removed.

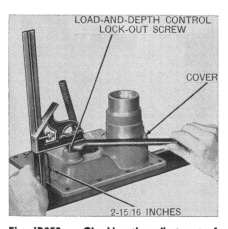

Fig. JD853 — Checking the adjustment of the load and depth control lock-out screw on Deere model 40S, 320S, 330S, 420S and 430S transmission case rear cover when equipped with single cylinder "Touch-O-Matic".

1. Lock nut
2. Adjusting nut
3. Stud
4. Pivot pins
5. Anchor yoke bushings
6. Anchor yoke
7. Load and depth control spring
8. Lock out screw
9. Cover
10. Set screw
11. Lock nut
12. Stud
13. Power shaft bushing
14. Oil seal
15. Hollow dowels
16. Control arm

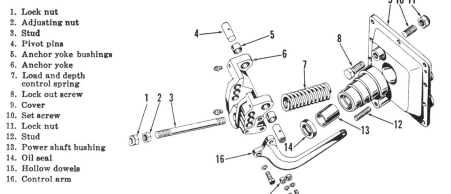

Fig. JD852—Exploded view of Deere model 40S, 320S, 330S, 420S and 430S transmission rear cover when equipped with single cylinder "Touch-O-Matic". Bushings (5 and 13) must be reamed after installation.

inches as shown. Reinstall set screw (10—Fig. JD852) and turn the screw in until same bottoms against lockout screw (8). Tighten jam nut (11) and recheck the 2 15/16 inches meas-

urement. Reinstall the rear cover by reversing the removal procedure and adjust the load and depth control anchor yoke as in the following paragraph.

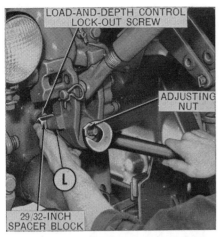

Fig. JD854—Adjusting the load and depth control anchor yoke on Deere model 40S, 320S, 330S, 420S and 430S when equipped with single cylinder "Touch-O-Matic". A 29/32-inch spacer block is used to gage the adjustment.

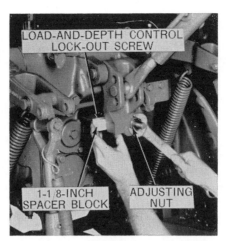

Fig. JD854B—Adjusting the anchor yoke on models 320S, 330S, 420S and 430S when equipped with dual "Touch-O-Matic." A 1⅛-inch spacer block is used to gage the adjustment.

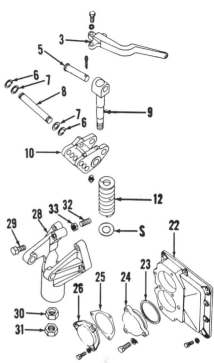

Fig. JD854C — Exploded view of models 40U, 40W, 320U, 330U, 420U and 430U transmission rear cover and anchor yoke assembly.

3. Control arm
5. Tension link pin
6. Snap rings
7. Spacer washers
8. Pivot pin
9. Tension link
10. Anchor yoke
12. Spring
22. Rear cover
23. Oil seal
24. Power shaft cover
25. Gasket
26. Input shaft cover
28. Spring housing
29. Lock out screw
30. Adjusting nut
31. Lock nut
32. Set screw
33. Lock nut
S. Shim

75B. Before attempting to adjust the load and depth control anchor yoke, first make certain that lock-out screw (8—Fig. JD852) is turned completely in against set screw (10). Then, place a 29/32 inch spacer block against head of lock-out screw and tighten the adjusting nut as shown in Fig. JD854 until the anchor yoke stop lug (L) just touches the spacer block. Reinstall and tighten the lock nut. Check the lift linkage as outlined in paragraph 127.

Models 320S-330S-420S-430S (Dual Touch-O-Matic)

76. **R & R AND OVERHAUL.** Drain transmission, disconnect upper lift link assembly from anchor yoke (3—

Fig. JD854A) and remove the power shaft guard. Remove lock and adjusting nuts from the load control spring stud (7). Remove control arm retaining cap screws and remove control arm (9). Remove snap ring from anchor yoke pivot pin, drift out pivot pin (2) and remove anchor yoke and spring (8). Remove cap screws from support (18) and remove support. Loosen the "Touch-O-Matic" mounting bolts and remove the cap screws retaining rear cover to transmission. Raise the "Touch-O-Matic" unit slightly and remove the rear cover.

The procedure for overhauling the rear cover is evident after an examination of the unit and reference to Fig. JD854A.

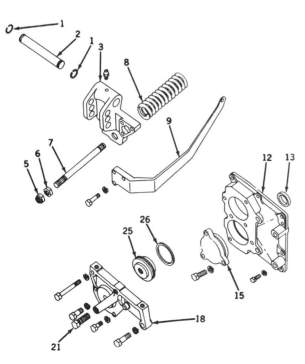

Fig. JD854A — Exploded view of models 320S, 330S, 420S and 430S transmission rear cover and anchor yoke assembly when equipped with dual cylinder "Touch-O-Matic".

1. Snap rings
2. Pivot pin
3. Anchor yoke
5. Lock nut
6. Adjusting nut
7. Stud
8. Spring
9. Control arm
12. Rear cover
13. Oil seal
18. Support
21. Lock out screw
25. Support pilot
26. Sealing ring

With the rear cover and components installed on transmission, adjust load and depth control linkage as follows: Turn lockout screw (21) inward until it bottoms, then turn the spring adjusting nut (6) either way as required until 1⅛ inches clearance is obtained between anchor yoke and head of lockout screw (21). Tighten lock nut (5). See Fig. JD854B.

Models 40U-40W-320U-330U-420U-430U

77. **R & R AND OVERHAUL.** Drain transmission, remove drawbar and disconnect upper link assembly from anchor yoke (10—Fig. JD854C). Remove power shaft guard if tractor is so equipped. Disconnect control link from control arm (3). Remove spring housing retaining cap screws and remove spring housing assembly. Unbolt and remove rear cover. The procedure for overhauling the rear cover is evident after an examination of the unit and reference to Fig. JD-854C.

If anchor yoke and spring housing was disassembled, refer to Fig. JD-

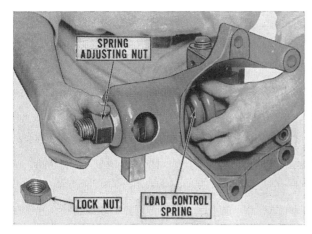

Fig. JD854D — To adjust the load control spring on models 40U, 40W, 320U, 330U, 420U and 430U, tighten adjusting nut until all end play (finger tight) is removed from spring, then tighten the nut an additional one-fourth turn.

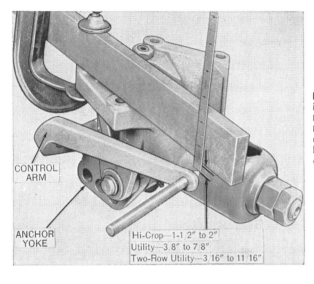

Fig. JD854F—View showing method of checking the position of the control arm on those models equipped with a spring housing. Note the different limits for different models.

Hi-Crop—1-1/2" to 2"
Utility—3/8" to 7/8"
Two-Row Utility—3/16" to 11/16"

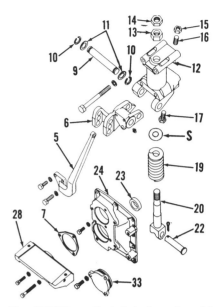

Fig. JD854G — Exploded view of models 40H, 420H and 430H transmission rear cover and anchor yoke assembly. Note support (28) used on these models.

5. Control arm	16. Set screw
6. Anchor yoke	17. Lock out screw
7. Gasket	19. Spring
9. Pivot pin	20. Tension link
10. Snap rings	22. Pin
11. Spacer washers	23. Oil seal
12. Spring housing	24. Rear cover
13. Adjusting nut	28. Support
14. Lock nut	33. Output shaft cover
15. Jam nut	S. Shim

854C to reassemble, be sure to reinstall the same number of shims (S) as were removed, then adjust spring tension as follows: Install adjusting nut (30) with counterbore toward spring and tighten the adjusting nut until spring can just be rotated with the fingers as shown in Fig. JD854D; then, tighten the nut an additional 1/4 turn and install lock nut. Clamp straight edge (or bar) on spring housing mounting surface as shown in Fig. JD854F. Distance from center of hole in control arm to mounting surface of spring housing should be 3/8-7/8-inch for models 40U-320U-330U-420U-430U and 3/16-11/16-inch for model 40W. If the measured dimension is less than specified, remove a shim (S—Fig. JD854C), readjust the spring tension and recheck. If the dimension is more than specified, add a shim, readjust the spring tension and recheck.

With preceding adjustments made, turn lockout screw (29) either way as required to obtain 9/16-inch clearance between anchor yoke and head of lockout screw (29). Then, install set screw (32) until it bottoms against

lockout screw (29) and tighten lock nut (33). Reinstall rear cover and spring housing and connect linkage.

Models 40H-420H-430H

78. **R & R AND OVERHAUL.** Drain transmission, disconnect upper link assembly from anchor yoke (6—Fig. JD854G) and remove power shaft guard if tractor is so equipped. Disconnect control link from control arm, remove spring housing retaining screws and withdraw spring housing. Remove cap screws from support bracket (28), then remove rear cover.

The procedure for overhauling the rear cover is evident after an examination of the unit and reference to Fig. JD854G.

If anchor yoke and spring housing was disassembled, refer to Fig. JD854G to reassemble, be sure to reinstall same number of shims (S) as were removed then adjust spring tension as follows: Install adjusting nut (13) with counterbore toward spring and tighten the adjusting nut (13) until spring can just be rotated with the fingers in a manner similar to that

shown in Fig. JD854D; then tighten the nut an additional 1/4 turn and install lock nut. Clamp straight edge, (or bar), on spring housing mounting surface as shown in Fig. JD854F. Distance from center of hole in control arm to mounting surface of spring housing should be 1 1/2-2 inches. If the measured dimension is less than specified, add a shim (S—Fig. JD854G), readjust the spring tension and recheck. If the dimension is more than specified, remove a shim, readjust spring tension and recheck.

After making the preceding adjustments, turn lockout screw (17) either way as required to obtain 3/4-inch clearance between anchor yoke and head of lockout screw (17), then install set screw (16) until it bottoms against lockout screw (17) and tighten lock nut (15). Reinstall rear cover and spring housing and connect linkage.

Models 420I-420W-430W-440I

79. **R & R AND OVERHAUL.** Drain transmission, disconnect upper link assembly from anchor yoke (7—Fig. JD854H) and remove power shaft attachment if tractor is so equipped. Remove lock nut (15) and adjusting nut (14) from spring stud (5). Disconnect lift arm return spring. Dis-

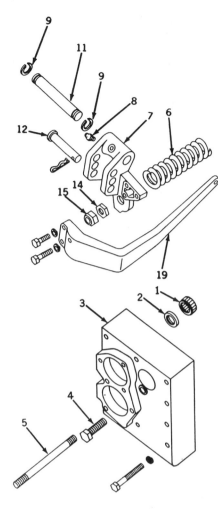

Fig. JD854H — Exploded view of models 420I, 420W, 430W and 440I transmission rear cover and anchor yoke assembly.

1. Needle bearing	8. Grease fitting
2. Oil seal	9. Snap ring
3. Rear cover	11. Pivot pin
4. Lock out screw	12. Upper link pin
5. Stud	14. Adjusting nut
6. Spring	15. Lock nut
7. Anchor yoke	19. Control arm

connect control link from control arm. Remove snap ring from pivot pin (11), drift out pivot pin and remove anchor yoke. Loosen "Touch-O-Matic" unit and block-up same. Remove rear cover.

The procedure for overhauling the rear cover is evident after an examination of the unit and reference to Fig. JD854H. Lip of seal faces front of tractor and if a new bearing is installed, it should be pressed in until end is $\frac{1}{16}$-inch below counterbore.

With the rear cover and components installed on transmission, adjust the load and depth control linkage, as follows: Turn lockout screw (4) inward until it bottoms against rear cover, then turn spring adjusting nut (14) either way as required until 1⅜ inches clearance is obtained between anchor yoke and head of lockout screw

Fig. JD854J—Adjusting the load and depth control anchor yoke on models 420I, 420W, 430W and 440I. A 1⅜-inch spacer block is used to gage the adjustment.

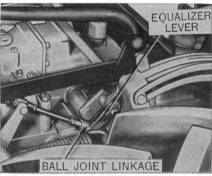

Fig. JD855—Dual "Touch-O-Matic" control linkage as used on Deere 40T, 420T and 430T.

(4). Install lock nut (15) and tighten. See Fig. JD854J.

Models 40T-40V-420T-420V-430T-430V

80. **R & R AND OVERHAUL.** Disconnect upper lift link assembly from anchor yoke and remove the power shaft guard. Disconnect the ball joint

linkage from the equalizer lever (Fig. JD855) and remove the four cap screws retaining the "Touch-O-Matic" unit to transmission case. Raise the "Touch-O-Matic" unit approximately two inches and block in the raised position. Unbolt and remove the rear cover from the transmission case.

The procedure for disassembling the removed unit is evident. Ream bushing (14—Fig. JD856) after installation to an inside diameter of 1.378-1.379. Ream the anchor yoke shaft bushings (8) to an inside diameter of 1.126-1.127 for the right bushing, 0.923-0.924 for the left bushing. Install seal (15) with lip of same facing front of tractor.

Reassemble the cover, using one shim (3). Install the load spring adjusting nut (12) with counterbore in nut toward spring and tighten the nut until end play is just removed from the spring; then, tighten the nut an additional ¼ turn and install the lock nut. Clamp a steel plate, measuring about 6 inches wide and 18 inches long, to cover as shown in Fig. JD857 and measure the distance from front face (gasket face) of the cover to the center of the hole in the control lever as shown. If the measured dimension is less than 4¼ inches, it will be necessary to add an additional shim (3—Fig. JD856). If the dimension is more than 4⅝ inches, remove the shim (3). Note: If shims are added or removed, it will be necessary to readjust the load control spring as previously outlined.

After the 4¼-4⅝ inch adjustment has been obtained, place a ¾ inch thick spacer block against head of lock-out screw (23—Figs. JD856 & 858) and turn the lock-out screw **out** until the spacer block just touches the anchor yoke stop lug (L—Fig. JD858). Turn the set screw (16) **in** as

1. Anchor yoke	
2. Pin	
3. Shim	
4. Tension link	
5. Spring	
6. Cover	
8. Bushing	
11. "O" ring	
12. Tension link nut	
13. Jam nut	
14. Bushing	
15. Oil seal	
16. Set screw	
17. Lock nut	
18. Control arm	
19. Roll pin	
20. Shaft	
21. Roll pin	
23. Stop screw	
24. Stud	

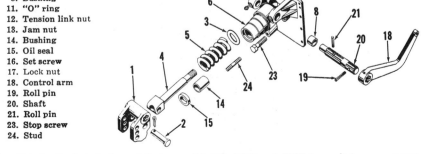

Fig. JD856—Exploded view of Deere model 40T, 420T and 430T transmission rear cover. Bushings (14 and 8) require final sizing after installation.

shown until the set screw bottoms on the lock-out screw. Install the set screw lock nut and recheck the adjustment.

Reinstall cover by reversing the removal procedure and tighten the four "Touch-O-Matic" retaining cap screws to a torque of 60 Ft.-Lbs. Reconnect the linkage to the equalizer lever and check the lift linkage as in paragraph 127A.

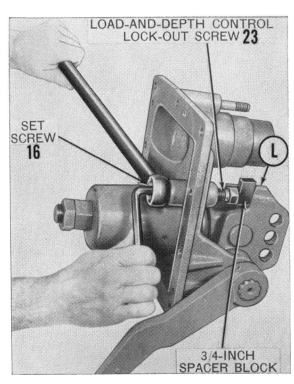

Fig. JD858—Adjusting the stop for load and depth control lock-nut screw on Deere model 40T, 420T and 430T.

R & R TRANSMISSION

All Models

81. To remove transmission from tractor and from final drive units, first disconnect transmission from center frame as outlined in paragraph 70 and proceed as follows: Support transmission housing and remove drawbar assembly and rear wheels. Remove platforms, disconnect brake linkage and remove the left brake lever. On models 40T, 420T and 430T, remove the brake shaft. Unbolt final drive units from transmission case and remove the units.

OVERHAUL SHIFTER RAILS

All Models

82. To overhaul the shifter rails remove top cover, which on standard and utility type tractors, requires the loosening and tilting of the "Touch-O-Matic" unit. On 40 series, 320 series, 330 series and 420 series prior to serial number 108922, slip rubber boot (2—Fig. JD860) off shift lever. Place lever in neutral position and remove snap ring (3), upper bearing (4) and lever. On 420 series

tractors after ser. no. 108921, 430 series and 440 series, remove lever knob. Then unbolt and remove shifter guide (25—Fig. JD860A and 860B). Slip rubber boot (2) off shift lever. Place lever in neutral position and remove snap ring (3), upper bearing (4) and shift lever.

On all models, shift second and reverse fork (8—Figs. JD860, 860A and 860B) against boss of cover and the other two forks in neutral position; then bump pin out of fork (8). Use

a soft drift and bump rail (12) out of fork and cover. Remove detent ball and spring and interlocking balls and pin. Follow same procedure with first and fifth speed fork (7) and with third and fourth speed fork (9). Length of new interlocking pin (16) is 0.527-0.532.

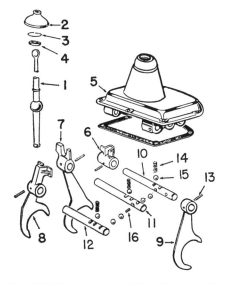

Fig. JD857—Checking the load and depth control anchor yoke adjustment on Deere model 40T, 420T and 430T. Notice how the steel plate is clamped to the cover.

Fig. JD860—Deere 40, 320, 330 and 420 series (prior serial number 108,922), transmission cover and shifter assembly.

1. Shift lever	10. 3rd shift rail
3. Snap ring	11. 1st & 4th shift rail
4. Lever bearing	12. 2nd & reverse shift
5. Cover	rail
6. 3rd shifter guide	13. Fork pin
7. 1st & 4th fork	14. Detent spring
8. 2nd & reverse fork	15. Detent ball
9. 3rd fork	16. Interlock pin

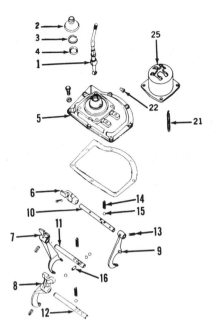

Fig. JD860A—Transmission top cover and shifter assembly for four speed transmission used on 420 series tractors, serial number 108,922 and up, 430 and 440 series tractors. Refer to legend of Fig. JD860.

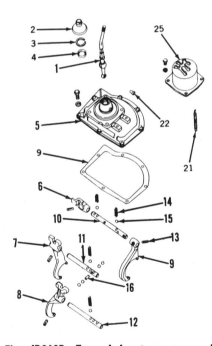

Fig. JD860B—Transmission top cover and shifter assembly for five speed transmission used on 420 series tractors, serial number 108,922 and up, 430 and 440 series tractors. Refer to legend of Fig. JD860.

OVERHAUL OUTPUT SHAFT

Series 40-320-330-420-430-440 (Four-Speed)

83. To remove the output shaft (33—Fig. JD861), which is integral with the main drive bevel pinion, first remove the transmission assembly as

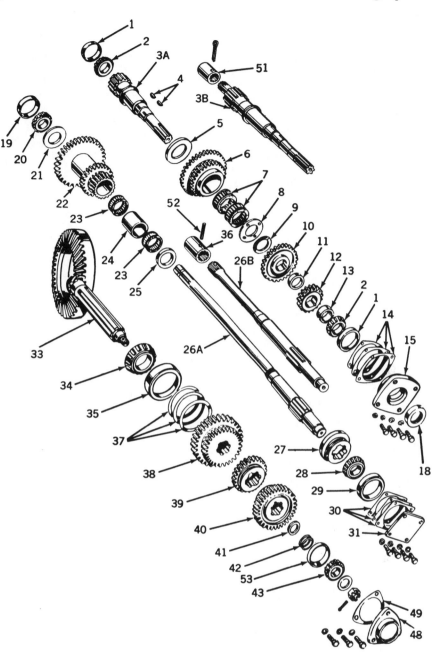

Fig. JD861—Exploded view of the four speed transmission used in tractors serial numbers 100,001 through 125,000. Output shaft (33) and the main drive bevel ring gear are available only as a matched set. Items (14), (15), (18), (31), (48) and (49) are not used on those models fitted with the early type direction reverser. Models prior to serial number 100,001 are similar.

1. Bearing cup	10. Fourth speed gear	26A. Power shaft (S, T and V)	38. Second and reverse sliding gear
2. Bearing cone	11. Spacer	26B. Power shaft (H, I, U and W)	39. First and fourth sliding gear
3A. Input shaft (H, S, T and V)	12. Third speed gear	27. Coupling	40. Third speed sliding gear
3B. Input shaft (I, U and W)	13. Spacer	28. Bearing cone	41. Spacer washer
4. Woodruff keys	14. Shims (0.002, 0.005 and 0.010)	29. Bearing cup	42. Shims (0.002, 0.005 and 0.010)
5. Rear washer	15. Quill	30. Shims (0.002, 0.005 and 0.010)	43. Bearing cone
6. First speed cluster gear	18. Oil seal	31. Cover	48. Cover
7. Bearing	19. Bearing cup	33. Output shaft	49. Gasket
8. Front washer	20. Bearing cone	34. Bearing cone	53. Bearing cup
9. Washer spring	21. Spacer washer	35. Bearing cup	
	22. Cluster gear	37. Shims (0.002, 0.005 and 0.010)	
	23. Bearing		
	24. Spacer		
	25. Front washer		

outlined in paragraph 81 and the differential as outlined in paragraph 90. Remove the transmission top cover and on models with direction reverser prior to ser. no. 131,309, remove re-

verser unit and mounting plate as in paragraph 71. On models without direction reverser, or with the differential type reverser, remove the output shaft front bearing cover (48).

49

On all models, remove cotter pin and nut from forward end of output shaft. Pull shaft rearward and remove front bearing cone (43) and shims (42). Withdraw shaft through rear of case and gears from above. Output shaft is supplied only as a matched set with the bevel ring gear.

Shims (37) between rear bearing cup (35) and transmission housing wall control mesh position of main drive bevel pinion gear. Be sure to install the same number of shims as were removed.

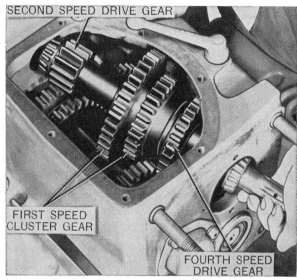

Fig. JD861B — Removing Deere 40, 320, 330, 420, 430 and 440 series four speed transmission input shaft. The shaft can be removed without disturbing the output shaft on some early models.

When reassembling, install bearing spacer (41) with chamfered edge toward shoulder on shaft. Vary thickness of shims (42) between spacer (41) and front bearing cone to remove all bearing play but permitting shaft to rotate freely; then remove 0.005 thickness of shims to obtain correct bearing pre-load of 0.004-0.006.

With the bearings adjusted, refer to paragraph 92B and check, and adjust if necessary, the mesh position of the main drive bevel pinion.

Refer to paragraph 71B when installing the early type direction reverser.

Series 420-430-440 (Five-Speed)

84. To remove output shaft (48—Fig. JD862), which is integral with the main drive bevel pinion, first remove the transmission assembly as outlined in paragraph 81 and the differential as outlined in paragraph 90. Remove the transmission top cover.

On models with no continuous-running power take-off or direction reverser, remove cover from front end of output shaft. On models with continuous-running power take-off, remove coupling from input shaft then unbolt and remove power take-off housing (30—Fig. JD863). On models with direction reverser prior to 131,809, remove reverser unit and mounting plate as in paragraph 71.

On all models remove nut (65—Fig. JD862) and washer (64) from output shaft (48). Pull shaft rearward and remove front bearing cone (63) and preload shims (61). On models prior to ser. no. 125,001, disengage snap ring (54) from its groove and slide snap ring toward front of shaft. Pull shaft rearward and remove third speed

Fig. JD861A—Exploded view of the four speed transmission used in tractors serial number 125,001 and up. This unit is similar to the unit used prior to 125,001 with the exception of detail changes in the input shaft and the output shaft and gears. Items (16), (17), (20), (35), (48) and (49) are not used on models that are equipped with the early type direction reverser.

2. Coupling	15. Spacer	30A. Power shaft (S, T and V)	40. Shims (0.002, 0.005 and 0.010)
3. Bearing cup	16. Shims (0.002, 0.005 and 0.010)	30B. Power shaft (H, I, U and W)	41. Second and reverse sliding gear
4. Bearing cone	17. Quill	31. Coupling	42. First and fourth sliding gear
5. Input shaft	20. Oil seal	32. Bearing cone	43. Third speed sliding gear
6. Woodruff keys	21. Bearing cup	33. Bearing cup	44. Spacer washer
7. Rear washer	22. Bearing cone	34. Shims (0.002, 0.005 and 0.010)	45. Shims (0.002, 0.005 and 0.010)
8. First speed cluster gear	23. Spacer washer	35. Cover	48. Gasket
9. Bearing	24. Cluster gear	37. Output shaft	49. Cover
10. Front washer	25. Bearing	38. Bearing cone	
11. Spring washer	26. Spacer	39. Bearing cup	
12. Fourth speed gear	27. Front washer		
13. Spacer	29. Coupling		
14. Third speed gear			

gear (59), drive member (58) and coupling (57). Continue to work shaft rearward and snap ring forward and remove fourth speed gear (56), thrust washer (55) and snap ring (54). Pull output shaft (48) from transmission case and remove fifth speed gear (53) and second and reverse gear (52).

On models ser. no. 125,001 and up, the snap ring has been discontinued and the output shaft can be moved rearward after the front bearing cone has been removed. Refer to Fig. JD-862A.

Output shaft is supplied only as a matched set with the bevel ring gear.

Shims (51) between rear bearing cup (50) and transmission housing wall control mesh position of main drive bevel pinion gear. Be sure to install the same number of shims as were removed.

When reassembling, refer to Fig. JD862. Be sure thrust washer (60) is installed with protruding shoulder facing front bearing cone. Vary the thickness of shims (61) under front bearing cone to remove all bearing play but permitting shaft to rotate freely; then remove 0.003 thickness of shims to obtain correct bearing preload of 0.002-0.004.

With the bearings adjusted, refer to paragraph 92B and check, and adjust if necessary, the mesh position of the main drive bevel pinion.

On models (prior to 131,309) so equipped, install the direction reverser as outlined in paragraph 71B.

OVERHAUL INPUT SHAFT
Series 40-320-330-420-430-440 (Four-Speed)

85. To remove the input shaft (3A or 3B—Fig. JD861), it is necessary to first disconnect transmission from center frame as outlined in paragraph 70 and remove the top cover.

NOTE: On models 40U, 40W, 420I, 420U, 420W, 430U, 430W and 440I a coupling (51) is fitted to rear end of input shaft to provide for high speed (engine speed) power take-off. To remove the input shaft on these models, first remove the transmission rear cover as outlined in paragraph 77 or 79 and remove the coupling (51).

On models with direction reverser prior to 131,809, remove the reverser unit and mounting plate as in paragraph 71. On models without direction reverser, or with differential type reverser, remove input shaft front bearing quill (15) from transmission front wall which on models 100,001 and up requires removal of a coupling

from input shaft. On all models, remove the power shaft front bearing cover (31) (or bracket) and tap rear end of power shaft to move the power shaft front bearing cup forward about ¼-inch. Move the input shaft forward through opening in front of housing, and lift rear end of shaft up and out through opening in top of case as shown in Fig. JD-861B.

NOTE: Two types of transmission housings have been used on the 420 series tractors as shown in Figs. JD861C and 861D. On the models fitted with the type shown in Fig. JD861D, the transmission output shaft will have to be removed before the input shaft can be removed as outlined above.

To disassemble the input shaft, use a suitable puller and remove front bearing cone. Pull or press third speed drive gear (12—Fig. JD861) off shaft and remove Woodruff key. Remove spacer (11), fourth speed gear (10) and Woodruff key. Remove spring washer (9), spacer washer (8), cluster gear (6), bearings and rear spacer washer (5).

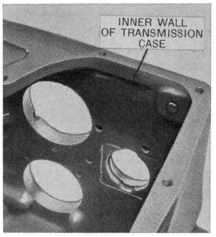

Fig. JD861C—View of inside wall of early type transmission case.

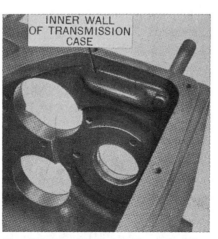

Fig. JD861D—View of inside wall of late type transmission case.

The washer (5) should be installed with chamfered side facing shoulder on shaft. Washer (8) should be installed with chamfered side toward cluster gear. Install spring washer (9) with concave (cupped) side nearest plain washer (8). Install fourth speed gear with chamfered end of teeth toward rear. Install spacer (11) with chamfered side toward fourth speed drive gear. Install third speed gear with chamfered end of teeth toward front. Install spacer (13) with chamfered side toward third speed drive gear, then install the front bearing cone and input shaft assembly. Bump the power shaft rearward and reinstall bearing cover (or bracket) (31). On models without direction reverser or with the differential type reverser, vary thickness of shims (14) between front bearing quill (retainer) and case to provide 0.002-0.004 shaft end play. Install oil seal (18) in quill with lip of seal facing toward rear. On models with direction reverser prior to serial number 131,309, install the reverser mounting plate and adjust the input shaft end play as in paragraph 71. Then assemble and install the reverser mechanism as in paragraph 71B.

Series 420-430-440 (Five-Speed)

86. To remove the input shaft (5 or 5A—Fig. JD862), it is necessary to first disconnect transmission from center frame as outlined in paragraph 70 and remove top cover.

NOTE: On models 420I, 420U, 420W, 430U, 430W and 440I, a coupling (2) is fitted to rear end of input shaft to provide for high speed (engine speed) power take-off. To remove the input shaft on these models, first remove the transmission rear cover as in paragraphs 77 and 79 and remove coupling (2).

On models with no continuous-running power take-off or direction reverser of the early type, or models after serial number 131,808 equipped with the late (differential) type direction reverser, remove power shaft front cover (45), coupling from input shaft and input shaft front bearing quill (7).

On models with continuous-running power take-off, remove coupling from input shaft then remove power take-off housing assembly as shown in Fig. JD863. Remove front bearing cone (19) from power shaft and withdraw driven gear (41). Remove input shaft front bearing quill (7—Fig. JD862).

On models prior to serial number 131,809, equipped with direction reverser, remove direction reverser housing assembly and mounting plate as in paragraph 71; then remove the

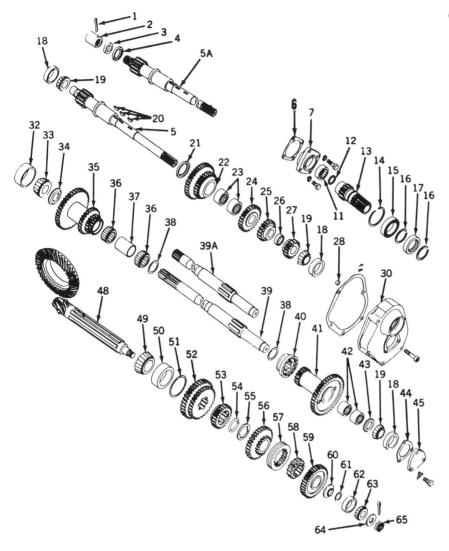

Fig. JD862—Exploded view of 420 series, prior to serial number 125,001, five speed transmission shafts and gears. View shown includes component parts of the optionally available continuous-running power take-off assembly which bolts to front wall of transmission housing. Refer to Fig. JD862A for exploded view of the five speed transmission from serial number 125,001 and up.

1. Cotter pin	19. Bearing cone	39. Power shaft
2. Coupling (models	20. Woodruff key	(models S, T
I, U and W)	21. Washer	and V)
3. Spacer*	22. First speed	39A. Power shaft
4. Oil seal*	cluster gear	(models H, I, U,
5. Input shaft	23. Needle bearings	W, some S)
5A. Input shaft	24. Fifth speed gear	40. Coupling
(models I, U and W)	25. Fourth speed gear	41. Power shaft
6. Shims	26. Spacer	driven gear*
7. Bearing quill	27. Third speed gear	42. Needle bearing*
11. Needle bearing*	28. Dowel*	43. Spacer washer*
12. Oil seal*	30. Housing*	44. Shims
13. Power shaft	32. Bearing cup	45. Cover
drive gear*	33. Bearing cone	48. Output shaft
14. Snap ring*	34. Spacer washer	49. Bearing cone
15. Ball bearing*	35. Cluster gear	50. Bearing cup
16. Snap ring*	36. Roller bearing	51. Snap ring
17. Oil seal	37. Spacer	52. Second and re-
18. Bearing cup	38. Washer	verse sliding gear

53. First and fifth	
sliding gear	
54. Snap ring	
55. Thrust washer	
56. Fourth speed	
driven gear	
57. Coupling	
58. Drive member	
59. Third speed	
driven gear	
60. Thrust washer	
61. Shims	
62. Bearing cup	
63. Bearing cone	
64. Washer	
65. Nut	

*Used on models with
continuous pto only.

control shaft bracket from front of power shaft.

On all models, tap power shaft (39 or 39A) forward about ¼-inch, then move input shaft forward through opening in front wall of transmission case, lift rear end of shaft up and out through opening in top of case.

To disassemble input shaft assembly use a bearing puller and remove front bearing cone. Press or pull third speed gear (27) off shaft and remove Woodruff key. Remove spacer (26) and using a knife edge puller, remove fourth speed gear (25) and Woodruff key. Press or pull fifth speed gear

(24) from shaft and remove Woodruff key.

CAUTION: Do not attempt to press off fourth speed gear (25) and fifth speed gear (24) together as keyways in input shaft may not be aligned and if attempted, either Woodruff key, fifth speed gear, or both could be damaged.

Remove first speed cluster gear (22), bearings (23) and rear thrust washer (21).

To reassemble, refer to Fig. JD862 and proceed as follows: Install rear thrust washer (21) on shaft. Install first speed cluster gear (22) and bearings with smaller diameter gear toward front. Install Woodruff key, fifth speed gear (24) and press gear on until it bottoms. Install Woodruff key, fourth speed gear (25) with shoulder toward front and press gear on shaft until it bottoms. Install spacer (26) with chamfered edge toward fourth speed gear. Install Woodruff key, third speed gear (27) with shoulder toward rear and press gear on until it bottoms; then install front bearing cone and input shaft assembly.

On models with no continuous-running power take-off or direction reverser, vary the thickness of shims (6) between bearing quill and transmission housing to obtain the desired input shaft end play of 0.002-0.004. Bump power shaft rearward and install cover (45).

On models with continuous-running power take-off, vary the thickness of shims (6) between bearing quill and transmission housing to obtain the desired input shaft end play of 0.002-0.004. Bump power shaft rearward and install driven gear (41) and bearing cone (19); then install pto housing assembly (30—Fig. JD863).

On models prior to serial number 131,809, with direction reverser install the reverser mounting plate and adjust the input shaft end play as in paragraph 71. Then, assemble and install the reverser mechanism as in paragraph 71B.

OVERHAUL POWER SHAFT

Series 40-320-330-420-430-440 (Four-Speed)

87. With output and input shafts removed as outlined in preceding paragraphs, bump rear end of shaft (26A or 26B—Fig. JD861) forward and remove front bearing cup. Remove rear bearing cone from shaft by bumping same with cluster gear (22). Withdraw shaft through front of trans-

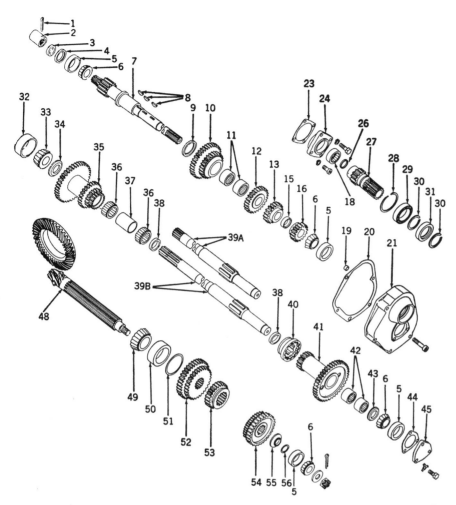

Fig. JD862A—Exploded view of the five speed transmission used in tractors serial number 125,001 and up. Basic difference between this transmission and that shown in Fig. JD862 is in the output shaft gears. Output shaft and gears are similar to those in four speed transmission. View shown includes the component parts used for models equipped with continuous-running power take-off.

1. Cotter pin
2. Coupling
3. Spacer (I, U and W)*
4. Oil seal*
5. Bearing cup
6. Bearing cone
7. Input shaft
8. Woodruff keys
9. Washer
10. First speed cluster gear
11. Needle bearing
12. Fifth speed gear
13. Fourth speed gear

15. Spacer
16. Third speed gear
18. Needle bearing*
19. Hollow dowel*
20. Gasket*
21. Housing*
23. Shim
24. Quill
26. Oil seal*
27. Power shaft drive gear*
28. Snap ring*
29. Ball bearing*
30. Snap ring*
31. Oil seal
32. Bearing cup

33. Bearing cone
34. Spacer washer
35. Cluster gear
36. Roller bearing
37. Spacer
38. Washer
39A. Power shaft (H and U)
39B. Power shaft
40. Coupling
41. Power shaft driven gear*
42. Needle bearing*
43. Spacer washer*
44. Shims (0.002, 0.005 and 0.010)

45. Cover*
48. Output shaft
49. Bearing cone
50. Bearing cup
51. Shims (0.002, 0.005 and 0.010)
52. Second and reverse sliding gear
53. First and fifth sliding gear
54. Third and fourth driven gear
55. Spacer washer
56. Shims (0.002, 0.005 and 0.010)

*Used on models with continuous pto only.

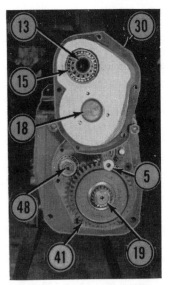

Fig. JD863—View showing five speed transmission with power take-off housing removed.

5. Input shaft
13. Power shaft drive gear
15. Ball bearing
18. Bearing cup
19. Bearing cone
30. Housing
41. Power shaft driven gear
48. Output shaft

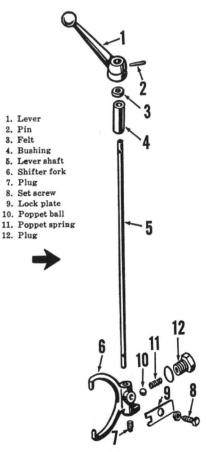

1. Lever
2. Pin
3. Felt
4. Bushing
5. Lever shaft
6. Shifter fork
7. Plug
8. Set screw
9. Lock plate
10. Poppet ball
11. Poppet spring
12. Plug

Fig. JD864—Deere 40 series power (PTO) shaft shifter assembly. Pre-sized bronze bushing (4) extends ¼-inch above top of case. Series 320, 330, 420, 430 and 440 are similar except for lever (1).

mission housing and remove gears, bearings, spacers and coupling from above.

When reassembling, install spacer washer (21) between rear bearing cone and cluster gear with beveled edge of spacer toward the rear. Vary the thickness of shims (30) between bearing cover (31) (or bracket) and transmission wall to remove all shaft end play but permitting shaft to rotate freely; then remove 0.003 thickness of shims to obtain correct bearing pre-load of 0.002-0.004.

Series 420-430-440 (Five-Speed)

88. With output and input shafts removed as in paragraphs 84 and 86, bump rear end of shaft (39 or 39A—Fig. JD862) forward and on those models without continuous-running power take-off, remove front bearing cup (18).

NOTE: On models with continuous pto, the bearing cup (18) was removed with the pto housing as shown in Fig. JD863.

Continue to bump shaft forward and remove rear bearing cone (33—Fig. JD862). Withdraw shaft through front of transmission housing and remove gears, spacers and coupling from above.

When reassembling, install spacer washer (34) between rear bearing cone and cluster gear with beveled edge toward the rear. Vary thickness of shims (44) under bearing cover

(45) to remove all shaft end play, but permitting shaft to rotate freely; then remove 0.003 thickness of shims to obtain correct bearing pre-load of 0.002-0.004.

OVERHAUL PTO SHIFTER CONTROLS
All Models

89. To remove power shaft shifter

control, unscrew plug (12—Fig. JD-864) from case and extract poppet (detent) assembly (10 and 11) from left side of case. Remove set screw (8) from shifter fork (6) and remove lever shaft (5) from fork. Top of bushing (4) should extend ¼-inch above top of case. Bushing is presized and if carefully installed, requires no final sizing.

DIFFERENTIAL AND FINAL DRIVE

DIFFERENTIAL
All Models

90. **REMOVE & REINSTALL.** To remove the differential, first remove the final drive assemblies as outlined in paragraph 95, 100 or 102A. Remove the transmission case rear cover as outlined in paragraphs 75 through 80. On all models except 420W serial numbers 80,001 through 100,000, remove both differential bearing quills (1—Figs. JD870 and 871). Working through transmission housing rear opening, remove bevel ring gear and differential unit.

On 420W tractors serial number 80,-001 through 100,000, the power shaft must be partially removed from transmission. On these tractors proceed as follows: Split transmission from center frame and remove power shaft front bearing cover. Drive power shaft forward to free front bearing cup from transmission housing and rear bearing cone from power shaft. Slide power shaft forward to provide clearance for removal of differential. Remove both bearing quills. Remove the sealing cup from the rear bore in right side of transmission housing, then remove bevel ring gear and differential assembly.

On models 40T, 420I, 420T, 420W, 430T, 430W and 440I, shims (4) interposed between bearing cones (3) and differential case (9) control differential bearing adjustment and backlash of main drive bevel ring gear and pinion.

On all models except 40T, 420I, 420T, 420W, 430T, 430W and 440I, shims (15) interposed between bearing quills (1) and bearing cups (3) control bearing adjustment and backlash of main drive bevel ring gear and pinion. Recommended bearing adjustment is 0.002-0.005 preload; recommended backlash is 0.006-0.008. Refer to MAIN DRIVE BEVEL GEARS section for adjustment procedure. Install oil seals (2) in quills with lip of seal facing differential.

91. **OVERHAUL.** Neither the bevel

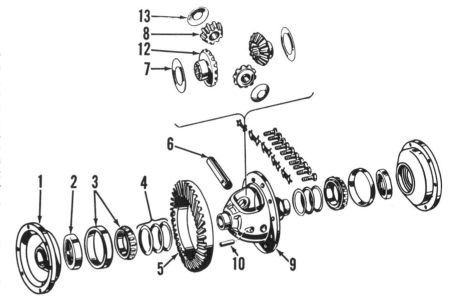

Fig. JD870—Deere 40T, 420I, 420T, 420W, 430T, 430W and 440I differential assembly. Bevel ring gear (5) is retained to differential case (9) by cap screws. Shims (4) are for adjustment of differential bearings and backlash of main drive bevel gears.

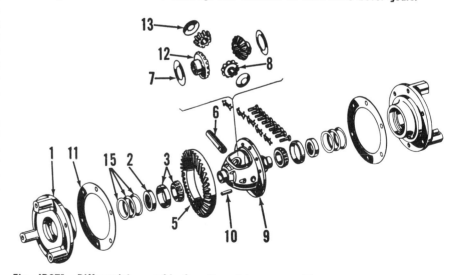

Fig. JD871—Differential assembly for all models except 40T, 420I, 420T, 420W, 430T, 430W and 440I. Shims (15) are for adjustment of differential bearings and backlash of main drive bevel gears. Refer to Fig. JD870 for legend.

pinion nor ring gear are furnished separately, but only as a matched pair. Bevel ring gear (5) is attached to differential case by cap screws. Note: Two dowels have been added to differential cage at serial number

133,693. To disassemble the differential unit, remove bevel ring gear and bump pinion shaft retaining (Groov) pin (10) out of differential case. Remove differential pinion shaft, pinions, side gears and thrust washers.

Reassemble in reverse order and stake differential case to prevent pinion shaft retaining (Groov) pin from backing out.

After reassembly, check trueness of ring gear back face by mounting the unit in its normal operating position. Total runout should not exceed 0.003.

MAIN DRIVE BEVEL GEARS

All Models

92. **GEAR MESH.** The main drive bevel pinion (output shaft) fore and aft position in the transmission case is adjustable (by means of shims) to provide the correct cone center distance or mesh position of the pinion with the main drive bevel ring gear. To adjust the bevel pinion mesh position, it is necessary to first remove the transmission from the tractor and then remove the differential unit.

The main drive bevel gears are available only as a matched pair. If new gears are being installed, or if the original gears require adjustment, do so as outlined in the following paragraphs.

92A. SETTING THE PINION. With differential removed and transmission removed from tractor, first step in setting the pinion position is to make certain that pinion (output) shaft bearings are correctly adjusted as outlined in paragraphs 83 or 84.

92B. After bearing pre-load is adjusted, install special pinion setting gage John Deere AM452-T in transmission case as shown in Fig. JD872.

NOTE: Gage set AM452-T is for all models except 40T, 420I, 420T, 420W, 430T, 430W and 440I and when used for these models, it requires two adapter rings (No. M1550-T) which are available from the factory.

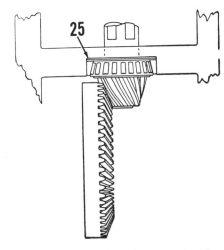

Fig. JD873—Simplified view of main drive bevel gears. Shims (25) control the fore and aft position of the main drive bevel pinion.

Note cone center distance etched on rear face of pinion. Measure the distance between end of gage and bevel pinion rear face with a feeler gage as shown. The thickness of the feeler gage plus 2.782 should equal the cone center distance as etched on the rear face of the pinion. Move the pinion in or out as required by adding or removing shims (25—Fig. JD873), until the feeler gage measurement when added to 2.782 exactly equals the cone distance (such as 2.815) etched on end of pinion. To vary the shims (25) which are located between the rear bearing cup and wall of the transmission housing, it will be necessary to remove the pinion shaft from the transmission.

If John Deere gage is not available, the bevel pinion adjustment can be obtained with an inside micrometer by measuring the distance between the gear end of the bevel pinion and a closely fitting mandrel installed in the differential quill bores. In this case, the micrometer reading plus one-half of the mandrel diameter should equal the measurement as indicated on the pinion.

93. **BACKLASH AND DIFFERENTIAL CARRIER BEARING ADJUSTMENT.** On models 40T, 420I, 420T, 420W, 430T, 430W and 440I, vary thickness of shims (4—Fig. JD870) between bearing cones (3) and differential case (9) to remove all end play and yet permit differential to rotate freely; then remove 0.004 of shims to obtain correct bearing preload of 0.002-0.005. After bearings are adjusted, adjust gear backlash to 0.006-0.008 by transferring shims (4) from left side of case to right side of case to increase gear backlash or from right side of case to left side of case to reduce backlash.

Procedure on all other models is the same except that backlash and differential bearing shims (15—Fig. JD-871) are located between bearing quills and bearing cup as shown.

94. **RENEWAL OF PINION & RING GEAR.** Neither the bevel pinion or the bevel ring gear are sold separately. They must be renewed as a matched pair. To renew the bevel pinion it is first necessary to remove the transmission from the tractor and from the final drive units as outlined in paragraph 81, and to remove the differential. The pinion shaft can now be renewed by following the procedure given in paragraph 83 or 84. To adjust mesh position of the new pinion shaft follow the procedure given in paragraph 92B.

The bevel ring gear and differential unit was removed from the tractor in removing the bevel pinion. The bench procedure for removal of the bevel ring gear is self-evident. To readjust the differential carrier bearings and the tooth backlash of the bevel gears, follow the procedure given in paragraph 93.

FINAL DRIVE UNITS

When a final drive unit has been removed for any reason, inspect for oil leakage from the differential. If there are signs of even slight leakage be sure to renew the seal (2—Figs. JD870 and 871) located in differential bearing quill on each side of differential.

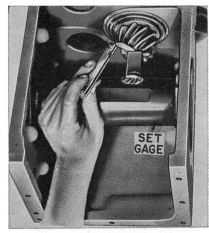

Fig. JD872—Setting the main drive bevel pinion mesh position with a John Deere special gage AM-452-T. Refer to text for alternate method.

Fig. JD874 — View showing method of mounting a dial indicator to check bevel gear backlash. Refer to text for adjusting procedure. Backlash is 0.006-0.008 for all models.

Models 40T-420T-430T

95. REMOVE AND REINSTALL. To remove a final drive unit, support rear portion of tractor and remove drawbar assembly and rear wheel. Unhook the brake pedal return springs, remove platforms and fenders and disconnect the brake adjusting yokes. Remove the left brake pedal and withdraw the brake shaft. Unhook the lift arm return springs and remove the lift arms. Remove the respective brake unit. Remove the two uppermost cap screws which retain the final drive housing to transmission case and insert two long studs (about 6 inches) in their place. Remove the remaining cap screws and using a hoist, slide final drive unit off studs.

96. R&R DRIVE SHAFT, BEARINGS AND/OR BULL PINION. To remove the bull pinion shaft (3—Fig. JD875) and gear (6), first remove the respective final drive unit as outlined in paragraph 95. Shaft, bearings and bull pinion can be bumped

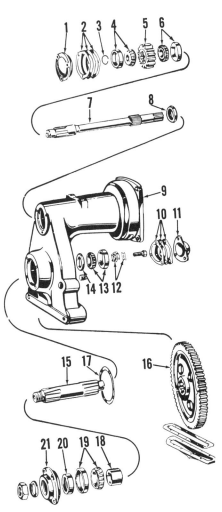

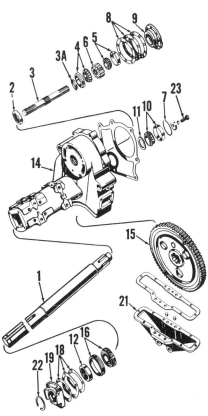

Fig. JD875—Exploded view of Deere 40T, 420T and 430T final drive. Shims (8) and (18) are used for adjusting the drive shaft and wheel axle bearings.

1. Wheel axle shaft	11. Washer
2. Oil seal	12. Oil seal
3. Drive shaft	14. Final drive housing
3A. Oil slinger	15. Drive (bull) gear
4. & 5. Bearing	16. Bearing
6. Drive (bull) pinion	18. Shims
7. Snap ring	19. Oil seal quill
8. Shims	21. Housing oil pan
9. Bearing quill	22. Snap ring
10. Bearing	23. Cap screw

Fig. JD876—Exploded view of Deere 40S final drive. Other models except Hi-Crop, Row-Crop Utility, Special Utility and Industrial (440I) are similar.

1. Bearing quill	12. Washers
2. Shims	13. Bearing
3. Snap ring	14. Washer
4. Bearing	15. Axle shaft
5. Bull pinion	16. Bull gear
6. Bearing	17. Gasket
7. Bull pinion shaft	18. Spacer
8. Oil seal	19. Bearing
10. Shims	20. Oil seal
11. Bearing cover	21. Bearing quill

out after removing bearing quill (9). The procedure for further disassembly will be evident after an examination of the unit.

Install oil seal (2) with lip of same facing the bull pinion gear. When installing oil slinger (3A), make certain that flange of slinger is toward bearing cone and 4 5/16 inches from beveled end of shaft.

Reassemble the unit by reversing the disassembly procedure and vary the number of shims (8) to obtain the desired bearing pre-load of 0.000-0.002.

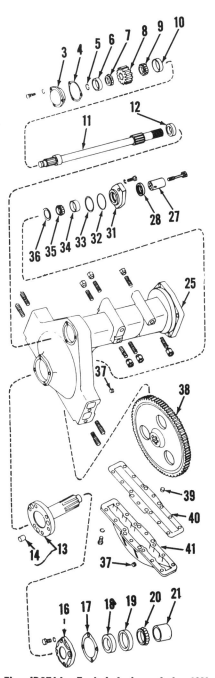

Fig. JD876A—Exploded view of the 40H, 420H and 430H final drive and component parts.

3. Quill	20. Bearing cone
4. Shim	21. Spacer
5. Snap ring	25. Final drive housing
6. Bearing cup	27. Planter drive shaft
7. Bearing cone	28. Oil seal
8. Pinion gear	31. Cover
9. Bearing cone	32. Sealing ring
10. Bearing cup	33. Shim
11. Final drive shaft	34. Bearing cup
12. Oil seal	35. Bearing cone
13. Axle shaft	36. Washer
14. Dowel	37. Pipe plug
16. Quill	38. Final drive gear
17. Gasket	39. Dowel
18. Oil seal	40. Gasket
19. Bearing cup	41. Oil pan

97. R&R WHEEL AXLE SHAFT, BEARINGS AND/OR BULL GEAR. To remove the wheel axle shaft (1—Fig. JD875) and gear (15), first re-

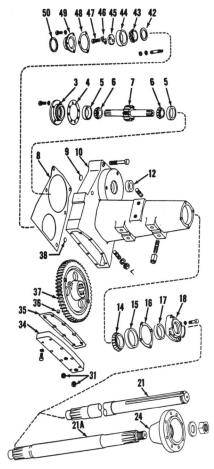

Fig. JD876B—Exploded view of 420W final drive and component parts. 420I, 430W and 440I are similar.

3. Quill	24. Hub (power
4. Shim	adjusted wheels)
5. Bearing cup	31. Pipe plugs
6. Bearing cone	34. Oil pan
7. Pinion shaft	35. Gasket
8. Gasket	36. Dowel
9. Expansion plug	37. Final drive gear
10. Final drive	38. Expansion plug
housing	42. Washer
12. Oil seal	43. Bearing cone
14. Bearing cone	44. Bearing cup
15. Bearing cup	45. Retainer
16. Shim	46. Lock
17. Oil seal	47. Cap screw
18. Quill	48. Gasket
21. Axle shaft	49. Cover
21A. Axle shaft (power	50. Sealing ring
adjusted wheels)	

move the respective final drive unit as outlined in paragraph 95, then remove the bull gear cover pan (21). Remove quill (19) and remove cap screw (23) from inner end of shaft. Using a reaction type pusher, push the shaft toward the wheel end and out of bull gear (15). The need and procedure for further disassembly is evident.

When reassembling, vary the number of shims (18) to provide the desired axle shaft bearing pre-load of 0.004-0.007.

Other Models Except 420I-420W-430W-440I

100. REMOVE AND REINSTALL. To remove a final drive unit, support rear of tractor and remove wheel. Disconnect lift arm return spring and remove the drawbar frame. Unhook the brake pedal return springs and remove the respective brake pull rod. Remove fender and foot rest. Unbolt and remove the final drive unit from transmission case.

When reinstalling the unit, adjust the brakes as outlined in paragraph 110.

101. R&R DRIVE SHAFT, BEARINGS AND/OR BULL PINION. To remove the drive shaft (7—Fig. JD-876 or 11—Fig. JD876A) or gear (5 or 8), support rear of tractor and remove wheel. Remove quill (1 or 3) and withdraw shaft, bearings and gear. Gear and bearing cones can be pressed off shaft after removing snap ring (3 or 5). Inner bearing cup can be removed with a suitable puller. Install oil seal (8 or 12) with lip of same facing out toward bull pinion.

When reassembling, vary the number of shims (2 or 4) to provide from 0.001 pre-load to 0.002 end play for the shaft bearings.

102. R&R WHEEL AXLE SHAFT, BEARINGS AND/OR BULL GEAR. To remove shaft (15—Fig. JD876 or 13—Fig. JD876A) or gear (16 or 38), support rear of tractor and remove wheel. Remove the housing oil pan and unbolt quill (21 or 16). Remove inner bearing cover (11 or 31) and cap screw and washers from inner end of shaft. Drive or press axle out of inner cone and gear and out through outer opening in housing. When shaft is being driven or pressed out, support hub of gear with a length of 3½ inch pipe slipped over outer end of axle.

When reassembling, vary the number of shims (10 or 33) to obtain the desired bearing pre-load of 0.004-0.007.

Models 420I-420W-430W-440I

102A. REMOVE AND REINSTALL. To remove final drive unit, support rear portion of tractor and remove lift arm spring and drawbar. Remove rear wheel. Unhook brake return spring and remove platform and fender, if so equipped. Disconnect brake adjusting yoke and remove respective brake unit. Unbolt and remove final drive unit from transmission case.

103. R & R DRIVE SHAFT, BEARINGS AND/OR BULL PINION. To remove shaft and pinion (7—Fig. JD-876B), first remove the respective final drive unit as outlined in paragraph 102A. With final drive unit removed, shaft and pinion and bearings can be removed after removing bearing quill (3). The procedure for further disassembly will be evident after an examination of the unit and reference to Fig. JD876B.

Reassemble the unit by reversing the disassembly procedure. Install oil seal (12) with lip facing inward and vary the number of shims (4) to obtain the desired bearing pre-load of 0.000-0.002.

104. R & R WHEEL AXLE SHAFT, BEARINGS AND/OR BULL GEAR. To remove the wheel axle shaft (21 or 21A — Fig. JD876B) and/or gear (37), first remove the respective final drives unit as outlined in paragraph 102A, then remove the bull gear cover (34). Remove quill (18) and slide quill from axle shaft being careful not to damage oil seal (17). Remove cover (49), then remove cap screws (47) and retainer (45) from inner end of axle shaft. Drive or press axle shaft from inner cone and gear and out through outer opening in housing. The need and procedure for further disassembly is evident.

When reassembling, vary the number of shims (16) to provide the desired axle shaft bearing pre-load of 0.004-0.007.

BRAKES

All Models

110. ADJUSTMENT. The only adjustment provided on the brakes is for pedal free travel. To adjust the brakes on models 40T, 420T and 430T, loosen the lock nut and turn the adjusting nut (Fig. JD880) until each pedal has a free travel of 1-1⅛ inches when measured between pedal and pedal stop on platform. On all other models loosen the lock nut and turn the adjusting nut (Fig. JD881) on each brake until each pedal has a free travel of 11/16-13/16 inch when measured between stop on pedal and edge of platform.

111. R&R AND OVERHAUL. To remove the brakes on models 40T, 420I, 420T, 420W, 430T, 430W and 440I, remove platforms and disconnect the brake pull yokes. Remove the brake housing retaining cap screws and remove assembly. To remove the brakes on all other models, remove the final drive housing assembly as outlined in paragraph 100 and withdraw the brake mechanism. The procedure for disassembling and overhauling the removed brakes is evident after an examination of unit and reference to Figs. JD882 or 883. Brake pedal and brake shaft bushings should be reamed to an inside diameter of 1.129-1.130. The pedal shaft bushings are located in final drive housings on 40T, 420T and 430T, and in transmission case on all other models.

BELT PULLEY

All Models

The belt pulley unit, Fig. JD885, is mounted over and driven by the PTO output shaft. The procedure for removing and reinstalling the unit is self-evident.

113. **OVERHAUL.** To disassemble the unit, remove pulley and Woodruff key. Unbolt and remove the drive

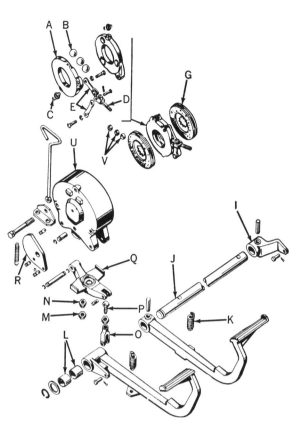

Fig. JD882—Exploded view of models 40T, 420I, 420T, 420W, 430T, 430W and 440I brakes.

A. Actuating disc
B. Ball
C. Spring
D. Actuating rod
E. Links
G. Lined disc
I. Left brake lever
J. Shaft
K. Pedal return spring
L. Bushings
M. Lock nut
N. Adjusting nut
O. Adjustable yoke
P. Eye bolt
Q. Operating lever
R. Cam
U. Brake housing
V. Hollow dowels

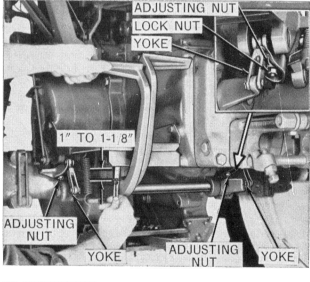

Fig. JD880—Adjusting the brakes on Deere models 40T, 420T and 430T. Brake pedals should have a free travel of 1-1 1/8 inches as shown.

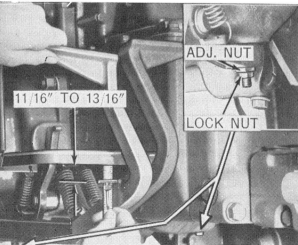

Fig. JD881—Adjusting the brakes on all models except tricycle. The pedals should have a free travel of 11/16-13/16-inch as shown.

shaft bearing quill (4—Fig. JD885) from the pulley case. Withdraw drive gear (8) and lift bearing cone (11) and shims (12) from bearing shaft (13). Using a lead hammer, bump pinion shaft (16) out of case and withdraw the remaining parts.

Examine all parts and renew any which are excessively worn. When reassembling the unit, proceed as follows: Install bearing cup (20), leaving out the shims (21) which are normally located behind the cup. Install bearing cone (19) on shaft (16) and position shaft (16) in the case. Position the case so that shaft (16) hangs in a vertical position and install the special set gage as shown in Fig. JD-886. Note: The special gage is available from John Deere. Using a feeler gage, measure the gap between the set gage and the gear as shown and remember the measurement. Remove the special set gage, shaft (16—Fig. JD885) and bearing cup (20).

Examine the gear end of shaft (16) for a number. If no number is found, then install shims (21) of a total thickness equal to the feeler gage measurement. If a number is found on the gear end of the shaft, install shims (21) of a total thickness equal to the feeler gage measurement minus the number which is stamped on the shaft.

Reinstall shaft (16) and vary the number of shims (23) to give the taper roller bearings a pre-load of 0.000-0.003. One method of determining this pre-load is to install more than enough shims (23), install pulley and tighten the retaining nut securely. Measure the shaft end play, then remove shims (23) of a total thickness equal to the measured end play plus 0.002. For example, if the measured end play was 0.010, remove 0.012 thickness of shims.

When installing the drive gear and shaft assembly (8), vary the number of shims (12) to obtain a bevel gear backlash of 0.004-0.006. One method of checking the backlash is to install the special John Deere set gage and use a dial indicator as shown in Fig. JD887. The indicator contact button should be located approximately ¼ inch from end of long tongue of gage as shown. After the correct backlash has been established, vary the number of shims (5) to give the drive gear and shaft (8) an end play of 0.002-0.004.

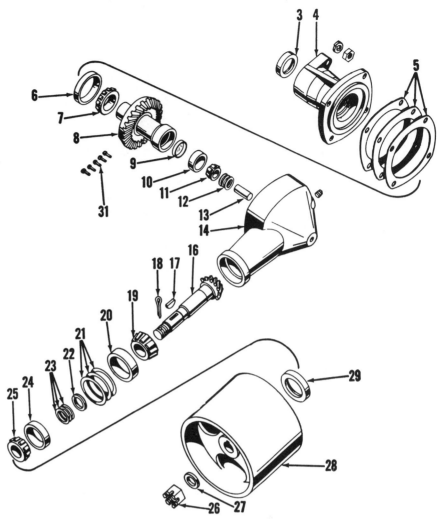

Fig. JD885—Exploded view of belt pulley attachment used on all models. The unit is driven by the power take-off shaft.

3. Seal	9. Cup plug	16. Pinion shaft	23. Shims
4. Bearing quill	10. Bearing cup	17. Woodruff key	24. Bearing cup
5. Shims	11. Bearing cone	18. Cotter pin	25. Bearing cone
6. Bearing cup	12. Shims	19. Bearing cone	26. Nut
7. Bearing cone	13. Bearing shaft	20. Bearing cup	27. Washer
8. Drive gear and	14. Housing	21. Shims	28. Pulley
shaft		22. Spacer	29. Oil seal

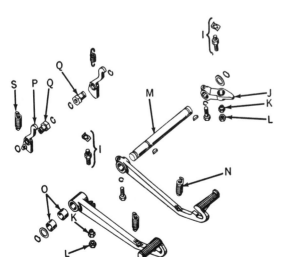

Fig. JD883—Exploded view of brakes used on all models except 40T, 420I, 420T, 420W, 430T, 430W and 440I

A. Actuating disc
B. Ball
C. Spring
D. Actuating yoke
E. Links
G. Lined discs
I. Rod and lock
J. Left brake lever
K. Adjusting nut
L. Lock nut
M. Shaft
N. Pedal return spring
O. Bushings
P. Brake lock
Q. Brake lock shaft

SPECIAL SET GAUGE

Fig. JD886—John Deere special set gage and a feeler gage can be used to determine the required thickness of the pinion shaft inner bearing shim pack.

Fig. JD887—Using John Deere special set gage and a dial indicator to measure the bevel gear backlash.

POWER TAKE-OFF

All Models

116. All series 40, 320, 330 and 440 tractors are equipped with a transmission driven power take-off shaft

Fig. JD888—View showing front end of five speed transmission with continuous-running power take-off unit installed.

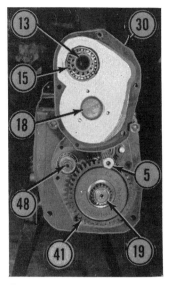

Fig. JD888A — View showing five speed transmission with PTO drive gear and housing removed.

5. Input shaft	19. Bearing cone
13. Power shaft drive gear	30. Housing
15. Ball bearing	41. Power shaft driven gear
18. Bearing cup	48. Output shaft

whereas the 420 and 430 series may be equipped with either a transmission driven or a continuous-running (engine driven) power take-off shaft.

In either case, the power shaft is an integral part of the transmission and the overhaul procedure is covered in paragraphs 87 and 88 of the transmission section. On some models, the transmission power shaft is the pto output shaft; whereas on other models, the output shaft is a part of the pto attachment shown in Fig. JD-889A.

For information pertaining to the pto clutch disc on models equipped with continuous-running power take-off, refer to paragraphs 63 through 65.

For information pertaining to the continuous-running power take-off hollow clutch shaft refer to paragraph 67A.

To overhaul power take-off drive gear and housing, proceed as follows:

PTO DRIVE GEAR AND HOUSING
Series 420-430
With Continuous PTO

116A. **R & R AND OVERHAUL.** With transmission separated from center frame as shown in Fig. JD888, remove housing as shown in Fig. JD-888A. With housing removed, refer to Fig. JD862 and proceed as follows: Remove snap ring (14) and push drive gear and bearing assembly from housing (30). Remove oil seal (12) from inner bore of drive gear and oil seal (17) from housing. Inspect and renew if necessary the needle bearing (11) in transmission input shaft quill. Remove bearing cone (19) from forward end of power shaft and withdraw driven gear (41) and needle bearings (42).

Carefully inspect all other parts and renew those showing wear or damage. Use care when reassembling to prevent splines on drive gear (13) from damaging oil seal (17). Also, be sure to install the same number of shims (6 and 44) as were removed.

POWER SHAFT ATTACHMENT
All Models So Equipped

116B. On models which are equipped with the power shaft attachment, remove same by removing cap screws from housing and withdrawing unit from transmission rear cover.

The procedure for disassembly and overhauling is evident after an examination of the unit and reference to Fig. JD889A.

Fig. JD889A — Exploded view of Power Shaft Attachment and associated parts. View shown is for H and U models, however other models are similar.

1. Oil seal
2. Needle bearing
3. Oil seal
5. Housing
6. Ball bearing
7. Snap ring
8. Snap ring
9. Shaft
10. Coupling
11. Roll pin
13. Screw plug

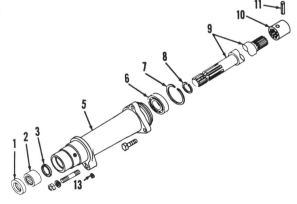

POWER LIFT UNIT

The maintenance of absolute cleanliness of all parts is of utmost importance in the operation and servicing of the hydraulic system. Of equal importance is the avoidance of nicks or burrs on any of the working parts.

SYSTEM TROUBLE SHOOTING

All Models

117. TROUBLE. Lift arms move up and down when control levers are stationary.

CAUSE AND CORRECTION. On the dual cylinder "Touch-O-Matic", this trouble could be caused by selector valves not seating properly. This condition can sometimes be corrected by tapping the selector lever firmly to the rear position. On both the single and dual cylinder "'Touch-O-Matic", the trouble could be caused by too many shims under the check valve (or valves). This condition can be corrected by adjusting the check valves or installing a two-stage check valve as in paragraph 126 or 126A.

TROUBLE. Loaded lift arms slowly drop from raised position.

CAUSE AND CORRECTION. This trouble could be caused by foreign material under check valve (or valves) and can sometimes be corrected by flushing the system as outlined in paragraph 123. This trouble could also be caused by a worn or damaged check valve and/or spring.

TROUBLE. System will not raise implement.

CAUSE AND CORRECTION. Trouble could be caused by improper adjustment of relief valves or a faulty hydraulic pump. Check and adjust the relief valves as in paragraph 124.

PUMP

The three section pump, shown in Fig. JD890, is mounted on the front face of the engine timing gear cover and receives its drive from a coupling on the front end of the engine camshaft. The pump is of the self-priming type and delivers approximately 8 gallons per minute when operating at the rated engine speed.

CAUTION: The engine should never be run when the hydraulic lines are disconnected. The pump will fail quickly when run without a supply of oil.

120. REMOVE AND REINSTALL. To remove the hydraulic pump, first drain cooling system and remove radiator as

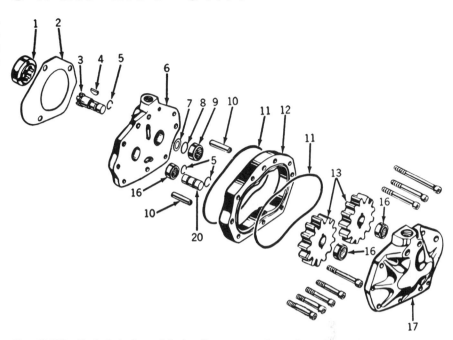

Fig. JD890—Exploded view of hydraulic pump used on all models. The pump is driven from the engine camshaft.

1. Coupling	5. Snap ring	9. Needle bearing	13. Gears
2. Gasket	6. Pump body	10. Dowel pin	16. Needle bearings
3. Drive shaft	7. Leather washer	11. Sealing rings	17. Cover
4. Woodruff key	8. "O" ring	12. Center section	20. Idler shaft

outlined in paragraph 48A, 49 or 49A. Drain the hydraulic system, disconnect the hydraulic lines from the pump and remove the three socket head pump retaining cap screws. When reinstalling the pump, tighten the retaining cap screws to a torque of 24-28 Ft.-Lbs.

121. OVERHAUL. To disassemble the pump, remove the assembly cap screws and using a lead hammer, tap the pump cover from the locating dowels. Remove the snap ring retaining the drive gear to the drive shaft as shown in Fig. JD891 and withdraw the drive gear and idler gear and shaft.

Lift pump center section from dowels and remove the drive gear Woodruff key. Remove drive shaft from pump body.

Inspect all parts and renew any which are damaged. Needle bearings can be collapsed for removal or can be removed with a suitable puller as shown in Fig. JD892. When installing needle bearings, use a piloted driver against end of bearing which has name stamped on it and press bearings into

Fig. JD891 — Removing snap ring from hydraulic pump drive shaft.

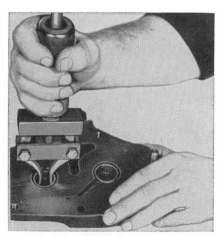

Fig. JD892—Pulling needle bearings from hydraulic pump. The bearings can also be removed by collapsing.

cover and pump body, allowing 0.060-0.065 clearance between machined surface and top of needle bearings in front cover and idler side of pump body, and 0.015-0.020 clearance between machined surface of pump body and drive shaft needle bearing. Refer to Fig. JD893. Install a new leather washer (7—Fig. JD890), with smooth side next to casting, into counterbore on drive shaft side of pump body; then, insert new "O" ring (8) into counterbore between leather washer and needle bearing. Reassemble the remainder of the pump by reversing the disassembly procedure and tighten the cover retaining cap screws to a torque of 24-28 Ft.-Lbs.

"TOUCH-O-MATIC" UNIT

All Models

"Touch-O-Matic" units are of the single or dual cylinder types. They are similar in construction and operate on the same principle.

The servicing procedures for the two systems are basically similar and the following paragraphs can be applied to either unit. In general, however, the dual cylinder system will require double the work which is listed.

122. **CHECKING SYSTEM BEFORE REMOVAL.** Before removing the "Touch-O-Matic" unit from tractor, visually inspect the system for external leaks. To test for internal leaks on the single cylinder system or internal leaks on the right half of the dual cylinder system, proceed as follows: On the dual cylinder system, move the selector lever firmly to the rear and attach a heavy implement to outer lift arms only. On the single cylinder system attach a heavy implement to lift arms. On both systems, raise the implement and shut off the engine. If lift arms do not move down after standing for a period of 15 to 20

minutes, it is reasonable to assume that there are no leaks within the system. The same check can be made on the left side of the dual cylinder system by attaching the implement to the inner lift arm.

If the lift arms move down during the 15-20 minute check, it may be caused by any one of several leaks within the system, and the "Touch-O-Matic" unit should be removed and overhauled. The following is a list of locations in the order in which the internal leaks are most likely to occur:

1. Selector crossover valves (dual cylinder system only)
2. Check valve, or valves
3. Piston sealing rings
4. Pipe plugs in oil passages
5. Relief valves in cylinder head, or heads.

123. **FLUSHING SYSTEM.** Drain hydraulic oil from the system and refill with distillate or kerosene. On the dual cylinder system move the selector lever to the rear position and remove lock pin from left lift arms. On both systems, operate the engine about three minutes and cycle the system several times. Shut off engine and drain flushing oil from system. Crank engine with starting motor NOT MORE THAN 30 SECONDS to remove flushing oil from pump and lines. If possible, raise front of tractor about 2 or 3 feet to make certain that all flushing oil is drained.

Refill system with recommended oil, start engine to fill pump and lines and add the required amount of oil.

124. **OPERATING PRESSURE—CHECK AND ADJUST.** To check and adjust the relief valve operating pressure on the single cylinder system, or to check and adjust the relief valve operating pressure for the right hand cylinder on the dual cylinder system, proceed as follows:

On the dual cylinder system, move the selector lever to the rear position and remove lock pin from lift arms as shown in Fig. JD895. Raise draft links until lift arms (outer lift arms on dual cylinder system) are approximately parallel to floor and block up under draft links to support them in this position. Move the manual control lever forward, remove the ½ inch pipe plug from cylinder (right cylinder on dual cylinder system) and install a ⅜ inch pipe plug into the tapped passage directly below the plug just removed. Install a 2000 psi pressure gage into the opening from which the ½ inch pipe plug was removed. See Figs. JD896 and 897.

With the manual control lever fully forward, start engine and run until the hydraulic system oil is at normal operating temperature. Move the engine speed control lever to wide open position, and while observing the pressure gage, slowly pull the "Touch-O-Matic" control lever (right control lever on dual cylinder system) to the rear and note the gage reading when the relief valve opens. Opening of the

Fig. JD895—Rear view of dual cylinder "Touch-O-Matic" unit, showing the selector lever in rear position and lift arm lock pin removed.

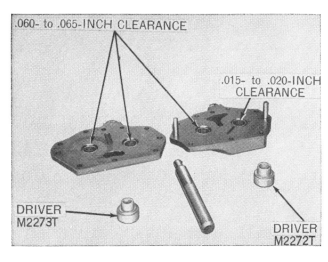

Fig. JD893 — Installation dimensions for needle bearings in hydraulic pump. The bearings should be driven in with a piloted type driver.

Fig. JD896—Pressure gage installation for checking the hydraulic system relief valve operating pressure on single cylinder "Touch-O-Matic".

Fig. JD897—Pressure gage installation for checking the hydraulic system relief valve operating pressure on dual cylinder "Touch-O-Matic".

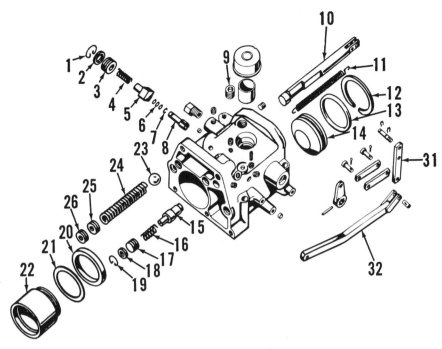

Fig. JD898—Exploded view of the single cylinder "Touch-O-Matic" unit as used on 40S, 330S, 420S and 430S. Other models so equipped are similar. Refer to legend under Fig. JD899.

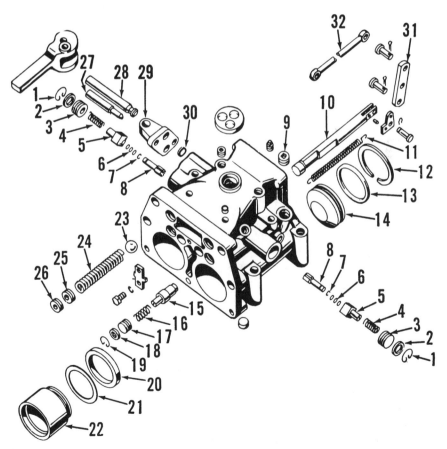

Fig. JD899—Exploded view of the dual cylinder "Touch-O-Matic" unit as used on 40T and 420T, prior to tractor serial number 125,001. Other models so equipped are similar.

relief valve can be detected by an audible "Whir" and the pressure gage should read 1040-1080 psi. If the operating pressure is not as specified, it will be necessary to reset the relief valve opening pressure as follows.

Drain the hydraulic system oil and remove the rockshaft and housing assembly. Loosen the relief valve locking screw (26—Fig. JD898 & 899) and turn the adjusting screw **in** to **increase** the pressure and **out** to **decrease** the pressure. One complete turn of the adjusting screw will change the operating pressure approximately 120 psi. A special wrench (John Deere No. AM 2048T) is available from John Deere for setting this pressure.

Note: If the recommended operating pressure cannot be obtained, check for a faulty hydraulic pump.

On dual cylinder systems, the relief valve operating pressure for the left hand cylinder can be checked in a similar manner.

When the adjustment is complete, remove the 3/8 inch plug, reinstall the 1/2 inch pipe plug and tighten the plug securely.

125. REMOVE AND REINSTALL. To remove the 40S, 320S, 330S 420S and 430S single cylinder "Touch-O-Matic" unit, drain the system, disconnect lift links from lift arms and unhook the lift arm return spring. Refer to Fig. JD900. Remove seat and tail light and disconnect the wire lead from the auxiliary light socket. Disconnect the load and depth control ball joint link from control arm and drive the roll pins from the load and depth control yoke pivot pins. Using

2. Seal	10. Control valve	18. Seal	26. Lock screw
3. Plug	11. Spring	19. Snap ring	27. Bottom crossover
4. Spring	12. Snap ring	20. Piston ring	valve
5. Check valve	13. Seal	21. Washer	28. Top crossover valve
6. Shims	14. Cylinder head	22. Piston	29. Selector bracket
7. Snap ring	15. By-pass valve	23. Ball	30. Oil seal
8. Push rod	16. Spring	24. Spring	31. Equalizer
9. 1/2 inch pipe plug	17. Plug	25. Adjusting screw	32. Control rod or link

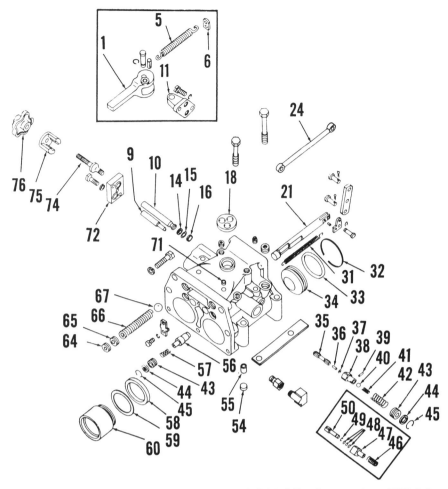

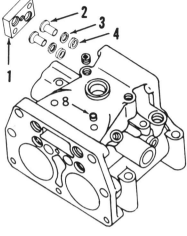

Fig. JD899B—View showing the plugs (2) which replace the cross-over valves in the "Touch-O-Matic" unit as used on the 440I series tractors.

1. Valve plate   3. Back-up washers
2. Plugs   4. Oil seals

Fig. JD899A—Exploded view of the dual "Touch-O-Matic" unit as used on 420T (after tractor serial number 125,000) and 430T series tractors. Selector lever and parts shown in upper left insert were replaced by handwheel (76) and associated parts at tractor serial number 127,415. Check valve assemblies (single stage), shown in lower right insert, were replaced by the two-stage check valve assembly at tractor serial number 125,001.

1. Selector lever	24. Control rod	43. Plug	58. Piston ring
5. Spring	31. Spring	44. Seal	59. Washer
6. Attaching clip	32. Snap ring	45. Snap ring	60. Piston
9. Bottom cross-over	33. Seal	46. Check valve spring	64. Lock screw
valve	34. Cylinder head	47. Check valve	65. Adjusting screw
10. Top cross-over valve	35. Push rod	48. Shims (0.002, 0.005	66. Spring
11. Bracket	36. Ball valve push pin	0.010)	67. Ball
14. Retainer washer	37. Shims (6 used)	49. Snap ring	71. Cup plug
15. Back-up washer	38. Check valve	50. Push rod	72. Valve plate
16. Oil seal	39. Roll pin	54. Cup plug	74. Special screw
18. Breather	40. Ball	55. Restrictor bushing	75. Cover
21. Control valve	41. Ball spring	56. By-pass valve	76. Handwheel
	42. Check valve spring	57. Spring	

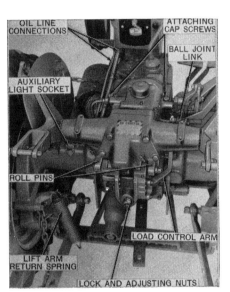

Fig. JD900—Rear view showing the single cylinder "Touch-O-Matic" installation on model 420S. Notice that lift arm return spring is disconnected. Other models so equipped are similar.

a short length of pipe and a ⅜ inch cap screw, pull the pivot pins as shown in Fig. JD901. Remove the load and depth control spring lock nut and adjusting nut and lift off the spring and yoke. Disconnect the hydraulic lines, remove the attaching cap screws and lift the hydraulic unit from tractor.

When reinstalling the "Touch-O-Matic" unit, tighten the retaining cap screws to a torque of 60 Ft.-Lbs. and adjust the load and depth control anchor yoke spring as outlined in paragraph 75.

Procedure for the 40S, 420I, 420W, 320S, 330S, 420S, 430S and 440I when

equipped with dual "Touch-O-Matic" is similar, except that a one piece pivot pin is used and pin can be removed after removing snap ring from end of pin.

On models 40U, 420U and 430U the removal of the "Touch-O-Matic" requires that the load and depth control spring housing be unbolted from the rockshaft housing.

125A. To remove the 40H, 40T, 40V, 40W, 420H, 420T, 420V, 430H, 430T and 430V "Touch-O-Matic" unit, drain the system and remove seat assembly. Disconnect the load and depth control linkage at equalizer lever and hydraulic lines at lower side of instrument

panel housing. Disconnect lift links from lift arms, remove the housing retaining cap screws and lift the "Touch-O-Matic" unit from tractor.

Install the "Touch-O-Matic" unit by reversing the removal procedure and tighten the retaining cap screws to a torque of 60 Ft.-Lbs.

126. DISASSEMBLE AND OVERHAUL. On the dual cylinder "Touch-O-Matic", remove the quadrant. On both systems, remove the cap screws retaining rockshaft housing to valve housing and separate the housings far enough to remove hair pin from anchor pin on crank pin. Slip control rod (or rods) from anchor pins and

remove rockshaft and housing assembly. On the dual cylinder "Touch-O-Matic", unhook the selector lever spring (11—Fig. JD899) from the clip, remove the cap screws retaining front cover to valve housing and separate the housings. Remove hair pins and disconnect control valves from equalizers as shown in Fig. JD902. On the single cylinder "Touch-O-Matic", remove the valve housing front cover and disconnect linkage from control valve (10—Fig. JD898). CAUTION: On both systems, do not attempt to withdraw control valves until after check valves have been removed, or parts may be damaged.

Withdraw piston (or pistons) from bore. If necessary to remove cylinder head, remove snap ring (12—Figs. JD898 & 899) and using a suitable press against piston side of cylinder head, remove cylinder head. Note: It is not necessary to remove cylinder head unless a new valve housing, cylinder head or cylinder head seal is to be installed.

Remove the check valve plug retaining snap ring (1), screw a ⅜ inch standard cap screw into check valve plug (3) and using a twisting motion,

remove the plug. Withdraw the check valve spring and check valve (or valves on dual cylinder systems) assembly. Now, the control valve (or valves) (10) can be withdrawn from its bore.

Note: The check valve (or valves) need not, and should not, be disassembled unless a new control valve, new check valve or a new check valve push rod is being installed.

On the dual cylinder systems prior to tractor serial number 127,415, remove the selector lever and bracket (29) and withdraw the cross-over valves (27 and 28). NOTE: On tractors serial number 127,415 and up (except 440I), the lever and bracket (29) have been replaced with a hand wheel (76—Fig. JD899A), cover (75) and valve

plate (72). On series 440I, the crossover valves are not used and plugs are substituted as shown in Fig. JD899B. Remove the by-pass valve plug retaining snap ring (19 — Fig. JD899), screw a ⅜-inch standard cap screw into plug (17) and using a twisting motion, withdraw the plug. Withdraw the by-pass valve spring (16) and valve (15). Remove locking screw (26), adjusting screw (25), noting how many turns is required to withdraw the adjusting screw. Remove the relief valve spring (24) and relief valve ball (23).

On the dual cylinder system, remove snap rings (27—Fig. JD903), screw a ⅜ inch standard cap screw into plugs (28) and using a twisting motion, withdraw the plugs. Shuttle valve (30)

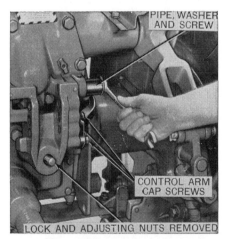

Fig. JD901—Pulling the anchor yoke pivot pin on Deere model 420S. Notice that the load control spring lock and adjusting nuts have been removed. Other models so equipped are similar.

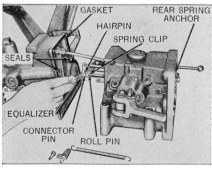

Fig. JD902 — Disconnecting dual cylinder hydraulic system control valves from equalizer linkage.

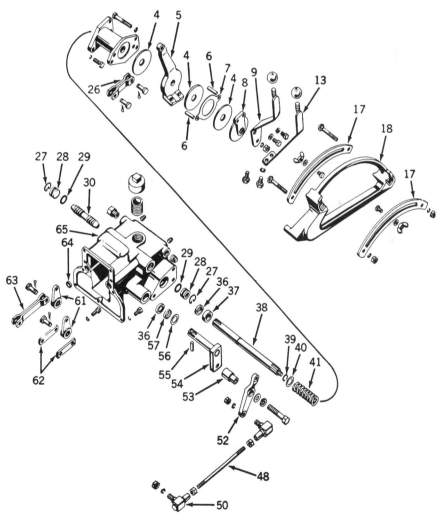

Fig. JD903—Dual cylinder "Touch-O-Matic" system valve housing front cover and quadrant. Other models so equipped are similar.

4. Friction disc	18. Quadrant	38. Left control shaft	53. Bushing
5. Right lever hub	27. Snap ring	39. Snap ring	54. Right control shaft
6. Roll pin	28. Plug	40. Retainer	55. Roll pin
7. Steel friction plate	29. Seal	41. Spring	56. Washer
8. Left lever hub	30. Shuttle valve	48. Load and depth	57. Retainer
9. Left lever	36. Oil seal	control linkage	61. Control crank
13. Right lever	37. Hollow dowel	50. Ball joint	62. Short equalizer link
17. Segment		52. Equalizing lever	63. Long equalizer link

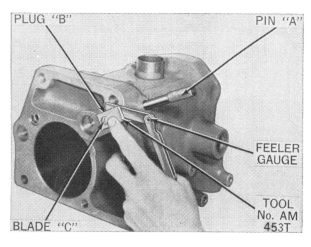

Fig. JD904—Determining the position of the by-pass valve passage on single cylinder "Touch-O-Matic".

milled in the head of the valve for this purpose. The seating surface of the by-pass valve can also be reconditioned by lapping. Make certain, however, that the by-pass bleeder hole is open and clean. When reassembling, renew all gaskets and seals.

To set the check valve or valves on system with single-stage check valves, proceed as follows:

Note: This need not be done unless the valve housing, control valve, check valve or check valve push rod has been renewed.

1. Set up John Deere special gage No. AM453T with pin (A) in hole and plug (B) in the control valve bore and contacting pin (A) as shown in Fig. JD904 and 905.

2. Using a feeler gage, measure and record the gap between the blade (C) and rear face of valve housing as shown.

3. Withdraw pin (A) and install the control valve into housing in operating position as shown in Fig. JD906 and 907.

4. Using about 0.030 of shims (6—Figs. JD898 & 899) assemble the check valve and push rod, leaving off the snap ring and install the assembly in the housing.

can be removed at this time. Renew the shuttle valve if it has any burrs, scratches or roughness that might prevent valve from operating freely. Shuttle valve must operate freely in its bore. Also, on the dual cylinder system, check the cross-over valves for burrs or scratches. The valves must operate freely in their bores.

Thoroughly clean all parts and examine same for excessive wear. Be

sure to clean all passages in the valve housing. Inspect the ramp on the control valve where the check valve push rod contacts the control valve during movement. A shiny line at this point is normal; if, however, the line is worn to a groove, renew the control valve. Examine the seating surface of the check valve. If the surface is not excessively worn, it can be reconditioned by lapping. A screw driver slot is

5. Hold check valve snugly against its seat. At this time, the check valve push rod must not contact the flat on the control valve. If it does, it will be necessary to remove some of the shims (6) between the check valve and push rod.

6. Move the control valve forward until ramp on same has just contacted the check valve push rod, but has not raised the check valve from its seat.

7. While holding the control valve in this position, slide plug (B) of gage

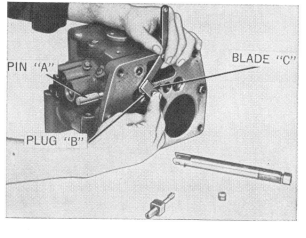

Fig. JD905—Determining the position of the by-pass valve passage on dual cylinder "Touch-O-Matic".

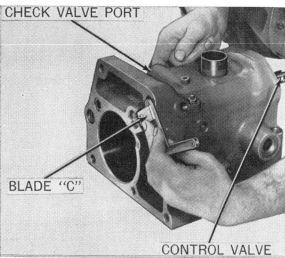

Fig. JD906 — Determining the setting of single cylinder "Touch-O-Matic" check valve.

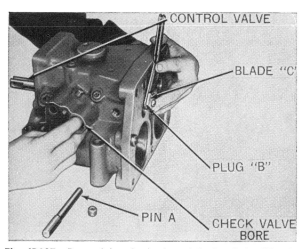

Fig. JD907—Determining the setting of dual cylinder "Touch-O-Matic" check valve.

against end of control valve as shown in Figs. JD906 and 907.

8. Again record the feeler gage reading between blade (C) and rear face of valve housing.

9. Subtract the reading obtained in step 8 from reading recorded in step 2.

10. If the reading obtained in step 9 is 0.021-0.030, the check valve adjustment is satisfactory. If reading is less than 0.021, it will be necessary to remove some shims (6—Figs. JD898 & 899) and recheck. If the reading is more than 0.030, add shims and recheck. After the proper number of shims (6) has been established, install snap ring on check valve push rod and reassemble in check valve. Withdraw control valve and set aside with check valve for later assembly.

Note: The same procedure can be used for setting the other check valve on dual cylinder systems.

126A. To set the check valve (or valves) on systems with two-stage check valves, proceed as follows:

1. Set up John Deere special gage No. AM453T with pin (A) in hole and plug (B) in the control valve bore and contacting pin (A) as shown in Fig. JD904 and 905.

2. Using a feeler gage, measure and record the gap between the blade (C) and the rear face of valve housing as shown.

3. Withdraw pin (A) and install the control valve into housing in operating position as shown in Fig. JD906 and 907.

4. Place a ball valve push-pin in the end of push rod and place assembly in

Fig. JD908—Checking Deere models 40S, 320S and 420S lift arms for free travel. When the lift arms are in the raised position, there should be a free travel of one inch as shown. Other models (including 330 and 430) with single cylinder "Touch-O-Matic" systems are checked in the same manner.

the check valve as shown in Fig. JD907A.

5. Place the check valve assembly in its operating position in the housing and hold same snugly against its seat. The check valve push rod must not contact the flat on the control valve. If it does, install a shorter ball valve push pin.

6. Move the control valve forward until ramp on same has just contacted the check valve push rod, but has not raised the check valve ball from its seat.

7. While holding the control valve in this position, slide plug (B) of gage against end of control valve as shown in Figs. JD906 and 907.

8. Again record the feeler gage reading between blade (C) and rear face of housing.

9. Subtract the reading obtained in step 8 from reading recorded in step 2.

10. If reading obtained in step 9 is 0.021-0.030, the check valve adjustment is satisfactory. If reading is less than 0.021, substitute a shorter ball valve push-pin or grind a slight amount from same. If reading is more than 0.030, use a longer ball valve push-pin.

11. After correct length of ball valve push-pin has been established, place the complete check valve assembly in a vertical position (push rod down) and push down on check valve until check valve ball is pushed as far off its seat as it will go. This travel represents the distance the first stage of the check valve opens before the check valve itself unseats. This travel should measure 0.020-0.040. If travel is less than 0.020 it will be necessary

to grind the check valve end of the push rod. If travel is more than 0.040, add shims at check valve end of push rod. See Fig. 907A.

12. Remove control valve and set aside with check valve for later assembly.

When installing relief valve adjusting screw (25—Figs. JD898 or JD899), turn the screw in the same number of turns as were used in removal unless a new relief valve spring was installed. If a new relief valve spring was installed, turn the adjusting screw into housing eight complete turns after the screw has first contacted the spring. Reinstall by-pass valve making certain that valve is free in bore. Install control valve and linkage, then install check valve. Install leather washer next to shoulder on piston head and install compression ring with flat side next to leather washer. Install piston. Reassemble the remaining parts and after the unit is installed on tractor, check and adjust if necessary the relief valve operating pressure as in paragraph 124.

LIFT LINKAGE
Single "Touch-O-Matic"

127. ADJUST. To adjust the single cylinder system linkage, start engine and pull the control lever to the extreme rear position; then, when lift arms have stopped movement at the top, lift up on lower draft link to check for free travel of lift arm which should be 1 inch as shown in Fig. JD-908. If the free travel is not as specified, adjust the ball joint link rod as shown in Fig. JD909.

Dual "Touch-O-Matic"

127A. To adjust the dual cylinder system lift linkage, move the selector lever to the extreme rear position and unpin the lift arms. Start engine and pull the left control lever to the

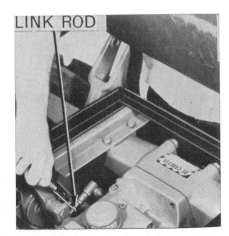

Fig. JD909—Adjusting the single cylinder system lift linkage to obtain the free travel shown in Fig. JD908.

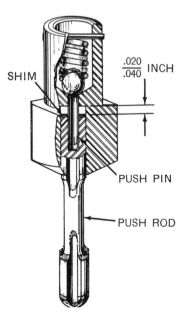

Fig. JD907A—Cut-away view of the two-stage check valve which is used to replace the earlier single-stage check valve.

SHIM

.020 / .040 INCH

PUSH PIN

PUSH ROD

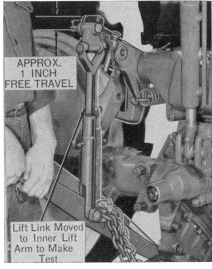

Fig. JD910 — Checking Deere models 40T, 420T and 430T inner lift arms for free travel. When the arms are in the raised position, there should be a free travel of one inch as shown. Other models with dual cylinder "Touch-O-Matic" systems are checked in the same manner.

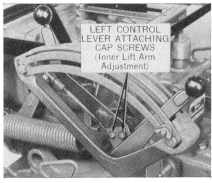
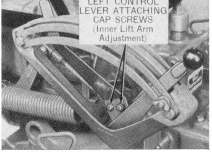

Fig. JD911—On Deere 40T, 420T and 430T dual cylinder systems, the free travel shown in Fig. JD910, can be obtained by shifting the left control lever on its hub. Other models so equipped are adjusted in the same manner.

Fig. JD912—Deere dual cylinder "Touch-O-Matic" lift linkage. After obtaining the desired inner lift arm free travel as shown in Fig. JD910, the outer and inner lift arms can be paralleled by adjusting the linkage as shown.

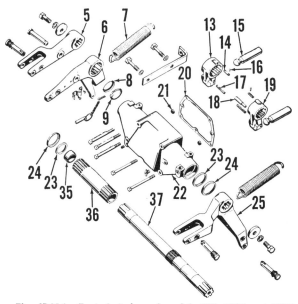

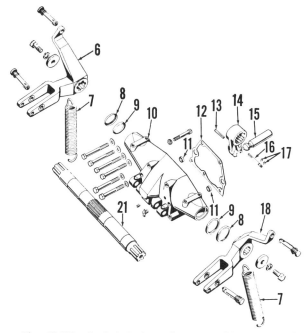

6. Left lift arm
7. Spring
8. Retainer
9. Oil seal
10. Housing
11. Hollow dowel
12. Gasket
13. Groov pin
14. Crank arm
15. Connecting rod
16. Groov pin
17. Hairpins
18. Right lift arm
21. Rockshaft

Fig. JD913—Exploded view of models 40S, 320S, 330S, 420S and 430S rockshaft and associated parts when equipped with single cylinder "Touch-O-Matic".

5. Left lift arm
6. Inner left lift arm
7. Spring
8. Retainer
9. Oil seal
13. Left crank arm
14. Groov pin
15. Connecting rod
16. Hairpin
17. Groov pin
18. Spacer
19. Right crank arm
21. Hollow dowel
22. Housing
23. Seal
24. Retainer
25. Right lift arm
35. Bushing
36. Hollow rockshaft
37. Rockshaft

extreme rear position; then, when inner lift arm has stopped movement at the top, lift up on draft link to check for free travel of lift arm which should be 1 inch as shown in Fig. JD910. If free travel is not as specified, it will be necessary to shift the position of the left control lever on its hub, which can be accomplished by loosening the cap screws shown in Fig. JD911.

After the free travel of the inner lift arm is satisfactorily adjusted, pull both control levers to the extreme rear position; at which time, the inner and outer lift arms should be parallel. If the lift arms are not parallel, adjust the position of the outer lift arm by changing the length of the link rod as shown in Fig. JD912.

Fig. JD914—Exploded view of models 40T, 420T and 430T rockshaft and associated parts. Models 40W, 40H, 40V, 420H, 420V, 430H and 430V are similar.

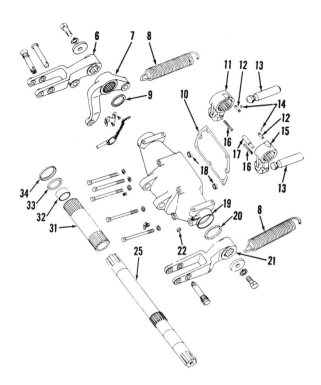

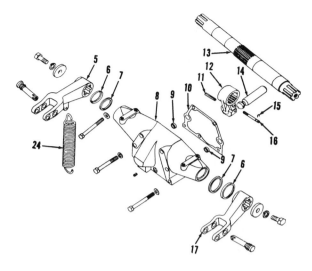

Fig. JD916—Exploded view of 40U, 320U, 330U 420U and
430U rockshaft and associated parts.

5. Left lift arm	12. Crank arm
6. Retainer	13. Rockshaft
7. Oil seal	14. Connecting rod
8. Rockshaft housing	15. Hair pin
9. Dowel	16. Groove pin
10. Gasket	17. Right lift arm
11. Roll pin	24. Spring

Fig. JD915—Exploded view of models 320S, 330S, 420S and
430S rockshaft and associated parts when equipped with dual
cylinder "Touch-O-Matic". Models 420I, 420W, 430W and
440I are similar.

6. Left outer lift arm	17. Spacer
7. Left inner lift arm	18. Dowel
8. Spring	19. Rockshaft housing
9. Oil seal	20. Oil seal
10. Gasket	21. Right lift arm
11. Left crank arm	22. Dowel
12. Groove pin	25. Rockshaft
13. Connecting rod	31. Hollow rockshaft
14. Hair pin	32. Bushing
15. Right crank arm	33. Oil seal
16. Roll pin	34. Retainer

Fig. JD917—Series 440 wiring diagram with key switch.
Series 320, 330, 420 and 430 wiring diagram with key
switch are similar except the oil pressure gage and sending unit are not used. On models without key switch, use
dashed circuit (X).

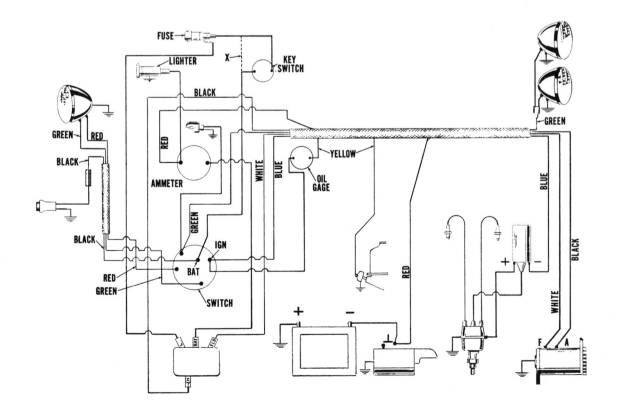

REMOTE HYDRAULIC CONTROL SYSTEM

A remote hydraulic control attachment is available as optional equipment for all models except the 420 Special (I). The remote hydraulic system is connected to the Touch-O-Matic system through a control valve mounted at the left of the seat. The control valve provides pressure for a double acting remote cylinder and is controlled by a lever at the right of the seat as shown in Fig. JD918.

CONTROL VALVE

All Models So Equipped

128. REMOVE AND REINSTALL. Drain the Touch-O-Matic system, then remove the seat and seat frame from tractor. Remove shield at front of control valve, disconnect linkage from pistons and remove support bracket from control valve housing. Remove the control shaft bracket from the right front side of the Touch-O-Matic front cover. Remove the lines between the control valve and break-away coupling. Loosen the control valve attaching cap screws, then remove the pressure and by-pass lines from control valve and Touch-O-Matic.

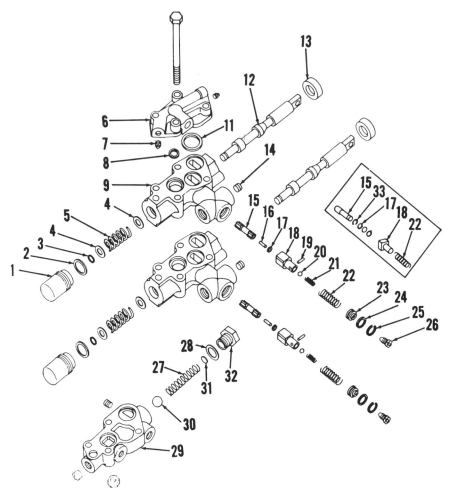

Fig. JD920—Exploded view of the control valve used with the remote hydraulic control system.

1. Cap	9. Housing	19. Roll pin	27. Spring
2. Copper washer	11. Seal (8 used)	20. Ball	28. Gasket
3. Snap ring	12. Piston	21. Spring	29. Base
4. Washer	13. Oil seal	22. Spring	30. Ball
5. Spring	14. Pipe plug	23. Plug	31. Shim
6. Cover	15. Push rod	24. Seal	(0.010 & 0.027)
7. Pipe plug	16. Push pin	25. Snap ring	32. Plug
8. Oil seal	17. Shim	26. Thermal relief	33. Snap ring
	18. Check valve	valve	

Fig. JD918—View showing location of the remote hydraulic system control valve and lever.

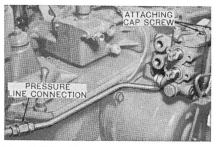

Fig. JD919—Disconnect lower pressure line at base of instrument panel and remove line and control valve together.

Disconnect pressure line at base of instrument panel as shown in Fig. JD919, then remove the attaching cap screws and lift the control valve and attached line from tractor.

129. OVERHAUL. With control valve assembly removed from tractor, refer to Fig. JD920 and proceed as follows: Tap check valve plugs (23) into housings and extract snap rings (25). Use caution during this operation not to damage the thermal relief valves (26) on models so equipped. Extract the check valve plugs (23) by using a ⅜-inch puller screw on

models without thermal relief valves, or on models so equipped, by grasping the thermal relief valves with a pair of pliers. Extract all parts of the single stage or two-stage check valves. Unscrew caps (1) and withdraw piston and spring assemblies.

Remove relief valve plug (32) and extract shims (31), spring (27) and ball (30). Remove assembly bolts and separate housings.

Carefully clean all parts in a suitable solvent, paying particular attention to seating surfaces and fluid passages. Renew any damaged or questionable parts. Inspect face of check valve (18) and head of push rod (15) for wear. If control valve is fitted with a two-stage check valve, examine push pin (16) and shims (17) for wear. Drive roll pin (19) from valve, remove ball and spring (20 & 21) and renew same if rusted or pitted. Examine valve seats for evidence of leaking. If valves are renewed, lap same to their respective seats, identify the valve with its seat for reinstallation and be sure to flush out all lapping compound.

130. When reassembling, renew all seals and gaskets and proceed as follows: Place all seal rings (11) in their respective counterbores and join housings with the long through bolts. Insert relief valve ball (30), spring (27) and the originally removed shims (31); then install gasket (28) and tighten plug (32). Install and thoroughly lubricate oil seals (13). Assemble washers (4), spring (5) and retainer (2) to pistons; then, using a slight rotary motion, install pistons into housings and through the oil seals. Push pistons into housings as far as they will go.

Now, if the original check valve and push rod (15 & 18) are being installed, use the same shims (17) and/or push pin (16) as were originally removed. If a new valve or push rod are used, use several shims as a starting point to set the valves.

131. **Adjust Single Stage Check Valve.** While holding valve against its seat by finger pressure as shown in Fig. JD921, check the piston travel from its neutral position to the point where the piston ramp just begins to raise the check valve from its seat. If piston is moved more than 0.040 before valve begins to raise, add a shim between check valve and push rod and recheck. If valve begins to raise before piston has moved 0.020, remove a shim and recheck. After the specified 0.020-0.040 piston movement is obtained,

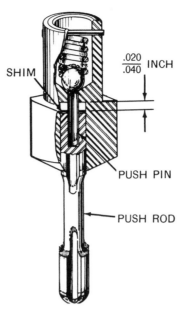

Fig. JD922—Sectional view of the assembled two-stage check valve used on models serial number 134,998 and up.

rotate piston 180 degrees and again try the valve action to be sure the valve is not held off its seat in any neutral position. Check and shim adjust, if necessary, the other check valve, using the same procedure previously outlined.

132. **Adjust Two-Stage Check Valve.** While holding valve against its seat by finger pressure as shown in Fig. JD921, check the piston travel from its neutral position to the point where

the piston ramp just begins to raise the first stage check valve ball (Fig JD922) from its seat. If piston is moved more than 0.040 before valve begins to raise, install a longer push pin. If valve begins to raise before piston has moved 0.020, install a shorter push pin. After the specified 0.020-0.040 piston movement is obtained, withdraw the complete check valve assembly and stand same in a vertical position as shown in Fig. JD922. Push down on check valve until ball is pushed as far off its seat as it will go. This total downward movement represents the distance the first stage of the valve opens before the check valve itself is raised off its seat. This desired movement of 0.020-0.040 is adjusted by adding or removing shims between push rod and valve. With all shims removed, if valve movement is still less than 0.020, it will be necessary to grind the necessary amount of material from check valve end of push rod. Check and adjust, if necessary, the other valve, using the same procedure previously outlined.

133. When all check valves are adjusted, install the valves, springs (22 —Fig. JD920), retainer plugs (23) with seals (24) and snap rings (25). On models so equipped, be careful not to damage the thermal relief valves (26).

After the control valve unit is installed. fill and bleed the system as outlined in paragraph 138.

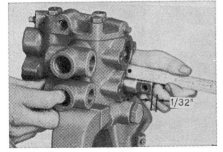

Fig. JD921—View showing one method of checking control valve piston travel when adjusting check valves.

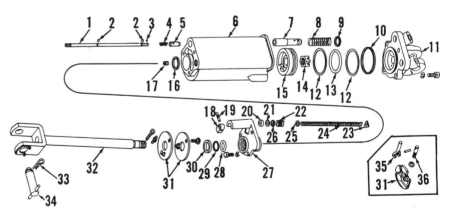

Fig. JD923—Exploded view of hydraulic stop type remote control cylinder. Lower right insert shows late type stop.

1. Stop rod	13. "O" ring	25. Stop rod washer
2. Snap rings	14. Nut	26. Packing adapter, male
3. Ball	15. Piston	27. Piston rod guide
4. Bleed valve spring	16. Piston rod guide gasket	28. "O" ring
5. Bleed valve	17. Pipe plug	29. Packing washer
6. Cylinder	18. Stop rod arm	30. Wiper seal
7. Stop valve	19. Groov-pin	31. Rod stop
8. Stop valve spring	20. Packing adapter, female	32. Piston rod
9. Gasket	21. "V" packing	33. Locking pin
10. Gasket	22. Packing spring	34. Attaching pin
11. End cap	23. Stop rod washer	35. Stop lever
12. Packing washer	24. Stop rod spring	36. Stop lever screw

REMOTE CYLINDER
(Hydraulic Stop Type)
All Models So Equipped

134. **OVERHAUL.** To disassemble the unit, remove oil line and end cap (11—Fig. JD923). Remove stop valve (7) and bleed valve (5) by pushing stop rod (1) completely into cylinder. Withdraw stop valve from bleed valve, being careful not to lose the small ball (3). Remove nut from piston rod, being careful not to distort the rod and remove piston and rod. Push stop rod (1) all the way into cylinder and drift out Groov-pin (19). Remove piston rod guide (27).

Examine all parts for being excessively worn and renew all seals. Wiper seal (30) should be installed with sealing lip toward outer end of bore. Install stop rod ("V" seal assembly) (20, 21, 22 & 26) with sealing edge toward cylinder. Complete the assembly by reversing the disassembly procedure and tighten the end cap bolts to a torque of 85 Ft.-Lbs. Install the piston rod stop as outlined in paragraph 136.

OPERATING ADJUSTMENT
All Models

135. **ADJUST RELIEF VALVE.** To check the relief valve pressure, proceed as follows: Connect a suitable pressure gage (at least 2000 psi) to the right hand breakaway coupling and install a break away coupling plug into left coupling.

Start engine and run same until hydraulic fluid is at normal operating temperature. Move the engine speed

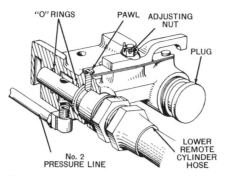

Fig. JD924 — Sectional view showing the breakaway coupling and associated parts. Refer to text for adjustment.

control lever to the high idle position and while observing the pressure gage, actuate the control lever and observe the gage pressure as the relief valve opens. Relief valve opening point can be detected by an audible "whirr".

If pressure is not within the limits of 1040-1080 psi, remove the relief valve plug (32—Fig. JD920) and add or remove shims (31) between plug and relief valve spring as required. One 0.010 thick shim will change the relief valve opening pressure approximately 30 psi. One 0.027 thick shim will change the relief valve opening pressure approximately 75 psi.

If the addition of shim washers will not correct a low pressure reading, it will be necessary to overhaul the hydraulic pump as outlined in paragraph 121.

136. **REMOTE CYLINDER STOP ADJUSTMENT.** To adjust the work-

ing stroke on cylinders with a cast stop plate, lift the piston rod stop lever, slide the adjustable stop along piston rod to the desired position and press the stop lever down. If clamp does not hold securely, lift stop lever, rotate it ½-turn clockwise and press in place. Make certain, however, that the adjustable stop is located so that the stop rod contacts one of the flanges on stop.

On models with sheet metal stop, loosen wing nut, slide stop assembly along piston rod to desired position and tighten the wing nut. Make certain that wing nut is directly opposite the stop rod arm.

137. **BREAKAWAY COUPLING RELEASE TENSION.** Install hoses and lock same in breakaway coupling, then check the amount of pull required to disconnect one hose from the coupling. If pull is less than 150 lbs., tighten the nut (Fig. JD924) and recheck. If pull is more than 200 lbs., loosen the nut and recheck. After the specified 150-200 pounds pull is obtained with both hoses in place, lock the nut with a cotter pin.

138. **BLEEDING.** To bleed the remote system, proceed as follows: Attach remote cylinder to breakaway coupling, start engine and allow to run until hydraulic fluid has reached operating temperature. Position cylinder with hose end up on floor, or some other support, and actuate control lever so cylinder piston will be extended and retracted about six or eight times.

Recheck oil level in reservoir and add oil if necessary.

MARVEL-SCHEBLER Calibration Data

Model	†Float Setting	Repair Kit	Gasket Set	Inlet Needle and Seat	Idle Jet or Tube	Nozzle	Economizer Plug, Jet or Idle Plug	Power Jet or Needle
TSX245	¼	286-765	16-654	233-558	49-285	47-288	49-218	.....
TSX530	¼	286-1005	16-654	233-543	49-285	47-391	49-219	43-622
TSX562	¼	286-1047	16-654	233-543	49-285	47-410	49-218	43-669
TSX641	¼	286-1107	16-654	233-543	49-285	47-451	49-258	43-699
TSX678	¼	286-1122	16-654	233-543	49-285	47-455	49-207	43-707
TSX756	¼	286-1246	16-654	233-543	49-101L	47-451	49-258	43-699
TSX768	¼	286-1247	16-654	233-543	49-101L	47-455	49-258	43-678
TSX777	¼	286-1262	16-654	233-543	49-101L	47-473	49-437	43-729

JOHN DEERE
(PREVIOUSLY JD-18)

Models ■ 435D ■ 440ID

SHOP MANUAL
JOHN DEERE
MODELS 435D-440ID

Tractor serial number is stamped on plate which is affixed to bottom of instrument panel.

★NOTE★

This shop manual provides methods for servicing the 435D and 440ID (diesel) tractors which, except for the engine and accessories covered in this manual, are serviced in the same manner as the similar 430W and 440I (non-diesel) tractors covered in the first section of this shop manual (previously JD15). In the index listed below, the first column lists the tractor components, the two columns under the tractor model list the beginning paragraph number and the manual in which the desired paragraph containing the service information will be found.

INDEX

INDEX CONT.

CONDENSED SERVICE DATA

GENERAL	435D	440ID
Engine Make	GM	GM
Cylinders	2	2
Bore—Inches	3⅞	3⅞
Stroke—Inches	4½	4½
Displacement—Cubic Inches	106.2	106.2
Compression Ratio	17.0:1	17.0:1
Pistons Removed From?	Above	Above
Main Bearings, Number of	3	3
Main & Rod Brgs., Adjustable?	No	No
Cylinder Sleeves?	Yes	Yes
Generator, Regulator & Starting Motor—Make	Delco-Remy	Delco-Remy
Electrical System—Voltage	12	12
Battery Ground Polarity	Positive	Positive
Forward Speeds	4 or 5	4 or 5

TUNE-UP	435D	440ID
Compression—Gage Pounds	500(1)	500(1)
Tappet Gap	0.009 Hot	0.009 Hot
Exhaust Valve Face Angle	30°	30°
Exhaust Valve Seat Angle	30°	30°
Injector Timing	See Par. 143	See Par. 143
Nozzle Opening Pressure—PSI	450-850	450-850
Engine High Idle—RPM	1975	1975
Engine Rated Speed—RPM	1850	1850
Belt Pulley High Idle—RPM	1356	1356
Belt Pulley Rated Speed—RPM	1270	1270
Power Shaft High Idle—RPM	597	597
Power Shaft Rated Speed—RPM	560	560

(1) Compression check made with engine running at 600 rpm.

SIZES—CAPACITIES—CLEARANCES	435D	440ID
Crankshaft Main Journal Dia	2.999-3.000	2.999-3.000
Crankpin Diameter	2.499-2.500	2.499-2.500
Camshaft Journal Diameter	2.1820-2.1825	2.1820-2.1825
Balance Shaft Journal Diameter	2.1820-2.1825	2.1820-2.1825
Piston Pin Diameter	1.3746-1.3750	1.3746-1.3750
Valve Stem Diameter—Exhaust	0.3095-0.3105	0.3095-0.3105
Main Bearing Running Clearance	0.002-0.004	0.002-0.004
Rod Bearing Running Clearance	0.0044-0.0084	0.0044-0.0084
Camshaft Bearing Running Clearance	0.0045-0.006	0.0045-0.006
Balance Shaft Bearing Running Clearance	0.0045-0.006	0.0045-0.006
Crankshaft End Play	0.004-0.011	0.004-0.011
Camshaft End Play	0.008-0.015	0.008-0.015
Balance Shaft End Play	0.008-0.015	0.008-0.015
Piston Skirt Clearance	0.0037-0.0074	0.0037-0.0074
Cooling System—Gallons	2½	2½
Crankcase Oil—Quarts	9	9
Fuel Tank—Gallons	10½	10½
Transmission & Differential—Qts	9	9
Final Drives (Each)—Quarts	1¾	1¾
Hydraulic System—Quarts	12	12
Belt Pulley—Pints	½	½

TIGHTENING TORQUES (Ft.-Lbs.)	435D	440ID
Cylinder Head Bolts	170-180	170-180
Main Bearing Bolts	120-130	120-130
Rod Bearing Bolts	45-50	45-50
Rocker Arm Bolts	50-55	50-55
Flywheel Cap Screws	130-140	130-140
Balance Weight Nuts	300-325	300-325
Timing Gear Cover Cap Screws	See Par. 96	See Par. 96
Injector Clamp Bolt	20-25	20-25

FRONT SYSTEM

SWEPT-BACK FRONT AXLE

The 435D models are available with a swept-back front axle as shown in Fig. JD101. To service the swept-back front axle assembly, proceed as outlined in the following paragraphs. To service the adjustable front axle for other 435D models and 440ID models, refer to paragraphs 2, 3 and 4 in the first section of this manual.

STEERING KNUCKLES

Model 435D

68. R&R AND OVERHAUL. To remove the knuckles (67—Fig. JD101), first support front of tractor and remove wheel and hub assemblies. Remove cap screws (57) and bump steering arms (55) from knuckles. Knuckles can now be withdrawn from below and needle bearings (69) can be driven from the axle extensions.

69. New needle bearings should be pressed or driven into position using a special driver such as John Deere tool number AM-457T, or its equivalent. Apply the driver against the end of the bearing which has the manufacturers' number stamped on it. When installing the steering arms (55), vary the number of 0.010 and 0.027 thick shims (54) to give an end play of 0.000-0.030 and tighten cap screws (57) to a torque of 60-65 Ft.-Lbs.

When installing front wheels, tighten the bearing adjusting nuts to a torque of 35-40 Ft.-Lbs., then back off the nut one castellation and insert the cotter pin.

TIE-RODS AND TOE-IN

Model 435D

70. The procedure for removing the tie-rods and/or tie-rod ends is evident. When reassembling, vary the length of each tie-rod an equal amount to obtain the recommended toe-in of $\frac{1}{8}$-$\frac{1}{2}$ inch.

NOTE: Steering gear must be in its mid-position and the front wheels must be pointing straight ahead when making the toe-in adjustment.

AXLE PIVOT SHAFT AND BUSHINGS

Model 435D

71. To renew the axle pivot shaft and bushings, first remove axle assembly from tractor as follows: Disconnect tie-rod ends from center

steering arm and remove one axle extension and wheel assembly. Loosen jam nut and set screw at right rear of front support, drive pivot shaft (53) forward and out of front support, then slide front axle assembly out of front support.

72. Check pivot shaft (53) and pivot shaft bushings (86) against the values which follow:

Pivot shaft O. D..........1.374-1.375
Pivot shaft bushings I. D...1.377-1.379
Recommended clearance ..0.002-0.005

If bushings and/or pivot shaft are excessively worn, install new bushings using a closely piloted arbor and ream them to the desired inside diameter, if necessary.

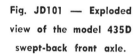

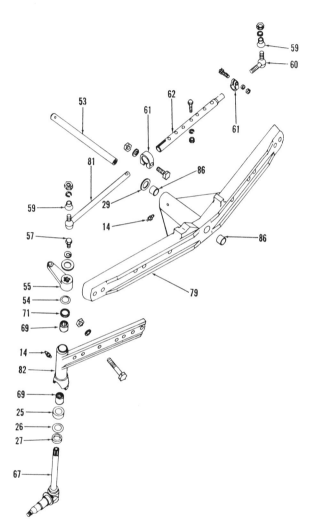

Fig. JD101 — Exploded view of the model 435D swept-back front axle.

14. Grease fitting
25. Thrust bearing
26. Retainer
27. Seal
29. Shim washer (0.030 & 0.060)
53. Pivot shaft
54. Shim (0.010 & 0.027)
55. Steering arm
57. Cap screw
59. Dust cover
60. Tie-rod end
61. Clamp
62. Tie-rod tube
67. Spindle and knuckle
69. Needle bearing
71. Seal
79. Axle center member
81. Adjustable tie-rod end
82. Axle extension
86. Bushing

POWER STEERING

Two types of pumps are used to provide the pressurized fluid for power steering operation. The type of pump used is determined by whether the tractor is, or is not, equipped with Touch-O-Matic.

When the tractor is equipped with power steering only, the pump shown in Fig. JD102 is used. When the tractor is equipped with power steering and Touch-O-Matic, the pump shown in Fig. JD182 is used.

Except for pumps, the power steering and Touch-O-Matic equipment remains the same as that outlined for 430 and 440 tractors in the first section of this manual.

For information on the pump used when tractor is equipped with Touch-O-Matic only, refer to paragraph 176.

PUMP

73. **REMOVE AND REINSTALL.** To remove either type pump, discon-

nect lines from pump and allow oil to drain, then remove the cap screws which retain pump bracket to engine and remove pump and bracket.

74. OVERHAUL (POWER STEERING ONLY). Remove pulley retaining nut, then remove pulley and Woodruff key from pump shaft. Remove pump mounting bracket from pump housing. Remove the rear mounting stud (20—Fig. JD102) and the flared union (21); then remove pump reservoir (18). Use a small punch in hole provided in housing (3), depress snap ring (16) and pry same out with a screw driver. Remove end plate (15). Remove the two pressure plate springs (13) and pressure plate (11). Pull the rotor (8) and vanes (7) assembly out of pump ring (9) about ½-inch and place a rubber band around assembly to keep vanes

from falling out, then remove the rotor and vanes assembly. Remove pump ring (9) from dowels and note location of arrow and oil relief notch. Remove thrust plate (5). Pull pump shaft (4) from reservoir end of housing. Dowels (12) can be removed, if necessary. Depress the flow control valve plunger retainer plug (26) and remove snap ring (24). Use caution when releasing the pressure on retaining plug to prevent parts from flying. Remove retaining plug (26), flow control valve assembly (27) and spring (28). Remove filter assembly from threaded hole in housing which accepts the flared tube union.

75. Clean all parts in a suitable solvent and inspect as follows: Inspect inside contour of pump ring for gouges, chatter marks or other undue

wear. If pump ring is damaged or worn, renew pump ring and vanes. Inspect rotor and vanes and if rotor and/or vanes are worn until vanes tilt in rotor, renew rotor and vanes. Inspect faces of thrust plate and pressure plate for scoring. Light scoring can be eliminated by lapping faces on a perfectly flat surface; however, if heavy scoring is present, renew parts. Inspect pump drive shaft for worn splines and/or grooving of oil seal seating surface and renew shaft if either condition is found. Inspect oil seal for signs of leakage, and if seal is renewed, install same with lettering toward outside. Be sure to pack the cavity between lips of seal with Lubri-Plate, or equivalent, prior to installing pump shaft. Inspect the flow control valve for wear or damage and make sure valve is free to move in its bore. NOTE: The pressure relief valve can be removed from the flow control valve for cleaning by removing the hex-head plug. However, if after cleaning, any damage is found, the complete flow control and relief valve assembly must be renewed as separate parts are not catalogued. The pressure relief valve is pre-set to open at 1000 psi. Renew the filter assembly if same is plugged or damaged.

76. Reassembly of pump is the reverse of disassembly, however, keep the following points in mind: When installing the pump ring, be sure the arrow points in the direction of pump rotation and the notch in outer face of pump ring aligns with the oil passage in pump housing. When installing rotor vanes, be sure rounded edges are toward pump ring. Either use all new "O" rings or be absolutely sure original "O" rings are in a satisfactory condition.

77. OVERHAUL (POWER STEERING AND TOUCH-O-MATIC). For information on this combined power steering and Touch-O-Matic pump, refer to paragraph 177 in the Power Lift System section.

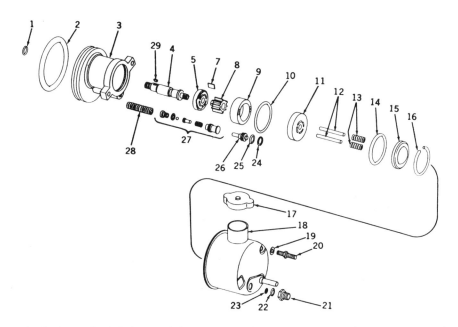

Fig. JD102 — Exploded view of the power steering pump used on models 435D and 440ID when equipped with power steering only.

1. Drive shaft seal	11. Pressure plate	22. "O" ring
2. "O" ring	12. Dowels	23. "O" ring
3. Housing	13. Pressure plate springs	24. Retaining ring
4. Drive shaft	14. "O" ring	25. "O" ring
5. Thrust plate	15. End plate	26. Plunger retainer ring
7. Rotor vane	16. Retaining ring	27. Flow control & relief valve
8. Rotor	18. Body (reservoir)	28. Flow control valve spring
9. Pump ring	19. Seal	29. Woodruff key
10. "O" ring	20. Rear mounting stud	
	21. Union	

ENGINE AND COMPONENTS

The John Deere 435D and 440ID tractors are fitted with a General Motors 2-53 series diesel engine. This 2-cylinder, 2-cycle engine has a displacement of 106.2 cubic inches; a bore of 3.875 inches; a stroke of 4.5 inches and a compression ratio of 17 to 1. The Roots-type blower mounted to the right side

of the cylinder block furnishes intake air to the cylinders via the ports in the cylinder liner (sleeve). Refer to Fig. 103 for schematic views showing the combustion cycle and to Fig. 104 for an exploded view of the cylinder block assembly.

NOTE: At the time this manual was printed, due to the relatively short time this

engine has been in the field, only standard size parts are available and the various oversize and undersize parts are not yet catalogued. However; they will be available in the near future and should the need for such parts arise, contact your nearest General Motors parts distributor.

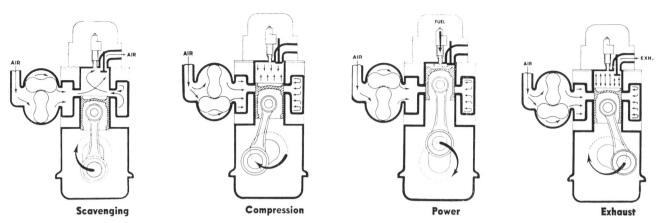

Scavenging   Compression   Power   Exhaust

Fig. 103—Schematic view showing the combustion and power strokes of the series 2-53 General Motors engine. This sequence is typical of all two cycle engines.

TUNE-UP

Normally, a tune-up on an engine in service will involve only making sure that the various adjustments have not changed. However, if it is necessary to remove and/or renew the cylinder head, injectors or the governor, it will be necessary to make a preliminary (cold) tune-up (setting) of the engine to permit it to be started and brought up to the recommended 160-185 degrees F. operating temperature; at which time the exhaust valve clearance must be rechecked, as well as the injector timing and linkage adjustments rechecked.

78. ADJUST EXHAUST VALVES (cold). To adjust the clearance for exhaust valves on number one cylinder,

proceed as follows: Pull fuel shut-off control to the "no-fuel" position and turn engine in the normal direction of rotation until the number one injector follower is fully depressed. Loosen lock nuts on push rods and rotate push rods as required until clearance between valve stems and rocker arms is 0.011 cold. Repeat the above operation for number two cylinder. NOTE: Exhaust valve clearance must be rechecked and adjusted, if necessary, to 0.009 after engine has been started and brought to operating temperature.

79. TIME FUEL INJECTORS. To time the fuel injector on number one cylinder, proceed as follows: Pull fuel shut-off control to the "no-fuel" posi-

tion and turn engine in the normal direction of rotation until both exhaust valves for the number one cylinder are fully depressed. Place the small end of the injector timing gage (General Motors tool J1242) in the hole provided in top of injector body. Loosen jam nut on push rod and rotate push rod until shoulder of timing gage will just pass over the top of injector follower as shown in Fig. 105. With injector set to this dimension (1.484 inches), hold push rod and tighten jam nut. Repeat the above operation for number two cylinder.

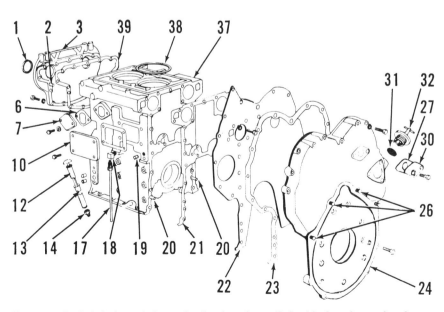

Fig. 104—Exploded view of the early diesel engine cylinder block and associated parts. Production engines serial number 2D-4010 and up, no longer use item number 10.

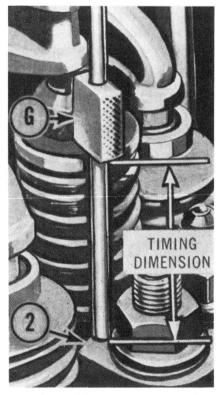

Fig. 105 — Injectors are correctly timed when shoulder of timing gage J1242 will just pass over top of injector follower as shown. Timing dimension is 1.484 inches.

G. Timing gage   2. Injector body

1. Oil seal	12. Dipstick	19. Plug	32. Oil filler cap
3. Upper front cover	13. Dipstick guide	22. Rear end plate	37. Cylinder block
7. Water hole cover	14. Dipstick adapter	24. Timing train cover	38. Compression
10. Hand hole cover	17. Air box drain tube	31. Strainer	gasket

80. **ADJUST INJECTOR RACK CONTROLS.** The injector control racks are adjusted so that each injector will inject an equal amount of fuel into the cylinders. Adjust injector racks as follows: Disconnect the vertical link from the governor rocker shaft lever. Loosen all four of the rack control lever adjusting screws and be sure control levers are loose on control tube. Now turn down the two adjusting screws of the rear cylinder control lever until they are equal in height and tight against control tube, as shown in Fig. 106. Rotate control tube to "full-fuel" (injector rack in) position and check to see that there is at least 1/16-inch clearance between the fuel rod and the boss of cylinder head as shown in Fig. 106. If clearance is not as specified, obtain same by loosening one screw and tightening the other. Be SURE both screws are tight after the above operation is complete. Hold the rear injector control rack in the "full-fuel" position and turn down the inner adjusting screw until the front injector control rack has moved to the "full-fuel" position. Now turn down outer adjusting screw until it bottoms lightly on control tube, then alternately tighten the two adjusting screws. Recheck the rear injector rack to be sure it has remained snug on the rack control lever. If rack of rear injector has loosened, back-off the inner adjusting screw of front injector slightly and tighten the outer adjusting screw. Check entire adjustment by moving injector control racks to the "full-fuel" position and

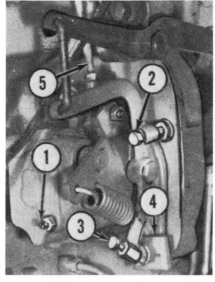

Fig. 107—View showing governor and governor linkage after oil filter support has been removed.

1. Buffer screw
2. Low idle screw
3. High idle screw
4. High idle stop
5. Vertical link

pushing straight down on injector racks with a screwdriver. Injector racks should spring back to their original positions when screwdriver is removed.

81. **ADJUST GOVERNOR LINKAGE.** Loosen lock nut and back-out buffer screw (1—Fig. 107). Move hand throttle to high idle position, then disconnect the vertical link from rocker shaft lever. Hold control racks in the "full-fuel" position by lifting

up on the vertical link (5) and adjust the ball and socket assembly on bottom of vertical link until stud will just enter the rocker shaft lever.

NOTE: Be sure that shut-off control (6—Fig. 108) is not interfering with vertical link. With hood and tappet cover removed and the hand throttle in high idle position, the injector racks should be checked for a maximum of 1/64-inch end play. If the end play is less than 1/64-inch, shorten the vertical link until it is obtained.

82. **ADJUST LOW IDLE SPEED.** Refer to Fig. 107, loosen lock nut and back-out buffer screw (1). Start engine and place hand throttle in low idle position. Loosen lock nut and adjust low idle screw (2) to obtain an engine low idle of 600 rpm, then tighten lock nut. With the engine operating at low idle rpm, turn buffer screw (1) in until engine surge (roll) is eliminated. DO NOT raise the engine low idle more than 20 rpm with the buffer screw. Tighten the buffer screw lock nut.

83. **ADJUST HIGH IDLE SPEED.** Start engine and move hand throttle to the high idle position and use caution not to overspeed engine. Refer to Fig. 107 and loosen lock nut on high idle adjusting screw (3). Turn the high idle adjusting screw to obtain an engine high idle of 1975 rpm, then tighten lock nut.

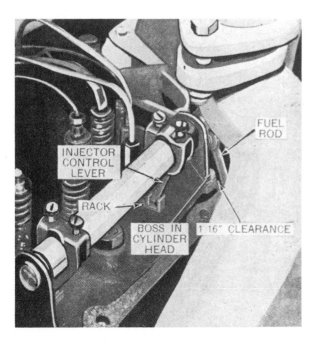

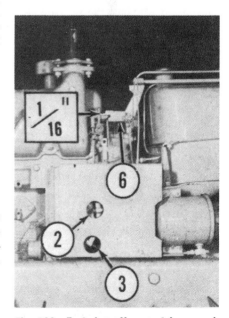

Fig. 106 — View showing the injector control tube and control levers. Refer to text for adjustment procedure.

Fig. 108—Fuel shut-off control is properly positioned when there is approximately 1/16-inch clearance between control arm and the special nut on vertical link with throttle in high idle position.

2. Low idle screw
3. High idle screw
6. Shut-off control

84. ADJUST SHUT-OFF CONTROL. Position control cable in its clip so that loom does not protrude above clip. Loosen swivel screw and position shut-off control (6—Fig. 108) so there is at least $\frac{1}{16}$-inch clearance between shut-off control arm and the special nut on top end of vertical linkage when the hand throttle is in high idle position.

R&R ENGINE WITH CLUTCH

85. Disconnect battery ground cable, shut off fuel and drain cooling system. Remove muffler and air cleaner cap, then proceed as follows:

On models 435D, loosen fasteners and remove hood and grille.

On models 440ID, remove grille screen, then unbolt and remove hood. Unbolt and remove side rails, leaving grille supports attached to side rails. Disconnect headlight wires from connector near generator and pull wires from hole in right baffle. Attach hoist to grille, then remove cap screws and dowels which retain grille to grille support and remove grille. If dowels are stuck, use a cap screw and a hammer type puller to remove same.

On all models equipped with power steering, disconnect lines from power steering cylinder and steering valve. Be sure front wheels are in the straight-ahead position, mark the location of the steering gear worm shaft in relation to the steering gear housing, then unbolt the steering gear assembly from steering gear support and remove steering gear and front steering shaft.

On models without power steering, remove roll pin from coupling at aft end of steering shaft, loosen socket head screw at front end of steering shaft, then drive steering shaft rearward. Unbolt steering gear housing from steering gear support, then rotate steering gear clockwise until steering gear worm shaft will allow radiator to clear.

On all models, disconnect upper and lower radiator hoses from radiator and unbolt fan shroud. Unbolt radiator from lower supports and upper radiator supports from engine, then lift radiator from tractor.

Support tractor under clutch housing with a floor jack, attach hoist to the front support and axle assembly, then unbolt front support from engine and pull support forward until dowels clear. Raise engine until crankshaft pulley clears the front support, then move front support and axle assembly away from tractor.

85A. Remove tube between blower and air cleaner and plug inlet of blower to prevent any foreign objects from entering blower. Unclip, then disconnect fuel return line at union located near fuel tank front support. Remove the fuel line running between primary fuel pump and fuel filter. Disconnect wire from temperature sending unit (bulb). Disconnect fuel supply line from fuel shut-off valve. Remove governor shield, disconnect wires from the hour meter and oil pressure switches and pull wires from hole in oil filter mounting bracket. Disconnect oil lines from oil filter. Disconnect governor adjusting (vertical) rod from governor control rod and the fuel shut-off control from shut-off arm. Unbolt fuel tank front support from engine, then install engine hangers and attach hoist to same. Unbolt engine and remove same from clutch housing.

Reinstall by reversing the removal procedure and tighten the engine to clutch housing nuts and cap screws to 57-63 Ft.-Lbs.

CYLINDER HEAD

86. **REMOVE AND REINSTALL.** To remove the cylinder head, first remove the hood which on 440ID models requires removal of grille screen so the elastic stop nuts can be held. Drain cooling system. On models equipped with power steering, disconnect lines from power steering cylinder and steering valve. Be sure front wheels are in the straight-ahead position, mark the location of the steering gear worm shaft in relation to the steering gear housing, then unbolt the steering gear assembly from steering gear support and remove steering gear and front steering shaft.

On models without power steering, remove roll pin from coupling at aft end of steering shaft, loosen socket head screw at front end of steering shaft and move steering shaft out of the way. Remove exhaust manifold and disconnect fuel lines at the cylinder head fuel manifolds. Loosen radiator hoses from thermostat housing and remove housing. Remove valve rocker cover and disconnect control link from injector control tube lever. Unbolt and remove the control tube and brackets as an assembly.

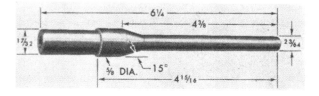

Fig. 109—View showing cylinder head removed from engine. Note injector tips (2) and cam followers (1).

1. Cam followers	4. Exhaust valves
2. Injector tips	W. Water holes
3. Cam follower guide	R. Oil return hole

On early models, disconnect oil by-pass line at right rear of cylinder head. Remove the cylinder head retaining bolts and lift cylinder head from engine. See Fig. 109. Note the location of the cylinder liner compression seals, the oil and water ring seals and the oil seal in the groove around the perimeter of the cylinder block.

CAUTION: When placing cylinder head on work bench, protect cam followers and injection nozzle tips by resting valve side of cylinder head on wooden blocks.

When reinstalling cylinder head, use all NEW compression gaskets and seals and proceed as follows: Place cylinder liner compression gaskets on cylinder liners, flange side down. Place the ring seals (7 used) in their counterbores in cylinder block and the oil seal in the groove around the perimeter of the cylinder block. To insure against disturbing seals when installing head, guide studs such as those shown in Fig. 110 must be used.

Fig. 110 — When installing cylinder head use two guide studs J7786, or equivalent, to insure against disturbing seals.

Place guide studs diagonally opposite in end cylinder head bolt holes in block, then clean under side of cylinder head and install the cylinder head. Check and BE SURE all seals are in place before finally positioning cylinder head. Install the cylinder head hold-down cap screws with the flat headed cap screws on the left (balance shaft) side. Start with the cap screws on right (camshaft) side of engine and tighten cap screws to bring cylinder head into position evenly. Beginning at the center and working toward ends of cylinder head, torque hold-down cap screws to 170-180 Ft.-Lbs. in ½-turn increments.

Install the injector control tube assembly and be sure that control arms are engaged in slots of injector control racks. After tightening bracket retaining cap screws, rotate control tube, until injectors are in full fuel position (injector racks inward), then release control tube. Control tube should return to the off position (injector racks outward) by spring pressure. If any binding exists, loosen brackets and shift the assembly slightly. Injector racks must return to the off position by the return spring tension only. Do not bend return spring to bring about this condition.

Balance of reinstallation is evident, however, perform an engine tune-up as outlined in paragraphs 78 through 81 before starting engine. After start-ing engine, check for fuel leaks and after engine has reached operating temperature, retorque the cylinder head cap screws to 170-180 Ft.-Lbs. Recheck the tune-up operations given in paragraphs 78 through 84.

87. **OVERHAUL.** Given herein are procedures for cylinder head leakage (cracks) inspection, cam follower hole inspection and renewal of injector hole tubes.

88. INSPECTION. Seal off the water holes in cylinder head using steel plates and suitable rubber gaskets held in place by bolts. Install dummy or scrap injectors to seat injector hole tubes and torque clamp bolts to 20-25 Ft.-Lbs. torque. Drill and tap one of the water hole plates and install an air hose, then apply 80-100 psi of air pressure to cylinder head. Place cylinder head in water that has previously been heated to 180-200 degrees F. and leave immersed for approximately twenty minutes. Observe water for bubbles indicating cracks or leaks.

After cylinder head has been pressure inspected, dry with compressed air. Renew cylinder head if same is cracked. Renew leaking injector hole tubes as outlined in paragraph 89.

Inspect cam follower holes in cylinder head for wear and/or scoring. Light scoring can be taken out by using crocus cloth wet with fuel oil.

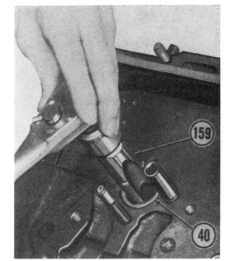

Fig. 112—The first step in removing an injector tube is to cut approximately ¾-inch of threads in tube using tool J5286-2, or equivalent, as shown.

40. Injector tube   159. Tool J5286-2

Bores in new cylinder head are 1.0620-1.0630. If bores are worn or scored until clearance between bore and cam follower exceeds 0.006, renew cylinder head. Recommended clearance between cam follower and cylinder head is 0.001-0.003.

89. RENEW INJECTOR HOLE TUBES. Refer to Fig. 111. Each injector is inserted into a thin-walled copper tube (22) which passes through the water jacket in the cylinder head. The copper tube is flared-over at the lower end, and it is sealed at the upper end with a neoprene ring (23). Effectiveness of the injector tube flared-over seal and neoprene seal can be checked by subjecting the cylinder head water jacket to 80-100 psi air pressure and submerging the head in water which has been heated to 180-200 deg. F.

Injector tubes can be removed as follows: Refer to Fig. 112. With injector removed, thread a tap (159), GM tool J5286-2 or equivalent, into upper end of injector tube (40) until ¾-inch of threads have been cut in tube. If necessary to prevent injector tube from turning while cutting the threads, use a tube holder tool, GM J5286-1 or equivalent, driven into lower or small end of tube. After cutting ¾-inch of threads in tube, remove tube holder tool. Then, using a small diameter rod, GM tool J5286-3

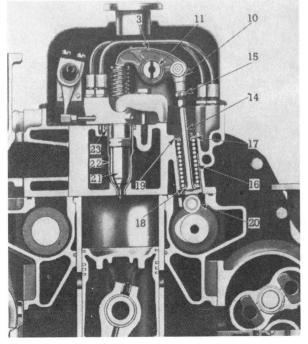

Fig. 111 — Cross sectional view of the diesel engine showing details of push rod, rocker arm, injector and injector tube.

3. Injector rocker arm
10. Push rod clevis
11. Rocker arm shaft
14. Push rod
15. Lock nut
16. Push rod spring
17. Upper spring seat
18. Lower spring seat
19. Spring seat retainer
20. Cam follower
21. Injector
22. Injector tube
23. Neoprene seal

Fig. 113—Drive-out injector tube by bumping on end of thread tap using a small diameter rod, or tool J5286-3, as shown.

Fig. 115 — Upset (flare) end of injector tube by applying 30-Ft.-Lbs. torque to up-setting die (J5286-6), or equivalent.

Fig. 117—Ream bevel seat in injector tube to receive the injector nut using tool J5286-9, or equivalent.

or equivalent, and inserted as shown in Fig. 113 to contact end of thread tap (159—Fig. 112), bump injector tube out of cylinder head.

To install an injector tube proceed as follows: Refer to Figs. 114 and 115. Place a new sealing ring in counterbore of cylinder head and using installing tool and pilot, GM tools J5286-4 and J5286-5 or equivalent, bump injector tube in place. Using injector tube installing tool and upsetting die, GM tools J5286-4 and J5286-6 or equivalent, upset (flare) lower end of injector tube by applying 30 Ft.-Lb. torque to upsetting die.

After installing injector tube, it must be finished in three operations: Hand reamed as shown in Fig. 116 to receive injector body; spot faced at the flared-over end; and hand reamed to provide a sealing surface for the bevel seat on lower end of injector nut as shown in Fig. 117.

As shown in Fig. 116, ream injector tube, using GM tool J5286-7 or equivalent, until lower shoulder of reamer contacts injector tube.

Remove excess stock from flared-over end of injector tube using GM tool J5286-8 or equivalent, so that lower end of tube is from flush to 0.010 below finished surface of cylinder head.

The third step as shown in Fig. 117, reaming bevel seat in injector tube, should be performed with care as this operation controls the location of the injector spray tip relative to the surface of the cylinder head. Use an injector as a gage to check for the desired flush condition between cylinder head and the shoulder on the injector spray tip as shown in Fig. 118. Note: If pre-finished injector tubes are used (those which have a narrow land machined at the beveled

Fig. 114—View showing method of installing injector tube with driver (J5286-4) and pilot (J5286-5), or their equivalents.

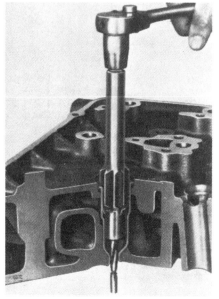

Fig. 116—Ream injector tube with tool J5286-7, or equivalent, to receive injector body and spray tip.

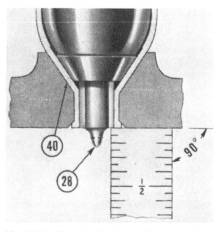

Fig. 118—Showing flush condition of injector spray tip shoulder relative to machined surface of cylinder head.

Fig. 119 — View showing the two exhaust valves (2) used for each cylinder. Note the injector tubes (1) between the exhaust valves.

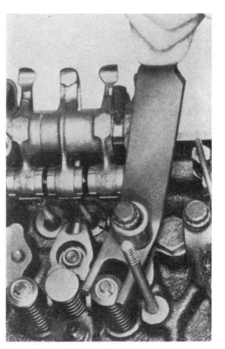

Fig. 121—View showing method of renewing valve spring without removing cylinder head. Piston is at TDC. Use tool J1227-01, or equivalent.

seat as distinguished from tubes having a straight bevel seat) exercise care during the final reaming operation to prevent removal of too much metal from the thin wall of the tube.

Check effectiveness of the flared-over seal and neoprene seal on injector tube by subjecting the cylinder head water jacket to 80-100 psi air pressure and submerging the head in water heated to 180-200 deg. F.

89A. CYLINDER HEAD WARPAGE. The fire deck (bottom) of cylinder head should be checked each time head is removed. The amount of longitudinal and/or transverse warpage should not exceed 0.004 and not more than 0.020 of metal should ever be removed from the fire deck. The distance from top edge of cylinder head and bottom (fire deck) of cylinder head must not be less than 4.376 inches. After a cylinder head has been refaced, it wil be necessary to correct the specified protrusion dimensions of valve inserts, valves and injector spray tips. Refer to paragraph 89 and 90. If injector hole tubes are renewed, it is recommended that the cylinder head be pressure checked as outlined in paragraph 89. If cylinder head is refaced, stamp fire deck on a non-sealing surface to indicate the amount of removed metal.

VALVES, SEATS AND GUIDES

90. Two exhaust valves, located in the cylinder head, are provided for each cylinder, Fig. 119. Inlet ports are in the cylinder liner. Valve head must not protrude more than 0.014 beyond the surface of cylinder head as shown in Fig. 120. Limiting the protrusion to a maximum of 0.014 will prevent the valve head from striking the top of the piston. Renew valve and/or seat when valve is more than 0.002 below surface of cylinder head.

91. Exhaust valves seat on renewable inserts in the cylinder head. Refer to Fig. 120 for reconditioning limits of valve and insert.

Shoulderless valve guides should be installed so that top or threaded end of guide projects 0.010-0.030-inch above machined face of valve tappet cover gasket surface of cylinder head. Service guides, requiring a press fit of 0.0005-0.0035, are prefinished and do not require final sizing.

Valve Face Angle..............30°
Valve Seat Angle..............30°
Valve Tappet Gap—Hot........0.009
Valve Stem Diameter...0.3095-0.3105
Valve Clearance in Guide..0.002-0.004
Renew If Clearance Exceeds..0.006
Guide Height Above Tappet
Cover Gasket Surface of
Cylinder Head—Inches ..0.010-0.030

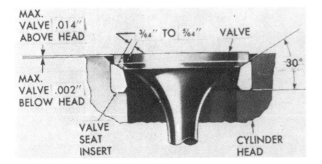

Fig. 120 — Valves and valves seats must be within the limits indicated. Renew valve and/or seat to maintain this relationship.

Valve Seat
Counterbore Diameter ..1.439-1.440
Valve Seat
Counterbore Depth 0.298-0.302

VALVE SPRINGS

92. Any exhaust valve spring can be renewed, as shown in Fig. 121, without removing cylinder head. To renew a valve spring, first position the piston at TDC (TDC is indicated when injector plunger has traveled downward approximately $\frac{3}{16}$-inch). Then, remove the injector fuel lines and cap screws attaching rocker arms shaft brackets to cylinder head.

Valve springs have a free length of 2½ inches and should require 141 plus or minus 8 lb. to compress spring to a length of 1 29/32 inches. Renew spring when less than 128 pounds will compress spring to 1 29/32 inches.

CAM FOLLOWERS AND PUSH RODS

Cam followers, operating in unbushed bores located in the cylinder head, are of the roller type.

93. Refer to Fig. 122. A cam follower (20), push rod (14), push rod spring (16), spring seats (17 and 18) can be removed from the cylinder head without removing the cylinder head as follows: Remove the fuel lines, the rocker arms shaft brackets and shaft. Unscrew rocker arm from push rod to be removed. Compress push rod spring with a tool similar

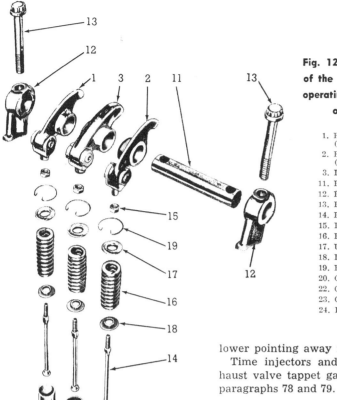

Fig. 122—Exploded view of the injector and valve operating mechanism for one cylinder.

1. Exhaust valve rocker (left)
2. Exhaust valve rocker (right)
3. Injector rocker
11. Rocker arm shaft
12. Rocker shaft bracket
13. Bracket cap screw
14. Push rod
15. Lock nut
16. Push rod spring
17. Upper spring seat
18. Lower spring seat
19. Retainer
20. Cam follower
22. Guide
23. Cap screw
24. Lock washer

Fig. 123—Removal of a push rod and cam follower can be accomplished without removing cylinder head by using tool J3092-01, or equivalent, as shown.

14. Push rod
15. Lock nut
19. Upper seat retainer
39. Tool J3092-01
40. Flat washer

lower pointing away from the valves.

Time injectors and adjust the exhaust valve tappet gap as outlined in paragraphs 78 and 79.

ROCKER ARMS AND SHAFTS

95. Three rocker arms are provided for each cylinder; two for the exhaust valves and one for the injector. Refer to Fig. 126.

Injector rocker arms are fitted with bushings for the rocker arm shaft as well as for the push rod clevis pin. Exhaust valve rocker arms have no rocker arm shaft bushings.

Diameter of new rocker arm shaft is 0.8735-0.8740. Inside diameter of new injector rocker arm bushing and rocker arm shaft bore in exhaust valve rocker arm is 0.8750-0.8760.

to the one as shown in Fig. 123, and remove spring seat retainer 19. Pull push rod assembly and cam follower out through top of cylinder head.

Push rods and cam followers can be removed from combustion chamber side of cylinder head as follows: Remove cylinder head and disconnect push rod from rocker arm. Then remove two cap screws attaching cam follower guide (3—Fig. 124) to cylinder head and withdraw cam follower assembly.

94. Cam followers, available in standard size only, have a running clearance in bores of 0.001-0.003. Renew the followers and/or cylinder head if clearance exceeds 0.006. Cam follower roller bushing to roller pin clearance should not exceed 0.010. Side clearance between roller and follower should not be less than 0.015 nor more than 0.023 as shown in Fig. 125. Roller, bushing and roller pin are available as service items.

Install cam followers in cylinder head with oil hole in bottom of fol-

Fig. 124 — Cam followers (1) can be removed from lower side of cylinder head after removing cam follower guide (3).

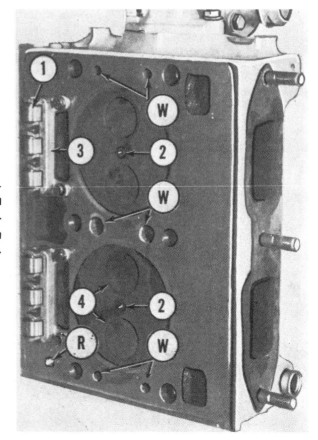

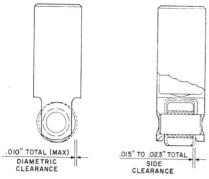

Fig. 125—View showing cam roller wear and clearance limits. Roller, bushing and roller pin are available as service items.

Running clearance of rocker arms on shaft is 0.001-0.0025 with a maximum allowable clearance of 0.004.

Clearance between the steel clevis pin bushing and the bronze bushing at push rod end of rocker arms is 0.0015-0.003. The side clearance between clevis and rocker arm is 0.008-0.017.

Contact face of the rocker arms can be refinished providing that not more than 0.010 material is removed. Maintain the original contour and keep contact face parallel with axis of the large diameter bore of rocker arm. Adjust exhaust tappet gap as in paragraph 78 and injector timing as in paragraph 79.

GEAR TRAIN COVER

The gear train shown in Fig. 131 is located at the flywheel end of the engine and is housed by the gear train cover.

96. **REMOVE AND REINSTALL.** To remove the gear train cover, first disconnect engine from center frame as outlined in paragraph 174. Remove engine clutch and flywheel. Remove fuel pump. Remove oil filter and support, governor and blower. Either completely remove oil pan, or remove the four rear cap screws, then loosen the remaining cap screws and allow oil pan to drop. Remove the cap screws attaching gear train cover to rear of engine.

CAUTION: Note the location of the attaching cap screws as they are removed. Various sizes are used and in addition, the six which pass through the idler gear hub and the spacer are a self-locking type.

Install pilot studs, tap gear cover free from dowels and remove cover. See Fig. 127.

If necessary, the rear oil seal can be renewed at this time.

Prior to reinstalling gear train cover, lubricate gear train teeth and lip of rear oil seal. Install new gasket on gear train cover. If there are shims

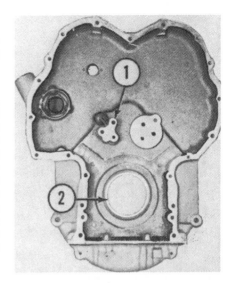

Fig. 127—Inside view of gear train cover after removal. Note spacer (1) which is integral with cover. Item (2) is crankshaft rear oil seal.

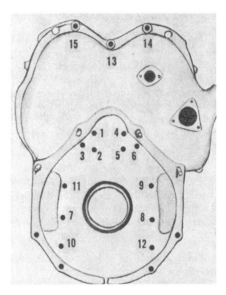

Fig. 128—During the initial tightening sequence, tighten cap screws until they are snug in the order shown.

Fig. 126 — Top side of cylinder head showing one push rod and exhaust rocker arm assembly removed. Note injector between exhaust valves.

1. Exhaust valves
2. Injector
3. Injector control rack
4. Injector fuel caps
5. Exhaust valve rocker arms
6. Injector rocker arm

between spacer and engine end plate, hold same in position with heavy grease. Place oil seal expander J7540, or equivalent, on end of crankshaft and if pilot studs were removed, reinstall them. Place gear train cover over crankshaft and while holding seal expander against end of crankshaft, slide gear train cover into position. Install the attaching cap screws and refer to Fig. 128 for the initial tightening sequence. Refer to Fig. 129 for the second tightening sequence. Torque the $\frac{3}{8}$-inch cap screws to 30 Ft.-Lbs. and the $\frac{5}{16}$-inch cap screws to 25 Ft.-Lbs. during the second sequence.

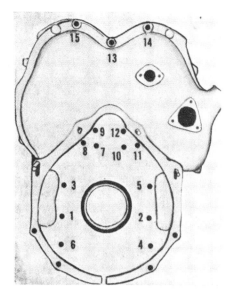

Fig. 129—During the second tightening sequence, torque the ³/₈-inch cap screws to 30 Ft.-Lbs. and the 5/16-inch cap screws to 25 Ft.-Lbs. in the order shown.

ENGINE UPPER FRONT COVER

97. This cover is mounted at the upper front of the engine and should require no servicing other than the possible renewal of oil seals. See Fig. 130. Renewal of the oil seals is evidenced by oil leaks below the balance weights.

To remove the engine upper front cover, remove hood, radiator, fan and bracket assembly and the balance weights. Unbolt and remove cover. Seals can now be pressed or driven from cover.

When installing new seals, use a non-hardening sealing compound on outer edge of seals and lubricate lips of seals with cup grease. Install seals with lips of same facing inward. Also inspect spacers on camshaft and bal-

ance shaft and if seal contacting surfaces are grooved or rough, renew the spacers. Tighten the cover retaining cap screws to 35 Ft.-Lbs. torque and the balance weight nuts to 300-325 Ft.-Lbs. torque.

TIMING GEARS

Refer to Fig. 131 for a view of the timing gear train located at the rear of engine. The governor drive gear is driven by the balance shaft gear and the upper rotor gear of the blower is driven by the camshaft gear. Normal backlash of all gears, except blower gears, is 0.003-0.005 with a maximum allowable backlash of 0.007. Normal backlash of blower gears is 0.0005-0.0025 with a maximum allowable backlash of 0.0035.

98. **GEAR TIMING MARKS.** The four basic gears of the timing gear train are marked as follows: The crankshaft gear is stamped with the symbols "IROVL" and "ILAVR." The idler gear has two markings of a triangle within a circle stamped 180 degrees apart. The balance shaft gear is stamped with two "O" marks. The camshaft gear is stamped with one "O" and a triangle.

NOTE: When timing the gear train, disregard all markings except those involving an "O". The blower gears are also marked with a small "O"; however, these are used to time the blower rotors and need not be considered at this time.

Time the four basic gears as fol-

lows: Mate an "O" mark on idler gear with the "O" of the "IROVL" mark on crankshaft gear. Position balance shaft gear so one "O" will mate with "O" on idler gear and the remaining "O" will be at a three o'clock position. Mate "O" mark on camshaft with the "O" mark on balance shaft gear.

99. **CRANKSHAFT GEAR.** The crankshaft gear is keyed and pressed on the crankshaft. To remove the crankshaft gear, remove the engine as outlined in paragraph 85 and the crankshaft as outlined in paragraph 112.

NOTE: In some cases it may be possible to remove the crankshaft gear without removing the crankshaft from engine; however, the interference fit is usually such that the crankshaft must be removed in order to remove the gear.

When reinstalling the gear, press or drive gear on crankshaft as shown in Fig. 132 until gear bottoms on shoulder of shaft.

100. **IDLER GEAR AND HUB.** To remove the idler gear, idler gear hub and thrust washers, remove the timing train cover. Remove rear thrust washer and idler gear from idler hub. Remove cap screw (5—Fig. 133) from counterbore of idler hub and remove hub and front thrust washer. See Fig. 134.

Fig. 130—Inside view of engine upper front cover. Note engine hanger at upper left of cover.

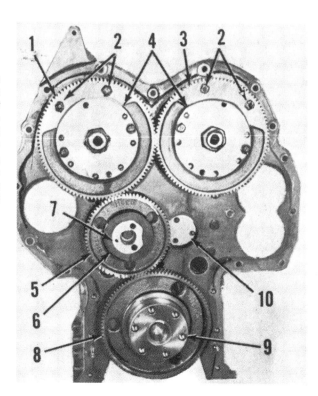

Fig. 131 — View of the four basic gears in the timing gear train of early models. Late model production engines do not use balance screws (2) or shaft nut retainers (4), however, service parts require that balance screws be installed. Refer to text for timing marks.

1. Balance shaft gear
2. Balance screws
3. Camshaft gear
4. Shaft nut retainers
5. Idler gear
6. Thrust washer
7. Idler gear hub
8. Crankshaft gear
9. Crankshaft
10. Shim

Fig. 132—Drive crankshaft gear on crankshaft until it bottoms using tool J7557, or equivalent.

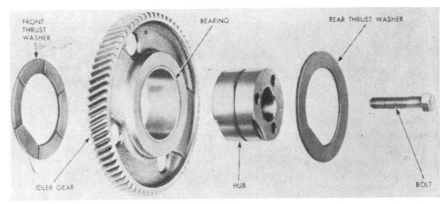

Fig. 134—Exploded view of idler gear, hub and thrust washers. Normal operating clearance of hub in bearing is 0.003-0.005.

Wash parts and inspect for wear and/or scoring. Idler gear and bearing must be renewed as an assembly as bearing is not catalogued separately. New idler gear bearing inside diameter is 2.187-2.188. New idler gear hub outside diameter is 2.183-2.184. Thrust washer thickness is 0.119-0.121. Normal gear backlash is 0.003-0.005 with a maximum allowable backlash of 0.007.

When reinstalling idler gear and hub assembly, proceed as follows: Place front thrust washer on idler hub with grooved side toward gear and position hub on engine end plate with flats of hub at the three o'clock position. Install three of the previously removed $\frac{5}{16}$-inch cap screws through idler hub to hold same in position, then insert the cap screw in counterbore of hub and tighten to 30-35 Ft.-Lbs. torque. Remove the $\frac{5}{16}$-inch cap screws, install the idler gear and align timing marks as outlined in paragraph 98. Place rear thrust washer on hub with grooves facing gear.

101. BALANCE SHAFT GEAR. To remove the balance shaft gear, first remove the balance shaft as outlined in paragraph 104. Remove the nut retainer plate (early engines) and nut. Place the balance shaft and gear in a press and carefully support gear. Use a short piece of ¾-inch brass rod between end of balance shaft and press ram and press shaft from gear. Remove Woodruff key and thrust plate.

When reinstalling gear, support balance shaft under rear bearing journal and use a hollow sleeve between gear and ram of press, to install gear.

NOTE: Do not disturb the two balance cap screws located on outer perimeter of gear if so equipped. Late production engines have gears which have no balance screws, however, service parts require that balance screws be installed.

Reinstall shaft and gear in engine with timing marks in register as outlined in paragraph 98.

102. CAMSHAFT GEAR. To remove the camshaft gear, first remove the camshaft as outlined in paragraph 103. Remove the nut retainer plate (early engines) and nut. Place camshaft and gear in a press and carefully support gear. Use a short piece of ¾-inch brass rod between end of

shaft and press ram and press shaft from gear. Remove thrust washer, Woodruff key and spacer from camshaft.

When reinstalling, place spacer on gear end of camshaft and install Woodruff key and thrust plate. Support camshaft under rear bearing journal and use a hollow sleeve between gear and ram of press to install gear.

NOTE: Do not disturb the two balance cap screws located on outer perimeter of gear is so equipped. Late production engines have gears which have no balance screws, however, service parts require that balance screws be installed. See Fig. 135.

Reinstall shaft and gear in engine with timing mark in register as outlined in paragraph 98.

CAMSHAFT

The camshaft, located at the upper right of the cylinder block, rotates in three renewable bearings (bushings). Oil is supplied under pressure to the end bearings and from the end bearings through hollow passages in the camshaft, to the center bearing. Refer to Fig. 136.

103. To remove the camshaft, first remove the cylinder head as outlined

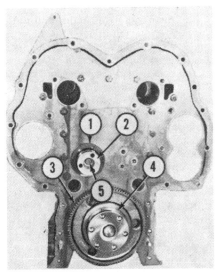

Fig. 133 — Idler hub (2) and forward thrust washer (1) can be removed after cap screw (5) has been removed.

Fig. 135—Balance screws such as those shown are installed in the outer perimeter of the camshaft gear as well as the balance shaft gear on early production engines.

in paragraph 86 and the gear train cover as outlined in paragraph 96.

Use a clean shop towel, or a piece of soft wood and wedge same between camshaft gear and balance shaft gear. Remove nut which retains balance weight to front of camshaft and pry off balance weight. Remove Woodruff key and balance weight spacer. Remove nut retainer plate (11) on early engines, then working through holes in camshaft gear, unbolt thrust plate from cylinder block and pull camshaft and gear assembly from engine.

Clean all parts and be sure all oil passages in camshaft and bearings are open and clean. Inspect all parts for wear, scoring or other damage and check same against the values which follow:

Journal Diameters 2.1820-2.1825
Bearings Inside Diameter..2.187-2.188
Running Clearance 0.0045-0.006
Maximum Allowable
    Running Clearance 0.0008
End Play 0.008-0.015
Maximum Allowable
    End Play 0.019
Maximum Allowable
    Run-Out 0.002
Thrust Plate Thickness....0.208-0.210

If camshaft bearings are to be renewed it will be necessary to perform the additional work of removing the radiator, fan assembly and the upper front cover. NOTE: Effective at engine serial number 2D-9278, an oil slinger has been added to the cam-

shaft and is positioned behind spacer (4—Fig. 136). Before, removing old bearings note the position of same in the bore with respect to the notch in the bearings, and install new bearings in same position. Bearings can be removed and reinstalled, using a suitably piloted driver. End bearings are installed flush with ends of cylinder block. Center bearing rear edge should measure 5.54 inches from rear face of engine block.

When reinstalling camshaft, torque thrust plate cap screws to 30-35 Ft.-Lbs. and the balance weight retaining nut to 300-325 Ft.-Lbs. and be certain that timing marks are in register as outlined in paragraph 98.

BALANCE SHAFT

The balance shaft, located at the upper left of the cylinder block, rotates in two renewable bearings (bushings). Oil is supplied under pressure to the two bearings. Refer to Fig. 136.

104. To remove the balance shaft, first remove the gear train cover as outlined in paragraph 96.

Use a clean shop towel, or a piece of soft wood and wedge same between balance shaft and camshaft gear. Remove nut which retains weight to front of balance shaft and pry off balance weight. Remove Woodruff key and balance weight spacer. Remove nut retainer plate (11) on early en-

gines, then working through holes in balance shaft gear, unbolt thrust plate from cylinder block and pull balance shaft and gear assembly from engine.

Clean all parts and be sure oil passages in bearings are open and clean. Inspect all parts for wear, scoring or other damage and check same against the values which follow:
Journal Diameters 2.1820-2.1825
Bearing Inside Diameter...2.187-2.188
Running Clearance 0.0045-0.006
Maximum Allowable
    Running Clearance 0.008
End Play 0.008-0.015
Maximum Allowable
    End Play 0.019
Thrust Plate Thickness....0.208-0.210

If balance shaft bearings are to be renewed, it will be necessary to perform the additional work of removing the radiator, fan assembly and the upper front cover. NOTE: Effective at engine serial number 2D-9278, an oil slinger has been added to the balance shaft and is positioned behind spacer (4—Fig. 136). Before removing the old bearings note the position of same in the bore with respect to the notch in the bearings and install new bearings in same position. Bearings can be removed and reinstalled, using a suitable piloted driver. Bearings are installed flush with ends of cylinder block.

When reinstalling balance shaft, torque thrust plate cap screws to 30-35 Ft.-Lbs. and the balance weight retaining nut to 300-325 Ft.-Lbs. and be certain that timing marks are in register as outlined in paragraph 98.

CONNECTING ROD AND PISTON UNITS

105. Piston and connecting rod units are removed from above after removing hood, cylinder head and oil pan. Remove carbon from inner surface of liner and if necessary, install cylinder liner hold-down clamps, then remove top ridge with a ridge cutter. Leave the cylinder hold-down clamps installed to prevent liners from working upward during crankshaft rotation or when removing the piston and rod units. Remove the oil pump screen and intake pipe assembly, then remove rod caps and push piston and rod assemblies out top side of engine. Refer to Fig. 137.

When reinstalling, numbered sides of connecting rods and connecting rod caps should be in register and face toward camshaft side of engine. Torque the connecting rod nuts to 45-50 Ft.-Lbs.

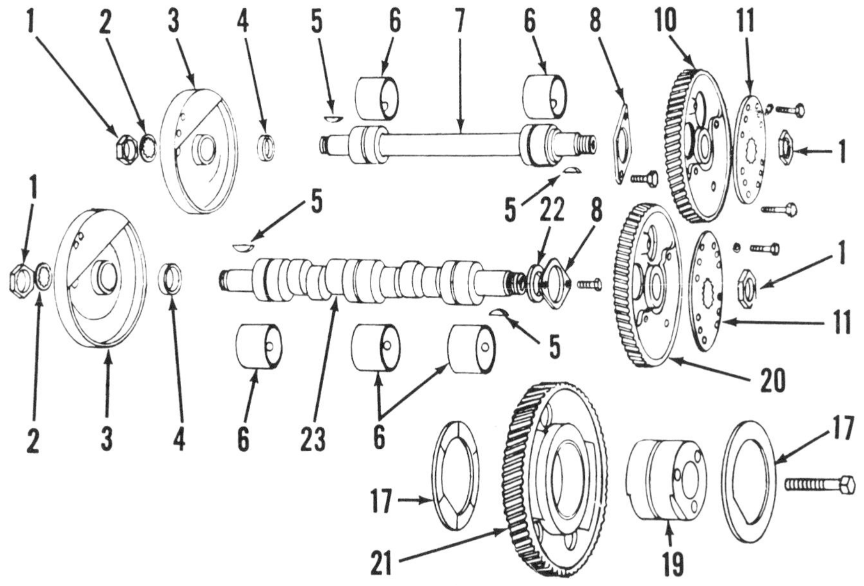

Fig. 136—Exploded view of the camshaft, balance shaft, idler shaft, gears and component parts. Late production engines are not fitted with nut retainer plates (11). Effective at engine serial number 2D-9278, oil slingers have been added to the balance shaft and camshaft and are positioned behind spacers (4).

1. Nut	5. Woodruff key	10. Balance shaft gear	20. Camshaft gear
2. Lock washer	6. Bushing	11. Nut retainer plate	21. Idler gear
3. Balance weight	7. Balance shaft	17. Thrust washer	22. Spacer
4. Spacer	8. Thrust plate	19. Idler gear shaft	23. Camshaft

PISTONS, LINERS AND RINGS

Trunk-type malleable iron pistons are tin plated and are equipped with four steel, chrome plated compression rings and two, three piece oil control rings. Both oil control rings are located below the piston pin. See Fig. 138.

The wet type cylinder liners are a hardened cast iron alloy and are a slip fit in the cylinder block.

The cylinder liners contain eighteen oval shaped ports which are cut at a 25 degree angle and are spaced equidistantly around the liner at approximately mid-length. See Fig. 139.

106. **PISTONS.** Pistons are supplied in standard size only with a skirt diameter of 3.8693-3.8715.

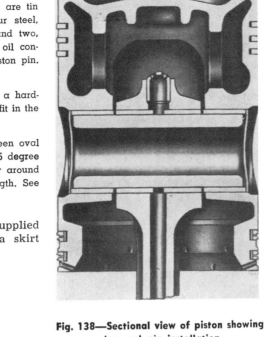

Fig. 138—Sectional view of piston showing ring and pin installation

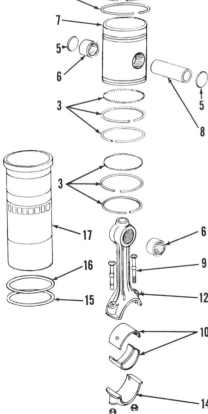

Fig. 137—Exploded view of piston, connecting rod, cylinder liner and component parts. Item 15 is not used on late production engines.

3. Oil ring	12 Connecting rod
5. Pin retainers	14. Cap
6. Bushing	15. Lower cylinder
7. Piston	seal
8. Piston pin	16. Upper cylinder seal
10. Bearing	17. Liner

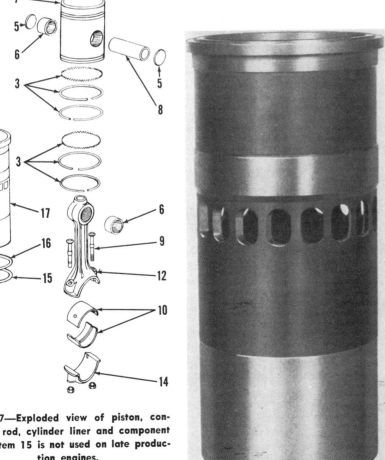

Fig. 139—Cylinder liner has eighteen ports spaced equidistantly around its circumference. Ports are cut at a 25 degree angle.

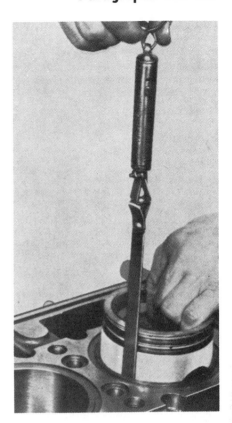

Fig. 140—Check piston to liner clearance with a ½-inch feeler and spring scale as shown. Refer to text.

Desired piston skirt clearance is 0.0037-0.0074 with a maximum wear limit of 0.009.

Clearance between piston and liner can be checked as shown in Fig. 140 by using a ½-inch wide feeler ribbon and a spring scale as follows: With liner in position in cylinder block, insert piston in liner with the feeler ribbon positioned between piston and liner. Now withdraw feeler ribbon with the spring scale and measure the force required to withdraw feeler ribbon. When this force is approximately six pounds, the clearance between piston and liner will be 0.001 greater than the thickness of the feeler ribbon used. For example: When actual clearance is 0.005, a spring scale pull of approximately six pounds will be required to withdraw a 0.004 feeler ribbon.

107. **LINERS.** Cylinder liners are a slip fit in the cylinder block and can be removed, in most cases, without the use of a puller. However, where difficulty is encountered, liner can be removed by driving on a hardwood block positioned against lower end of liner.

Cylinder liners are available in the standard size only, however, used liners can be honed to remove top ridge and/or glaze providing that the values given below and in paragraph 106 are not exceeded.

Inside Diameter 3.8752-3.8767
Maximum Out-of-Round
    (New) 0.002
Maximum Out-of-Round
    (Used) 0.003
Maximum Taper (New) 0.001
Maximum Taper (Used) 0.002
Depth of Liner Flange
    Below Block 0.0465-0.050
Outside Diameter (Upper
    Seal Surface) 4.480-4.481
Outside Diameter (Lower
    Seal Surface 4.355-4.356

When reinstalling liners, be sure counterbore in cylinder block is clean and free of foreign material. Counterbore in cylinder block should measure 0.300-0.302 in depth and must not vary more than 0.001 around its circumference. Lubricate the liner seal rings and install them in the cylinder block with the smallest, which is identified by white marking compound, in the bottom groove. See Fig. 141. Note: Late production engines no longer use the lower sealing ring. Install liner, which should slide into place freely when installed by hand. Tap liner lightly with a soft

hammer to be sure it is properly seated in counterbore, then mount a dial indicator as shown in Fig. 142 and measure the distance between top of liner flange and top of cylinder block. This distance should be 0.0465-0.050 below top of cylinder block and should not vary more than 0.0015 between the two cylinders. If the above limits are not met, switch liners and recheck, or use new parts.

108. RINGS. There are four compression rings and two oil control rings on each piston. Refer to Fig. 143.

Check ring end gap and side clearance against the values which follow:

End Gap
    Compression Rings 0.020-0.046
    Oil Control Rings.......0.010-0.025
Side Clearance
    Top compression 0.008-0.010
    2nd Compression 0.007-0.010
    3rd & 4th Comp. 0.005-0.008
    Oil Control Rings.....0.0015-0.0055

PISTON PINS AND BUSHINGS

The hollow type piston pin which is of the full floating type, rotates in bushed bores of the piston and in similar bushings located in the small end of the connecting rod.

109. Piston pins are retained in piston by stamped metal retainers (5—

Fig. 137). Pry out the retainer by punching a hole through its center.

After installing new piston pin retainers using GM tool J-4895-01 or equivalent, check same for leakage as follows: Fill inside of piston with fuel oil until pin boss is submerged. Let piston and rod assembly set for approximately 15 minutes, then carefully inspect around both pin retainers. If fuel oil leaks are present, install new retainers.

The 1.3746-1.3750 diameter piston pin has a normal operating clearance in piston bushings of 0.0025-0.0034 and a normal operating clearance in rod bushings of 0.000-0.0019. Maximum allowable limit of piston pin clearance in either bushing is 0.010.

The split type bushings should be installed in piston with split side at bottom of piston and with inner end of bushing flush with edge of pin boss. Rod bushings should be installed in rod with split side at top of rod, as shown in Fig. 143, and with outer edge of bushings flush with edge of rod.

New bushings should be sized after installation to provide a pin clearance of 0.0025-0.0034 in piston bushings, and 0.0010-0.0019 clearance in rod. Connecting rod center to center length is 8.799-8.801.

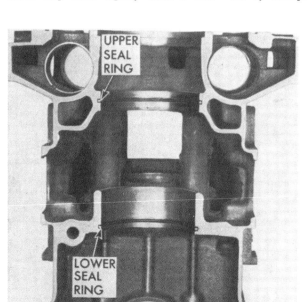

Fig. 141—Cross sectional view of cylinder block showing location of liner seals of early engines. Bottom (smallest) seal is identified by a white marking. Late production engines do not use lower sealing ring.

Fig. 142—Top of liner should be 0.0465-0.050 below top of cylinder block when measured with a dial indicator as shown.

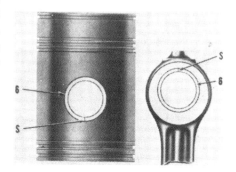

Fig. 143—The split type piston pin bushings are installed in the piston with the split at bottom side and in the connecting rod with the split at top side.

CONNECTING RODS AND BEARINGS

110. The shimless, precision type rod bearings are renewable from below without removing the rods from the engine.

Check crankpins and bearings for wear, scoring and out-of-round conditions.

Crankpin Diameter 2.499-2.500
Taper, Maximum 0.003
Out-of-Round (Max.) 0.003
Bearing Running
  Clearance 0.0044-0.0084
  Renew if Clearance Exceeds .. 0.0104
Bearing I.D.
  (Vertical Axis) 2.5044-2.5074
Bearing Thickness (90 Deg.
  From Parting Line) ... 0.1245-0.1250
Minimum Allowable Bearing
  Thickness 0.123
Rod Side Play 0.006-0.012
Rod Nut Torque .... 45-50 Ft.-Lbs.
Rod, C to C Length 8.799-8.801

CRANKSHAFT AND BEARINGS

111. The induction hardened crankshaft rotates in three precision type bearings which are renewable from below without removing the crankshaft. A special bolt or pin inserted in the crankshaft main journal oil hole will facilitate removal of upper shells for the two front main journals. Rear main journal upper bearing shell can be removed with a small diameter curved rod as shown in Fig. 144. Identification numbers on main bearing caps face camshaft side of engine.

Fig. 145—Crankshaft end play of 0.004-0.011 can be measured with a dial indicator as shown.

112. Crankshaft can be removed after removing the engine from tractor chassis, gear train cover, oil pan, rod caps, engine lower front cover, and main bearing caps.

Crankshaft end play (Fig. 145) of 0.004-0.011 is controlled by thrust washer halves (9—Fig. 146), located on either side of the rear main bearing journal. Grooved face of thrust washers contact thrust surface on crankshaft. If thrust surfaces of crankshaft are worn or ridged excessively, they should be reground. For data on crankshaft front and rear oil seals, refer to paragraphs 114 and 115.

113. Check shaft and main bearings for wear, scoring and out-of-round conditions.

Main Journal Diameter 2.999-3.000
Running Clear., Mains 0.002-0.004
  Renew if Clear-
    ance Exceeds 0.006
Bearing I.D.
  (Vertical Axis) 3.002-3.003
Bearing Thickness (90 deg.
  from part line) 0.1245-0.1250
Minimum Allowable
  Bearing Thickness 0.123
Crankpin Diameter 2.499-2.500
Running Clearance,
  Crankpins 0.0044-0.0084
  Renew If Clear-
    ance Exceeds 0.0104
Main & Rod Journal
  Out-of-Round, Max. 0.003
Main & Rod Journal
  Taper, Max. 0.003
Crankshaft End Play,
  Recommended 0.004-0.011
Crankshaft End Play
  Control .... No. 3 Brg.
Main Brg. Bolt Tightening
  Torque—Ft.-Lb. 120-130

CRANKSHAFT OIL SEALS

114. REAR OIL SEAL. The lip type oil seal for the rear end of the crankshaft is located in the gear train cover and can be renewed after gear train cover is removed as outlined in paragraph 96.

Fig. 144—Upper half of rear main bearing shell can be removed with crankshaft installed by using a small diameter curved rod as shown.

Fig. 146—Crankshaft end play of 0.004-0.011 is controlled by thrust washers on either side of rear main journal.

9. Thrust washers
10. Oil pump drive gear
13. Dowels
14. Rear main bearing cap

Rear oil seal can also be removed in the following manner: Detach engine from center frame as outlined in paragraph 174, unbolt clutch from flywheel and flywheel from crankshaft. Drill two holes 180 degrees apart in metal case of seal, install self threading screws fitted with washers, then pry out seal.

When reinstalling seal, use a nonhardening sealing compound on outer edge of seal and lubricate lip of seal with cup grease. Lip of seal faces toward engine.

NOTE: Use caution when installing the seal over crankshaft. Seal is soft and flexible and if not well lubricated, lip of seal will tend to roll.

If gear train cover was not removed, place seal over crankshaft and start in counterbore. Use a hardwood block as a driver and drive seal into place.

If gear train cover was removed, use pilot studs and a seal expander tool when reinstalling same as outlined in paragraph 96.

115. FRONT OIL SEAL. The lip type oil seal for front end of crankshaft is located in the engine front lower cover, and can be renewed after cover is removed as outlined in paragraph 118.

The front oil seal can also be removed in the following manner. Remove hood, grille, radiator and pump (if so equipped), front stabilizer and crankshaft pulley. Drill two holes 180 degrees apart in the metal case of the seal, install self threading screws fitted with washers, then pry out seal.

When reinstalling seal, use a nonhardening sealing compound on outer edge of seal and lubricate lip of seal with cup grease. Lip of seal faces toward engine.

If front lower cover was not removed, place seal over crankshaft and

start in counterbore. Use a hardwood block as a driver and drive seal into place.

If front lower cover was removed, reinstall same as outlined in paragraph 118.

FLYWHEEL, RING GEAR AND PILOT BEARING

116. To remove flywheel, first detach engine from center frame as outlined in paragraph 174, then unbolt and remove clutch assembly.

Clutch surface of flywheel can be refaced providing that not more than 0.020 of metal is removed and all radii are maintained. When reinstalling, tighten cap screws to 130-140 Ft.-Lbs. torque. Total run-out of flywheel at clutch surface should not exceed 0.005.

Ring gear can be renewed after flywheel is removed. To install a new ring gear, heat same to not more than 400 degrees F. and install with chamfer end of teeth facing rearward.

Inspect pilot bearing before reinstalling clutch and if defective, remove same from flywheel using a suitable puller. Fill grease retainer with a good grade of bearing grease and when installing pilot bearing, use a driver that will contact only the outer race. Shielded side of bearing is toward rear of tractor.

OIL PUMP AND ENGINE LOWER FRONT COVER

117. A rotor type oil pump, Fig. 147, is incorporated in the engine lower front cover.

The inner rotor is driven by a gear pressed on the front end of the crankshaft. The outer rotor is driven by the inner rotor and rotates in a bore in the pump body which is eccentric

(off-center) to the crankshaft and inner rotor.

118. R&R LOWER FRONT COVER. Disconnect battery ground cable, shut off fuel and drain cooling system. Remove muffler and air cleaner cap, then proceed as follows:

On models 435D, loosen fasteners and remove hood and grille.

On models 440ID, remove grille screen, then unbolt and remove hood. Unbolt and remove side rails, leaving grille supports attached to side rails. Disconnect headlight wires from connector near generator and pull wires from hole in right baffle. Attach hoist to grille, then remove cap screws and dowels which retain grille to grille support and remove grille. If dowels are stuck, use a cap screw and a hammer type puller to remove same.

On all models equipped with power steering, disconnect lines from power steering cylinder and steering valve. Be sure front wheels are in the straight-ahead position, mark the location of the steering gear worm shaft in relation to the steering gear housing, then unbolt the steering gear assembly from steering gear support and remove steering gear and front steering shaft.

On models without power steering, remove roll pin from coupling at aft end of steering shaft, loosen socket head screw at front end of steering shaft, then drive steering shaft rearward. Unbolt steering gear housing from steering gear support, then rotate steering gear clockwise until steering gear worm shaft will allow radiator to clear.

On all models, disconnect upper and lower radiator hoses from radiator and unbolt fan shroud. Unbolt radiator from lower supports and upper radiator supports from engine, then lift radiator from tractor.

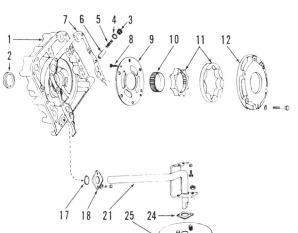

Fig. 147—Exploded view of the crankshaft lower front cover and oil pump assembly.

1. Front cover
2. Front oil seal
3. Plug
4. Gasket
5. Spring
6. Pressure regulator valve
8. Drive screw
9. Pump cover plate
10. Drive gear
11. Rotor assembly
12. Pump body
17. "O" ring
21. Inlet tube
25. Screen assembly

Fig. 148—Lower front cover being installed. Notice pilot studs and oil seal expander.

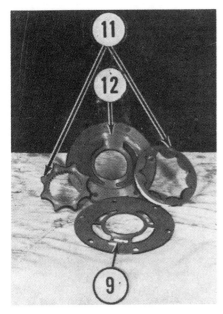

Fig. 149—View of the component parts of the oil pump assembly. Refer to Fig. 147 for legend.

Support tractor under clutch housing with a floor jack, attach hoist to the front support and axle assembly, then unbolt front support from engine and pull support forward until dowels clear. Raise engine until crankshaft pulley clears the front support, then move front support and axle assembly away from tractor.

On all models, remove fan belt and crankshaft pulley, which on models with front mounted hydraulic pump, requires removal of hydraulic pump drive coupling. Note: The very early crankshaft pulleys were fitted with two tapered cones, whereas later engines were fitted with only one tapered cone. Beginning with engine serial number 2D-3783, both tapered cones were eliminated and the pulley now incorporates a press fit which requires the use of pulley installer set GM J-7773, or its equivalent, to service pulley. Drain and remove oil pan. Unbolt and remove oil pump intake pipe and screen assembly, then unbolt and remove the lower front cover from engine. See Fig. 148.

Reinstall by reversing the above procedure.

119. **PUMP.** The oil pump contained in the engine lower front cover can be disassembled after removing the lower cover as outlined in paragraph 118. Remove the six cap screws which retain the pump body in the lower front cover and lift out pump. Drift out the two drive screws (8—Fig. 147) which hold cover plate (9) to pump body (12) and lift out the inner and outer rotors (11). See Fig. 149.

Clean all parts in a suitable solvent. Inspect lobes and faces of rotors for scratches and/or scoring. Inspect surfaces of pump body for scoring. Small scratches or scoring marks can be removed with an emery stone. Renew parts if heavily scored. Inspect splines of inner rotor and drive gear on crankshaft and renew damaged parts. Drive gear can be removed from crankshaft using a suitable puller. When reinstalling gear, use tool J7558-1 and spacer J7558-2, or equivalent as shown in Fig. 150. Gear is correctly positioned when inner face is 0.440 from front face of main bearing journal.

When reassembling pump, be sure that the marking "UP RH" is on top side when installing pump body. If

Fig. 150—Install oil pump drive gear as shown. Gear is correctly installed when inner face measures 0.440 from front face of main bearing journal.

crankshaft front oil seal is renewed at this time, coat outer edge with a non-hardening sealing compound and install seal with lip facing inward. Center inner rotor as near as possible prior to installing the lower front cover to engine. Use new seal for inlet pipe flange.

120. **OIL PRESSURE REGULATOR.** An oil pressure of 40-50 psi is maintained within the engine at all speeds, regardless of oil temperature, by means of a regulator (6—Fig. 147) located in the engine lower front cover. Regulator valve can be removed after plug, gasket and spring are removed.

Wash parts and inspect. Regulator valve must move freely in its bore. Measure the regulator spring and replace the early 2 1/64 and 2 1/16 inch springs with the latest 2 1/4 inch spring.

121. **OIL PAN.** The oil pan can be removed in the conventional manner.

DIESEL ENGINE BLOWER

The Roots-type blower, Fig. 151, mounted on right side of engine, supplies the filtered air needed for combustion and scavenging. Two hollow, double lobe rotors rotate in a hollow housing. The blower rotors are pinned to steel shafts which are carried by aluminum end plates. Helical gears located on the splined ends of the rotor shafts provide the proper spacing and timing of the rotor lobes. Refer to Fig. 152 for an exploded view of blower.

122. **REMOVE AND REINSTALL.** Loosen air intake tube clamps and remove tube. Cap intake of blower housing to prevent entry of foreign objects. Remove the six through bolts

Fig. 151 — Blower is bolted to engine end plate on right side of engine.

which run lengthwise through blower, special washers (32—Fig. 152) and the reinforcement plates (30). Note the position of the two shorter (8-inch) bolts. Remove the front end plate cover (29) and gasket. On early models, disconnect hose from oil pipe at top of blower rear end plate. Remove the four blower to cylinder block bolts and lift blower from engine.

When reinstalling blower to engine, use a new gasket between blower and cylinder block and use a non-hardening gasket cement on block side of gasket only. Position end plate cover (29) and gasket (28) on forward end of blower, then install the reinforcements (30), special washers (32) and the six long through bolts (31). NOTE: While gaskets (1 & 28—Fig. 152) appear to be the same, the thickness of rear gasket (1) is 0.026-0.036 while that of the front gasket is 0.054-0.070. DO NOT intermix the two gaskets as oil leaks could result. Be sure the two shorter bolts (33) are installed in the same holes from which they were removed. Position a new blower to end plate gasket and use a non-hardening gasket cement on engine end plate side of gasket only. Mount blower to engine and tighten the through

bolts finger tight. Install the blower to cylinder block bolts and torque to 10-15 Ft.-Lbs. Start with the two center through bolts and torque all through bolts to 20-25 Ft.-Lbs., then retorque the blower to cylinder block bolts to 46-50 Ft.-Lbs. Backlash between upper rotor gear and camshaft should be 0.003-0.005, with a maximum backlash of 0.007.

123. **OVERHAUL.** To overhaul blower, first remove blower from engine as outlined in paragraph 122, then proceed as follows: Wedge a clean cloth between rotors of blower and remove the cap screws and washers which retain the blower gears to their shafts. Use two pullers and remove both gears from shafts at the same time. Note that upper gear has two counterbores 180 degrees apart as shown in Fig. 153. Remove shims (7—Fig. 152) and spacers (8) and keep same with their mating gears. Remove cap screws from thrust plate (26) and lift off thrust plate and the three spacers (22). Unbolt and remove the thrust washers (23). Set blower on wooden blocks with aft end down and remove the two screws which retain forward end plate (19) to blower housing, then use a plastic hammer

Fig. 153—View showing gear end of early model blower. Note counter bores in upper rotor gear. Oil by-pass pipe has been eliminated from late type blower.

and tap forward end plate from blower housing. Use caution not to damage the mating surfaces. Remove blower rotors from housing. Remove seal washers (11), seals (12), seal retainers (13) and springs (14) from each end of rotor shafts. Invert housing on blocks, and remove the rear end plate from blower housing in the same manner as the front end plate.

124. Thoroughly clean and dry all parts. Inspect rotors, end plates and blower housing for nicks, burrs and signs of scoring. Small defects can be removed with an oil stone. Inspect bearing surfaces of end plates and rotor shafts for undue wear and/or scoring and renew parts as necessary. Rotors are supplied in sets only. Check blower gears and splines on shafts for wear, chipping or other damage. While seal washers, seals, seal retainers and springs can be reused if found to be satisfactory, it is recommended that the seals and their component parts be renewed during reassembly.

125. To reassemble blower, first place the front end plate on wooden blocks with inside surface on top side as shown in Fig. 154. Place springs (14—Fig. 152), retainers (13), seals (12) and seal washers (11) into cavities in forward ends of rotors, then install rotors in front end plate as shown in Fig. 154. NOTE: Lubricate seals when installing and if the original seals are being reused, be sure to assemble them on the same shafts from which they were removed and with the same side facing the blower end plates. Place blower housing over rotors, then install the seals and their component parts in cavities in aft end of rotors. Be sure tangs of seal

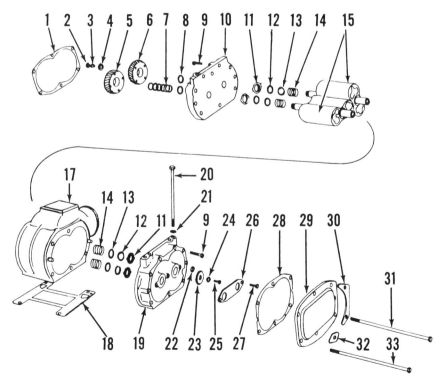

Fig. 152—Exploded view of the Roots type blower used on the GM 2-53 series engine. Rotors (15) are available in sets only.

5. Rotor gear (L. H. Helix)	8. Spacer	15. Rotors	29. End plate cover
6. Rotor gear (R. H. Helix)	10. Rear end plate	17. Housing	30. Reinforcement (large)
7. Shims (0.002, 0.003, 0.004, 0.005)	11. Seal washer	19. Front end plate	31. Through bolt (8¼")
	12. Seal	22. Spacer	32. Reinforcement
	13. Retainer	23. Thrust washer	33. Through bolt (8")
	14. Spring	26. Thrust plate	

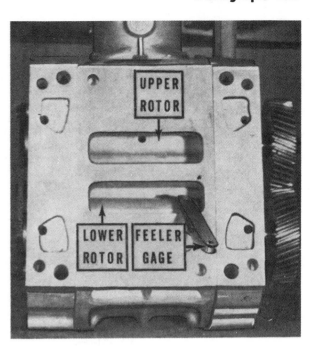

Fig. 156 — View showing method of checking clearance between rotors. Use same method to check clearances between rotors and housing.

Fig. 154 — Support front end plate on blocks when installing rotors and shafts. Be sure seal assemblies are in place.

washers are in their slots, install the rear blower end plate, then install the screws which hold both end plates to the blower housing.

Working at front end of blower, install the thrust washers (23—Fig. 152) and tighten the retaining cap screws to 25-30 Ft.-Lbs. torque. Install the three spacers (22) and thrust plate (26) and torque cap screws to 7-9 Ft.-Lbs. Clearance between thrust plate and thrust washer is 0.001-0.003. Renew thrust washers if necessary to obtain this clearance. Invert blower and position rotor shafts so the missing serrations (blind splines) are positioned 90 degrees from each other as indicated in Fig. 155. Install spacers and the original number of shims (7— Fig. 152) on ends of rotor shafts, then with the "O" marks on gears in regis-

ter, start gears on shafts. Be sure the gear having the two counterbores is on upper shaft and that counterbores in gear point toward rear end plate. Install plain washers on the cap screws used to retain gears on rotor shafts, then install cap screws. Insert clean rag between rotor lobes and pull the gears on shafts by tightening the cap screws until they bottom. Now remove the gear retaining cap screws and remove the plain washers. Install the gear washers (4) on the cap screws, then reinstall cap screws and torque same to 25-30 Ft.-Lbs.

Normal backlash between blower gears is 0.0005-0.0025 with a maximum allowable backlash of 0.0035. Renew gears if backlash exceeds 0.0035. Backlash can be measured with a dial indicator.

126. TIMING BLOWER ROTORS. After blower is assembled and prior to installation to engine, the blower rotors may require timing to insure the proper operating clearances. Clearance between rotor lobes and blower housing and end plates should also be checked.

Rotors are positioned in relation to each other by adding or removing shims from behind either of the gears. Adding shims behind upper gear moves the gear outward and the upper rotor will be moved counter-clockwise. Adding shims behind lower gear will move the lower rotor in a clockwise direction. Shims are available in thicknesses of 0.002, 0.003, 0.004 and 0.005.

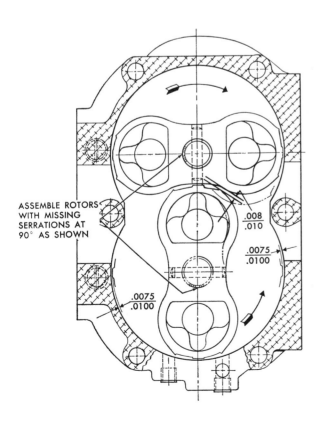

ASSEMBLE ROTORS WITH MISSING SERRATIONS AT 90° AS SHOWN

.008
.010

.0075
.0100

.0075
.0100

Fig. 155 — When assembling blower, position rotors so missing serrations are 90 degrees apart as shown. Note the clearances between rotors and between rotors and housing. See Fig. 156.

Check rotor timing as follows: Refer to Figs. 155 and 156, rotate gears until rotor lobes are at their closest relative position and measure the clearance between the lobes. This clearance should be 0.008-0.010 at any point along the length of the lobes. If clearances between lobes is not between the 0.008-0.010 limits, determine which rotor is to be repositioned, then add or subtract shims behind rotor gear; however, bear in mind that both gears should be removed together as outlined in paragraph 12.3

127. To check clearance between rotor lobes and blower housing turn the rotors until clearance between rotor lobes and housing can be measured as shown in Fig. 155. Clearance at any point along rotor lobe should be 0.0075-0.0100. Measure all four lobes and if clearance is not between 0.0075-0.010, renew end plates and/or rotors.

128. Measure clearance between ends of rotors and the blower end plates. Clearance between rear (gear) end plate and end of rotors should be 0.006-0.012. If clearances are not as end plate and end of rotors should be 0.006-0.012. If clearance are not as specified, renew thrust washers (23—Fig. 152) and/or the front end plate.

Install the front end plate cover to blower and the blower to the engine as outlined in paragraph 122.

DIESEL FUEL SYSTEM

The main components of the diesel fuel system are: A gear type fuel pump, mounted on the gear train cover and driven by the upper blower shaft gear; fuel filter; unit type fuel injectors and an injector rack control tube which transfers the governor action to the injector racks.

A restrictor (orifice) is fitted in the cylinder head fuel return passage outlet to maintain pressure in the fuel system.

TROUBLE SHOOTING

The following data should be helpful in locating trouble on the GM 2-53 series engine.

129. **HARD STARTING.** Cranking speed, engine compression and engine air supply are okay.

Check for no fuel which could be caused by air leaks, faulty pump, or obstruction in fuel line. Also could be caused by injector racks not in full fuel position.

130. **NO FUEL OR INSUFFICIENT FUEL.** Check for air leaks in fuel system, fuel flow obstruction, faulty fuel pump or missing restriction fitting located in fuel return line. Check fuel flow as outlined in paragraph 185.

131. **UNEVEN RUNNING OR FREQUENT STALLING.** Check for below normal coolant temperature, no fuel or insufficient fuel, improper injector timing, incorrect injector rack control setting, leaking injector spray tips, faulty governor adjustments or binding in injector rack control tube.

132. **DETONATION.** The trouble can be caused either by oil in the inlet supply of air for the engine, by low engine coolant temperature or by faulty injection. Oil in the engine air supply can be traced either to restricted air box drains, to a defective gasket between blower and engine, to defective blower oil seals, or to excessive oil supply in air cleaner oil cups. Check injectors for incorrect timing, for leaking check valve, for enlarged spray tip holes or for a damaged spray tip.

133. **LACK OF POWER.** Check governor, injector rack control, injector timing and exhaust valve clearance for correct adjustment. Also check for insufficient fuel or insufficient air supply.

134. **EXHAUST SMOKE ANALYSIS.** Black or gray color of the exhaust smoke indicates incomplete burning of fuel. Check for insufficient air supply, excess fuel or irregular fuel distribution, lugging engine and incorrect grade of fuel.

Blue color of the exhaust smoke indicates fuel or lube oil not being burned in the cylinder; that is, fuel or lube oil is being blown through the cylinder during the scavenging period. Check for internal fuel or lube oil leaks and for excessive oil supply in air cleaner cups.

White color of the exhaust smoke usually indicates a mis-firing cylinder. Check for faulty injectors, low compression and low cetane fuel.

INJECTORS

The 35 cubic MM injectors (Fig. 161) used in the series 2-53 General Motors engine are of the unit type. The unit type injector combines in a single unit all of the parts necessary to meter, pressurize and inject fuel to the cylinders. Excess fuel is by-passed through the injector body and serves as a coolant to the injector as well as a means of eliminating air pockets.

142. **LOCATING A MISFIRING CYLINDER.** If one engine cylinder is misfiring, it is reasonable to suspect a faulty injector. Generally, a faulty injector can be located by making each injector inoperative. As in checking spark plugs in a spark ignition engine, the faulty injector is one which, when it is made inoperative, least affects the running of the engine.

With engine running at idle speed, make the injectors inoperative, one at a time, by holding down the injector follower with a screw driver as shown in Fig. 162.

A misfiring cylinder because of dirt under the spray tip valve may be corrected by flushing the injector as follows: With engine operating, set throttle in idle position. Loosen, approximately six turns, both adjusting screws on the rack control lever of the injector to be flushed. Move the injector rack control lever into the full fuel position. This will cause a maximum amount of fuel to be forced through the injector. During this time the engine will detonate, but do not allow detonation to continue for more

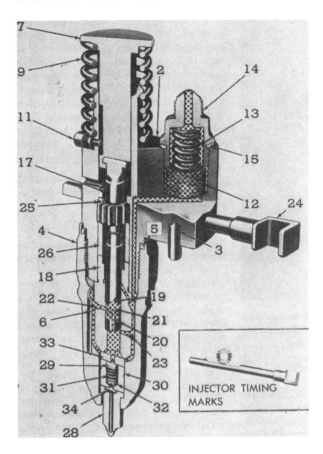

Fig. 161 — View showing details of the fuel injector assembly.

2. Injector body
3. Dowel pin
4. Injector nut
5. Seal ring
6. Spill deflector
7. Follower
9. Plunger spring
11. Stop pin
12. Filter element
13. Filter spring
14. Filter cap
17. Plunger
18. Plunger bushing
19. Upper helix
20. Lower helix
21. Meter recess
22. Bushing upper port
23. Bushing lower port
24. Control rack
25. Gear
26. Gear retainer
28. Spray tip
29. Valve
30. Valve cage
31. Valve spring
32. Valve stop
33. Valve seat
34. Check valve

INJECTOR TIMING MARKS

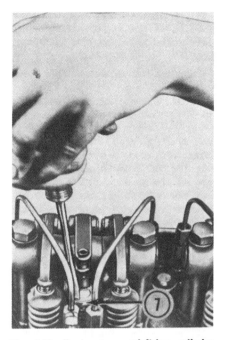

Fig. 162—To locate a misfiring cylinder, make the injector inoperative by holding down the injector follower (7) with a screw driver.

than a few seconds. If detonation does not occur, continued flushing of the injector will not correct a misfiring cylinder.

Remove and test the condemned injector as outlined in paragraphs 144 through 152.

143. **TIMING INJECTORS.** This adjustment synchronizes the no fuel position of each injector by positioning the top surface of each injector follower a certain distance above the machined face of the injector body. This measurement is 1.484 and is made with timing gage J1242.

To time the injectors proceed as follows: Pull fuel shut-off control to the no fuel (off) position. Rotate engine crankshaft until both exhaust valves are opened fully. Position injector timing gage so that small end of gage enters the drilled hole located in top surface of injector body. Loosen push rod lock nut and adjust injector rocker arm by lengthening or shortening rocker arm push rod until bottom face of timing gage head will just pass over the top of injector follower. Tighten push rod lock nut and recheck the timing dimension. Refer to Fig. 105.

Check and adjust the timing of the other injector in a similar manner.

144. **R&R INJECTORS.** To remove one or both injectors, proceed as follows: First remove tractor hood and engine rocker arms cover. Remove fuel lines from both the injector and fuel connectors. Cover each fuel fitting with a shipping cap to prevent dirt from entering the injector and fuel lines.

Rotate engine crankshaft so that the clevis pins of the three rocker arms for one cylinder are on the same plane. Remove rocker arms shaft support bolts, then swing rocker arm assembly over and away from the valves and injector.

Remove injector hold-down clamp bolt and washer. Using GM tool J1227-01 or equivalent placed under the injector body, free the injector from its seat. Lift injector from its seat while disengaging the injector rack from the control lever.

After installing the injector, torque tighten the injector clamp bolt to 20-25 Ft.-Lb. Tighten the rocker arm support bolts to 50-55 Ft.-Lbs. torque and the fuel line connections to 12-15 Ft.-Lbs. torque.

Time injectors as outlined in preceding paragraph 143, and adjust injector rack control tube levers as outlined in paragraph 80.

145. **INJECTOR BENCH TEST.** The following bench tests can be made on injector units providing the shop is equipped with an injector testing and popping fixture (J7509) and an injector comparator (J7041), or their equivalents. An injector which fails to pass one or more of the following tests should be disassembled and overhauled, however, all tests should be completed before disassembling the injector to correct any one condition.

CAUTION: The extreme pressure of the injector spray can cause the fuel to penetrate the flesh which could result in blood poisoning. Avoid this source of danger by directing the spray away from your person when testing injectors.

146. **INJECTOR RACK AND PLUNGER FREENESS.** To check whether plunger works freely in its bushing and whether the rack moves freely back and forth, proceed as follows: Place injector follower against bench, then slowly depress follower to the bottom of its stroke and at the same time move the rack back and forth through its range of travel.

Failure of the rack to have freedom of travel, or sticking of the plunger indicates dirty or damaged internal parts. Disassemble and clean, or overhaul injector as outlined in paragraph 153.

147. VALVE OPENING PRESSURE TEST. This test is performed to determine the valve opening (popping) pressure.

Place injector in tester J7509 or its equivalent and connect line to injector inlet. Operate tester to purge all air from test fixture, then connect outlet line. Place injector rack in the "full fuel" position, then operate the tester with smooth even strokes and note the valve opening (pop) pressure as the injector starts to spray. This pressure should be 450-850 psi. If pressure is not within these limits, disassemble, clean and/or overhaul injector as outlined in paragraph 153.

148. VALVE HOLDING PRESSURE. The valve holding pressure test will determine whether the lapped sealing surfaces within the injector are sealing properly.

With injector mounted in tester J7509, or equivalent, operate tester until pressure rises to a point just below the injector popping pressure (approximately 450 psi.), then close the shut-off valve and observe the time it takes for the pressure to drop from approximately 450 psi to 250 psi. This time must be NOT LESS than 40 seconds. If the injector pressure drops from 450 psi to 250 psi in less than 40 seconds check injector as follows prior to disassembly: Completely dry injector with compressed air, then open the fixture shut-off valve and operate test fixture to maintain approximately 450 psi.

Check for a leak at the injector rack opening and if this occurs, a poor bushing-to-body fit is indicated.

Check for leaks around the spray tip or seal ring which indicates a loose injector nut, a damaged seal ring, or a damaged (brinnelled) surface on the injector nut or spray tip.

Check for a leak at the filter cap which indicates a loose cap or a damaged gasket.

Check for a "dribble" at the spray tip orifices which indicates a leaking valve assembly due to dirt or damaged surfaces. An injector that "dribbles" will cause pre-ignition in the engine. NOTE: A drop or two of fuel appearing at the spray tip is an indication of fuel being trapped at the beginning of the test and can be disregarded if the time required for the pressure drop from 450 psi to 250 psi is not less than 40 seconds.

149. HIGH PRESSURE TEST. The high pressure test is performed to discover any fuel leaks at the filter caps,

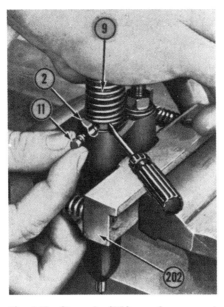

Fig. 163—Stop pin (11) can be removed by depressing follower and holding spring up with a screw driver.

body plugs, seal ring and internal lapped surfaces which may not have appeared during the valve holding pressure test. It also indicates whether the plunger and bushing clearances are satisfactory.

Completely dry injector with compressed air and check all injector and test fixture connections for leaks. Place injector rack in "full-fuel" position and lock the popping handle of tester in position with handle lock. Operate tester to build-up and maintain pressure, then use the adjusting screw in popping handle to depress the injector plunger to a point where both ports in the injector bushing are closed. This point is easily determined by the fact that the injector spray will decrease appreciably and a pressure rise will occur.

At this time, the condition of the plunger and bushing can be determined. If clearance between plunger and bushing is excessive, it will not be possible to raise the pressure above the normal valve opening (popping) pressure and renewal of plunger and bushing, as outlined in paragraph 153, will be required.

If the plunger and bushing assembly will hold pressure, continue to operate tester and maintain a pressure of 1400-2000 psi and inspect injector for leaks at the filter caps, body plugs, nut seal ring area and injector rack hole. If disassembly, cleaning and/or overhaul are indicated, refer to paragraph 153.

150. SPRAY PATTERN. With the injector rack in the "full-fuel" position, operate tester to maintain pressure at a point just below the valve opening (popping) pressure, then pop the injector several times and observe the spray as it emerges from the spray tip. An equal amount of fuel should be discharged from each orifice and spray pattern should be uniform.

If above conditions are not met, clean orifices during overhaul as outlined in paragraph 153.

151. VISUAL INSPECTION OF PLUNGER. An injector which has passed all of the preceding tests outlined in paragraphs 146 through 150 should, in addition, have the plunger visually checked under a magnifying glass. There are small areas on the lower helix that may be chipped and still not be indicated in the above tests.

To remove the plunger, mount injector in holding fixture as shown in Fig. 163. Compress the follower spring, insert a screw driver between lower end of spring and injector body then, raise spring and remove stop pin. Release follower slowly to prevent parts from flying. Remove injector from holding fixture, and in order to prevent entry of dirt, invert injector and catch follower, spring and plunger in hand.

Carefully inspect plunger and if found to be chipped, renew plunger and bushing assembly as outlined in paragraph 153.

152. FUEL OUTPUT TEST. When making the fuel output test, set the comparator J7041 to 1200 strokes per minute, then using adaptor J7041-88, seal the injector firmly in place. Start comparator and place injector rack to the "full-fuel" position. Allow comparator to run for at least 30 seconds to purge the system of any air that might be present, then push the fuel flow start button on comparator. The fuel flow will automatically stop after the pre-determined (1200) number of strokes. When the full flow has stopped, move the injector rack to the "no-fuel" position, turn the comparator off, then observe the reading on the vial. The volume of fuel should register between a minimum of zero and a maximum of 6 in the vial of the comparator. If the volume of fuel in the vial does not meet the above specification, either install a new injector or overhaul the existing injector as outlined in paragraph 153.

NOTE: The fuel output test can be used to select a set of injectors which will inject the same amount of fuel and thus result in a smooth running engine.

Should a comparator not be available, and the injector is checked with other equipment, bear in mind that the unit should inject 35 cubic millimeters of fuel per stroke with control rack in wide open position.

153. **OVERHAUL INJECTOR.** With injector removed as outlined in paragraph 144, place same in fixture J6868, or equivalent, and remove filter caps (14—Fig. 164), gaskets, springs and the filter elements. Discard the filter elements. Compress the follower spring, insert a screw driver between lower end of spring and injector body, then raise spring and remove stop pin as shown in Fig. 163. Release spring slowly to prevent parts from flying, then remove plunger follower, spring and plunger as an assembly. Invert injector in fixture and remove the injector nut (4—Fig. 165). Use caution when removing nut not to spill out the valve parts and spray tip. Remove

valve assembly parts, spill deflector (6) and seal ring (5) from nut. NOTE: If injector has been in use for sometime, the spray tip may be stuck in the nut, in which case, support nut on a wood block and drive the spray tip out as shown in Fig. 166 by using tool J1291-01 or a hollow brass tube of the proper size. Remove plunger bushing (18—Fig. 165), gear retainer (26) and gear (25) from injector body, then withdraw the injector control rack (24).

Since most injector troubles can be traced to dirt and carbon, it is essential that a

clean working area be provided. It is particularly important to have a clean area on which to place the injector parts after they have been cleaned, inspected and/or lapped.

Wash and clean all parts in clean fuel oil or other suitable cleaning solvent as follows: NOTE: DO NOT use waste or rags to clean injector parts. Use reamer J1243, or equivalent, and clean inner chamber of spray tip as shown in Fig. 167. Turn reamer in a clockwise direction only. Clean orifices of spray tip using a pin vise fitted with 0.004 diameter wire (probe) as shown in Fig. 168, however, hone tip end of wire before using to free it of any burrs. Be sure that the orifices are not enlarged. Wash spray tip and dry with compressed air.

Clean carbon from spray tip seat in injector nut by installing reamer

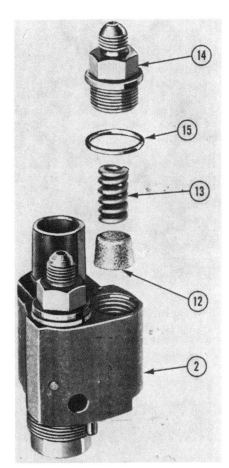

Fig. 164 — Details of injector filter, spring and cap.

2. Injector body
12. Filter
13. Spring
14. Filter cap
15. Gasket

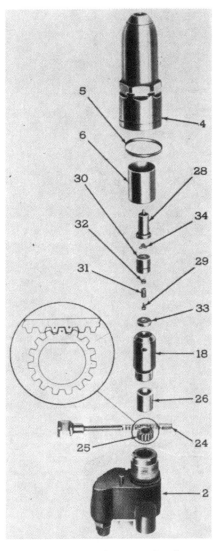

Fig. 165—View showing control rack, gear, spray tip, valve assembly and component parts in their relative positions. Note timing marks on control rack and gear.

2. Injector body
4. Nut
5. Ring seal
6. Spill deflector
18. Plunger bushing
24. Control rack
25. Gear
26. Gear retainer
28. Spray tip
29. Valve
30. Valve cage
31. Valve spring
32. Valve stop
33. Valve seat
34. Check valve

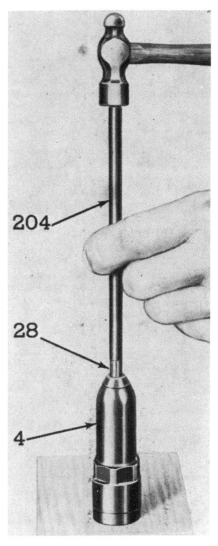

Fig. 166—If spray tip (28) is stuck in nut (4), drive tip out using tool J1291-01, or a hollow brass tube of proper diameter.

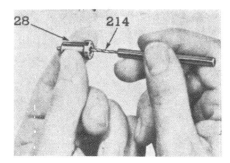

Fig. 167—Ream inner chamber of spray tip using tool J1243, or its equivalent.

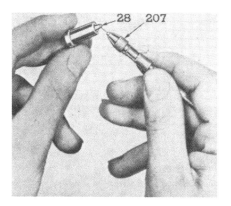

Fig. 168—Spray tip orifices can be cleaned by using 0.004 wire probes held in a pin vise.

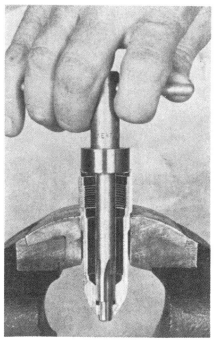

Fig. 169—Clean carbon from spray tip seat in injector nut with tool J4986-1, or equivalent.

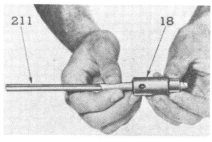

Fig. 170—Clean bore of plunger bushing as shown by wrapping tissue paper around tool J1291-01, or its equivalent.

J4986-1, or equivalent, in the nut and turning reamer in a clockwise direction as shown in Fig. 169. Use caution during this operation not to remove any metal or raise any burrs on the spray tip seat. Wash injector nut and dry with compressed air.

When handling the injector plunger, do not touch the polished surfaces with the fingers. Final cleaning of plunger bore in bushing may be accomplished by wrapping tissue paper around tool J1291-01, or its equivalent, and cleaning the bore as shown in Fig. 170. Keep plunger and bushing together as they are mated parts.

Use hole cleaning brushes and clean all passages in injector body. Blow out and dry passages with compressed air.

After washing and cleaning, submerge parts in a receptacle of clean fuel oil.

154. With parts washed and cleaned as outlined above, inspect, recondition and/or renew parts as follows:

Inspect the teeth on injector control rack and control rack gear. Inspect bore of gear. Renew parts which show signs of excessive wear, scoring or other damage.

Inspect plunger for scoring, erosion and chipping or wear at helix. Also check the portion of the plunger which slides in control gear and if any rough edges are found, remove same with a 500 grit stone. Inspect plunger bushing for scoring, chipping and undue wear. Place plunger in bushing and check for freedom of

movement. If any damage was found which renders either part unserviceable, both parts must be renewed as they are available only as mated parts.

Inspect plunger spring for fractures and distortion. The spring has a free length of 1.659 inches and should require 53-59 (56 desired) pounds to compress it to a length of 1.028 inches. Renew spring when less than 48 pounds will compress same to the 1.028 inches length.

Inspect the injector valve spring for fractures and distortion. Spring should require 4¾-5¾ pounds to compress it to a length of 0.240 inch. Renew spring when less than 4¼ pounds will compress same to the 0.240 inch length.

Inspect both ends of spill deflector for burrs and sharp edges which could mar or peel metal from injector body or nut during assembly. Defects can be removed from deflector by using a medium stone.

Inspect the spray tip seating surface in the injector nut for nicks, burrs or other damage. Reseat surface if not badly damaged, however, if severely damaged, renew injector nut.

Refer to Fig. 171 for sealing surfaces of parts which may require lapping. Use a magnifying glass and carefully inspect the parts for burrs, nicks, erosion, cracks, chipping and excessive wear. Renew those parts found to be unserviceable and if lapping is required on the parts found to be serviceable, proceed as follows for all parts except the valve seat (33—Fig. 165). Clean lapping blocks using compressed air only. DO NOT use cloth for this purpose. Use a 600 grit dry lapping powder on one of the lapping blocks. place part to be lapped on block and using only enough pressure to hold part flat, move part in a figure eight pattern. After making four or five passes, clean part by drawing it across clean tissue placed on a flat surface and inspect. Repeat this operation until part is lapped flat. Wash part and dry with compressed air. Ap-

2. Injector body
18. Plunger bushing
28. Spray tip
29. Valve
30. Valve cage
33. Valve seat
34. Check valve

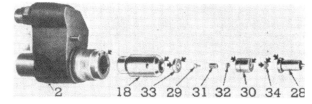

Fig. 171—Heavy arrows indicate the injector sealing surfaces which may require lapping.

ply lapping powder to second lapping block, apply dry part and make several passes using the figure eight pattern. This will impart a smooth finish to part. Again wash part and dry with compressed air. Place dry part on third lapping block and DO NOT use the lapping powder. Hold part flat and make several passes using the figure eight pattern. This will produce the "mirror" finish required for perfect sealing.

Only the edge of the hole in the valve seat contacts the valve, this edge must be a perfect circle and present an unbroken surface. Examine edge under a magnifying glass and if hole shows small irregularities, lap same as follows: Install tool J7174, or equivalent, in an electric drill mounted in a vise as shown in Fig. 172. Place a small amount of lapping powder and oil mixture on the tool. Place valve seat over pilot of tool, start drill and while holding valve seat with fingers, lightly touch the valve seat against the rotating tool. After lapping the valve seat in this manner, flat lap the face on a lapping block, then clean and inspect the width of the edge (seat). Width of seat should be 0.002-0.005. If width of seat exceeds 0.005, renew valve seat as a seat width of more than 0.005 will lower the pop pressure of the injector.

155. Reassemble the injector as follows: Position injector body right side up; install new filter elements. Make sure that dimple in filter elements is on bottom side. Place a spring above each filter, use new gaskets and install filter caps. Torque filter caps to 65-75 Ft.-Lbs. and install shipping caps to prevent entry of foreign material. Invert injector body and slide control rack into body until the two marked teeth can be seen when looking into the injector bore. Hold rack in this position and install control gear with marked tooth engaged with the space between the two marks on the control rack as shown in Fig. 165. Place gear retainer (26) on top of gear, then align the locating pin of plunger bushing with slot in injector body and install plunger bushing (18). Place ring seal (5) on shoulder of injector body, slide spill deflector (6) over plunger bushing, then place valve seat (33) on end of bushing with seat side up. Insert stem of valve (29) through spring (31) and position valve stop (32) at opposite end of spring. Place valve cage (30) over valve assembly, then position the

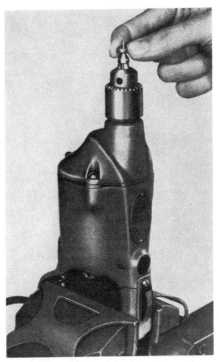

Fig. 172 — Use tool J7174, or equivalent, to lap the edge of hole in valve seat. Seat width must not exceed 0.005.

Fig. 173—Checking concentricty of injector spray tip to injector nut with tool J5119, or its equivalent.

valve and cage assembly on valve seat. Position check valve (34) on end of valve cage, then place the spray tip over the check valve and against cage. Lubricate threads of nut (4)

and pilot the nut over shoulder of spray tip. Tighten the nut only hand tight at this time. Invert the injector (filter caps up) and push the control rack all the way in. Slide head of plunger into follower, then insert this assembly through plunger spring. Start the stop pin in the injector body so that bottom coil of plunger spring is resting on the narrow flange of stop pin. With the slot in the follower aligned with stop pin hole and the flat side of plunger aligned to engage flat of control gear, push down on top of follower and at the same time, push inward on stop pin. The stop pin will slide into position when the slot in follower aligns with stop pin hole in injector body. Plunger spring will hold the stop pin in position. Invert injector and torque injector nut to 55-65 Ft.-Lbs.

With injector assembly same in gage J5119, or equivalent, as shown in Fig. 173, and check the concentricity of spray tip to nut. The units must be concentric within 0.008. If the total runout exceeds 0.008, loosen nut, re-center spray tip and recheck. If, after several attempts, the spray tip cannot be satisfactorily positioned, recheck the complete injector assembly.

Before installing injector in tractor, test same as outlined in paragraphs 146 through 152. In addition, all carbon deposits must be removed from the beveled seat of the injector tube in cylinder head. Use reamer J5286-9, or its equivalent, and make certain that no metal is removed during the reaming operation. Use grease in flutes of reamer to retain the removed carbon.

Install injector in engine as outlined in paragraph 144.

NOTE: In order to preclude any possibility of scoring the internal parts of the injector at installation due to lack of lubrication, any trapped air should be bled from the injector. Do this by placing the control rack in "no-fuel" position and cranking the engine briefly prior to tightening the fuel connections.

INJECTOR HOLE TUBES

156. To insure efficient injector cooling, each injector is inserted into a thin-walled copper tube which passes through the water jacket in the cylinder head. The copper tube is flared-over at the lower end and sealed at top with a neoprene ring.

To renew the tubes which involves removal of the cylinder head, refer to paragraph 89.

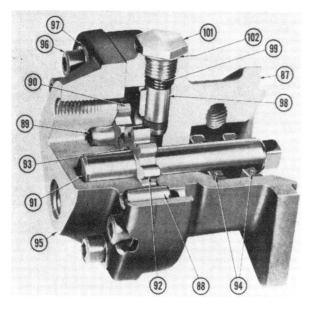

Fig. 174—Cut-away view of the primary fuel pump.

87. Pump body	93. Gear retaining ball
88. Dowel	94. Oil seals
89. Driven shaft	95. Pump cover
90. Driven gear	98. Relief valve
91. Drive shaft	99. Spring
92. Drive gear	101. Plug

Fig. 175—Exploded view of the primary fuel pump showing the relative position of component parts.

87. Pump body	98. Relief valve
90. Driven gear	99. Spring
91. Drive shaft	100. Pin
92. Drive gear	101. Plug
94. Oil seals	103. Coupling
95. Pump cover	

PRIMARY FUEL SUPPLY PUMP

The GM 2-53 series Diesel engine is equipped with a gear type primary fuel supply pump, Fig. 174, which is mounted on the gear train cover and is driven by the upper rotor shaft of the blower. This pump provides a continuous flow of fuel from the supply tank through the fuel chambers within the injectors and then, back to the tank. A restriction placed in the cylinder head fuel return passage outlet provides sufficient resistance to maintain a fuel pressure of 50-70 psi @ 1800 engine rpm throughout the fuel system.

157. SEAL LEAKAGE. A tapped hole located in lower side of pump body carries off fuel oil which passes the pump shaft seals. If leakage exceeds one drop per minute when checked at fuel pump body drain hole, the seals (94—Fig. 174) must be re-newed. Pump shaft oil seals are installed with their lip facing drive end of pump.

158. FUEL FLOW CHECK. To check fuel flow, disconnect fuel return line at supply tank. Operate engine at 1200 rpm and measure the fuel flow for a period of one minute. At least one half gallon of fuel should flow from the return line in one minute.

If fuel flow is insufficient, check for air leaks in fuel system by immersing end of fuel return line in clean fuel oil. Air bubbles indicate an air leak on the suction side of the pump. Also check for a restricted fuel line or an inoperative pressure relief valve which is located in the pump body. Pressure relief valve can be removed without removing the pump assembly.

159. REMOVE & REINSTALL. To remove fuel pump proceed as follows: Disconnect suction and pressure lines from pump. Remove three cap screws which attach pump to gear train housing and remove pump. Copper type flat washers are used with the three attaching cap screws.

Non-adjustable pump pressure relief valve can be removed without removing the pump assembly from gear train housing. Pressure relief valve opens at 65-75 psi.

160. DISASSEMBLE & OVERHAUL. Refer to Figs. 174 and 175. Disassembly of the pump is self-evident after an examination of the unit.

Pump shaft (91) rotates in unbushed bores of the pump body. Install pump driven gear (90) with slot in face of gear facing pump cover. Install the two lip type oil seals (94) with lips facing drive end of pump. The pump body and cover are assembled without a gasket.

DIESEL ENGINE GOVERNOR

A flyweight, variable speed centrifugal governor is mounted on left side of the engine. The governor is driven by a gear that extends through the engine rear end plate which meshes with the balance shaft gear of the engine gear train.

ADJUSTMENT

Before adjusting the governor be sure no binding occurs in the linkage and that the engine shut-off arm does not interfere when the speed control lever is placed in the high idle position.

161. ADJUST LOW IDLE. Start engine and allow to run until engine has reached operating temperature. Refer

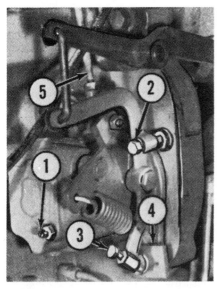

Fig. 176—View showing governor and governor linkage after governor shield has been removed.

1. Buffer screw
2. Low idle screw
3. High idle screw
4. High idle stop
5. Vertical link

1. Expansion plug
2. Washer
3. Snap ring
4. Bearing
5. Body
6. Lock nut
7. Buffer screw
8. Spring
9. Gasket
10. Rocker shaft fork
11. Thrust bearing
12. Bearing races
13. Retainer ring
14. Riser
15. Weight
16. Shaft and carrier
17. Spacer
18. Snap ring
19. Bearing
20. Mounting flange
21. Spacer
22. Drive gear
24. Weight pin
25. Clip
26. Rocker shaft
27. Oil seal
28. Seal retainer
29. Rocker shaft lever
30. Speed control lever
31. Shoulder bolt
33. Low idle screw
34. Governor spring
35. Pin
36. High idle screw

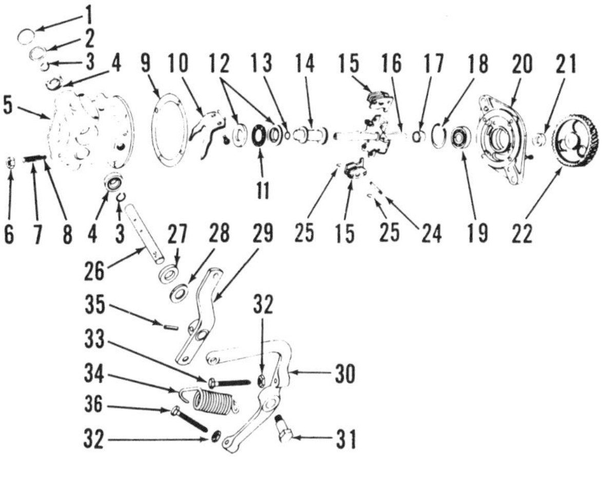

Fig. 178—Exploded view of diesel engine governor showing component parts and their relative positions.

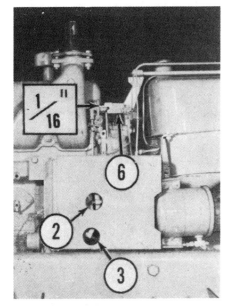

Fig. 177—Fuel shut-off control is properly positioned when there is approximately 1/16-inch clearance between control arm and the special nut on vertical link with throttle in high idle position.

2. Low idle screw
3. High idle screw
6. Shut-off control

to Fig. 176 and back-out the buffer screw (1); then, with the speed control lever in idle position, turn the idle speed adjusting screw (2) until

engine operates at 600 rpm and tighten lock nut. Now turn the buffer screw in until the engine roll (surge) is eliminated and tighten the buffer screw lock nut. NOTE: Do not raise the engine speed more than 20 rpm with the buffer screw.

162. **ADJUST HIGH IDLE.** With engine running, move the speed control lever to the high idle position and use caution not to over-speed the engine. Turn the high idle speed adjusting screw (3) until engine operates at a high idle speed of 1975 rpm. Tighten lock nut.

163. **ADJUST ENGINE SHUT-OFF CONTROL.** The engine shut-off control must be adjusted to eliminate any interference with the governor control linkage when the speed control lever is in the high idle position. To adjust the engine shut-off control, proceed as follows: Loosen screw holding control wire in the swivel, then position the cable so that end of conduit (loom) does not extend above its clip. Place the engine speed control lever in the high idle position, then adjust the engine shut-off control until there is at least 1/16-inch between the shut-off control arm and the special nut on the vertical linkage as shown in Fig. 177. Tighten screw in swivel.

R&R AND OVERHAUL

164. Remove governor shield, disconnect the linkage from speed control lever (30—Fig. 178) and the injector control linkage from rocker shaft lever (29). Unbolt and remove the governor assembly from engine. NOTE: In some cases on early models, it may be necessary to remove the air box cover plate before the governor assembly can be removed. Late production engines no longer have the air box cover plate.

To disassemble the governor, proceed as follows: Place a scribe mark on the governor body and bearing housing to facilitate reassembly, then remove the four counter-sunk screws from bearing housing and separate governor. Balance of disassembly is evident after an examination of the unit and reference to Fig. 178.

Inspect all parts and renew as necessary. When reassembling, install oil seal (27) with lip of same facing bearing (4). When pressing bearing (19) on shaft (16), press on bearing inner race only. Press gear (22) on shaft (16) until gear bottoms.

When reinstalling governor to engine, be sure the high idle stop (4—Fig. 176) is installed in a horizontal position under the lower mounting cap screw as shown.

Adjust governor as outlined in paragraphs 161 through 163.

DIESEL ENGINE COOLING SYSTEM

RADIATOR

165. To remove the radiator, remove muffler and air cleaner cap, then proceed as follows:

On models 435D, loosen fasteners and remove hood and grille.

On models 440ID, remove grille screen, then unbolt and remove hood. Unbolt and remove side rails, leaving grille supports attached to side rails. Disconnect headlight wires from connector near generator and pull wires from hole in right baffle. Attach hoist to grille, then remove cap screws and dowels which retain grille to grille support and remove grille. If dowels are stuck, use a cap screw and a hammer type puller to remove same.

On all models equipped with power steering, disconnect lines from power steering cylinder and steering valve. Be sure front wheels are in the straight-ahead position, mark the location of the steering gear worm shaft in relation to the steering gear housing, then unbolt the steering gear assembly from steering gear support and remove steering gear and front steering shaft.

On models without power steering, remove roll pin from coupling at aft end of steering shaft, loosen socket head screw at front end of steering shaft, then drive steering shaft rearward. Unbolt steering gear housing from steering gear support, then rotate steering gear clockwise until

steering gear worm shaft will allow radiator to clear.

On all models, disconnect upper and lower radiator hoses from radiator and unbolt fan shroud. Unbolt radiator from lower supports and upper radiator supports from engine, then lift radiator from tractor.

Balance of disassembly is self-evident. Check the radiator filler cap which is set to release when pressure in cooling system reaches 3½-4½ pounds.

FAN

166. The fan is bolted to a combination hub and pulley. The hub and pulley is carried on a shaft and bearing assembly which is pressed into a bracket mounted to the front of the engine.

To remove the fan and bracket assembly, first remove radiator as outlined in paragraph 165. Loosen bolts in bracket, remove fan drive belt, then remove bracket and fan assembly from engine.

To disassemble, unbolt fan from hub (6—Fig. 179), then use a suitable puller and pull hub from the shaft and bearing assembly (5). Press the shaft and bearing assembly from bracket (4), using a pusher that will apply pressure to the outer race of bearing only.

Inspect shaft and bearing. The bearing is permanently lubricated and

sealed and if renewal is required, the bearing and shaft must be renewed as a unit.

When reassembling, press the shaft and bearing assembly into bracket until outer end of bearing is flush with outer end of bearing bore in bracket. Use a pusher that will apply pressure to outer race of bearing only. Support inner end of shaft and press the hub on shaft until the measured distance from the back of the hub and the back of the bracket is 2.46 inches. Reinstall fan blades to hub, then mount the entire assembly to engine. Adjust fan drive belt to allow ½-inch deflection when measured midway between pulleys.

THERMOSTAT

167. The cooling system is equipped with a thermostat which starts to open at 170 deg. F. and is fully opened at 190 deg. F.

The thermostat is located in a housing which is bolted to the front of cylinder head. To renew the thermostat, remove the hood, drain coolant to a point below the thermostat housing, then disconnect outer flange from housing and remove thermostat.

WATER PUMP

168. The water pump is located at the lower right hand front portion of the engine and is driven by the fan belt.

169. **REMOVE AND REINSTALL.** Drain cooling system, then loosen and remove generator and fan belt. Disconnect by-pass line at top of pump, then unbolt and remove pump from mounting adaptor.

Reinstall by reversing the above procedure and adjust belt to allow a ½-inch deflection when measured midway between pulleys.

170. **OVERHAUL.** With water pump removed as outlined in paragraph 169, use a suitable puller and remove pulley (22—Fig. 179). Remove end cover (12) and gasket (13) from impeller end of pump. Place pump in a press with hub end up and using a pusher which will apply pressure to the bear-

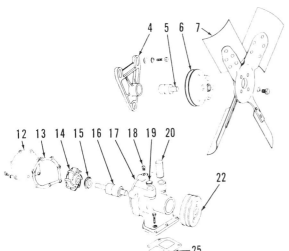

4 5 6 7

12 13 14 15 16 17 18 19 20

22

4. Bracket
5. Shaft and bearing
6. Fan hub
7. Fan
12. Pump cover
13. Gasket
14. Impeller
15. Seal
16. Shaft and bearing
17. Pump body
18. Plug
19. Cup-plug
20. By-pass connector
22. Pulley
25. Gasket

25

Fig. 179—Exploded view of the fan and water pump assemblies.

ing (16) outer race only, push the shaft, bearing and impeller (14) assembly out of housing (17). Support impeller on seal surface and press shaft out of impeller.

Wash all parts except the shaft and bearing assembly (16) in a suitable solvent and inspect. Pay particular attention to sealing surfaces of seal (15) and impeller (14). If bearing is faulty, bearing and shaft must be renewed as a unit.

Reassemble pump as follows: Place pump body (17) in a press, hub end up. Press bearing (16) and shaft into body until outer end of bearing is flush with end of housing. Use a pusher that will apply pressure to outer race of bearing only as shown in Fig. 180. Coat outside diameter of

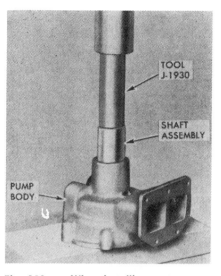

Fig. 180 — When installing water pump shaft and bearing, use a pusher which will contact only the outer race of bearing.

seal (15—Fig. 179) with sealing compound, then while applying pressure only on the outer flange of seal, press same into position until outer flange bottoms against pump body. Clean sealing surfaces of seal and impeller; then, while supporting hub end of shaft, press impeller on shaft until rear of impeller and end of shaft are flush. With impeller properly installed, there should be a clearance of 0.010-0.022 between ends of impeller vanes and pump body. Place pulley (22) on bed of press and using a suitable rod between press ram and shaft, press shaft into pulley until end of shaft is flush with front face of pulley. Install gasket (13) and cover (12).

DIESEL ELECTRICAL SYSTEM

GENERATOR

171. Delco-Remy generators are used and the specifications are as follows:
1102102

Brush spring tension—oz....	28
Field draw—volts	12
amps..........	1.48-1.62
Output cold, volts	14.0
amps..........	25
rpm	2000

1100367

Brush spring tension—oz........	28
Field draw—volts	12
amps	1.58-1.67
Output cold—volts	14
amps	20
rpm	2300

REGULATOR

172. Delco-Remy regulators are used and the test specifications are as follows:

1118955

Ground polarity	Pos.
Cut-out relay	
air gap—inches	0.020
point gap—inches	0.020
closing voltage range.....	11.8-13.5
adjust to—volts	12.6
Voltage Regulator	
air gap—inches	0.075
setting voltage range.....	13.8-14.8
adjust to—volts	14.3
Current Regulator	
air gap—inches	0.075
setting current range........	23-27
adjust to—amps.	25

1118779

Ground polarity	Pos.
Cut-out relay	
air gap—inches	0.020
point gap—inches	0.020
closing voltage range......	11.8-14.0
adjust to—volts	12.8
Voltage Regulator	
air gap—inches	0.075
setting voltage range......	13.6-14.5
adjust to—volts	14

STARTING MOTOR

173. Delco-Remy starters are used and the specifications are as follows:
1108654

Voltage	12
Brush spring tension—oz...	24
No load test	
volts	11.8
amperes	40-70
rpm	6800-9200
Lock test	
volts	5.85
amperes	615
torque—Ft.-Lbs.	29

1108656

Voltage	12
Brush spring tension—oz........	24
No load test	
volts	11.8
amps	40-70
rpm	6800-9200
Lock test	
volts	5.85
amps	615
torque—Ft.-Lbs.	29

CLUTCH

Models 435D may be equipped with either a single or a dual plate clutch assembly, whereas models 440ID are available only with the single plate clutch assembly.

Clutches can be serviced after removal by following the information given in paragraphs 62 through 65 in the first section of this shop manual. Keep in mind however; that individual repair parts are now available for the single plate clutch pressure plate assembly.

The clutch assemblies can be removed for servicing by following the information given in paragraph 174.

TRACTOR SPLIT
(Engine From Center Frame)

174. Remove hood, which on 440ID models will require removal of grille screen. On 440ID models, remove side rails. On all models, disconnect power steering lines. Disconnect wires from generator. Remove tube between blower and air cleaner and cap off blower intake. Unclip and disconnect fuel return line at union located at fuel tank front support. Shut off fuel and disconnect fuel supply line at shut-off valve. Remove fuel line between fuel primary pump and fuel filter. Remove the governor shield, disconnect wires from hour meter and oil pressure switches and pull wires from hole in oil filter bracket. Disconnect oil lines from oil filter. Disconnect governor control rod from hand control lever. Disconnect fuel shut-off control from shut-off arm. Disconnect wire from temperature sending unit, then unbolt fuel tank front support from engine. Install engine hangers, attach hoist to same, then unbolt engine from clutch housing (center frame) and roll engine and front wheels assembly away from tractor.

POWER LIFT SYSTEM

Except for hydraulic pumps, the hydraulic system of the 435D and 440ID tractors is identical to that outlined in the first section of this shop manual. For service information on hydraulic components other than pumps, refer to the first section of this shop manual.

When tractor is equipped with Touch-O-Matic only, a pump such as the one shown in Fig. JD181 is used. When tractor is equipped with Touch-O-Matic and power steering, a pump such as that shown in Fig. JD182 is used. For service information on the two types of pumps, refer to the following paragraphs.

PUMP

175. **REMOVE AND REINSTALL.** To remove either type pump, disconnect lines from pump and allow oil to drain, then remove the cap screws which retain pump bracket to engine and remove pump and bracket.

176. **OVERHAUL (TOUCH-O-MATIC ONLY).** With pump removed as outlined in paragraph 175, clean pump and place a scribe mark across front plate, pump body and backplate to facilitate reassembly. Remove pulley, then unbolt and remove mounting bracket. Refer to Fig. JD181 and remove key from drive shaft, remove through bolts, separate pump and remove gears and shafts. If pump body (5) sticks to either front plate (13) or back plate (4), loosen same by tapping with a plastic hammer. Pry diaphragm (7) from front plate and remove the two springs (11) and steel balls (12). Remove the phenolic gasket (8), protector gasket (9), molded "V" seal (10) and shaft oil seal (14) from front plate.

Clean all parts in a suitable solvent and inspect. Small nicks and burrs can be removed with emery cloth or a fine stone. Check both shafts for wear at bearing surfaces and roughness at seal areas. If shafts measure less than 0.6850 at bearing surfaces, renew shaft and gear. Note: One gear and shaft assembly may be renewed separately, however; shafts and gears are available as assemblies only. Inspect gear faces for wear and scoring. Inspect snap rings to see that they are in grooves of shafts and not worn. Inspect bearings in front and back plates and if inside diameter is more than 0.691, renew plates as bearings are not available separately. Bearings in front and back plates have the oil grooves in line with dowel pin holes and the bearings in front plate should be flush with islands in groove pattern. Check wear on face of back plate and if wear exceeds 0.0015, renew back plate. Check gear pockets of pump body for wear and scoring. If inside diameter of gear pockets exceeds 1.719, renew pump body.

When reassembling, use new diaphragm (7), phenolic gasket (8), "V" seal (10), protector gasket (9) and shaft oil seal (14). With open part of "V" seal (10) toward front plate (13),

Fig. JD181—Exploded view of the hydraulic pump used on models 435D and 440ID when equipped with Touch-O-Matic only.

4. Back plate	8. Phenolic gasket	13. Front plate
5. Body	9. Protector gasket	14. Oil seal
6. Drive gear & shaft	10. Molded "V" seal	15. Idler gear & shaft
7. Diaphragm	11. Spring	16. Snap ring
	12. Steel ball	

work same into grooves of front plate using a dull tool if necessary. Press protector gasket (9) and phenolic gasket (8) into "V" seal. Install the two steel balls (12) and springs (11), then install diaphragm (7) with bronze face on gear side. Note: The diaphragm must fit inside the raised rim of the "V" seal. Dip gear assemblies (6 and 15) in oil and install in front plate. Place a thin coat of heavy grease on both finished sides of pump body (5) and install body with the half-moon cavities facing away from rear plate. Note the small drilled hole in one of the cavities. This hole must be on pressure side of pump. Install rear plate (4) over gear shafts, install through bolts and torque same to 25 Ft.-Lbs. Lubricate shaft oil seal and install same. Use caution not to damage the rubber sealing lip. Tap into position with a plastic hammer. Check pump rotation. Pump will have a slight amount of drag after assembly but should turn freely after a short period of use.

177. OVERHAUL (POWER STEERING AND TOUCH-O-MATIC). With pump removed as outlined in paragraph 175, first remove pulley, then unbolt and remove pump mounting bracket. Remove oil seal (pilot) plate (2—Fig. JD182) and seal (4). Oil seal (3) can be renewed at this time. Remove through bolts, separate pump sections and mark gears so they can be reinstalled in the same position. Remove drive gear (22) and Woodruff key (9) from pump shaft (8), then press shaft and bearing (11) from body (17). Remove snap ring (5) and washer (6), then press bearing (11) from shaft (8). Remove bearing (18) from pump body. Remove snap ring (19), then remove driven (idler) gear (23) and Woodruff key (21) from idler shaft (20). Remove cap (30), adjusting screw (29), relief valve spring (28) and relief valve (27). Remove cap (33) and flow control valve spring (31), then remove cap (36) and flow control valve (35). Remove bearings (18) from end cover (26). Dowels (24) can be removed, if necessary.

Clean and inspect all parts carefully. Pay close attention to oil seal (3), oil seal contacting surface of shaft (8) and all bearings. Be sure all mating surfaces are free of burrs. Flow control valve (35) should fit snug yet move freely in its bore. Renew any parts which are damaged or show undue wear. When installing new oil seal (3), install same with lip facing pumping gears.

Dip all parts in clean hydraulic oil prior to installing and reassemble the pump by reversing the disassembly procedure. Adjust power steering relief valve screw (29) to provide an operating pressure of 1000 psi for the steering circuit.

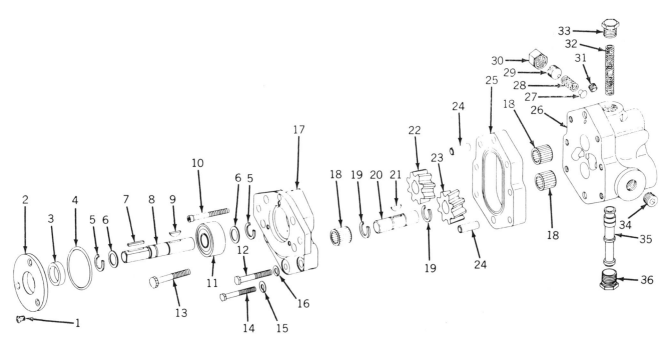

Fig. JD182 — Exploded view of the pump used when tractor is equipped with both power steering and Touch-O-Matic. Items 27, 28 and 29 control power steering operating pressure.

1. Flat head screw	7. Drive key	18. Roller bearing	24. Dowel	31. Plug
2. Oil seal (pilot) plate	8. Pump shaft	19. Snap ring	25. Gear plate	32. Flow control valve spring
3. Oil seal	9. Woodruff key	20. Idler shaft	26. Pump cover	33. Cap
4. "O" ring	10. Socket head screw	21. Woodruff key	27. Relief valve	34. Plug
5. Snap ring	11. Ball bearing	22. Drive gear	28. Relief valve spring	35. Flow control valve
6. Washer	13. Cap screw (12 pt)	23. Driven (idler) gear	29. Adjusting screw	36. Cap
	17. Pump body		30. Cap	

JOHN DEERE
(PREVIOUSLY JD-36)

Series ■ 820 (3 Cyl.) ■ 830 (3 Cyl.)

SHOP MANUAL

JOHN DEERE

820(3 Cyl.)—830(3 Cyl.)

Tractor serial number is located on right side of transmission case. Engine serial number is stamped on a plate at lower right front of engine cylinder block.

INDEX (By Starting Paragraph)

CONDENSED SERVICE DATA

GENERAL	820 Diesel	830 Diesel
Engine Make	Own	Own
Number Cylinders	3	3
Bore—MM	98	98
-Inches	3.86	3.86
Stroke—MM	110	110
-Inches	4.33	4.33
Displacement—cc	2490	2490
Cubic Inches	152	152
Compression Ratio	16.7:1	16.7:1
Battery Terminal Grounded	Neg.	Neg.
Forward Speeds	8	8

TUNE-UP	820 Diesel	830 Diesel
Firing order	1-2-3	1-2-3
Compression Pressure—PSI @ 200 RPM	300	300
Valve Clearance:		
Inlet—MM	0.35	0.35
Inch	0.014	0.014
Exhaust—MM	0.45	0.45
Inch	0.018	0.018
Timing Mark Location	Crankshaft Pulley	
Engine Low Idle—RPM	650	650
Engine High Idle—RPM	2500	2545
Working Range	1400-2100	....
PTO Horsepower		
@ 2100 RPM	29	....
@ 2400 RPM	....	35

SIZES—CAPACITIES— CLEARANCES	820 Diesel	830 Diesel
Crankshaft Journal Diameter MM	79.35	79.35
Inches	3.124	3.124
Crankpin Diameter MM	69.812	69.812
Inches	2.7485	2.7485
Piston Pin Diameter MM	30.168	30.168
Inches	1.1877	1.1877
Main Bearing Clearance MM	0.02-0.1	
Inches	0.001-0.0041	
Rod Bearing Clearance MM	0.03-0.11	
Inches	0.0012-0.0042	
Camshaft Journal Clearance MM	0.09-0.14	
Inches	0.0035-0.0055	
Crankshaft End Play MM	0.05-0.20	
Inches	0.002-0.008	
Camshaft End Play MM	0.06-0.22	
Inches	0.0025-0.0085	
Piston Skirt Clearance MM	0.07-0.13	
Inches	0.0028-0.0052	
Cooling System—Qts.	8	11
Crankcase (with filter)—Qts.	7	6
Fuel Tank—Gallons	16½	16½
Trans. & Hydraulic System—Gals.	7.9	7.9

TIGHTENING TORQUES— FT.-LBS.	820 Diesel	830 Diesel
Cylinder Head	110	110
Main Bearings	85	110
Con. Rod Bearings (Oiled)	70	70
Rocker Arm Assembly	35	35
Flywheel	85	85

FRONT SYSTEM

AXLE AND SUPPORT

1. **AXLE CENTER MEMBER.** Center axle unit (5—Fig. 1) attaches to front support by pivot bolt (6) and rear pivot pin (7). End clearance of pivot is controlled by shims (3). Five 0.015 shims are installed at factory assembly and recommended maximum end play is 0.015. Removing shims reduces the end play. Thrust washers (2) at front and rear of pivot bushing (4) are iden-tical and may be interchanged to compensate for wear when unit is removed. Rear pivot pin (7) is pressed into front support (1—Fig. 2).

On early models, steering bellcrank (13—Fig. 1) should have 0.001-0.006 clearance in the two bushings (12) pressed into center axle unit. Late models use needle roller bearings instead of bushings (12). Recommended end play of 0.010 is controlled by shims (11) which are 0.010 in thickness. Two shims are normally used, install additional shims when necessary providing snap ring (9) can still be installed.

2. **SPINDLES AND BUSHINGS.** Refer to Fig. 3 for exploded view.

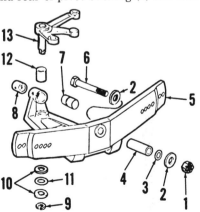

Fig. 1—Exploded view of front axle and associated parts.

1. Nut
2. Washer
3. Shim
4. Bushing
5. Axle
6. Pivot bolt
7. Pivot pin
8. Bushing
9. Snap ring
10. Washer
11. Shim
12. Bushing
13. Steering bellcrank

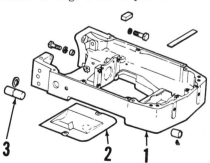

Fig. 2—View of front support casting and associated parts.

1. Front support
2. Pan
3. Rear pivot pin

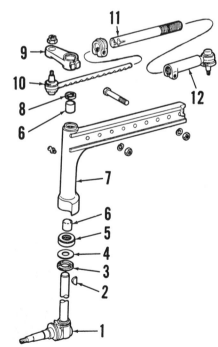

Fig. 3—Axle extension and associated parts of the type used on all models.

1. Spindle
2. Woodruff key
3. Lower seal
4. Washer
5. Thrust bearing
6. Bushing
7. Axle extension
8. Upper seal
9. Steering arm
10. Tie-rod end
11. Tube
12. Tie-rod end

3

Steering arm (9) is keyed to spindle and retained by a clamp screw. Spindle end play should be maintained at not more than 0.76 MM (0.030 inches) by repositioning steering arm on shaft. Tighten steering arm clamp screw to approximately 85 ft.-lbs. Bushings and associated parts are pre-sized.

TIE RODS AND TOE-IN

3. The recommended toe-in is 1/8-3/8 inch. Remove cap screw in outer clamp and loosen clamp screw in tie rod end (12—Fig. 3); then turn tie rod tube (11) as required. Both tie rods should be adjusted equally.

On manual steering models, stops on both spindles (1) should contact stops on axle extensions (7) at the same time. On power steering models, none of the spindle stops must contact at extreme turning position. Readjust tie rods if necessary until the correct condition is attained. Tighten tie rod clamp screws to a torque of 9.7 kg/m (70 ft.-lbs.) when adjustment is correct.

MANUAL STEERING GEAR

Steering gear is a recirculating ball nut type and the housing is mounted on top side of clutch housing. See Fig. 4 for a cross-sectional view of the manual steering gear unit.

LUBRICATION

All Manual Steering Models

4. Recommended steering gear lubricant is John Deere 303 Special Purpose Oil or Automatic Transmission Fluid, Type A. Fluid should be maintained at level of fill plug located on rear of steering column housing as shown in

Fig. 5. Drain plug is located on right hand side of transmission housing immediately below steering shaft cover.

ADJUSTMENT

All Manual Steering Models

5. Except for tie rods and toe-in (paragraph 3) no adjustment is provided. Excessive steering column shaft end and side play can be eliminated by loosening locknut (2—Fig. 6) and turning adjustor (4) until end play is eliminated and barrel bearings (7) slightly preloaded.

OVERHAUL

All Manual Steering Models

6. **REMOVE AND REINSTALL.** Either the steering wheel shaft unit or rockshaft assembly can be removed independently of the other if required for service. First drain steering gear as outlined in paragraph 4; then refer to the appropriate following paragraphs.

7. STEERING WHEEL SHAFT HOUSING. If tractor is equipped with foot throttle, disconnect pedal linkage and lift foot pedal up out of way. On all models, unbolt and remove steering shaft cover (14—Fig. 6). Refer to Inset, Fig. 7 and turn steering wheel until yoke pin is centered in opening as shown; then remove cap screw (18—Fig. 6), clip (18) and yoke pin (17). NOTE: Yoke pin has tapped hole which will accept one of the cover cap screws for pulling leverage.

Remove steering wheel using a suitable puller. Remove the cap screws securing steering shaft housing to clutch housing and dash, and lift unit from tractor.

Install by reversing the disassembly procedure. Make sure "R" mark on yoke (13) is toward cover opening when

pin is installed. Tighten lockplate screw (18) to 115 kg/cm (100 inch-pounds), steering wheel nut to 6.9 kg/m (50 ft.-lbs.) and housing cap screws to 4.8 kg/m (35 ft.-lbs.) Refill steering gear housing as outlined in paragraph 4.

8. STEERING CROSS SHAFT. The steering cross shaft (20—Fig. 6) can be removed without removing steering wheel shaft, or with wheel shaft and housing removed.

If steering wheel shaft housing is not removed, drain the unit, remove cover (14), cap screw (18) and yoke pin (17) as outlined in paragraph 7; and turn steering wheel until yoke (13) is withdrawn from shaft arm.

Refer to Fig. 7 and remove button plug from left side of clutch housing

Fig. 5–Steering gear lubricant should be maintained at level of filler plug as shown.

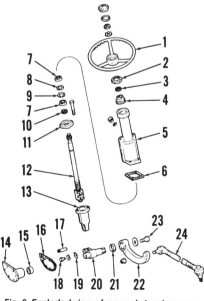

Fig. 6–Exploded view of manual steering gear.

1. Wheel	13. Yoke
2. Locknut	14. Cover
3. Oil seal	15. Bushing
4. Bearing adjuster	16. Gasket
5. Housing	17. Pin
6. Gasket	18. Cap screw
7. Bearing	19. Clip
8. Thrust washer	20. Cross shaft
9. Retainer	21. Seal
10. Oil seal	22. Steering arm
11. Retainer	23. Cap screw
12. Shaft & ball nut	24. Drag link

BARREL BEARINGS

OIL SEAL

STEERING WHEEL SHAFT HOUSING

STEERING WHEEL SHAFT WORM

BALL NUT

BALL GUIDE

YOKE

STEERING SHAFT

STEERING SHAFT ARM

STEERING DRAG LINK

Fig. 4–Cross-sectional view showing component parts of the manual steering gear.

and remove arm attaching cap screw (23—Fig. 6). Make sure cross shaft arm is still aligned with right cover opening as shown in Inset—Fig. 7 and bump steering shaft to right out of clutch housing and arm.

Install by reversing removal procedure. Tighten arm attaching cap screw (23—Fig. 6) fully, strike arm with a hammer, then retighten to 23.5 kg/m (170 ft.-lbs.) Complete the assembly by reversing disassembly procedure and fill steering gear as outlined in paragraph 4.

9. **OVERHAUL.** To disassemble the removed steering shaft housing, loosen locknut (2—Fig. 6) and remove bearing adjuster (4). Pull steering shaft upward until bearings (7) are exposed and remove upper bearing, retainer (9) and thrust washer halves (8). Lift off lower bearing (7) then withdraw steering shaft downward out of housing and oil seal (10). Yoke (13) can be removed from shaft ball nut by removing the two retaining cap screws. Do not disassemble ball nut as unit is available only as an assembly.

Install oil seal (10) if necessary, with lip toward inside and outer face flush with bottom of seal bore chamfer.

Reassemble by reversing the disassembly procedure. Tighten steering shaft bearing adjuster (4) until a pull of approximately 15 lbs. is required at end of yoke (13) to rock the shaft in barrel bearings (7).

Bushings (15) in cover (14) and clutch housing should be installed 0.030 below face of bushing bore if renewed. Normal clearance between shaft (20) and bushings is 0.13-0.23MM (0.005-0.009.)

Reinstall steering cross shaft if removed, as outlined in paragraph 8 and steering wheel shaft housing as in paragraph 7. Refill steering gear as in paragraph 4.

POWER STEERING SYSTEM

The power steering system is a factory option and consists of an open center valve type using a front-mounted tandem hydraulic pump, with one section providing power for steering only.

The power assist cylinder and control valve is built into main steering gear and uses the same mechanical steering linkage as the manual steering models.

TROUBLE SHOOTING

All Power Steering Models

10. Any problems that may develop in the power steering system usually appear as sluggish steering, power assist in one direction only, or excessive noise in power steering pump.

Sluggish steering can be caused by leaking "O" rings in valve body, improperly assembled valve body, worn pump or clogged fluid filter or lines. Power assist in one direction only is caused by a faulty or improperly adjusted power steering valve. Excessive noise is caused by a plugged filter, foaming operating fluid, malfunctioning pump and/or air leaks in intake line or shaft seal.

SYSTEM CHECKS

All Power Steering Models

To determine the overall condition of the power steering system or isolate a malfunction, make the following tests.

11. **LINKAGE TEST.** Jack up front end and, with engine stopped, turn steering wheel from lock to lock. Steering wheel should turn easily and smoothly without binding or catching at any point.

While front end is raised, turn front wheels straight ahead and start and run engine. Any tendency for front wheels to wander from the set position will indicate leakage, binding, malad-

justment or incorrect assembly of steering valve. With engine running, turn steering wheel fully in both directions. Steering effort and reaction speed should remain constant through full turn in both directions. A noticeable variation would indicate valve trouble.

12. **PUMP PRESSURE AND FLOW.** To check power steering pump pressure and flow, disconnect power steering pump pressure line (F—Fig. 8) from elbow in pump manifold and tee a flow meter in the line. Open control valve of flow meter, start and run engine at 2400 rpm. Slowly close flow meter valve until pressure rises to 1000 psi, then check pump flow which should be 4½-4¾ gpm. Continue to close flow meter valve until gage pressure reads 2000 psi. Pump flow should not drop more than ½ gallon per minute.

If pump pressure and flow are tested as outlined, it can be assumed that any trouble existing in the system must be due to power steering system relief valve or control valve. If pump test is not satisfactory, remove and overhaul the pump as outlined in paragraph 15.

13. **PRESSURE RELIEF VALVE.** To test the power steering relief valve pressure, install a suitable pressure gage in plug port (P—Fig. 8), or leave flow meter installed at pressure line fitting (F) as outlined in paragraph 12. Run engine at 2000 rpm. If flow meter is used, open flow meter pressure valve. Turn steering wheel fully to right or left and observe gage pressure which should be 2105-2130 psi.

If pressure is not as specified, loosen the locknut and turn power steering pressure adjusting screw (5—Fig. 9) in or out as required. Secure the adjustment by tightening locknut when adjustment is correct. If pressure cannot

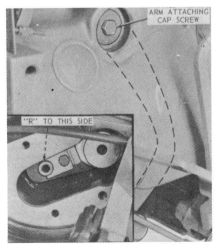

Fig. 7–Removing steering cross shaft. Inset shows yoke arm pin, which should be installed with "R" mark visible.

Fig. 8–Power steering pump flow can be checked by teeing a flow meter into pressure line fitting (F). Pressure can be checked at port plug (P).

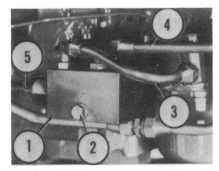

Fig. 9–Power steering pressure relief valve mounts on dump valve housing as shown.

1. Valve
2. Cap screw
3. Return line
4. Pressure line
5. Adjusting screw

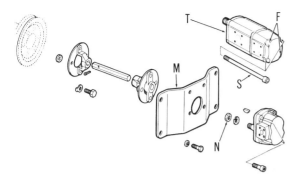

Fig. 10–Tandem pump (T) is used on power steering models, instead of regular hydraulic pump. Two opposing cap screws (F) fasten housing blocks together while the other two (long) screws (S) secure the pump to mounting bracket (M).

N. Shaft nut

be properly adjusted, overhaul relief valve assembly (1) as outlined in paragraph 18.

POWER STEERING PUMP

All Power Steering Models

14. **REMOVE AND REINSTALL.** The power steering pump is the forward section of the front mounted, gear-type pump shown installed in Fig. 8. To remove the pump, first remove right grille and air cleaner. Remove cotter pin from pump drive shaft and push shaft rearward out of front coupling. Disconnect the inlet and both pressure lines at pump. Remove the four retaining cap screws and lift off pump and mounting plate as an assembly.

Refer to paragraph 15 for removal of pump from mounting bracket and/or power steering (front) section of tandem pump from rear section. Install pump and mounting bracket by reversing removal procedure. Tighten screws securing pressure line adapters to pump body to a torque of 2.8 kg/m (20 ft.-lbs.) and screws securing mounting bracket to front support to a torque of 11.8 kg/m (85 ft.-lbs.)

15. **OVERHAUL.** If both sections of pump are to be disassembled, withdraw drive shaft from rear coupling and clamp solidly in a vise as shown in Fig. 11. Install front coupling on drive shaft and remove nut (N—Fig. 10) from

pump drive shaft, then remove front coupling from pump shaft as an assembly. Remove the two long socket head through-bolts (S) from diagonally opposite holes in front housing and lift pump off mounting bracket.

If only power steering (front) section of pump is to be disassembled, remove through-bolts (S) and the diagonally opposite slotted screws (F), then lift off front section only. Rear pump section will be loose on mounting bracket but the two parts cannot be separated until front coupling is removed.

With power steering pump section removed, disassemble as follows: Remove the two slotted head screws retaining cover (C—Fig. 12) and lift off the cover. Remove and save the "X" shaped spacer. Discard "O" rings and seals. Use a permanent type felt tip pen and mark the exposed pump bearing for correct reassembly, then remove pumping gears and bearings, bumping housing on a clean hardwood block to remove rear bearing.

Wash all parts in a suitable solvent and inspect for wear, scoring or other damage. Seals, packing and bearing plugs are available individually. If gears, bearings or housing are damaged or worn the complete pump section must be renewed as an assembly.

Prior to reassembly, dip all packings in clean oil and assemble wet, ob-

serving the assembly marks made when pump was disassembled, tightening slotted screws evenly and firmly. Reinstall the assembled pump as outlined in paragraph 14.

STEERING VALVE

All Power Steering Models

16. **REMOVE AND REINSTALL.** To remove the steering valve assembly, first remove cap screws from steering shaft cover (55—Fig. 13), remove cover and drain fluid from steering shaft compartment. Remove steering wheel using a suitable puller. Remove transmission housing shield. Disconnect pressure line at right side of steering housing. Remove foot throttle pivot screw and move throttle linkage out of the way.

Remove either side cowl, the two bolts securing steering valve housing (35) to dash and the four cap screws securing valve body to clutch housing. Working through opening in clutch housing where steering shaft cover (55) was mounted, remove the cap screw, clip and pin attaching connecting rod (52) to yoke shaft (62). NOTE: Pin is threaded for using a puller screw or slide hammer.

Steering valve and cylinder housing can now be lifted off of clutch housing. Install by reversing removal procedure, making sure housing gasket does not block oil passage. Tighten housing and side cover cap screws to a torque of 5 kg/m (35 ft.-lbs.).

17. **OVERHAUL.** To disassemble the steering valve, use Fig. 13 as a guide. Place unit in a vise, right side up. Remove lock nut (6), then using a suitable spanner, remove adjuster (8) and seal (7). Set steering wheel back on shaft splines and turn wheel clockwise until piston is retracted to top of cyl-

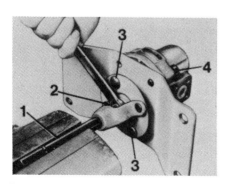

Fig. 11–Removing front coupling from hydraulic pump.

1. Drive shaft
2. Shaft nut
3. Coupling rivets
4. Mounting screw

Fig. 12–Exploded view of tandem hydraulic pump used on models with power steering.

C. Cover
H. Hydraulic system section
S. Steering pump section

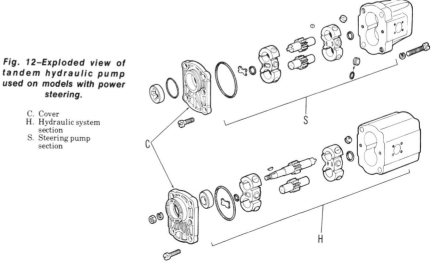

inder. Exert upward pressure on connecting rod (52) and turn steering wheel counter-clockwise until shaft (33) is threaded out of piston rod (47); then carefully lift steering shaft (33) and valve bodies out top of housing. Withdraw piston rod (47), piston and associated parts out bottom opening.

Support screw end of steering shaft (33) in a protected vise and apply down pressure to step washer (12) until top snap ring (11) can be removed. Lift off step washer, spring (13) and thrust washer (14); then remove lower snap ring (11). Remaining valve parts can now be removed as assemblies, from steering shaft.

NOTE: Upper valve body (21) and lower valve body (22) are factory matched and assembled; and factory adjusted by selection of shim packs (33). Keep the individual units separate from each other and properly identified, and disassemble

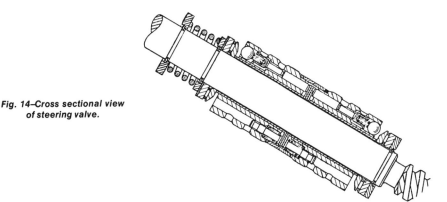

Fig. 14–Cross sectional view of steering valve.

only as required for cleaning or inspection.

Remove piston (44) and rod guide (51) from piston rod if service is indicated. Press retaining pin (48) from rod guide to remove connecting rod (52). Reassemble by reversing disassembly procedure, using new "O" rings and

seals. Make sure spanner wrench holes in piston face upward. Tighten piston (44) on piston rod to a torque of 34 kg/m (250 ft.-lbs.). Tighten adjuster (8) to a torque of 7 kg/m (50 ft.-lbs.) and lock nut (6) to 4.3 kg/m (30 ft.-lbs.).

Bushings (56) in side cover and clutch housing have a normal installed clearance of 0.02-0.11MM (0.001-0.0055 inch) on cross shaft (62). Install bushings flush with their bores if removed.

PRESSURE RELIEF VALVE

All Power Steering Models

18. **OVERHAUL.** Refer to Fig. 15 for exploded view. Front elbow (9) is connected by a tee to steering pressure line and rear elbow is connected by a return line to a fitting in transmission housing as shown in Fig. 9.

The power steering relief valve can be disassembled for checking without removal from its mounting place on dump valve, however if valve seat is damaged, the unit must be renewed as an assembly. Valve spring (7) should have a free length of 50.8MM (2.0 inches) and should test 53 kg (117 lbs.) when compressed to a height of 44.8MM (1.76 inches). The steering relief valve and all components are completely interchangeable with hydraulic system relief valve.

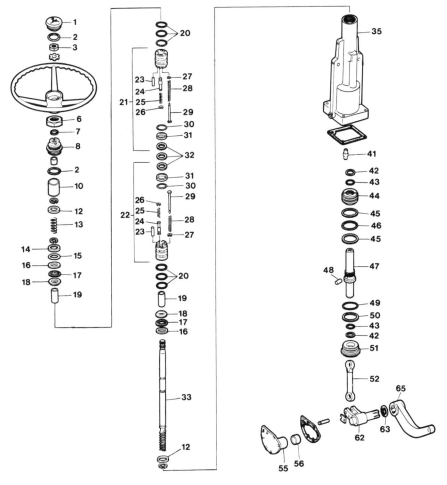

Fig. 13–Exploded view of power steering gear assembly with integral steering valve and cylinder.

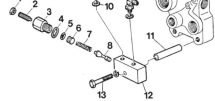

Fig. 15–Exploded view of power steering pressure relief valve.

1. Cap	17. Thrust bearing	29. Pressure pin	46. O-ring
2. O-rings	18. Thrust washer	30. Shim pack	47. Piston rod
3. Nut	19. Sleeve	31. Spacer	48. Pin
6. Lock nut	20. O-rings	32. Special washer	49. O-ring
7. Oil seal	21. Upper valve	33. Shaft	50. Backup ring
8. Adjuster	22. Lower valve	35. Housing	51. Rod guide
10. Spacer	23. Control pin	41. Sleeve	52. Connecting rod
12. Special washer	24. Spool valve	42. Backup ring	55. Cover
13. Spring	25. Spring	43. O-ring	56. Bushing
14. Washer	26. Snap ring	44. Piston	62. Steering shaft
15. Washer	27. Snap ring	45. Backup ring	63. Oil seal
16. Thrust washers	28. Spring		65. Arm

Pressure relief valve legend:
1. Locknut
2. Adjusting screw
3. Plug
4. Seal ring
5. O-ring
6. Tension ring
7. Spring
8. Valve body
9. Elbow
10. O-ring
11. Spacer tube
12. Housing
13. Cap screw

ENGINE AND COMPONENTS

All models are equipped with diesel engine only.

R&R ENGINE WITH CLUTCH

All Models

19. To remove engine and clutch as a unit, first drain cooling system and if engine is to be disassembled, drain oil pan. Remove front end weights if tractor is so equipped. Remove side grille screens, hood, tool box and side frames. Remove air intake pipe and radiator hoses. Disconnect fuel feed line, return line, fuel gage wire and air cleaner restriction warning switch wire. Remove hydraulic line clamps and separate hydraulic pump pressure and return lines, including power steering pressure line if so equipped.

Separate hydraulic pump drive coupling and disconnect steering drag link. Support tractor beneath clutch housing and attach a hoist to front support and axle assembly. Place wood wedges between axle and front support to prevent axle from tipping, remove cap screws securing front support to engine and roll front axle, front support, radiator and fuel tank as an assembly away from engine.

NOTE: Use care to keep front end unit from tipping forward when disconnected from engine. If an overhead hoist is not available, it may be necessary to drain or remove fuel tank before disconnecting front end unit.

Disconnect battery cables and remove batteries. Disconnect wiring harness, speed-hour meter cable and coolant temperature sending unit bulb. Disconnect ether starting unit pipe from manifold if so equipped. Disconnect throttle linkage and stop cable. If tractor has an underslung exhaust, disconnect exhaust pipe at manifold.

Install two JD-244 engine lifting eyes or equivalent to cylinder head and attach a hoist to lifting eyes. Keep en-gine horizontal with hoist, unbolt engine from clutch housing and pull engine forward until clutch clears clutch shaft.

Reassemble tractor by reversing disassembly procedure. It may be necessary to turn engine crankshaft or rear pto shaft to align both sets of clutch splines when rejoining the tractor. Alignment dowels or studs will assist in rejoining the housings. Make sure flywheel housing is snug against clutch housing before tightening the retaining cap screws. Tighten engine to clutch housing cap screws and front support to engine cap screws to a torque of 23.5 kg/m (170 ft.-lbs.).

CYLINDER HEAD

All Models

20. To remove cylinder head, first drain cooling system and remove hood. Disconnect battery ground straps. Remove air cleaner tube. Remove exhaust manifold from cylinder head and if tractor is equipped with underslung exhaust, the manifold can be left attached to the exhaust pipe. Disconnect injector leak-off line from fuel tank, injectors and injection pump and remove complete leak-off line. Disconnect pressure lines from injectors, remove hold-down clamps and spacers and withdraw injectors. Disconnect the cold starting aid from the inlet manifold elbow and remove elbow. Unbolt the fuel filters from the cylinder head. Plug all fuel openings. Remove vent tube from rocker arm cover, then remove cover and the rocker arm assembly. Identify and remove push rods and the valve stem caps. Disconnect fan baffle and water outlet elbow from cylinder head, then unbolt and remove cylinder head.

When reinstalling cylinder head, use a thin coat of No. 3 Permatex on both sides of head gasket and tighten head bolts to 15.2 kg/m (110 ft.-lbs.). torque in sequence shown in Fig. 16. Be sure oil holes in rear rocker arm shaft bracket and cylinder head are open and clean as this passage provides lubrication for the rocker arm assembly. Be sure the special, hardened flat steel washers are used beneath all bolt heads.

Head bolts should be retightened after engine has run about one hour at 2500 rpm under half-load. Loosen head bolts about 1/6-turn and retighten in sequence to 15.2 kg/m (110 ft.-lbs.). Valve tappet gap (cold) is 0.35 MM (0.014 inch) for intake valves and 0.45 MM (0.018 inch) for exhaust valves.

VALVES AND SEATS

All Models

21. Intake and exhaust valves seat directly in the cylinder head, however, hardened exhaust valve seats are available as a service item if required. Intake and exhaust valves are fitted with hardened steel stem caps.

Valves are normally recessed approximately 1.5MM (0.060 inch) below gasket surface of cylinder head. Valve should be renewed or a seat insert installed if valve recess exceeds 3.0MM (0.120 inch). See Fig. 17.

Recommended valve face angle is 43½° and recommended seat angle is 45°. Desired seat width is approximately 1.5MM (1/16 inch) and maximum allowable seat run-out is 0.002 inch. Standard valve stem diameter is 9.43-9.46MM (0.3715-0.3725 inch) and recommended clearance in cylinder head bore is 0.05-0.1MM (0.002-0.004 inch). Guide bores can be knurlized to reduce the clearance. Valves are also available with stem oversizes of 0.076, 0.381 and 0.762MM (0.003, 0.015 and 0.030 inch), for use in reamed valve guide bores.

TAPPET GAP ADJUSTMENT

All Models

22. The recommended cold tappet

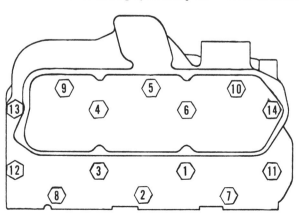

Fig. 16—Tighten cylinder head cap screws in sequence shown to a torque of 15.21 Kg/M (110 ft.-lbs.)

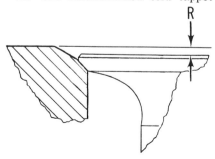

Fig. 17—Valve head should be recessed not more than 3MM (0.120 inch) as shown at (R).

gap setting is .35 MM (0.014 inch) for intake valves and .45 MM (0.018 inch) for exhaust. Cold (static) setting of all valves can be made from just two crankshaft settings using the procedure shown in Figs. 19 and 20. Proceed as follows:

Turn crankshaft by hand until No. 1 piston is at top dead center and a TDC timing dowel will enter timing hole in flywheel as shown in Fig. 18. If No. 1 piston is on compression stroke, rear valve (No. 3 Exhaust Valve) will be open and rocker arms on No. 1 cylinder will both be loose; adjust the four valves indicated in Fig. 19. If No. 1 piston is on the exhaust stroke, the two front valves will be both partially open (exhaust valve closing and intake valve commencing to open) and all other valves will be closed; adjust the two valves shown in Fig. 20.

With the indicated valves adjusted, turn crankshaft one complete turn until timing dowel will again enter timing hole in flywheel and adjust remainder of valves.

VALVE GUIDES

All Models

23. Valve guides are integral with cylinder head and have an inside diameter new of 9.51-9.53MM (0.3745-0.3755 inch), which provides 0.05-0.1MM (0.002-0.004 inch) operating clearance for valves. Maximum allowable valve stem clearance in guide is 0.15MM (0.006 inch) and when this value is exceeded, ream valve guide as required to fit next oversize valve stem. Valves are available with oversize stem of 0.076, 0.381 and 0.762MM (0.003, 0.015 and 0.030 inch).

VALVE SPRINGS

All Models

24. Intake and exhaust valve springs are interchangeable. Renew springs which are distorted, discolored, rusted

Fig. 18–Timing dowel (2) will enter a drilled hole in flywheel when No. 1 piston is at TDC.

1. Timing hole cover

or do not meet the test specifications which follow:

Free length (approx.) .... 2⅛ in.
Test lbs. at 1 13/16 in. 52-64
Test lbs. at 1 23/64 in. 129-157

ROCKER ARMS AND SHAFT

All Models

25. Rocker arms are interchangeable and bushings are not available. Inside diameter of shaft bore in rocker arm is 20.07-20.12MM (0.790-0.792 inch). Outside diameter of rocker arm shaft is 19.99-20.01MM (0.7869-07879 inch). Normal operating clearance between rocker arm and shaft is 0.05-0.12MM (0.0021-0.0051 inch). Renew rocker arm and/or shaft if clearance is excessive.

Valve stem contacting surface of rocker arm may be refaced but original radius must be maintained.

When reinstalling rocker arm assembly, be sure oil holes and passages are open and clean. Pay particular attention to the rear mounting bracket as lubrication is fed to rocker arm shaft through this passage. Oil hole in rocker arm shaft must face downward when installed on cylinder head.

CAM FOLLOWERS

All Models

26. The cylinder type cam followers

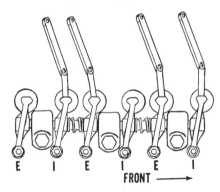

Fig. 19–The indicated valves can be adjusted when No. 1 piston is at TDC on compression stroke. Turn crankshaft one complete turn and adjust remaining valves; refer to Fig. 20.

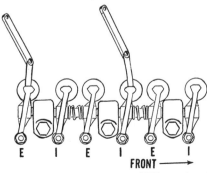

Fig. 20–When No. 1 piston is at TDC on exhaust stroke, the two indicated valves can be gapped. Refer also to Fig. 19.

(tappets) can be removed from below after camshaft has been removed. If necessary, they can aso be removed from above after cylinder head, rocker arm shaft and push rods are removed. Cam followers are available in standard size only and operate directly in machined bores in cylinder block.

It is recommended that new cam followers be installed if a new camshaft is being installed.

VALVE TIMING

All Models

27. Valves are correctly timed when timing mark on camshaft gear is aligned with timing tool (JD254) when tool is aligned with crankshaft and camshaft centerlines as shown in Fig. 23.

TIMING GEAR COVER

All Models

28. To remove timing gear cover, first remove the front axle and front support assembly as outlined in paragraph 19.

With front support assembly removed, remove fan, fan belt, alternator and water pump. Remove crankshaft pulley retaining cap screw, attach puller and remove pulley. Remove power steering pump, if so equipped. Remove the oil pressure regulating plug, spring and valve (See Fig. 21). Drain and remove oil pan, then unbolt and remove the timing gear cover. See Fig. 22.

Fig. 21–Oil pressure relief valve is located in front of timing gear cover as shown.

Fig. 22–View of timing gear train with front cover removed.

With timing gear cover removed, the crankshaft front oil seal can be renewed. To renew oil seal, coat outside diameter of seal with sealing compound and with seal lip toward inside, support timing gear cover around seal area and press seal into bore until it bottoms.

NOTE: Do not attempt to install the front oil seal in timing gear cover without providing support around seal area. Cover is a light cast aluminum alloy and could be warped or cracked rather easily.

CAMSHAFT

All Models

29. To remove camshaft, timing gear cover must be removed as outlined in paragraph 28. Remove vent tube, rocker arm cover, rocker arm assembly and push rods. Shut off fuel and unbolt fuel pump from cylinder block. Use wires with a 90 degree bend in end, pushed into push rod bore of tappet, to hold tappets away from camshaft lobes (wood doweling of proper size and spring type clothes pins can also be used). Turn engine until thrust plate retaining cap screws can be reached through holes in camshaft gear, then remove cap screws and pull camshaft and thrust plate from cylinder block.

NOTE: If upper idler gear is not being removed, mark camshaft gear and upper idler gear so camshaft can be reinstalled in its original position. If upper idler gear is being removed, turn engine to TDC and align camshaft gear timing mark as shown in Fig. 23 when installing the upper idler gear.

Support camshaft gear, press camshaft from gear and remove woodruff key.

If tachometer drive shaft at aft end of

camshaft requires renewal, thread exposed end and install a nut, then attach a puller to nut and remove the tachometer drive shaft from camshaft.

Camshaft is carried in three unbushed bores in cylinder block. When checking camshaft journal diameters, also check inside diameter of the camshaft journal bores using the following data.

Camshaft Journal O. D. 55.87-55.9MM
(2.2-2.201 inch)
Camshaft Journal bore
I. D. 55.99-56.01MM
(2.204-2.205 inch)
Normal operating
clearance 0.09-0.14MM
(0.0035-0.0055 inch)
Max. allowable
clearance 0.18MM
(0.007 inch)
Camshaft end play 0.06-0.22MM
(0.0025-0.0085 inch)
Max. allowable end play 0.38MM
(0.015 inch)
Thrust plate thickness
(new) 3.96-4.01MM
(0.156-0.158 inch)

When installing new camshaft gear, be sure timing mark is toward front and support camshaft under front bearing journal. When installing new tachometer drive shaft, be sure drive slot is toward rear and support camshaft under rear journal.

When installing camshaft in cylinder block be sure timing mark is aligned as shown in Fig. 23 and tighten thrust plate cap screws to a torque of 4.8 kg/m (35 ft.-lbs.).

IDLER GEARS

All Models

30. Upper and lower idler gears are bushed and operate on shafts attached to engine front plate by cap screws. Idler gear end play is controlled by thrust washers. Both idler gears are driven by crankshaft gear. Upper idler gear drives the timed camshaft and

injection pump drive gear; lower idler gear drives the oil pump.

Drive hubs are interchangeable but spring pins and retaining cap screws are not. The upper spring pin (P—Fig. 25) is 1⅛ inches long and should protrude 3.5-4.5MM (0.138-0.177 inch) from front face of hub as shown at (X). The lower spring pin is 1½ inches long and should protrude 5-7MM (0.197-0.275 inch) from front face of hub as shown at (Z). Lower spring pin extends through rear thrust washer and into front plate as shown in cross section, thus keeping hub from turning when rear retaining cap screw is tightened. Idler gear bushing clearance should not exceed 0.15MM (0.006 inch) and end play should not exceed 0.38MM (0.015 inch).

When reinstalling upper idler gear, make sure camshaft gear is correctly timed as outlined in paragraph 31, injection pump gear is correctly timed as outlined in paragraph 33 and No. 1 piston is at TDC. Tighten upper idler retaining cap screw to a torque of 9 kg/m (65 ft.-lbs.) and lower idler gear retaining cap screw to a torque of 13 kg/m (95 ft.-lbs.).

TIMING GEARS

All Models

31. **CAMSHAFT GEAR.** The camshaft gear is keyed and pressed on the camshaft. The fit of gear on camshaft is such that removal of the camshaft, as

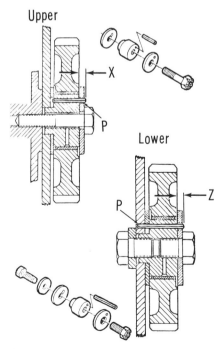

Fig. 25–Cross sectional view of timing gear train idler gears showing relative lengths of spring pins (P). Refer to paragraph 30 for details of installation.

X. 3.5—4.5MM
Z. 5.0—7.0MM

Fig. 23–With crankshaft turned so No. 1 piston is at TDC and timing tool positioned as shown, camshaft gear timing mark should be directly under edge of timing tool when idler.gear is installed. Refer also to Fig. 24.

Fig. 24–With crankshaft and camshaft gears positioned as shown in Fig. 23, turn injection pump drive gear until timing mark aligns with edge of timing tool; then install idler gear.

outlined in paragraph 29 is recommended. Camshaft is correctly timed when centerline of camshaft, timing mark on camshaft gear and centerline of crankshaft are aligned and crankshaft is at TDC-1 as shown in Fig. 23.

32. **CRANKSHAFT GEAR.** Renewal of crankshaft gear requires removal of crankshaft as outlined in paragraph 39. Gear is keyed and pressed on crankshaft. Support crankshaft under first throw when installing new gear. Installation of gear may be eased by heating gear in oil.

Crankshaft gear has no timing marks but keyway in crankshaft will be straight up when engine is at TDC.

33. **INJECTION PUMP GEAR AND SHAFT.** Type of gear and method of attachment will depend on what model injection pump is used; refer to Fig. 26.

On Model CDC injection pump, gear

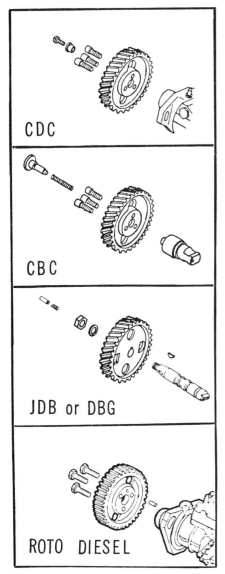

Fig. 26–Different injection pumps have been used requiring different gears and removal procedures. Refer to paragraph 33 for details.

can be removed after removing the four retaining cap screws. Gear shaft is an integral part of the pump and screws must be removed in pump removal. Tighten the three outer screws to a torque of 180 inch pounds and the one center screw to 100 inch pounds.

On Model CBC pump, gear and shaft unit can be withdrawn from timing gear housing and injection pump after timing gear cover is off. Gear is retained by three unevenly spaced cap screws which should be tightened to a torque of 180 inch pounds.

On Models JDB and DBG injection pump, gear and shaft unit can be withdrawn from timing gear housing and pump after cover is off. Gear is keyed to shaft and retained by washer and nut as shown. Tighten nut to a torque of 35 ft.-lbs.

On ROTO DIESEL injection pump, gear can be removed after removing the three retaining cap screws. Gear shaft is an integral part of pump and gear must be removed in pump removal. Dowel pin positions the gear in pump hub. Tighten gear retaining cap screws to a torque of 2.5 kg/m (18 ft.-lbs.).

On all models, injection pump drive gear is interchangeable. Two timing marks appear on the gear, each identified by a stamped "3" and "4". Use only the timing mark "3", when timing the gears as shown in Fig. 24.

34. **TIMING GEAR BACKLASH.** Excessive timing gear backlash may be corrected by renewing the gears concerned, or in some instances by renewing idler gear bushing and/or shaft. Refer to the following for maximum recommended backlash:

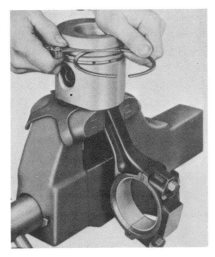

Fig. 27–Maximum allowable piston ring side clearance is 0.2MM (0.008 inch) when measured as shown.

Crankshaft gear to upper
idler 0.42MM
(0.0166 inch)
Camshaft gear to upper
idler 0.47MM
(0.0185 inch)
Injection pump gear to upper
idler 0.47MM
(0.0185 inch)
Crankshaft gear to lower
idler 0.47MM
(0.0185 inch)
Oil pump gear to lower
idler 0.40MM
(0.0157 inch)

ROD AND PISTON UNITS

All Models

35. Piston and connecting rod assemblies are removed from above after removing cylinder head and oil pan. Secure cylinder liners (sleeves) in cylinder to prevent liners from moving as crankshaft is turned.

The word "FRONT" is stamped on the head of each piston and embossed in the web of each connecting rod. Replacement rods are not marked and should be stamped with cylinder number at installation. Lubricate rod screws and tighten to a torque of 8.3-9.7 kg/m (60-70 ft.-lbs.).

PISTONS, RINGS AND SLEEVES

All Models

36. All pistons are cam-ground, forged aluminum-alloy and are fitted with three rings located above the piston pin. All pistons have the word "FRONT" stamped on the piston head and are available in standard size only.

Normal side clearance on all piston rings is 0.09-0.13MM (0.0035-0.0053 inch), with a maximum allowable side clearance of 0.2MM (0.008 inch). The two top compression rings are marked "TOP" for correct installation; oil control ring can be installed either side up. Piston ring end gap is correct as furnished and should not be altered.

The renewable wet type cylinder sleeves are available in standard size only. Sleeve flange at upper edge is sealed by the cylinder head gasket. Sleeves are sealed at lower edge by a square section packing and two "O"

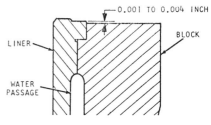

Fig. 28–Sleeve stand-out, without packing, should be 0.001-0.004 above cylinder block.

rings as shown in Fig. 29. Sleeves normally require loosening using a sleeve puller, after which they can be withdrawn by hand. Out-of-round or taper should not exceed 0.13MM (0.005 inch). If sleeve is to be re-used, it should be de-glazed using a normal cross-hatch pattern.

When reinstalling sleeves, first make sure sleeve and block bore are absolutely clean and dry. Carefully remove any rust or scale from seating surfaces and packing grooves, and from water jacket in areas where loose scale might interfere with sleeve or packing installation. If sleeves are being re-used, buff rust and scale from outside of sleeve.

Install sleeve without the seals and measure standout as shown in Fig. 28. Also check to be sure that sleeve will slip fully into bore without force. If sleeve cannot be pushed down by hand, recheck for scale or burrs. If necessary, select another sleeve. After matching sleeves to all the bores, mark the sleeves and keep any required flange shims with the sleeve. Install dry packing on cylinder sleeve, making sure packing is not twisted and that long cross section is parallel with sleeve as shown in Fig. 30. Install dry "O" rings in block grooves, making sure they are not twisted or damaged and that part of ring does not project into liner bore. Coat packing, lower step of sleeve and "O" rings in cylinder block liberally with clean engine oil and immediately install sleeve by pushing straight down into block bore. Work sleeve gently in by hand until flange enters block counterbore, then fully seat sleeve using a clean hardwood block and a hammer.

Specifications of pistons and sleeves are as follows:
Sleeve Bore 97.996-98.032MM
(3.8581-3.8595 inch)
Piston Skirt
   Diameter 97.726-97.752MM
(3.8475-3.8485 inch)

Piston Skirt Clearance . 0.07-0.13MM
(0.003-0.005 inch)
Maximum Allowable
   Clearance 0.26MM
(0.010 inch)

PISTON PINS AND BUSHINGS
All Models

37. The 30.163-30.173MM (1.1875-1.1879 inch) full floating piston pins are retained in pistons by snap rings. A pin bushing is fitted in upper end of connecting rod and bushing must be reamed after installation to provide a thumb press fit for the piston pin. Piston pin to piston clearance is 0.007-0.02MM (0.0003-0.0008 inch), with a maximum wear limit of 0.038MM (0.0015 inch).

CONNECTING RODS AND BEARINGS
All Models

38. The steel-backed, aluminum lined bearings can be renewed without removing rod and piston unit by removing oil pan and rod caps.

The connecting rod big end parting line is diagonally cut and rod cap is offset away from camshaft side as shown in Fig. 31. A tongue and groove cap joint positively locates the cap. Rod marking "FRONT" should be forward and locating tangs for bearing inserts should be together when cap is installed.

Connecting rod bearings are available in undersizes of 0.002, 0.010, 0.020 and 0.030, as well as standard. Refer to the following specifications:
Crankpin Diameter ... 69.8-69.82MM
(2.748-2.749 inch)
Diametral Clearance .. 0.03-0.106MM
(0.0012-0.0042 inch)
Cap Screw Torque—kg/m 8.3-9.7
   Ft.-Lbs. 60-70

CRANKSHAFT AND BEARINGS
All Models

39. The crankshaft is supported in four main bearings. Main bearing in-

serts can be renewed after removing oil pan, oil pump and main bearing caps. All main bearing caps except the rear are interchangeable and should be identified prior to removal so they can be reinstalled in their original position. Rear main bearing is flanged and controls crankshaft end play, which should be 0.05-0.2MM (0.002-0.008 inch). Install main bearing caps with previously affixed identification marks aligned and tighten the retaining cap screws to a torque of 11.7 kg/m (85 ft.-lbs.).

To remove the crankshaft, first remove engine from tractor as outlined in paragraph 19. Remove timing gear cover, timing gears, camshaft and engine front plate. Remove oil pan, oil pump, flywheel and flywheel housing. Be sure main bearing caps are identified for reinstallation, then remove rod and main bearing caps and lift crankshaft from cylinder block.

Crankshaft main journal diameter is 79.34-79.36MM (3.1235-3.1245 inches). Crankpin diameter is 69.80-69.82MM (2.748-2.749 inches). Recommended main bearing clearance is 0.03-0.10MM (0.001-0.004 inch), with a maximum allowable clearance of 0.15MM (0.006 inch). Main bearing inserts are available in undersizes of 0.05, 0.25, 0.51 and 0.76MM (0.002, 0.010, 0.020 and 0.030 inch) as well as standard.

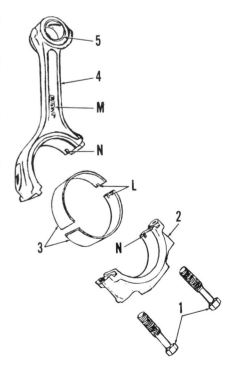

Fig. 31–Connecting rod assembly used on all models.

1. Cap screws	5. Pin bushing
2. Cap	L. Locating tangs
3. Inserts	M. "FRONT" marking
4. Rod	N. Locating notches

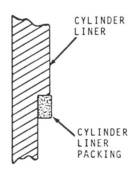

Fig. 29–Correct installation of square packing and O-rings used for sealing cylinder sleeves on some models. Refer to paragraph 36.

1. Square packing	B. Cylinder block
2. O-rings	S. Cylinder sleeve

Fig. 30–Cross section of cylinder sleeve, showing square section packing properly installed. Refer to paragraph 36.

CRANKSHAFT REAR OIL SEAL

All Models

40. A lip type crankshaft rear oil seal is contained in flywheel housing and a sealing wear ring is pressed on flywheel mounting flange or crankshaft. To renew the seal, first remove clutch, flywheel and flywheel housing. If wear ring on crankshaft is damaged, worn or scored, expand the ring by scoring it on sealing surface in several places using a dull chisel and hammer. Lift off the ring when it is loose. Install new ring with rounded edge to rear, being careful not to cock the ring. Ring should start by hand, and can be seated using a suitable driver such as JD251 which contacts entire edge of ring. Install seal in flywheel housing working from the rear, with sealing lip toward front. When properly installed, rear of seal should be flush with rear of housing bore.

FLYWHEEL

All Models

41. To remove the flywheel, first remove clutch as outlined in paragraph 85, then unbolt and remove flywheel from its doweled position on crankshaft flange.

To install a new flywheel ring gear, heat gear to approximately 500° F. and install with chamfered end of teeth toward front of flywheel.

When installing flywheel, tighten the retaining cap screws to a torque of 11.7 kg/m (85 ft.-lbs.).

FLYWHEEL HOUSING

All Models

42. The cast iron flywheel housing is secured to rear face of engine block by eight cap screws. Flywheel housing contains the crankshaft rear oil seal and oil pressure sending unit switch. The rear camshaft bore in block is open and the tachometer drive passes through flywheel housing. It is important therefore, that gasket between block and flywheel housing be in good condition and cap screws properly

tightened. Tighten all screws evenly to 3 kg/m (22 ft.-lbs.), then retorque to 4.8 kg/m (35 ft.-lbs.).

OIL PUMP AND RELIEF VALVE

All Models

43. To remove the oil pump, first drain and remove oil pan. Wedge a clean shop towel between teeth of oil pump drive gear and lower idler gear to keep gear from turning, remove the self-locking nut retaining oil pump drive gear to shaft; pull the gear, then unbolt and remove oil pump.

Refer to Fig. 32 for an exploded view. Pump drive shaft (4) is 15.99-16.02MM (0.6295-0.6305 inch) in diameter and should have 0.025-0.1MM (0.001-0.004 inch) clearance in housing bore. Gears should have 0.025-0.1MM (0.001-0.004 inch) radial clearance in pump housing with a maximum allowable clearance of 0.13MM (0.005 inch). End clearance of gears is 0.025-0.15MM (0.001-0.006 inch) with a maximum clearance of 0.2MM (0.008 inch).

Idler gear shaft (2) can be pressed from pump housing if renewal is indicated.

44. The oil pump pressure relief valve should be adjusted to maintain a pressure of 3.5-4.2 kg/cm² (50-60 psi) at 2400 engine rpm and normal operating temperature. Pressure is adjusted by adding or removing shim washers between spring and cap, located at front of timing gear cover as shown in Fig. 33.

The regulating valve spring has a free length of approximately 119MM

(4.7 inches) and should test 6.1-7.5 kg (13.5-16.5 lbs) when compressed to a height of 42.5MM (1.68 inch).

The relief valve seat is pressed into front face of cylinder block as shown in Fig. 34. When installing new bushing, use special driver (JD-248 or equivalent) which protects the raised seating surface of bushing from installation damage.

FUEL LIFT PUMP

R&R AND OVERHAUL

All Models

45. The fuel lift pump (Fig. 35) should maintain 3.5-4.5 psi pressure 16 inches above pump outlet at 900 engine rpm. A repair kit consisting of diaphragm, spring and gaskets is provided for service.

When installing the kit, mark cover and body before disassembly to assure proper positioning of inlet and outlet ports when reassembling. Install body screws loosely and actuate rocker arm to center and pull the necessary slack into diaphragm. Tighten body screws alternately and evenly.

Fig. 34—Oil pressure relief valve seat is located in front face of cylinder block and is renewable. Refer to paragraph 44.

Fig. 33—Oil pressure relief valve is located in front of timing gear cover as shown.

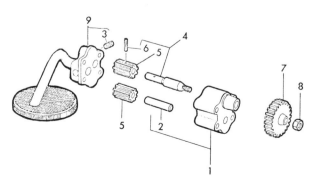

Fig. 32—Exploded view of engine oil pump showing component parts.

1. Housing
2. Idler shaft
3. Drive shaft and gear
4. Drive shaft and gear
5. Pump gears
6. Groove pin
7. Drive gear
8. Shaft nut
9. Cover and screen

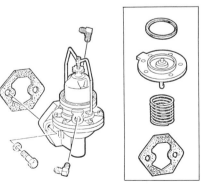

Fig. 35—Removed view of fuel pump. Parts enclosed in inset are available as a repair kit. Refer to paragraph 45.

DIESEL FUEL SYSTEM

FILTERS AND BLEEDING

All Models

46. **FILTERS.** Series 820 tractors are equipped with a ROTO-DIESEL Dual Stage, dual element filter of the type shown in Fig. 36. Series 830 tractors use the single element Two Stage Filter shown in Fig. 37.

Filter life depends primarily on care and cleanliness of fuel handling procedure. It is recommended that drain tap at bottom of fuel tank (Arrow—Fig. 38) and drain plug(s) on fuel filter(s) be loosened periodically and fuel allowed to run until any accumulated sediment, sludge or water is allowed to drain. Pump hand primer lever on fuel lift pump when draining fuel filters. Renew primary element of dual element system or element unit of single element type when contamination is suspected, or about once a year if fuel is clean. Renew final filter element of dual element type if primary element is contaminated, or at engine overhaul periods.

47. **BLEEDING.** Whenever fuel system has been run dry or when a line has been disconnected, air must be bled from fuel system as follows:

Be sure there is sufficient fuel in tank and that tank outlet valve is open. Loosen front bleed screw (B—Fig. 36) on dual element system or bleed screw (B—Fig. 37) on later models and actuate primer lever of fuel transfer pump until a solid, bubble-free stream of fuel emerges from bleed screw, then tighten the screw. On dual filter models, open rear bleed screw (B—Fig. 36) and continue pumping primer lever until fuel stream is bubble free; close bleed screw and continue pumping primer lever a few more strokes to fill pump and lines.

Injection pump is self-bleeding on models with Roosa Master pump. On models equipped with Roto-Diesel pump, refer to Fig. 39. Open bleed screw (1) on side of injection pump governor cover and continue to actuate primer lever until bubble-free fuel flows. Close bleed screw (1), open bleed screw (2) and repeat operation until air pocket is evacuated from governor cover.

NOTE: Air in governor cover removed by bleed screw (2) will not prevent tractor from starting and running properly; however, condensation in the trapped air can cause rusting of governor components and eventual pump malfunction. Do not fail to bleed governor cover even though tractor starts and runs properly.

If injectors or lines have been removed, loosen pressure line connections at injectors about one turn, open throttle and crank engine with starter until fuel flows from loosened connections, then tighten connections and start engine.

INJECTOR NOZZLES

All Models

WARNING: Fuel leaves the injection nozzles with sufficient force to penetrate the skin. When testing, keep your person clear of the nozzle spray.

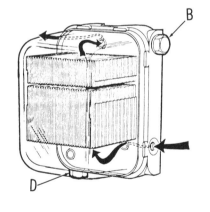

Fig. 37–Cross sectional view of fuel flow through two stage filter. Bleed plug is at (B), drain plug at (D).

48. **TESTING AND LOCATING A FAULTY NOZZLE.** If one engine cylinder is misfiring it is reasonable to suspect a faulty injector. Generally, a faulty injector can be located by running the engine at low idle speed and loosening, one at a time, each high pressure line at injector. As in checking spark plugs in a spark ignition engine, the faulty unit is the one that least affects the engine operation when its line is loosened.

Remove the suspected injector as outlined in paragraph 49. If a suitable nozzle tester is available, test injector as outlined in paragraphs 50 through 54 or install a new or rebuilt unit.

49. **REMOVE AND REINSTALL.** To remove an injector, remove hood and wash injector, lines and surrounding area with clean diesel fuel. Use hose clamp pliers to expand clamp and pull leak-off boot from injector. Disconnect high pressure line, then cap all openings. Remove cap screw from nozzle clamp and remove clamp and spacer. Pull injector from cylinder head.

NOTE: Unless the carbon stop seal has failed causing injector to stick, the injectors can be easily removed by hand. If injectors cannot be removed by hand, use John Deere nozzle puller JDE-38 and be sure to pull injector straight out of bore. DO NOT attempt to pry injector from cylinder head or damage to injector could result.

When installing injector, be sure nozzle bore and seal washer seat are clean and free of carbon or other foreign material. Install new seal washer and carbon seal on injector and insert injector into its bore using a slight twisting motion. Install and align locating clamp then install hold-down clamp and spacer and tighten cap screw to a torque of 2.8 kg/m (20 ft.-lbs.).

Bleed system as outlined in paragraph 47 if necessary.

50. **TESTING.** A complete job of nozzle testing and adjusting requires

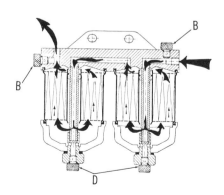

Fig. 36–Cross sectional view showing fuel flow through dual filter. Bleed plugs are at (B), drain plugs at (D).

Fig. 38–Arrow shows drain tap in bottom of fuel tank.

Fig. 39–On ROTO-DIESEL injector pump, open the two bleed screws in numerical sequence.

the use of an approved nozzle tester. Only clean, approved testing oil should be used in the tester tank. The nozzle should be tested for spray pattern, opening pressure, seat leakage and back leakage (leak-off). Injector should produce a distinct audible chatter when being tested and cut off quickly at end of injection with a minimum of seat leakage.

NOTE: When checking spray pattern, turn nozzle about 30 degrees from vertical position. Spray is emitted from nozzle tip at an angle to the centerline of nozzle body and unless injector is angled, the spray may not be completely contained by the beaker. Keep your person clear of the nozzle spray.

51. SPRAY PATTERN. Attach injector to tester and operate tester at approximately 60 strokes per minute and observe the spray pattern. A finely atomized spray should emerge at each nozzle hole and a distinct chatter should be heard as tester is operated. If spray is not symmetrical and is streaky, or if injector does not chatter, overhaul injector as outlined in paragraph 55.

NOTE: Some Series 820 units were equipped with an injector having five spray holes, equally spaced. All later models use a four hole nozzle which is the recommended service replacement for all five hole nozzle units.

52. OPENING PRESSURE. The correct opening pressure is 206.5-213.5 kg/cm² (2950-3050 psi) for all nozzles. If opening pressure is not correct, loosen locknut (2—Fig. 40) and back out valve lift adjusting screw (1) at least one full turn. Loosen locknut (4), actuate tester and adjust nozzle pressure by turning pressure adjusting screw (3) as required. With the correct nozzle opening pressure set and locknut (4) retightened, gently turn valve lift adjusting screw (1) in until it

bottoms, back screw out ½-turn and tighten locknut (2). Recommended tightening torque for locknut (4) is 70-80 inch-lbs.

53. SEAT LEAKAGE. To check nozzle seat leakage, attach injector to tester in a horizontal position. Raise pressure to approximately 170 kg/cm² (2400 psi), hold for ten seconds and observe nozzle tip. A slight dampness is permissible, but if a drop forms in the ten seconds, renew the injector or overhaul as outlined in paragraph 55.

54. BACK LEAKAGE. Attach

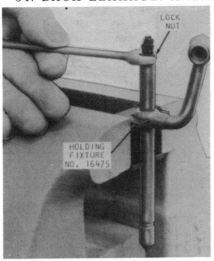

Fig. 41–Use holding fixture to secure injector for service.

Fig. 42–Use the special retractor as shown, to remove a sticking nozzle valve.

injector to tester with tip slightly above horizontal. Raise and maintain pressure at approximately 105 kg/cm² (1500 psi) and observe leakage from return (top) end of injector. After the first drop falls, back leakage should be at the rate of one drop every 3 to 10 seconds. If back leakage is excessive, renew injector or overhaul as outlined in paragraph 55. Back leakage from a new injector considerably slower than the indicated rate may indicate a potential sticking problem when the injector is put into service.

55. OVERHAUL. First wash the unit in clean diesel fuel and blow off with clean, compressed air. Remove carbon stop seal and sealing washer. Clean carbon from exterior of spray tip using a brass wire brush. Also, clean carbon or other deposits from carbon seal groove in nozzle body. DO NOT use wire brush or other abrasive on the Teflon coating on outside of nozzle body above carbon seal groove. Teflon coating can be cleaned with a soft cloth and solvent. Coating may discolor from use, but discoloration is not harmful.

Secure nozzle in vise using a suitable holding fixture as shown in Fig. 41. Loosen locknut (4—Fig. 40) and remove pressure adjusting screw (3), spring (5) and spring seat (6). If nozzle valve (7) will not slide from body when body is inverted, use special retractor (Fig. 42) or reinstall injector on tester with spring and lift adjusting screw (1 —Fig. 40) removed (pressure adjusting screw (3) loosely installed); and use hydraulic pressure from test pump to loosen the valve.

Nozzle and valve body are a matched set and should never be intermixed. Keep parts for one injector separate and immerse in clean diesel fuel in a compartmented pan, as unit is disassembled.

Clean all parts thoroughly in clean diesel fuel using a brass wire brush and lint-free wiping towels. Hard carbon or varnish can be loosened with a suitable non-corrosive solvent.

Clean spray tip orifices, first with a 0.2MM (0.008 inch) cleaning needle held in a pin vise as shown in Fig. 43, then follow up with a 0.25MM (0.010 inch) needle. Cleaning needle should

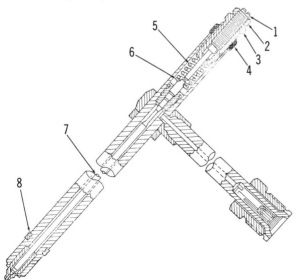

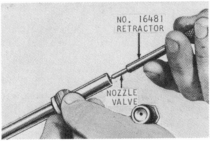

Fig. 40–Cross sectional view of ROOSA-MASTER injector nozzle used on late models. Early models are similar.

1. Lift adjusting screw
2. Locknut
3. Pressure adjusting screw
4. Locknut
5. Spring
6. Spring seat
7. Nozzle valve
8. Carbon seal

Fig. 43–Using a pin vise and cleaning needle to clean spray tip. Refer to paragraph 55 for cleaning needle size.

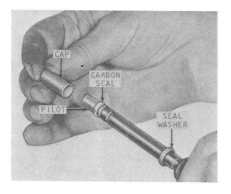

Fig. 45–Use the special pilot or a nozzle storage cap when installing a new carbon seal.

protrude only a slight amount (1/16 inch) from pin vise. Use light pressure and a twirling motion to penetrate the carbon. Moisten tip with diesel fuel or solvent if trouble is encountered.

Clean the valve seat using a valve tip scraper and light pressure while rotating scraper. Use a Sac Hole Drill to remove hard carbon from sac hole.

Piston area of valve and guide can be lightly polished by hand, if necessary, using Roosa Master No. 18649 lapping compound. Use the valve retractor to turn valve. Move valve in and out slightly while rotating, but do not apply pressure while valve tip contacts seat.

Thoroughly flush and wash the parts in clean diesel fuel to remove loosened carbon, lapping compound or other impurities; and assemble the parts while still immersed in clean diesel fuel.

Remove lift adjusting screw (1—Fig. 40) or back it out several turns, when adjusting opening pressure. Adjust as outlined in paragraph 52, increasing the opening pressure to 220-227 kg/cm² (3150-3250 psi) if a new spring (5) is installed. This allows for an initial pressure drop as the new spring takes a set.

INJECTION PUMP

All Models

A CAV Roto Diesel injection pump is used on all Model 830 tractors. Roosa Master CBC, CDC, DBG and JDB pumps have been used on Model 820. All pumps are flange mounted on left side of engine front plate and driven by upper idler gear of timing gear train. Pumps are not individually interchangeable and procedures for removal, installation and adjustment differ slightly. Refer to the appropriate following paragraphs for procedure.

NOTE: Injection pump service demands the use of specialized equipment and special training which is beyond the scope of this manual. This section, therefore, will cover only the information required for removal, installation and adjustment of the injection pump.

Model DBG Pump

Model DBGFC 331-17 DH injection pump was used on Series 820 tractors prior to serial number 047752.

56. **REMOVE AND REINSTALL.** To remove injection pump, first close fuel shut-off valve on fuel tank, then clean injection pump, line connections and the surrounding area with clean diesel fuel. Turn engine in direction of normal rotation until number one piston is starting up on compression stroke. Remove timing pin and cover from flywheel housing, reverse timing pin and reinsert it into threaded hole in flywheel housing. Press on timing pin and continue to turn engine until timing pin enters hole in flywheel. Engine is now at TDC.

NOTE: While it is not absolutely necessary, it is recommended that engine be set at TDC so timing can be checked and adjusted if necessary, when injection pump is reinstalled.

With engine set at TDC, disconnect fuel inlet line, fuel return line and throttle rod from injection pump. Disconnect pressure lines from injectors and pump and remove pressure lines. Remove pump mounting nuts and pull pump straight rearward off pump shaft.

When reinstalling pump, remove timing hole cover from pump and install timing window. Be sure timing lines on governor weight retainer and pump cam are aligned, start pump on pump shaft, and using seal compresser, slide pump into position and install the mounting nuts. Rotate pump back and forth, re-align timing marks and tighten mounting nuts.

NOTE: Use caution when sliding injection pump over shaft seals. If undue resistance is encountered, stop and examine rear seal. If lip has been rolled, renew the seal.

Removing timing pin, turn engine two complete revolutions, reinsert timing pin and again check, and align if necessary, the pump timing lines. The normal backlash of timing gears is enough to slightly affect injection pump timing.

Remove timing window, reinstall timing hole cover, throttle linkage and all lines. Bleed fuel system as outlined in paragraph 47.

57. **TIMING AND ADJUSTMENT.** To check injection timing with injection pump on tractor, first turn engine until number one piston is starting compression stroke, then remove timing pin and cover and reinsert timing pin into threaded hole in flywheel housing. Continue to turn engine until timing pin slides into hole in flywheel.

With engine set at TDC, shut off fuel and remove timing hole cover from injection pump. The timing lines on the governor weight retainer and pump cam should be aligned. If lines are not aligned, loosen pump mounting nuts and rotate pump either way as required to bring the lines into alignment. Hold pump in this position and tighten pump mounting nuts securely.

58. SPEED ADVANCE. The injection pump automatic advance unit is adjusted during production, however, due to variations in both fuel and operating temperatures, it may be necessary to make slight adjustments to provide optimum engine performance. The pump automatic advance can be checked, and adjusted if necessary, as follows:

NOTE: When working on injection pumps, take into consideration that pumps are sealed and breaking of seals by unauthorized personnel will void the injection pump warranty.

Shut off fuel, remove pump timing hole cover and install timing window (Roosa-Master No. 13366 or John Deere No. JD-259). Identify the window scribe line which is in line with pump cam ring line for future reference. Turn on fuel, start engine and bring to operating temperature. Run engine at 1400 rpm and read the cam advance at timing window. Cam scribe line should be advanced (moved up) 1½ marks on window which equals 3 pump degrees (6 crankshaft degrees) advance. If advance is not correct, use Fig. 46 as a guide and proceed as follows: Remove cap from the advance trimmer screw and loosen lock nut. Use an Allen wrench and turn trimmer screw clockwise to retard timing, or counterclockwise to advance timing.

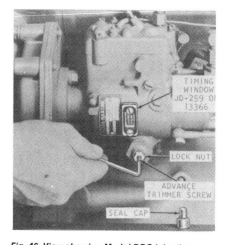

Fig. 46–View showing Model DBG injection pump cam advance trimmer screw. Refer to text for adjustment procedure.

To complete check, slowly increase engine speed to 2400 rpm and note the pump cam scribe line which should now have advanced (moved up) 3 marks on window which equals 6 pump degrees (12 crankshaft degrees) advance.

If injection pump will not meet both conditions outlined, either renew or overhaul the pump.

Model CDC Pump

Series 820 tractors, serial number 047753 to 054484 use Roosa Master Model CDC331-8DG injection pump. Recommended replacement pump for most models is the appropriate Model CBC pump plus attaching screws, thrust button and spring.

59. **REMOVE AND REINSTALL.** To remove injection pump, first close fuel shut-off valve on fuel tank, then clean injection pump, line connections and the surrounding area with clean diesel fuel.

Turn engine in the direction of normal rotation until number one piston is starting up on compression stroke, then remove flywheel timing hole cover and timing pin, reverse timing pin and reinsert in the threaded hole. Continue to slowly turn engine until timing pin enters hole in flywheel.

NOTE: If hole in flywheel goes past timing pin, turn engine counter-clockwise about ¼-turn, then again turn engine clockwise until timing pin and hole in flywheel are indexed. Engine is now at TDC.

Drain radiator, remove lower radiator hose, then remove access plate from front of timing gear cover. See Fig. 48. Remove pump drive gear mounting screw and the three gear retaining screws. Disconnect throttle rod, fuel supply line and fuel return line from pump. Remove the injector lines, then cap or plug all fuel openings. Support pump, remove the retaining nuts and washers and pull injection pump from engine. Pump

drive gear will be retained by timing gear cover.

When reinstalling pump, be sure engine is positioned at TDC. Remove injection pump timing pin from drive housing, reverse pin and reinsert it in the threaded hole. Turn pump drive shaft until timing pin drops into groove of pump drive shaft.

NOTE: Observe pin carefully as it drops only about 1/16-inch.

Hold pump in this position and mount pump on engine. Install pump drive gear, then recheck pump installation by loosening mounting nuts and turning pump counter-clockwise as far as it will go, then turning pump clockwise until timing pin indexes with pump drive shaft groove. Tighten mounting nuts and complete assembly by reversing disassembly procedure. Bleed fuel system as outlined in paragraph 47.

60. **TIMING AND ADJUSTMENT.** To check injection timing with injection pump on tractor, first turn engine until number one piston is starting compression stroke, then remove injection pump timing pin from drive housing, reverse pin and reinsert it in the threaded hole. Continue to turn engine in normal direction of rotation until timing pin drops into groove of pump drive shaft.

NOTE: Observe timing pin carefully as it will drop only about 1/16-inch when it indexes with groove in pump shaft. If pump shaft groove goes past timing pin, turn engine counter-clockwise about ¼-turn, then again turn engine clockwise.

With injection pump set as outlined, remove flywheel timing pin and timing hole cover, reverse pin and reinsert it

in threaded hole. Timing pin should enter hole in engine flywheel. If both timing pins index at the same time, injection timing is correct.

If both timing pins do not index at the same time, temporarily remove injection pump timing pin and index the flywheel timing pin. Loosen injection pump mounting nuts, rotate injection pump counter-clockwise on mounting studs and reinsert injection pump timing pin. Slowly turn injection pump clockwise until timing pin indexes with groove in pump drive shaft, then tighten mounting nuts.

61. **LOAD (CAM) ADVANCE.** The injection pump automatic advance unit is adjusted during production, however, due to variations in both fuel and operating temperatures, it may be necessary to make slight adjustments to provide optimum engine performance. The pump automatic advance can be checked, and adjusted if necessary, as follows: Be sure injection pump is static timed correctly as outlined in paragraph 60, then remove advance cam hole plug and install timing window (part number JD270). See Fig. 49.

NOTE: Notice that timing window has a bulls eye at the center with several concentric circles around it. Each circle is equal to two degrees.

Look through window and locate pump cam pin and align the side of pin nearest to engine with circular line of timing window. This point may vary between individual units and may not be at the bulls eye of timing window. Start engine and bring to operating temperature. Run engine at 800 rpm and observe pump cam pin which should still be at zero (original at rest) advance position. Advance engine speed to 1900 rpm and note cam advance in window. Cam pin should have advanced two marks (4 degrees) in window. If advance is not correct, re-

Fig. 47–Installed view of Model C injection pump showing points of adjustment.

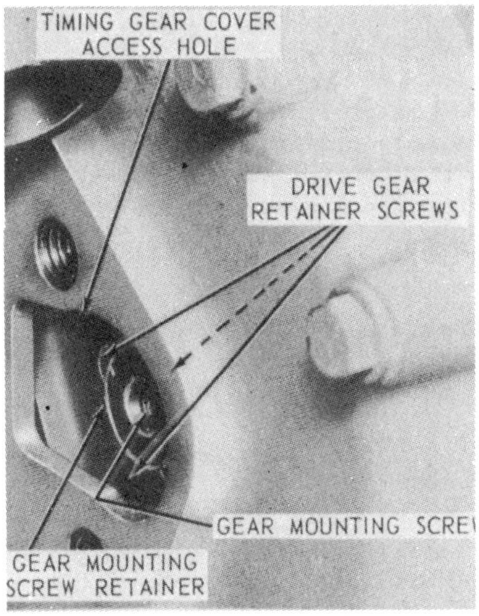

Fig. 48–Remove access hole cover from timing gear cover to gain access to the injection pump drive gear mounting screw and retaining screws on model CDC pump.

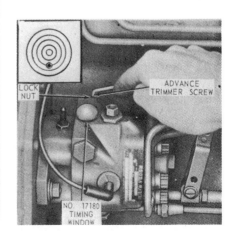

Fig. 49–Pump timing window installed to adjust intermediate advance. Concentric lines (inset) are two pump degrees apart.

move cap seal on the trimmer screw, located on engine side of injection pump, and adjust trimmer screw as required. See Fig. 49.

NOTE: When working on injection pumps, take into consideration that pumps are sealed and breaking of seals by unauthorized personnel will void the injection pump warranty.

Slowly advance engine speed to 2400 rpm and note the cam pin advance which should be 3 marks (6 degrees) in window. If pump does not advance 6 degrees at 2400 rpm, remove pump for test stand adjustment and/or service.

Model CBC Pump

Series 820 tractors, serial number 054485 to 082142, are equipped with a Roosa Master Model CBC 331-4AL injection pump. The appropriate model CBC pump is used as parts replacement for the earlier model CDC pump when complete pump is renewed, by also obtaining attaching cap screws, thrust button and spring.

62. REMOVE AND REINSTALL. To remove the injection pump, first close fuel shut-off valve on fuel tank, then clean injection pump, line connections and surrounding area with clean diesel fuel or solvent. Turn engine crankshaft until No. 1 piston is at TDC on compression stroke.

NOTE: Pump can be removed and reinstalled without regard to crankshaft timing position, however, positioning crankshaft at TDC-1 is recommended so timing can be properly checked and/or adjusted when pump is reinstalled. If timing is not to be checked, scribe timing marks on injection pump mounting flange and engine front plate which can be realigned when pump is reinstalled.

Disconnect or remove fuel inlet, return and pressure lines, throttle rod and solenoid wire from injection pump. Remove pump mounting stud nuts and pull pump straight to rear off pump shaft.

To install the pump without changing the timing, align mating timing marks (Fig. 50) on drive shaft and pump rotor and reinstall the pump with previously installed scribe lines in register.

To install and time the pump, check to be sure that No. 1 piston is at TDC on compression stroke. Remove timing pin (1—Fig. 51) from top of injection pump body and reinsert the pin long end down. Turn injection pump rotor until timing pin enters timing hole in rotor shaft, then install the pump.

On all installations, tighten injection pump retaining stud nuts to a torque of 4.2 kg/m (30 ft.-lbs.). Bleed pump and lines as outlined in paragraph 47. Adjust advance timing if necessary as in paragraph 64 and linkage as in paragraphs 70 or 71.

63. TIMING. To check injection pump timing without removing injection pump, turn engine until No. 1 piston is starting compression stroke, then remove engine timing pin and cover (Fig. 52). Insert timing pin into threaded hole in housing as shown and continue turning crankshaft until timing pin slides into timing hole in flywheel.

Invert pump timing pin (1—Fig. 51). Loosen mounting stud nuts (2) and rotate pump body in slotted holes if necessary until pump timing pin drops into hole in rotor. Tighten pump mounting nuts to 4.2 kg/m (30 ft.-lbs.) and reinstall and tighten both timing pins.

64. ADVANCE TIMING. First make sure static timing is properly set as outlined in paragraph 63. Shut off fuel, remove advance cam hole plug and install No. 17180 Timing Window as shown in Fig. 49. The timing window contains a series of concentric circles as shown in inset, and is designed so that all lines are two pump degrees apart. Note the location of hole in center of cam pin, which should be offset from center of bullseye as shown. Turn on fuel and start and warm engine, then note cam advance which should be 5 degrees at 1600 engine rpm. If it is not, loosen the locknut and turn advance trimmer screw until intermediate advance is correct. Maximum advance at full throttle should be 7°. If maximum advance is incorrect or intermediate advance cannot be properly adjusted, renew or overhaul the pump.

Model JDB Pump

Series 820 tractors after serial number 082143 are equipped with Roosa Master Model JDB33kAL2443 injection pump.

65. REMOVE AND REINSTALL. To remove the injection pump, first shut off fuel and clean injection pump, lines and surrounding area. Turn engine until No. 1 piston is at TDC on compression stroke.

NOTE: Pump can be removed and reinstalled without regard to crankshaft timing position, however, TDC-1 position is necessary if timing is to be checked. If timing is not to be changed, scribe timing marks on pump flange and engine front plate which can be realigned when pump is reinstalled.

Disconnect or remove fuel inlet, return and pressure lines, throttle rod and solenoid wire from injection pump. Remove mounting stud nuts and carefully slide pump straight to rear until clear of pump shaft and seals.

The pump shaft contains two soft plastic seals which are installed back to back as shown in Fig. 53. A special tool (Roosa Master 13369) is required

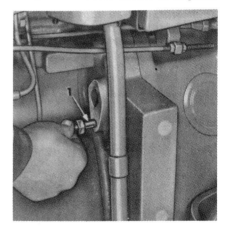

Fig. 52–Invert the threaded, flywheel timing pin (1) as shown to properly position crankshaft for injection pump installation.

Fig. 50–Mating timing marks on drive shaft and pump rotor must be aligned when CBC pump is installed.

Fig. 51–Invert the threaded timing pin on Model C pump to properly time the pump. Pin is normally installed as shown in inset.

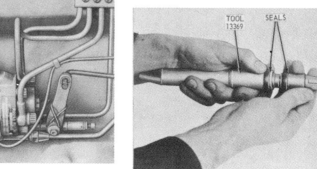

Fig. 53–A special tool is required to properly install shaft seals as shown.

to install seals as shown. Only the rear seal can be installed without removing shaft from tractor and gear from shaft. If both seals must be renewed, drain cooling system and remove lower radiator hose. Remove access plate from front of timing gear cover, back gear nut out until flush with end of shaft and jar shaft sharply with a drift and hammer to loosen gear, then withdraw shaft after removing the nut. Gear will remain in engagement with idler gear if timing gear cover is not removed.

To install new seals, first examine seal grooves in shaft carefully and remove any roughness or burrs. Coat seal liberally with Lubriplate and install from each end of shaft using the special installing tool. If shaft is removed and both seals renewed, reinstall shaft in pump before installing pump on engine. Reference mark (dot) on shaft tang and pump slot must align as shown in inset, Fig. 54.

The special Seal Installation Tool (Roosa Master 13371 or equivalent) must be used when installing pump (or shaft in pump). Also use extreme care. If resistance is felt, remove the pump and re-examine rear seal. If lip has been turned back, renew the seal.

If pump is not to be timed, realign the previously installed scribe marks and tighten stud nuts, then complete the assembly by reversing the removal procedure.

To time the pump, first be sure that crankshaft is at TDC-1. Remove timing cover from side of pump housing and turn the pump until governor weight timing line and cam timing line are in register as shown in Fig. 55. Tighten retaining cap screws to a torque of 35 ft.-lbs.

Bleed fuel system as outlined in paragraph 47 and if necessary, adjust throttle linkage as in paragraph 66.

66. **TIMING.** To check the timing without removing injection pump, turn crankshaft until No. 1 piston is coming up on compression stroke, then remove engine timing pin and cover as shown in inset, Fig. 55. Insert timing pin, long end first, into threaded hole in housing and continue turning crankshaft until pin slides into timing hole in flywheel.

Shut off fuel and remove timing cover from side of injection pump. With crankshaft at TDC-1, the timing scribe line on governor weight retainer should align with cam timing line as shown. If it does not, loosen pump mounting stud nuts and rotate pump in slotted holes until scribe lines are aligned. Hold pump in this position and retighten mounting stud nuts.

67. **ADVANCE TIMING.** The injection pump is provided with automatic speed advance which is factory set and will not normally need to be checked or reset. Minor adjustments can, however, be made without removal or disassembly of the pump. To check the advance mechanism, proceed as follows:

Shut off fuel, remove pump timing hole cover and install timing window as shown in Fig. 56. Turn on fuel and bleed fuel system, then start and run engine.

If maximum advance is not correct when tested at 2200 rpm, renew or overhaul the pump. If maximum advance was correct, reduce engine speed to 1200 rpm and set intermediate advance to 4° by loosening locknut and turning advance trimmer screw as shown. Tighten locknut and reinstall seal cap, then remove timing window and reinstall timing hole cover.

Roto Diesel Pump

All Series 830 tractors are equipped with a CAV Roto Diesel injection pump R3432 310.

68. **REMOVE AND REINSTALL.** To remove the injection pump, first shut off fuel and clean injection pump, lines and surrounding area. Pump can be removed and reinstalled without regard to crankshaft timing position and timing cannot be checked. The only critical requirement of the timing process is correct positioning of the injection pump drive gear as shown in Fig. 24.

Disconnect or remove fuel inlet, return and pressure lines, throttle rod and stop cable from injection pump. Drain radiator and remove lower radiator hose, then remove access plate from front of timing gear cover. Remove the three cap screws attaching drive gear to injection pump shaft flange. Support pump and remove the three mounting stud nuts, then pull pump from timing gear housing. Pump drive gear will be retained by timing gear cover.

When reinstalling pump, turn pump hub until timing slot in hub flange aligns with dowel pin in gear, and reverse removal procedure. Tighten gear mounting cap screws and flange mounting stud nuts to a torque of 2.5 kg/m (18 ft.-lbs.), making sure scribe marks (3 & 4—Fig. 57) are aligned as shown. Bleed the system as outlined in

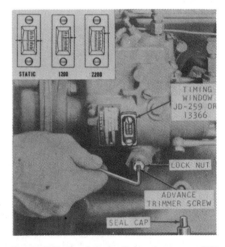

Fig. 56–Adjusting intermediate advance on JDB pump. Inset shows movement of scribe line on cam ring.

Fig. 54–Align reference marks (Inset) and use the special seal tool when installing pump.

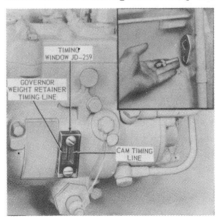

Fig. 55–With engine at TDC-1 when checked with timing pin (Inset), governor weight retainer timing line and cam timing line must register as shown. Timing window need only be installed to check intermediate advance as shown in Fig. 56.

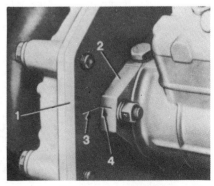

Fig. 57–Installed view of ROTO-DIESEL injection pump showing timing marks aligned. Pump will be in time with engine if gear timing marks are also aligned as shown in Fig. 24.

1. Front plate   3. Engine timing mark
2. Pump flange   4. Pump timing mark

paragraph 47 and adjust linkage as in paragraph 72.

SPEED AND LINKAGE ADJUSTMENT

Model DBG and JDB Pumps

69. To adjust throttle and governor linkage, first start engine and bring to operating temperature. Disconnect control rod (6—Fig. 59) from injection pump, move injection pump throttle arm to high idle position and check the engine high idle speed which should be 2545 rpm. If engine high idle is not as stated, break seal on pump fast idle adjustment (Fig. 58), turn screw as required and reseal. Move pump throttle arm to slow idle position and check engine speed which should now be 650 rpm. Turn pump slow idle adjustment (Fig. 58) as required.

NOTE: When adjusting high idle speed, take into consideration that breaking seal on adjustment screw may void the warranty if pump is still covered.

On DBG pump, be careful not to move throttle arm past Slow Idle position into Stop position which is controlled by the same arm. Slow idle position can be determined by the small area of free travel where engine speed ceases to decline.

With engine speeds adjusted at injection pump, reconnect control rod (6—Fig. 59). Raise cowl top door and remove right side cowl panel. Adjust stop screws (2 & 3) as required to allow slight overtravel of hand lever (1) in both directions.

Adjust foot throttle at yoke (4) until injection pump throttle arm contacts fast idle stop just before foot pedal contacts step plate.

Model CDC Pump

70. To adjust throttle and governor linkage, first start engine and bring to operating temperature. Disconnect control rod (6—Fig. 59) from injection pump. Move pump throttle arm AWAY from injection pump body to high idle

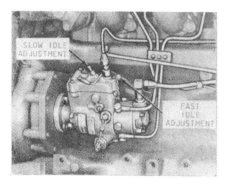

Fig. 58—Installed view of Model JDB injection pump showing slow and fast idle adjustment screws.

position and check engine high idle speed which should be 2545 rpm. If engine high idle speed is not as stated, remove slow idle adjusting screw from rear end of throttle control cap, insert a small blade screwdriver through screw hole and turn fast idle adjusting screw IN to increase, or OUT to decrease fast idle speed. DO NOT remove throttle control cap while engine is running.

With fast idle speed adjusted, reinstall slow idle adjusting screw and adjust speed to 650 rpm when throttle arm is moved fully toward injection pump.

Reconnect control rod (6—Fig. 59). Raise cowl top door, remove right side cowl panel and adjust stop screws (2 & 3) as required to allow slight overtravel of hand lever (1) in both directions.

Adjust foot throttle yoke (4) so that injection pump throttle arm contacts fast idle stop just before foot pedal contacts step plate.

Model CBC Pump

71. To adjust throttle and governor linkage, first start engine and bring to operating temperature. Disconnect control rod (6—Fig. 59) from injection pump. Move pump throttle arm (7—Fig. 61) away from injection pump as far as it will go and check engine high idle speed which should be 2545 rpm. If fast idle speed is not as stated, remove throttle cap (3) containing slow idle stop screw (1) and turn high speed adjusting nut (4) until specified high idle speed is obtained. Reinstall throttle cap (3) and check slow idle speed which should be 650 rpm.

Reconnect throttle rod (6—Fig. 59). Raise cowl top door, remove right side cowl panel and adjust stop screws (2 &

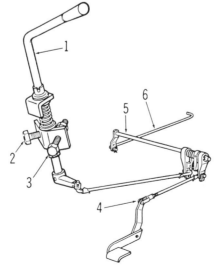

Fig. 59—Exploded view of throttle linkage showing points of adjustment.

1. Hand lever	4. Foot throttle yoke
2. Slow idle stop screw	5. Cross shaft
3. Fast idle stop screw	6. Control rod

3) as required to allow slight overtravel of hand lever (1) in both directions.

Adjust foot throttle yoke (4) so that injection pump throttle arm contacts fast idle stop just before foot pedal contacts step plate.

Roto Diesel Pump

72. To adjust throttle and governor linkage, start engine and bring to operating temperature. Disconnect control rod (6—Fig. 59) from injection pump and move throttle lever (3—Fig. 62) against fast idle adjusting screw (2). Engine speed should be 2545 rpm. If it is not, turn fast idle screw (2) in or out as required. Move injection pump throttle lever against slow idle adjusting screw (1); engine speed should be 650 rpm. If it is not, adjust slow idle screw. Reconnect throttle control rod (6 —Fig. 59) to injection pump.

Raise cowl top door, remove right side cowl panel and adjust stop screws (2 & 3) as required to allow slight overtravel of hand lever (1) in both directions.

Adjust foot throttle yoke (4) so that injection pump throttle arm contacts

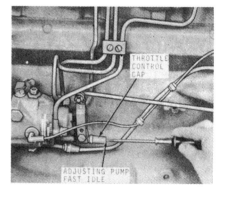

Fig. 60—Remove low idle adjusting screw and insert screw driver in end of throttle control cap to adjust the high idle screw on Model CDC pump.

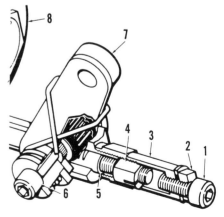

Fig. 61—Cross sectional view of governor adjustment on Model CBC injection pump; refer to paragraph 71.

1. Slow idle stop	
2. Jam nut	
3. Throttle cap	
4. High speed adjusting	5. Governor rod
nut	6. Clamp screw
	7. Lever arm
	8. Pump body

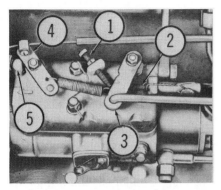

Fig. 62—Installed view of ROTO-DIESEL pump showing linkage adjustments.

1. Slow idle screw
2. Fast idle screw
3. Pump throttle arm
4. Stop lever
5. Lever stop

fast idle stop just before foot pedal contacts stop plate.

To adjust shut-off cable, completely push in stop knob and check to be sure stop lever (4—Fig. 62) contacts stop (5) on injection pump governor cover. If it does not, loosen cable clamp screw and reposition clamp on stop cable.

INJECTION PUMP SHUT-OFF

Models CDC, CBC, JDB

73. These models are equipped with a solenoid type shut-off which must be energized to provide fuel flow. The shut-off valve is spring loaded in the closed position and is opened when key switch is turned to "ON" or "START" position.

If tractor will not start, turn switch to "ON" and check continuity of solenoid lead.

COOLING SYSTEM

RADIATOR

All Models

74. **REMOVE AND REINSTALL.** To remove the radiator, first drain cooling system, remove grille screens and hood. Unbolt fan shroud from radiator and lay shroud back over fan. Disconnect upper and lower radiator hoses and upper radiator brace, then unbolt and remove radiator from tractor.

Install by reversing the removal procedure. Tighten top radiator brace as required to straighten radiator with cooling fan and lock in place with locknut.

WATER PUMP

All Models

75. **REMOVE AND REINSTALL.** To remove water pump, first remove

radiator as outlined in paragraph 74, then remove fan and fan belt. Disconnect by-pass hose from water pump, then unbolt and remove water pump from engine.

Reinstall by reversing removal procedure and adjust fan belt so a 25 lb. pull mid-way between pulleys will deflect belt ⅝-inch.

76. **OVERHAUL.** To disassemble water pump, use Fig. 63 as a guide and proceed as follows:

Support fan pulley hub in a press and, using a suitable mandrel, press shaft from pulley hub. Suitably support housing on gasket surface and press shaft, bearing, seal and impeller as a unit from housing.

Bearing outer race is a tight press fit in housing bore and is not otherwise secured. It is important therefore, that

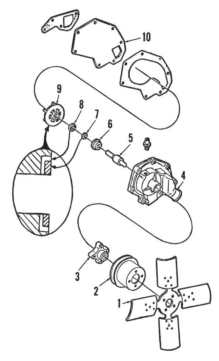

Fig. 63—Exploded view of water pump showing component parts.

1. Fan blades
2. Pulley
3. Hub
4. Body
5. Shaft & bearing
6. Seal
7. Insert
8. Cup
9. Impeller
10. Cover

Fig. 64—Exploded view of alternator unit showing component parts.

1. Nut
2. Pulley
3. Fan
4. Spacer
5. Drive end frame
6. Bearing
7. Snap ring
8. Rotor
9. Bearing
10. Stator
11. Negative heat sink
12. Positive heat sink
13. Insulators
14. Retainer
15. Slip ring end frame
16. Brush holder
17. Cover
18. Isolation diode

reasonable precautions be taken during assembly to prevent bearing movement during installation of impeller and pulley hub.

Coat outer edge of seal (6) with sealant and install in housing (4) using a socket or other driver which contacts only the outer flange of seal, then insert long end of bearing (5) through front of housing bore. Use tool No. JD-262 or a similar tool which contacts only outer race of bearing and press bearing into housing until front of bearing is flush with housing bore.

Install insert (7) in cup (8) with "V" groove of insert toward cup. Parts must be clean and dry. Dip cup and insert in engine oil then press cup and insert into impeller as shown in inset, until cup bottoms in impeller counterbore. Support front end of pump shaft and press impeller (9) on rear of shaft until fins are flush with gasket surface of housing.

Invert the pump assembly and support the unit on rear of shaft which is recessed into impeller hub. DO NOT support impeller or housing. With shaft suitably supported, press pulley hub (3) on front of shaft until front face of pulley hub is approximately flush with hub bore. Complete the assembly by reversing the disassembly procedure.

ELECTRICAL SYSTEM

ALTERNATOR AND REGULATOR

Motorola Alternator

77. **OPERATION AND TESTING.** Refer to Fig. 64 for an exploded view of alternator unit.

The isolation diode and brush holder can be renewed without removal or disassembly of alternator; all other alternator service requires disassembly.

The primary purpose of the isolation diode is to permit use of charging indicator lamp. Failure of the isolation diode is usually indicated by the indi-

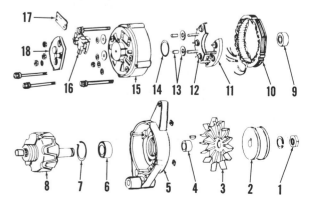

cator light; which glows with engine stopped and key switch off if diode is shorted; or with engine running if diode is open.

Failure of a rectifying diode may be indicated by a humming noise when engine is running, if diode is shorted; or by a steady flicker of charge indicator light at slow idle speed if diode is open. Either fault will reduce alternator output.

To check the charging system, refer to Fig. 65 and proceed as follows:

(1) With key switch and all accessories off and engine not running, connect a low reading voltmeter to terminals D-F. Reading should be 0.1 volt or less. A higher reading would indicate a short in isolation diode, key switch or wiring.

(2) Turn key switch on but do not start engine. Recheck voltmeter reading which should be 1-3 volts. A higher or lower reading may indicate a defective alternator, regulator or wiring.

(3) Start and run engine at approximately 1400 rpm and, with all accessories off, again check voltmeter reading which should be 15 volts. A lower reading could indicate a discharged battery or defective alternator.

(4) Move voltmeter lead from auxiliary terminal (D) to output terminal (E) and recheck voltage. Reading should drop one volt from reading in previous test (3), reflecting the 1-volt resistance built into isolation diode (18 —Fig. 64). If battery voltage (12 volts) is obtained, isolation diode is open and must be renewed.

(5) If a reading lower than the specified 15 volts was obtained when checked as outlined in test 3, stop engine and disconnect regulator plug (B). Connect a jumper wire between output terminal (E) and field terminal (C) on alternator brush holder. Connect a suitable voltmeter to terminals (D-F) on alternator. Start engine and slowly

increase engine speed while watching voltmeter. If a reading of 15 volts can now be obtained at 1300 engine rpm or less, renew the regulator. If a reading of 15 volts cannot be obtained, renew or overhaul the alternator.

CAUTION: DO NOT allow voltage to rise above 16.5 volts when making this test. DO NOT run engine faster than 1500 rpm with regulator disconnected.

78. **OVERHAUL.** The isolation diode and brush holder can be removed without removing alternator from tractor. Brush holder should be removed before attempting to separate the frame units.

To disassemble the removed alternator unit, remove through-bolts and separate slip ring end frame (15—Fig. 64) from drive end frame (5). Rotor (8) will remain with drive end frame and stator (10) with slip ring end frame. Be careful not to damage stator windings when prying units apart.

Examine slip ring surfaces of rotor for scoring or wear and field windings for overheating or other damage. Check bearing surfaces of rotor shaft for visible wear or scoring. Check rotor for grounded, shorted or open circuits using an ohmmeter as follows:

Refer to Fig. 66 and touch the ohmmeter probes to points (1-2) and (1-3); a reading near zero will indicate a ground. Touch ohmmeter probes to the two slip rings (2-3); reading should be 5.5 ohms. A higher reading will indicate an open field circuit, a lower reading a short.

Runout should not exceed 0.05MM (0.002 inch). Slip ring surfaces can be trued if runout is excessive or if surfaces are scored. Finish with 400 grit or finer polishing cloth until scratches or machine marks are removed.

Stator is "Y" connected (Fig. 67) and center connection need not be unsoldered to test for continuity. Each field winding uses two coils as shown. Conti-

nuity should exist between any two of the stator leads but not between any lead and stator frame. Because of the low resistance, shorted windings within a coil cannot be satisfactorily checked. Three positive diodes are located in slip ring heat sink (12—Fig. 64) and three negative diodes in grounded heat sink (11). Diodes should test at or near infinity in one direction when checked with an ohmmeter, and at or near zero when meter leads are reversed. Renew any diode with approximately equal meter readings in both directions. Diodes must be removed and installed using an arbor press or vise and a suitable tool which contacts only outer edge of diode. Do not attempt to drive a faulty diode out of heat sink, as shock may cause damage to other good diodes. If all diodes are being renewed, make certain the positive diodes (marked with red printing) are installed in positive heat sink (12) and negative diodes (marked with black printing) are in-

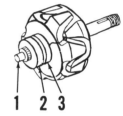

Fig. 66–Removed rotor assembly showing test points to be used when checking for grounds, shorts and opens.

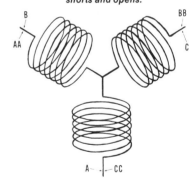

Fig. 67–Schematic view of typical "Y" connected stator. Center connection need not be unsoldered to check for continuity.

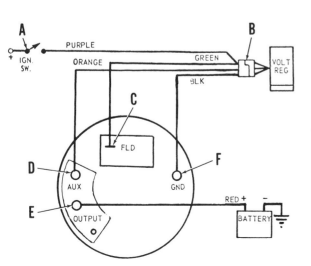

Fig. 65–Wiring diagram of charging system showing test points. Refer to paragraph 77 for test procedure.

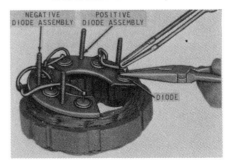

Fig. 68–Use needle nose pliers as a heat sink and an iron only, when soldering diode connections.

stalled in negative heat sink (11). Use a pair of needle nose pliers as a heat sink when soldering diode leads (See Fig. 68). Use only rosin core solder and an iron instead of a torch. Excess heat can damage a good diode while it is being installed.

Exposed length of brushes in removed brush holder should be ¼-inch or more. Brushes are available only in an assembly with the holder. Check for continuity between field terminal (A—Fig. 69) and insulated brush (C); and between brush holder (B) and grounded brush (D). Wiggle the brush and lead while checking, to test for poor connections or an intermittent ground.

NOTE: A battery powered test light can be used instead of an ohmmeter for all electrical tests except shorts in rotor winding. When checking diodes however, test light must not be more than 12 volts.

Bosch Alternator

79. OPERATION AND TESTING. Refer to Fig. 70 for exploded view of BOSCH 12 Volt, 28 Ampere alternator used on some tractors.

To check the charging system, connect a voltmeter to B+ (Output) terminal (1—Fig. 71) on alternator and to a suitable ground. With engine running at 1200 rpm, reading should be 13 volts or above. A lower reading could indicate a discharged battery or faulty alternator or regulator.

Fig. 69–Check brush holder for continuity between A-C; and B-D.

With engine not running, disconnect the three-terminal plug (2) from alternator and touch ammeter leads to DF (Green Wire) terminal and B+ terminal. Current draw should be approximately 2 Amperes. High readings are caused by shorts or grounds. Low readings may be caused by dirty slip rings or defective brushes.

Connect a jumper wire between B+ and DF terminals and a voltmeter between B+ terminal and ground. Start engine and increase engine speed until voltage rises to not to exceed 15.5 volts. If maximum obtainable reading is less than 14 volts and battery is fully charged, alternator is at fault.

Move jumper wire connection from B+ terminal to D+ terminal. Voltage reading should remain the same. If voltage drops, exciter diodes are defective. If voltage reading was normal with plug disconnected and jumper wire installed, but was low when tested with plug connected, renew the regulator.

80. OVERHAUL. Brush holder (1—Fig. 70) can be removed without removing alternator from the tractor. Brush holder should be removed before attempting to separate alternator frame units.

To disassemble the removed alternator, first remove capacitor and brush holder. Immobilize pulley and remove shaft nut, pulley and fan. Mark the brush end housing, stator frame and drive end housing for correct reassembly and remove through-bolts. Rotor will remain with drive-end frame and stator will remain with brush-end frame when alternator is disassembled.

Remove the two terminal nuts and three screws securing rectifier (2) to brush end frame and lift out rectifier and stator (3) as a unit. Carefully tag the three stator leads for correct rejoining then unsolder the leads from rectifier diodes, using an electric soldering iron and minimum heat. Be careful not to get solder on diode plates or overheat the diodes.

Check brush contact surface of slip rings for burning, scoring or varnish coating. Surfaces must be true to within 0.05MM (0.002 inch). Contact surface may be trued by chucking armature in a lathe. Polish the contact surface after truing using 400 grit polishing cloth, until scratches and machine marks are removed. Check continuity of rotor windings using an ohmmeter as shown in Fig. 73. Ohmmeter reading should be 4.0-4.4 Ohms between the two slip rings and infinity between either slip ring and rotor pole or shaft.

Stator is "Y" wound, the three individual windings being joined in the middle and two coils being used for each stator circuit. Test the windings using an ohmmeter as shown in Fig.

Fig. 71–Installed view of BOSCH alternator.

1. Output terminal
2. Terminal plug
3. D+ terminal

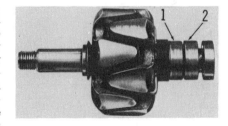

Fig. 72–A reading of 4.0-4.4 Ohms should exist between slip rings (1 & 2) when tested with an ohmmeter.

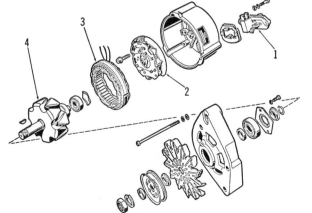

Fig. 70–Exploded view of BOSCH alternator used on some tractors.

1. Brush holder
2. Rectifier
3. Stator
4. Rotor

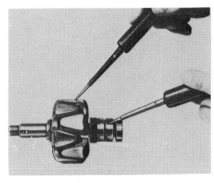

Fig. 73–No continuity should exist between either slip ring and any part of rotor frame.

74. Ohmmeter reading should be 0.4-0.44 Ohms between any two leads and infinity between any lead and stator frame.

Alternator brushes and shaft bearings are designed for 2000 hours of service life. New brushes protrude 10MM (0.4 inch) beyond brush holder when unit is removed; for maximum service reliability, renew BOTH the brushes and shaft bearings when brushes are worn to within 5MM (0.2 inch) of holder. Solder copper leads to allow for 10MM (0.004 inch) protrusion using

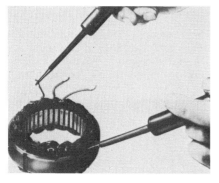

Fig. 74–No continuity should exist between any stator lead and stator frame.

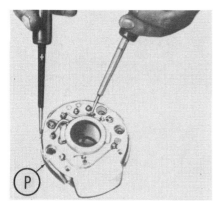

Fig. 75–A near infinity reading should be obtained when positive probe rests on positive heat sink (P) and negative probe touches diode leads as shown. Reverse the probes and reading should be near zero Ohms.

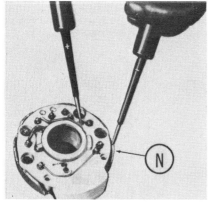

Fig. 76–With negative probe on negative heat sink (N) and positive probe touching diode leads, an infinity reading should be obtained. Reverse the probes and reading should be near zero.

rosin core solder only, and making sure that solder does not seep into and stiffen the wire lead.

The rectifier is furnished as a complete assembly and diodes are not serviced separately. The rectifier unit contains three positive diodes, three negative diodes and three exciter diodes which energize the rotor coils before engine is started. If any of the diodes fail, rectifier unit must be renewed.

To test the positive diodes, touch positive ohmmeter probe to positive heat sink as shown in Fig. 75, and negative test probe to each diode lead in turn. Ohmmeter should read at or near infinity for each test. Reverse the leads and repeat the series; ohmmeter should read at or near zero for the series.

Test negative diodes as shown in Fig. 76. Place negative test probe on negative heat sink and touch each diode lead in turn with positive test probe. Ohmmeter should read at or near infinity for the series. Reverse the leads and repeat the test; ohmmeter should read at or near zero for the series.

Test exciter diodes by using the D+ terminal as the base as shown in Fig. 77. Ohmmeter should read at or near infinity with positive test probe on terminal screw and at or near zero with negative test probe touching screw.

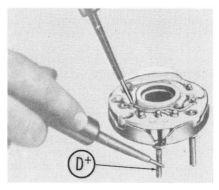

Fig. 77–D+ terminal is used to test exciter diodes, refer to paragraph 79.

When assembling alternator, tighten through-bolts to approximately 38-55 kg/cm (35-45 inch-pounds and pulley nut to 3.5-4.5 kg/m (25-30 ft.-lbs.).

STARTING MOTOR
All Models

81. A ROBERT BOSCH 0001 359 016 starting motor is used. To remove the starter on some models, it is first necessary to remove engine oil dipstick and tube (T—Fig. 78) from cylinder block. When reinstalling, make sure top of tube is adjusted to a height of 152.5MM (6 inches) from machined surface of cylinder block as shown in inset, Fig. 78.

To disassemble the removed starter unit, remove through-bolts from brush end, then remove brush end cover. Brushes can be examined or renewed at this time. Recommended wear limit of used brushes is 15.5MM (0.625 inch). Renew springs which are damaged or heat discolored.

To further disassemble starter, remove the two insulated brushes (B—Fig. 79) to free field coil leads, lift out "C" clip (C) and withdraw brush end frame as a unit, being careful not to lose thrust washer and shim pack (S—Fig. 80) which control armature end play. Disconnect starter lead from sole-

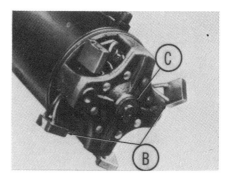

Fig. 79–To remove brush end frame from BOSCH starter, first remove insulated brushes (B), then "C" clip (C).

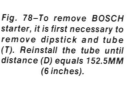

Fig. 78–To remove BOSCH starter, it is first necessary to remove dipstick and tube (T). Reinstall the tube until distance (D) equals 152.5MM (6 inches).

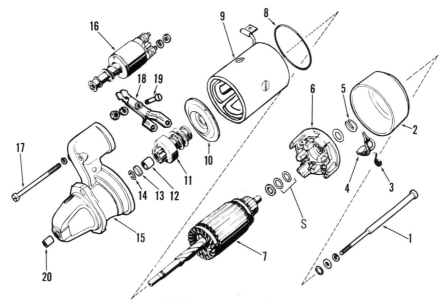

Fig. 80–Exploded view of BOSCH starting motor, showing component parts.

S. Shim pack	5. "C" clip	10. Guide plate	15. Drive end frame
1. Through-bolts	6. Brush end frame	11. Drive	16. Solenoid
2. Cover	7. Armature	12. Bushing	17. Through-bolts
3. Spring	8. Seal ring	13. Pinion stop	18. Shift lever
4. Brush	9. Center frame	14. Snap ring	19. Pivot bolt

noid (16) and remove solenoid through-bolts (17). Push starter drive (11) toward engaged position, then unhook and lift off solenoid unit. Remove shift lever pivot bolt (19), then withdraw drive end frame (15) from armature and drive unit, pulling shift lever out with starter drive as drive end frame is withdrawn.

To remove starter drive (11), push

pinion stop (13) back away from snap ring (14), then spread and remove snap ring. Drive parts can now be removed and armature withdrawn from guide plate (10).

Assemble by reversing the disassembly procedure. Pinion stop (13) must be pushed down over snap ring

(14) when drive unit is installed, to contain the ring. Armature should have 0.1-0.3MM (0.004-0.012 inch) end play in brush end frame (6) with "C" clip (5) installed. End play is controlled by thickness of shim pack (S). Shims are available in an armature assembly kit which also includes "C" clip (5), commutator end thrust washers, shims (S), pinion stop (13) and snap ring (14).

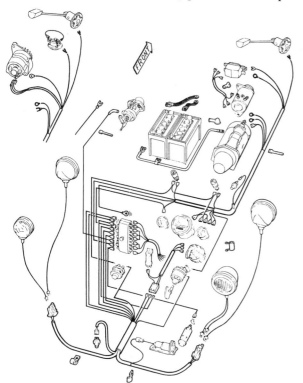

Fig. 81–Schematic view of wiring harness used on early 820 models.

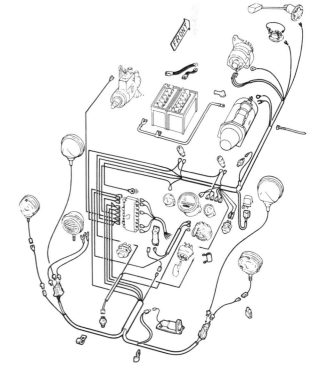

Fig. 82–Schematic view of wiring diagram used on late 820 models.

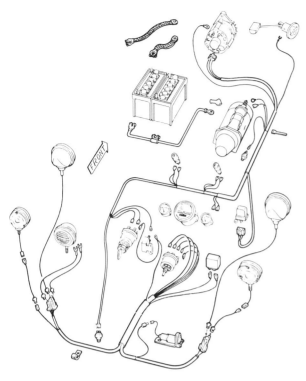

Fig. 83—Schematic view of wiring diagram used on 830 models.

ENGINE CLUTCH

ADJUSTMENT

All Models

82. **FREE PLAY AND PEDAL POSITION.** Clutch pedal free play should be one inch, and should be adjusted when free play decreases to ½-inch. Adjustment is made by disconnecting yoke (Fig. 84) from clutch pedal arm and shortening or lengthening clutch operating rod as required.

Clutch pedal is used in two positions, adjusted by pedal positioning cap screw shown in Fig. 85. When power take-off is being used, loosen cap screw and move pedal pad rearward until cap

screw contacts front of slot. If only the transmission clutch is being used, pedal pad may be moved forward until cap screw contacts rear of slot. CAUTION: PTO clutch will not release with pedal in forward position.

83. **RELEASE LEVERS.** Clutch release levers can be adjusted for wear without disassembly or removal of unit. To make the adjustment, disconnect clutch operating rod from clutch pedal arm (Fig. 84) and refer to Fig. 87. Remove access cover from clutch housing and back off clutch operating rod bolt nuts until operating lever contacts pto clutch plate pins for all three operating levers. Tighten inner nut until lever begins to pull away from pto clutch plate pin, tighten nut an additional 2½ turns and secure by tight-

ening locknut. Adjust clutch operating rod until, with yoke connected, release bearing just contacts operating lever. Turn flywheel until each of the other clutch levers are in position, and adjust the lever to lightly contact release bearing. Tighten all locknuts securely while maintaining the adjustment, then adjust clutch pedal free travel as outlined in paragraph 82.

TRACTOR SPLIT

All Models

84. To detach (split) engine from clutch housing, first drain transmission case and cooling system. Remove hood. Disconnect battery cables and remove batteries and the wood (insulator) block, then remove the two cap screws securing cowl frame to flywheel housing. Disconnect wiring and move wiring harness rearward. Disconnect and remove interfering hydraulic lines. Disconnect throttle control rod from injection pump and on models so equipped, disconnect stop cable. Disconnect drag link at either end. Place a suitable floor jack underneath clutch housing and attach a hoist to engine, then remove attaching cap screws and separate the tractor.

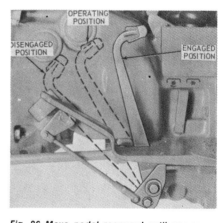

Fig. 86—Move pedal rearward until cap screw contacts front of slot (Fig. 85) to permit full disengagement of pto clutch.

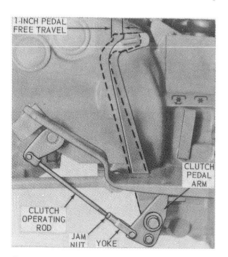

Fig. 84—External clutch adjustment is made by disconnecting rod yoke from pedal arm and adjusting length of operating rod.

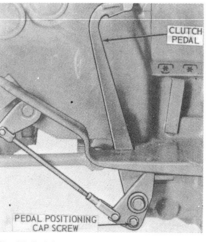

Fig. 85—Pedal position is adjusted by loosening cap screw and shifting pedal in slotted hole.

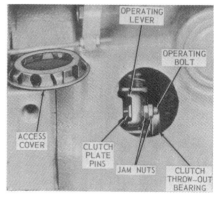

Fig. 87—Wear adjustment of dual clutch levers can be made as shown. Refer to paragraph 83.

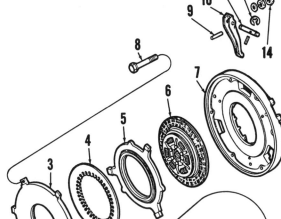
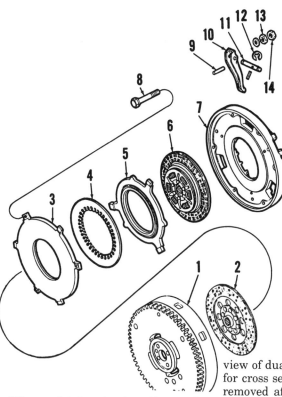

Fig. 88–Exploded view of dual clutch unit.

1. Flywheel
2. Transmission clutch disc
3. Front pressure plate
4. Diaphragm spring
5. Rear pressure plate
6. PTO clutch disc
7. Cover
8. Actuating bolt
9. PTO release pin
10. Release lever
11. Pivot pin
12. "E" ring
13. Adjusting nut
14. Jam nut

When rejoining tractor, it may be necessary to turn flywheel or input shafts until splines engage. Keep the parts in close alignment, using aligning studs if necessary. Make sure flywheel housing and clutch housing are butted together before tightening cap screws, then tighten evenly to a torque of 23.5 kg/m (170 ft.-lbs.).

R&R AND OVERHAUL

All Models

85. Refer to Fig. 88 for an exploded

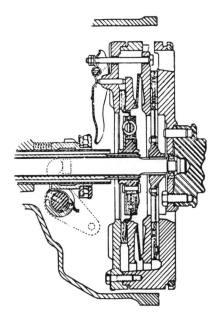

Fig. 89–Cross sectional view of dual clutch assembly showing clutch discs correctly installed.

view of dual clutch unit and to Fig. 89 for cross sectional view. Clutch can be removed after clutch split outlined in paragraph 84. When installing clutch, make sure long hub of transmission clutch disc (2—Fig. 102) is forward. Use a suitable alignment tool and tighten retaining cap screws to a torque of 4.8-5.5 kg/m (35-40 ft.-lbs.).

To disassemble the clutch cover and associated parts, remove locking nuts (14) and back off jam nuts (13) evenly until spring pressure is relieved. Mark the cover (7), rear pressure plate (5) and front pressure plate (3) with paint or other suitable means so balance can be maintained, and separate the units.

Inspect pressure plates (3 & 5) for cracks, scoring or heat discoloration and renew as necessary. Check diaphragm spring (4) for heat discoloration, distortion or other damage and renew if its condition is questionable. Renew transmission clutch disc (2) if facing wear approaches rivet heads, if hub is loose or splines are worn, or if disc is otherwise damaged. PTO disc (6) should be renewed if total thickness at facing area is 3/16-inch or less.

Assemble by reversing the disas-

Fig. 90–View of clutch housing showing brake reservoir supply line (L) and transmission pump relief valve (R) exploded.

S. Seal rings

sembly procedure, installing pto clutch disc with hub offset toward front as shown in Figs. 88 and 89. Adjust release levers after installation as outlined in paragraph 83 and clutch linkage as in paragraph 82.

CLUTCH SHAFT

All Models

86. To remove clutch shaft it is necessary to separate clutch housing from transmission housing. Clutch shaft can be removed from rear of clutch housing.

If tractor has continuous-running pto and there is evidence of oil seepage between clutch shaft and powershaft (pto shaft) separate shafts and inspect oil seal and pilot. Press new pilot (cup rearward) into bore of powershaft until it bottoms. Press oil seal in bore of powershaft, with lip rearward, until it contacts pilot.

When installing clutch shaft, or powershaft when tractor is equipped with continuous-running pto, be sure lugs on shaft align with slots in transmission oil pump drive gear.

CLUTCH RELEASE BEARING AND YOKE

All Models

87. The clutch release bearing can be removed after clutch housing is split from engine as outlined in paragraph 84. Release bearing and carrier is sold as an assembly only.

CLUTCH HOUSING

All Models

88. Clutch control linkage can be serviced after clutch housing is split from engine as outlined in paragraph 84.

Brake reservoir supply line (L—Fig. 90), clutch shaft and associated parts can be renewed after transmission split as outlined in paragraph 90.

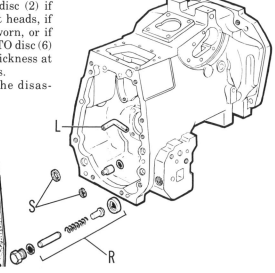

TRANSMISSION

Transmissions are constant mesh type using helical cut gears. Two shift levers are used, the left lever selecting high, low and reverse ranges as well as a park position. The right lever controls a four step gear arrangement. Thus with the two shift levers, eight forward speeds (four in high range and four in low range) and four reverse speeds are available. The four reverse speeds approximate in mph the four forward speeds obtained in low range.

TOP (SHIFTER) COVER

All Models

89. **REMOVE AND REINSTALL.** To remove the shifter cover, remove the shield and work it up over shifter lever boots. Disconnect connector from starting safety switch and bleed line from fitting on shift cover. If necessary, disconnect rear wiring harness, then unbolt and lift shifter cover from clutch housing.

Any further disassembly required will be obvious after examination of the unit. See Fig. 91.

Reinstall by reversing the removal procedure.

NOTE: Shifter rails and forks are an integral part of the transmission and can be serviced after transmission is split from clutch housing as outlined in paragraph 90.

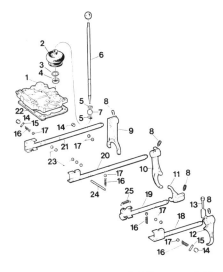

Fig. 91–Exploded view of shift cover, shifter shafts, forks and associated parts.

1. Cover	13. Safety switch pin
2. Boot	14. Caps
3. Snap ring	15. Spring pins
4. Retaining ring	16. Detent springs
5. Snap rings	17. Steel balls
6. Lever	18. Low/reverse shaft
7. Actuating ball	19. High/park shaft
8. Set screws	20. High speed shifter
9. Low speed shifter	shaft
fork	21. Low speed shifter
10. High speed shifter	shaft
fork	22. Gasket
11. High/park range fork	23. Interlock pin
12. Low/reverse range	24. Interlock balls
fork	

TRACTOR SPLIT

All Models

90. To split tractor between clutch housing and transmission case, disconnect battery ground straps, drain transmission case, remove shifter cover as outlined in paragraph 89, then remove the two clutch housing to transmission case cap screws located at rear of shifter cover opening under mounting flange. Remove left platform and unhook clutch return spring. Remove right platform, disconnect the two brake pressure lines from brake valve and the main hydraulic pressure line from tee or elbow near transmission oil filter. Remove plate retaining hydraulic pump inlet line and reservoir return line at lower right side of transmission case. Disconnect tail light wires and if so equipped, disconnect hydraulic outlet mid-couplers. Support transmission case, place a rolling floor jack under front section and block front axle to prevent front end from tipping. Remove remaining clutch housing to transmission cap screws and separate tractor.

NOTE: The cap screw located in front of the transmission filter cannot be completely removed unless filter is removed, however, cap screw can be unscrewed as tractor is split and left in casting hole.

When rejoining tractor, make sure sealing rings (S—Fig. 90) are in position in clutch housing flange. Make sure clutch housing flange is flush against transmission housing and tighten retaining cap screws to a torque of 11.8 kg/m (85 ft.-lbs.).

SHIFTER MECHANISM

All Models

91. **REMOVE AND REINSTALL.** To remove shifter shafts (rails) and forks, first split tractor between transmission and clutch housing as outlined in paragraph 90 and remove rockshaft housing as in paragraph 129. Loosen set screws securing forks to shift shafts.

Before attempting to remove any shift rail, refer to Figs. 91 and 92 and determine location and interaction of detent balls and interlock mechanism. When withdrawing shifter shafts, DO NOT turn shafts so that set screw holes will align with detent location as shown in Fig. 91. If detent ball drops into set screw hole, shaft will be impossible to remove.

Place all shift shafts in neutral and withdraw left hand shaft (18), removing fork (12) as it is free and retrieving detent ball and two interlock balls as shift shaft emerges from housing front wall. Move high SPEED shift shaft (20) into either gear to free interlock pin (24), then withdraw range shift shaft (19) removing fork and detent ball.

Pull inner speed shift shaft (20) forward until cross drilling in shaft emerges, and retrieve the three interlock balls (23) from cross drilling. Continue to withdraw the shaft, capturing detent ball and the FOUR interlock balls in two housing walls as shaft is removed. Remove the remaining shifter shaft, detent ball and shifter fork. If interlock pin (24) must be renewed, drive the spring pin (15) from either end of housing cross drilling and push cup plug (14) out with interlock pin.

Assemble shift mechanism by reversing removal procedure, using Figs. 91 and 92 as a guide for installation of interlock components. Check shifter mechanism in all gears before rejoining tractor, keeping the following points in mind. Outer (right) speed shifter shaft (21—Fig. 92) must be in either gear before inner range shifter shaft can be moved into PARK. When one speed shifter shaft or range shifter shaft is moved to an engaged position, the other shaft in that pair cannot be moved out of neutral detent.

COUNTERSHAFT

All Models

92. **R&R AND OVERHAUL.** To remove the countershaft, split tractor as outlined in paragraph 90 and remove shifter shafts and forks as outlined in paragraph 91. Remove the pto gear, or gears, from front of transmission. Remove cap screws from countershaft bearing support (13—Fig. 93). Remove snap ring (52) from its groove at rear end of countershaft, then use a screw driver and turn locking washer (48) until splines of washer index with splines of countershaft. Pry bearing support off dowels (12), pull assembly

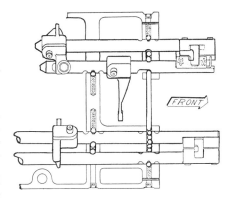

Fig. 92–Schematic view of shifter mechanism showing location and operation of detent assemblies and interlock balls and pin. Refer to Fig. 91 for exploded view.

forward and lift gears from transmission case as they come off shaft.

Inspect all gears, thrust washers and shift collar for broken teeth, excessive wear or other damage and renew as necessary. If support assembly bearings or shafts, or the snubber brake assemblies (19) require service, the shafts and bearings can be pressed out after removing retaining snap rings; however, the transmission drive gear must be removed before countershaft can be removed. Snubber brake springs (18) should test 28.5-34.9 kg/m (63-77 lbs.) when compressed to a length of 38.1MM (1½ inches). Needle bearing (54) can be removed from its bore after removing snap ring (53).

INPUT SHAFT

All Models

93. **R&R AND OVERHAUL.** To remove input shaft it is first necessary to remove countershaft as outlined in paragraph 92.

With countershaft removed, the transmission input shaft is removed as follows: Remove transmission oil cup and lines. Straighten lock plates and remove input shaft bearing quill (8—Fig. 93) and shims (7) from front of input shaft. Bump input shaft forward until front bearing cup (6) clears its bore, then move input shaft forward, lift rear end of shaft and remove input shaft from transmission case.

With input shaft removed, inspect all gears for chipped teeth or excessive wear. Inspect bearings and renew as necessary. Bump bearing cup (1) forward if removal is required. Inspect needle bearing (4) and renew if necessary.

Install and adjust end play of input shaft as follows: Be sure bearing cup (1) is bottomed in bore and place input shaft in position. Use original shim pack (7), or use a new shim pack approximately 0.75MM (0.030 inch) thick, install front bearing quill (8) and tighten cap screws to a torque of 4.8 kg/m (35 ft.-lbs.). Use a dial indicator to check the input shaft end play which should be 0.1-0.15MM (0.004-0.006 inch). Vary shims as required. Shims are available in thicknesses of 0.07, 0.13, 0.25, 0.051 and 1.0MM (0.003, 0.005, 0.010, 0.020 and 0.040 inch). Do not forget front oil line clamp when making final installation.

PINION SHAFT

All Models

94. **R&R AND OVERHAUL.** To remove the transmission pinion shaft, remove the input shaft as in paragraph 93 and the differential as in paragraph 98.

With differential removed, remove oil line, nut (43—Fig. 93), bearing (42), shims (40) and spacer (39). Use a screw driver and turn thrust washers until splines of thrust washers are indexed with splines of countershaft. Pull countershaft rearward and remove parts from transmission case as they come off shaft. Bearing cup (25) and shims (26) can be removed from housing by bumping cup rearward. Be sure to keep shims (26) together as they control the bevel gear mesh position. Bearing cup (41) can be removed from housing by bumping cup forward.

Check all gears and shafts for chipped teeth, damaged splines, excessive wear or other damage and renew as necessary. If pinion shaft is renewed, it will also be necessary to renew the differential ring gear and right hand differential housing as these parts are not available separately. Bearing (24) is installed with large diameter toward gear end of shaft.

NOTE: Mesh (cone point) position of the pinion shaft and main drive bevel pinion gear is adjusted by two 0.25MM (0.010 inch) shims (26) located between rear bearing cup (25) and housing.

Install pinion shaft and adjust shaft bearing preload as follows: Use Fig. 93 as a guide and with bearing (24) on pinion shaft, start shaft into rear of housing. With shaft about half-way into housing, place 1st and 5th speed gear (27) on shaft with teeth for shift collar (29) toward front. Place the thickest thrust washer (28) on shaft, then install coupling sleeve (30) and shift collar (29). Move shaft forward slightly and install 2nd and 6th speed gear (31) with teeth for shift collar toward rear. Place thrust washer with outer tangs (32) over shaft, then slide retaining washer (33) over thrust washer (32). Move shaft slightly forward and install 4th and 8th speed gear (34) on shaft with teeth for shift collar toward front. Place the thinnest thrust washer (35) on shaft and install shift collar sleeve (37) and shift collar (36). Install 3rd and 7th speed gear (38) on shaft with teeth for shift collar toward rear. Push shaft forward until rear bearing cone (24) seats in bearing cup (25) and use screw driver to turn thrust washers until splines on thrust washers lock with splines of pinion

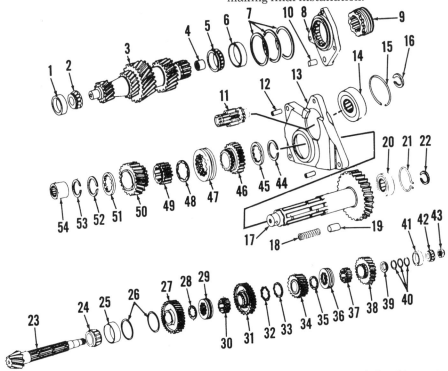

Fig. 93—Exploded view showing transmission shafts and gears. Note that transmission drive gear (11) is mounted in countershaft support (13).

1. Bearing cup	(thickest)	41. Bearing cup
2. Bearing cone	16. Snap ring	42. Bearing cone
3. Input shaft	17. Countershaft	43. Nut
4. Needle bearing	18. Brake spring	44. Snap ring
5. Bearing cone	19. Brake plug	45. Thrust washer
6. Bearing cup	20. Ball bearing	46. Reverse pinion
7. Shims	21. Snap ring	47. Shift collar
8. Bearing quill	22. Snap ring	48. Thrust washer
9. Shifter collar	23. Pinion shaft	49. Shift collar sleeve
10. Dowel	24. Bearing cone	50. Low range pinion
11. Drive gear	25. Bearing cup	51. Thrust washer
12. Dowel	26. Shims	52. Snap ring
13. Support	27. 1st & 5th gear	53. Snap ring
14. Ball bearing	28. Thrust washer	54. Needle bearing
15. Snap ring	29. Shift collar	
	30. Shift collar sleeve	
	31. 2nd & 6th gear	
	32. Thrust washer (outer tangs)	
	34. 4th & 8th gear	
	35. Thrust washer (thinnest)	
	36. Shift collar	
	37. Shift collar sleeve	
	38. 3rd & 7th gear	
	39. Spacer	
	40. Shims	

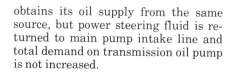

Fig. 94–View of transmission oil pump (P) with clutch shaft and pto powershaft removed.

I. Inlet line
O. Outlet line
P. Transmission pump

shaft. Install spacer (39), shims (40), bearing (42) and nut (43), then adjust pinion shaft bearing preload as outlined in paragraph 95.

95. PINION SHAFT BEARING ADJUSTMENT. The pinion shaft bearings must be adjusted to provide a bearing preload of 0.15MM (0.006 inch) (5-15 in. lbs. rolling torque). Adjustment is made by varying the number of shims (40—Fig. 93).

To adjust the pinion shaft bearing preload, proceed as follows: Mount a dial indicator with contact button on front end of pinion shaft and check for end play of shaft. If shaft has no end play, add shims (40) to introduce not more than 0.05MM (0.002 inch) shaft end play. NOTE: Excessive end play increases the possibility of inaccuracies due to parts shifting. If original shims (40) are not being used, install a preliminary 0.9MM (0.035 inch) thick shim pack. Shims are available in thicknesses of 0.05, 0.13 and 0.25MM (0.002, 0.005 and 0.010 inch). Tighten nut (43) to a torque of 22 kg/m (160 ft.-lbs.) and measure shaft end play, then remove shims from shim pack (40) equal to the measured shaft end play, PLUS an additional 0.13MM (0.005 inch). This will give the recommended bearing preload of 0.006. Retighten nut to torque of 22 kg/m (160 ft.-lbs.) and stake in position.

TRANSMISSION OIL PUMP
All Models

The transmission oil pump is a gear type pump mounted on rear wall of clutch housing and driven by the pto powershaft. See Fig. 94.

Pump capacity is 22 liters (5.8 gpm) at 2400 engine rpm. Fluid from the transmission oil pump flows through the transmission/hydraulic oil filter and past transmission pump relief valve (R—Fig. 90) then the entire transmission pump flow is delivered to intake side of main hydraulic pump. On models with power steering, the power steering section of main pump

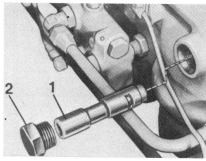

Fig. 95–Filter bypass valve (1) and plug (2) is located directly above filter unit.

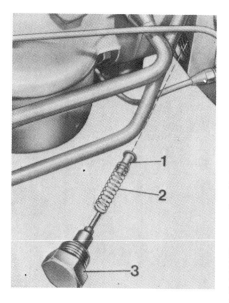

Fig. 96–Low pressure relief valve (1), spring (2) and plug (3) are located immediately forward of system filter as shown.

obtains its oil supply from the same source, but power steering fluid is returned to main pump intake line and total demand on transmission oil pump is not increased.

96. TESTING. To test the transmission pump pressure and flow, install a suitable flow test meter in oil filter bypass plug (2—Fig. 95) and place the meter return line in transmission filler plug opening. With engine speed at 2400 rpm and fluid at operating temperature, open meter load valve and check the flow, which should be 20-22 liters (5.3-5.8 gpm). Slowly close flow meter load valve and note the pressure at which flow decreases. Meter pressure reading should be 60 psi which is transmission pump relief valve pressure. If pressure is not as specified, remove the valve (Fig. 96) and renew valve spring, valve or seat as required. NOTE: The renewable valve seat is pressed into housing; refer also to Fig. 90.

97. R&R AND OVERHAUL. To remove the transmission oil pump, first split clutch housing from transmission housing as outlined in paragraph 90.

With clutch housing separated from transmission case, pull clutch shaft, or clutch shaft and pto power shaft, from clutch housing. Remove pump inlet and outlet lines from pump, then remove pump from wall of clutch housing and separate pump body from adapter. See Fig. 97.

Clean and inspect all parts for chipping, scoring or excessive wear. If bearing in pump body requires renewal, press new bearing in bore until it bottoms. Pump gears are available as a matched set only. Pump idler shaft is renewable and diameter of new shaft is 15.85-15.87MM (0.624-0.625 inch). Thickness of new pump gears is 12.89-12.94MM (0.508-0.510 inch).

When reassembling pump, coat gears with oil and tighten adapter mounting cap screws to a torque of 4.8 kg/m (35 ft.-lbs.). Align slots of clutch shaft, or pto powershaft, with lugs of pump drive gear when installing shafts and be sure seals are on ends of inlet and outlet tubes before rejoining tractor.

Fig. 97–View showing transmission oil pump separated. Note drive lugs in I. D. of pump drive gear.

DIFFERENTIAL AND FINAL DRIVE

DIFFERENTIAL

All Models

98. REMOVE AND REINSTALL. To remove differential, drain transmission case, then remove final drives as outlined in paragraph 101 and the rockshaft housing as outlined in paragraph 129.

If tractor is equipped with a differential lock, the assembly must be removed as follows: Remove clamp screw from lever (1—Fig. 98), and remove the square key (7). Hold yoke (14) in place, bump shaft (9) rearward and remove Woodruff key (8). Remove shaft (9) to the rear after bumping out plug (11).

Disconnect load control arm spring, slide pivot shaft to the left and lift out load control arm. Remove transmission oil cup and rear oil line. Support the differential from the top and remove both carrier bearing quills (2 & 16—Fig. 99), keeping shim packs (3) with the respective quills.

Shims (3) control differential carrier bearing preload which should be 0.05-0.13MM (0.002-0.005 inch), and back-

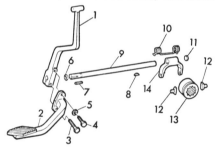

Fig. 98—Exploded view of differential lock actuating mechanism showing component parts.

1. Hand lever	8. Woodruff key
2. Pedal	9. Shaft
3. Cap screw	10. Spring
4. Cap screw	11. Plug
5. Lock washer	12. Shoes
6. O-ring	13. Collar
7. Key	14. Yoke

lash which should be approximately 0.3MM (0.012 inch). Refer to paragraph 100 for adjustment procedure.

99. OVERHAUL. The bevel ring gear (13—Fig. 99) is available only as a matched set which contains bevel pinion shaft (12) and right hand half (13) of differential case.

To disassemble the unit, remove the eight retaining cap screws (6) and separate differential case halves. Note that differential cross shaft (10) is located by a dowel pin (11). Additional splines on left differential case half (7) are used for hand parking brake on Model 820 and are unused on most 830 Models. Refer to Fig. 106 for exploded view of hand brake and to paragraph 112 for overhaul. On models with hand brake, left carrier bearing (5—Fig. 99) must be removed to remove brake sheave, and sheave must be installed before bearing cone. If differential pinions (9) or cross shaft (10) are damaged, all should be renewed. If axle (side) gears (8) are damaged or excessively worn, closely examine bores and seating surface in differential case for damage.

Reassemble by reversing the disassembly procedure. Tighten differential case retaining cap screws (6) to a torque of 4.8 kg/m (35 ft.-lbs.).

MAIN DRIVE BEVEL GEARS
All Models

100. ADJUSTMENT. If differential is removed for access to other parts and no defects in adjustment are noted, shim packs (26—Fig. 93 and 3—Fig. 99) should be kept intact and reinstalled in their original positions.

However, if new parts are installed, mesh position, carrier bearing preload and backlash should be adjusted as follows:

Mesh position of the main drive bevel pinion is controlled by the two 0.25MM (0.010 inch) shims (26—Fig. 93) located between bearing cup (25) and transmission housing wall as shown in Fig. 100. Make sure the two shims are present at reinstallation if bearing cup has been removed.

Carrier bearing preload and bevel gear backlash are both controlled by the thickness of shim pack (3—Fig. 99) located at each differential carrier bearing quill. The differential carrier bearings should have a preload of 0.05-0.13MM (0.002-0.005 inch) and adjustment is made as follows: Install differential and bearing quills with original shim packs, then check differential end play using a dial indicator.

NOTE: When making this adjustment, be positive that clearance exists between the main drive bevel ring gear and pinion shaft at all times.

If no differential end play exists, add shims under right bearing quill to introduce not more than 0.05MM (0.002 inch) end play. If more than 0.05MM (0.002 inch) end play existed on original check, subtract shims.

Measure end play of differential, then subtract shims equal to the measured end play PLUS an additional 0.07MM (0.003 inch) to give the desired 0.05-0.13MM (0.002-0.005 inch) bearing preload. Shims are available in thicknesses of 0.07, 0.05 and 0.25MM (0.003, 0.005 and 0.010 inch).

With differential carrier bearing preload set, adjust backlash between main drive bevel gear and pinion shaft to 0.15-0.30MM (0.006-0.012 inch) by transferring bearing quill shims from one side to the other as required. Moving shims from left to right will

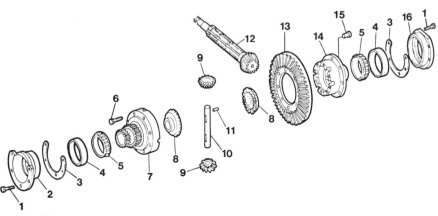

Fig. 99—Exploded view of main drive bevel gears, differential and associated parts.

1. Cap screw	5. Bearing cones	9. Differential pinions	13. Bevel ring gear
2. Bearing housing	6. Cap screws	10. Cross shaft	14. Case half
3. Shims	7. Differential case half	11. Dowel pin	15. Rivet
4. Bearing cups	8. Axle gear	12. Drive pinion	16. Bearing housing

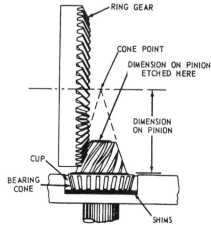

Fig. 100—Simplified view of main drive bevel gears showing location of shims which control gear mesh position. Refer to text for adjustment procedure.

decrease backlash. Do not remove any shims during backlash adjustment or the previously determined preload adjustment will be changed.

FINAL DRIVE
All Models

101. **REMOVE AND REINSTALL.** To remove final drive, support rear of tractor and remove wheel and tire. Disconnect fender lights and free wiring harness from clamp on final drive housing, then remove fender. If right hand final drive is being removed and tractor has selective control valve, disconnect pressure line, coupler lines and return hose between valve and rockshaft housing and remove control valve. Disconnect the brake line from final drive housing. Attach hoist to final drive, remove attaching cap screws and pull final drive from transmission case.

Reinstall by reversing removal procedure and tighten attaching cap screws to a torque of 11.7 kg/m (85 ft.-lbs.).

NOTE: If brake disc came off with final drive, install disc so that thickest facing is toward transmission case.

102. **OVERHAUL.** To overhaul the removed final drive unit, refer to Fig. 101 and proceed as follows:

Remove lock plate (25), cap screw (24) and retainer washer (23), then pull planet carrier assembly (22) from axle. Support outer end of final drive housing (7) so oil seal (2) will clear and press axle out of housing. The axle bearings, bearing cups and oil seals are now available for inspection or renewal. If outer bearing (4) is renewed, heat it to approximately 300°F. and drive it into place while hot. NOTE: If axle is flanged, be sure oil seal is on axle, metal side out, before installing outer bearing. Bearing cups are pressed in bores until they bottom. Seal cup (3) will be pushed out when outer bearing cup is removed. Be sure to reinstall seal cup after bearing cup is installed. If ring gear and/or final drive housing is damaged, renew complete unit.

To remove planet pinions (19), expand snap ring (21), lift it from groove of carrier (22) and pull pinion shafts (17). Check carrier, pinions and rollers for pitting, scoring or excessive wear and renew parts as required. If any of the planet pinion rollers are defective, renew the complete set.

Reassemble final drive and adjust axle bearings as follows: Coat bores of planet pinions with grease and position rollers (23 in each bore) in pinions. Place a thrust washer on each side of pinions, then place pinions in carrier and insert pinion shafts only far enough to retain rollers and thrust washers. Install snap ring (21) in slots of pinion shafts, then complete insertion of pinion shafts and be sure snap ring seats in groove in carrier. Coat inner seal (14) with grease and install axle in housing. Heat inner bearing (16) to approximately 300°F. and install bearing on inner end of axle. Place carrier assembly on axle, install retaining washer (23) and cap screw (24) and tighten cap screw until bearing is pulled into place and a small amount of axle end play remains. Now while bearing is still hot, check the amount of torque required to turn the axle with the existing axle end play, then tighten the cap screw to increase the rolling torque 10-35 in.-lbs. for old bearings, or 20-70 in.-lbs. if new bearings are used. Install lock plate (25). Fill axle outer bearing opening with multi-purpose grease and install oil seal with metal side out.

Use new gasket (13) when reinstalling final drive to transmission case. However, before installing final drive, pull final drive shaft (26) and brake disc and inspect. Brake disc is installed with thickest facing next to transmission case.

Refer to paragraph 110 for information on brake pressure plate and pressure ring.

BRAKES

The brakes on all models are hydraulically actuated and utilize a wet type disc controlled by a brake operating valve located on right side of clutch housing. See Fig. 102. Brake discs are splined to the final drive shafts and the brake pressure ring is fitted in inner end of final drive housing. Except for a pedal adjustment, no other brake adjustments are required.

BLEED AND ADJUST
All Models

103. **BLEEDING.** Brakes must be bled when pedals feel spongy, pedals bottom, or after disconnecting or disassembling any portion of the braking system.

To bleed brakes, start engine and run for at least two minutes at 2000 rpm to insure that brake control valve reservoir is filled.

NOTE: Brakes can also be bled without engine running if necessary. Remove right platform and fill brake control valve reservoir by removing filler plug (11—Fig. 103). Follow same bleeding procedure used when engine is running except brake valve reservoir will need to be refilled after each fifteen strokes of the brake pedal.

Attach a bleeder hose (preferably clear plastic) to brake bleed screw located on top side of final drive housing and place opposite end in filler hole of rockshaft housing. Slowly depress and release brake pedal until oil flowing from bleeder hose is completely free of air bubbles, then depress brake pedal and tighten bleed screw.

Repeat bleeding operation for opposite side brake.

104. **ADJUSTMENT.** Whenever brake control valve has been disassembled, a brake pedal and equalizing valve adjustment must be made to prevent mechanical interference between brake valve pistons and reservoir check valves.

Before making this adjustment,

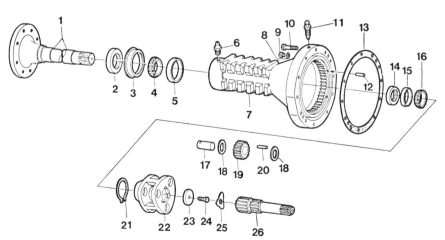

Fig. 101—Exploded view of final drive assembly showing component parts.

1. Wheel axle	8. O-ring	14. Oil seal	20. Needle roller
2. Oil seal	9. Plug	15. Bearing cup	21. Snap ring
3. Oil seal cup	10. Cap screw	16. Bearing cone	22. Planet carrier
4. Bearing cone	11. Bleed plug	17. Pinion shaft	23. Washer
5. Bearing cup	12. Dowel pin	18. Thrust washer	24. Cap screw
6. Grease fitting	13. Gasket	19. Planet pinion	25. Lock plate
7. Axle housing			26. Final drive shaft

Fig. 102–View showing brake control valve. Right platform has been removed.

A. Pedal adjusting screws
B. Brake control valve
C. Mounting cap screws
P. Brake valve pistons

bleed brakes as outlined in paragraph 103.

105. RIGHT PEDAL. Adjust right hand pedal stop screw so brake valve piston is fully extended and arm of brake pedal is snug against end of piston without piston being depressed (zero clearance). Apply a force of about 4.5 kg (10 lbs.) to LEFT brake pedal and if left brake pedal settles, turn pedal stop screw for right brake pedal counterclockwise about ⅓-turn at which time left brake pedal should stop settling. If left brake pedal does not stop settling, a leak in the braking system is indicated and must be isolated and corrected. Refer to paragraph 108.

106. LEFT PEDAL. Adjust left hand pedal stop screw so brake valve piston is fully extended and arm of brake pedal is snug against end of piston without piston being depressed (zero clearance). Apply a force of about 4.5 kg (10 lbs.) to RIGHT brake pedal and if right pedal settles, turn pedal stop screw for left brake pedal counterclockwise about ⅓-turn at which time right brake pedal should stop settling. If right brake pedal does not stop settling, a leak in the braking system is indicated and must be isolated and corrected. Refer to paragraph 108.

107. PEDAL HEIGHT. If brake pedal height is not aligned after equalization valves are adjusted as outlined in paragraphs 105 and 106, align pedals by turning stop screw on highest pedal about ⅛-turn counter-clockwise.

BRAKE TEST

All Models

108. PEDAL LEAK-DOWN. With a 27.2 kg (60 lb.) pressure applied continuously to each pedal for one minute, the pedal leak-down should not exceed one inch. Excessive brake pedal leak-down can be caused by air in the brake system, faulty brake control valve pistons and/or "O" rings, faulty brake pressure ring seals, or faulty brake control valve equalizing valves or reservoir check valves.

Brakes should always be bled as outlined in paragraph 103 before any checking or adjusting of braking system is attempted. Faulty brake control valve pistons or "O" rings will be indicated by external leakage around the brake control valve pistons.

Faulty brake pressure ring seals, or brake control valve, can be determined as follows: Isolate brake from brake control valve by plugging brake line. If leak-down stops, the brake pressure ring seals are defective. If leak-down continues, the brake control valve is faulty and can be checked further by depressing brake pedals individually, then simultaneously. If leak-down occurs in both cases, a defective reservoir check valve is indicated. If leak-down occurs during individual pedal operation but not on simultaneous pedal operation, a faulty equalizer valve is indicated.

Refer to paragraph 109 for brake control valve information and to paragraph 110 for brake pressure ring information.

OVERHAUL

All Models

109. BRAKE CONTROL VALVE. To remove brake control valve, remove right platform and thoroughly clean valve and surrounding area. Disconnect brake lines from rear of control valve, remove the mounting cap screws (C—Fig. 102) and remove control valve from clutch housing. Discard gasket located between control valve and clutch housing. Remove "E" ring (40—Fig. 103), pull shaft (39) and remove pedals from control valve. Remove connectors (23 & 27), check valve springs (20) and balls (19). Remove seats (21) and ball retainers (18), then push pistons (14) and springs (15) out rear of valve body. Remove filler plug (3) and cup plug (30), then using a screw driver with proper sized bit, remove reservoir check valve assemblies (9). Remove equalizer valve assemblies (items 10, 11, 12 and 13). "O" rings (16) and oil seals (17) can be removed from piston bores.

Fig. 103–Exploded view of brake valve, linkage and lines.

1. Valve housing
2. Gasket
3. Plug
4. O-ring
5. Check valve seat
6. O-ring
7. Check valve
8. Spring
9. Valve assembly
10. Plug
11. O-ring
12. Spring
13. Ball
14. Brake piston
15. Spring
16. O-ring
17. Seal ring
18. Retainer
19. Ball
20. Spring
21. Valve seat
22. O-ring
23. Adapter
24. O-rings
25. Brake line
26. Brake line
27. Fitting
28. Clamp
29. Cap screw
30. Plug
31. Pedal
32. Pedal
33. Bushing
34. Latch
35. Spring
36. Spring pin
37. Cap screw
38. Nut
39. Pedal shaft
40. Snap ring
41. Cap screw
42. Cap screw

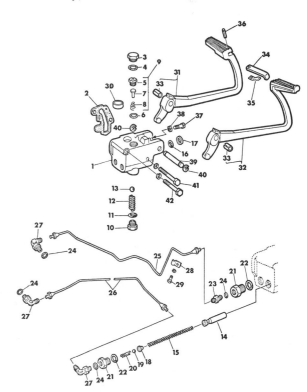

Clean and inspect all parts. Piston spring (15) should test 9.1 kg (20 lbs.) when compressed to a length of 146MM (5¾ inches). Renew housing (1) if seats for equalizer balls (13) are damaged. Oil seals (17) are installed with lips toward outside. Pay particular attention to area of reservoir check valve (7) where contact is made with valve piston and renew valve if any doubt exists as to its condition. Brake pedals are fitted with bushings for brake pedal shaft (39) and bushings and/or shaft should be renewed if clearance is excessive.

Lubricate lips of oil seals (17) and all other parts. Use a new cup plug (30) and reassemble by reversing disassembly procedure. Use a new gasket when installing valve on tractor. Bleed brakes as outlined in paragraph 103 and adjust pedals as outlined in paragraphs 105, 106 and 107.

110. **BRAKE PRESSURE PLATE, RING AND DISC.** To remove brake pressure plate, pressure ring and brake disc, remove final drive housing as outlined in paragraph 101. Pull final drive shaft from differential and remove brake disc from final drive shaft. See Fig. 104. Lift brake pressure plate from dowels in final drive housing. Remove brake pressure ring by prying it out evenly. If pressure ring is difficult to remove, attach a small hydraulic pump to brake line connections, be sure bleed valve is closed, then pump oil behind

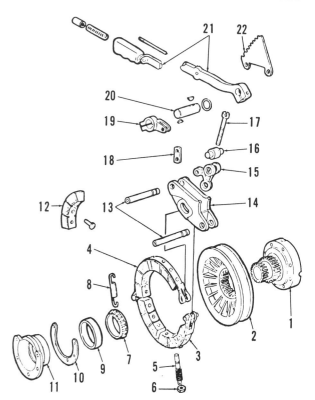

Fig. 106–Exploded view of hand brake unit used on some 820 tractors.

1. Differential case half
2. Brake drum
3. Lower band
4. Upper band
5. Support screw
6. Locknut
7. Carrier bearing
8. Return spring
9. Bearing cup
10. Shims
11. Bearing quill
12. Brake facing
13. Anchor pins
14. Anchor
15. Actuating arm
16. Toggle
17. Adjusting screw
18. Link
19. Actuating lever
20. Brake shaft
21. Hand lever
22. Quadrant

Fig. 104–Removing final drive shaft (1) and brake disc (2) from transmission housing.

Fig. 105–Pressure ring (1) and packings (2) removed from final drive housing. Dowels must align with housing bores.

pressure ring to force it from cylinder (groove). See Fig. 105. Dowels can be removed from final drive housing, if necessary.

Inspect brake disc for worn or damaged facing or damaged splines. If facings require renewal, renew complete disc assembly as facings are not available separately. Inspect pressure plate for scoring, checking, or other damage and renew if necessary. Remove and discard seals from pressure ring and inspect ring for cracks or other damage.

To reassemble brake assembly, proceed as follows: Place brake disc on final drive shaft so thickest facing is next to transmission case and insert final drive shaft into differential. Place new inner and outer seals on pressure ring and lubricate assembly liberally. Start pressure ring into its cylinder (groove) with flat chamfered side first and press into cylinder until it bottoms. Be absolutely sure that neither seal is cut or rolled during installation. Place pressure plate over dowels in final drive housing, hold in place if necessary, then install final drive housing.

Bleed brakes as outlined in paragraph 103.

HAND BRAKE

Some Series 820 tractors are factory equipped with a hand brake of the type shown exploded in Fig. 106. The brake

drum (sheave) is splined to left half of differential case and the contracting band wedges in groove of sheave to hold tractor.

ADJUSTMENT

Series 820 So Equipped

111. To adjust the brake, remove plug on top rear face of transmission case as shown in Fig. 107. With brake released, reach through plug hole with screwdriver as shown. Turn screw (17 —Fig. 106) until band is tight in sheave, then back off one turn. Loosen locknut (6) and turn support screw (5) finger tight, back off one turn and tighten locknut.

Fig. 107–Hand brake can be adjusted with a screwdriver after removing plug from rear, top face of transmission housing as shown.

OVERHAUL

Series 820 So Equipped

112. All hand brake components except brake drum (sheave) can be removed after removing hydraulic lift cover as outlined in paragraph 129. To remove brake drum, remove differential unit as in paragraph 98.

Refer to Fig. 106 for an exploded view of hand brake and associated parts. Brake facings (12) should be renewed if rivets are loose or when grooves in contact surfaces are worn smooth. Facings are available individually but must be renewed in sets of four. Adjust the brake as outlined in paragraph 111 after tractor is assembled.

POWER TAKE-OFF

When service is required on the pto system, the following should be taken into consideration. Work involving rear pto shaft can be done by working from rear of tractor. Work involving the pto driven gears, powershaft clutch shaft, mid-pto shaft or mid-pto shifter assembly will require that the clutch housing be separated from the transmission case. Work involving the rear pto shaft shifter assembly will involve removing the rockshaft housing, high and low range shifter shafts and countershaft assembly in addition to separating the clutch housing from transmission case.

REAR PTO SHAFT

All Models

113. **R&R AND OVERHAUL.** To remove rear pto shaft (12—Fig. 108), drain transmission and remove pto shield and shaft guard, if so equipped. Place rear pto shaft control lever in "OFF" position, then remove cap screws from pto shaft bearing quill (13) and pull shaft and quill assembly from transmission case. Be careful not to pull shift collar from front drive shaft. Bearing (15) can be renewed after removing snap rings (3 and 16). Press new oil seal (14) in quill with lips toward front until it bottoms. For service on remainder of pto assembly, split tractor as outlined in paragraph 90.

Reassemble by reversing removal procedure and mate splines of rear shaft with splines of shift collar as shaft is installed.

FRONT DRIVE SHAFT AND DRIVEN GEARS

All Models

114. **R&R AND OVERHAUL.** To remove pto front drive shaft (1—Fig. 108) and gear (4), first split tractor as outlined in paragraph 90. Note: the pto clutch shaft and engine clutch shaft can also be removed at this time by withdrawing them rearward out of clutch housing.

Prior to any disassembly, install rear pto shaft if removed, and place rear pto shaft shifter lever in "ON" position. This will position rear shift collar on splines of rear shaft and prevent it

from dropping to bottom of transmission case. If collar comes off, it will be necessary to remove rockshaft housing to retrieve it.

Remove snap ring from front of gear and lift off the gear, then withdraw pto drive shaft (1) from transmission. Inspect needle bearing (7) in front quill (5) and renew if damaged. Renew other parts as required.

Reassemble by reversing the disassembly procedure, using shim stock or other seal protector and extreme care to protect lip of seal (14) when reinstalling output shaft in rear quill (13).

REAR PTO SHAFT SHIFTER

All Models

115. **R&R AND OVERHAUL.** To remove rear pto shaft shifter assembly, remove the high and low range shifter shafts as outlined in paragraph 91 and the transmission countershaft as outlined in paragraph 92. Remove detent assembly located forward of and slightly below pto shift lever. Clip lock wire and remove shifter fork set screw. Remove snap rings from ends of shifter shaft, push shifter shaft out front of transmission case and lift out shifter fork. Pull front pto shaft out and remove shift coupler. Bump roll pin from shifter arm and pull shifter lever from shifter arm and transmission case.

Clean and inspect all parts and renew as necessary. Reinstall shifter mechanism by reversing removal procedure.

HYDRAULIC LIFT SYSTEM

All models are equipped with an open center type hydraulic system consisting of a crankshaft driven gear-type pump supplied with working fluid from the transmission housing by the transmission oil pump. System priority (routing of fluid) is as follows:

1. **Transmission reservoir**
2. **Transmission oil pump**
3. **Supply bypass valve**
4. **Oil Filter & filter relief valve**
5. **Low pressure relief valve**
6. **Main hydraulic pump**
7. **System relief valve**
8. **Selective control valve**
9. **Rockshaft control valves**
10. **Brake valves**
11. **Transmission & final drive lubrication.**

Supply bypass valve opens under negative pressure when transmission pump is inoperative (engine clutch disengaged,

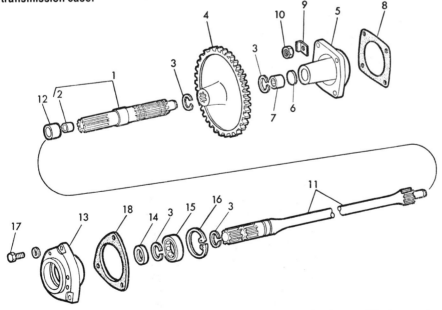

Fig. 108–Exploded view of pto output shaft, drive shaft, gear and associated parts of the type used.

1. PTO drive shaft
2. Bushing
3. Snap ring
4. Gear
5. Front quill
6. Thrust washer
7. Needle bearing
8. Gasket
9. Lock plate
10. Nut
11. Output shaft
12. Bushing
13. Rear quill
14. Oil seal
15. Bearing
16. Snap ring
17. Cap screw
18. Gasket

etc.), enabling main hydraulic pump to pick up fluid directly from transmission reservoir. The low pressure relief valve regulates inlet pressure of the main pump, normally bypassing a small excess flow from transmission pump when oil is warm and filter is in good condition.

On models with power steering, the power steering fluid is supplied by a second main pump section installed in tandem with the main hydraulic pump. Initial steering fluid is supplied by the transmission oil pump but return steering fluid is directed to the main pump inlet line, therefore the fluid supply available to the hydraulic system is not decreased by adding the power steering option. NOTE: The only exception to this statement would be complete failure of power steering relief valve, in which case the fluid pumped by the power steering pump would be returned to reservoir and the supply available for hydraulic functions would be reduced.

Transmission and final drive lubrication passage tees into main supply line ahead of brake valve, assuring continuing lubricant flow when brakes are being applied. An orifice in rockshaft dump valve supplies a continuing lubricant flow directly to lubricant passage when rockshaft valve is actuated, thus the only time lubricant flow is temporarily interrupted is during raising action of the single acting selective control valve.

Refer to the appropriate following paragraphs for testing, removal and overhaul of hydraulic system components.

TESTING

All Models

116. **PUMP PRESSURE AND FLOW.** Main pump system flow is 5.8

Fig. 109–View of pressure relief valve housing and lines.

1. Relief valve	5. Pressure line
2. Bracket	6. Mounting screws
3. Return tube	7. Outlet pressure line
4. Inlet elbow	8. Adjusting screw

gpm at 2400 engine rpm. Main system relief pressure should be 148-150 kg/cm^2 (2105-2130 psi). Pressure and flow can be checked by disconnecting rear pressure line (5—Fig. 109) from elbow (4) and teeing a suitable flow meter in the line.

Pressure only can be checked by installing a gage in port (4—Fig. 110) in rockshaft housing as shown, and immobilizing lift arms to test the pressure. Flow is correct if lift arms will move a load from fully lowered to fully raised position in 2½ seconds with engine running at 2100 rpm and fluid at operating temperature.

To adjust the pressure, loosen jam nut and turn adjusting screw (8—Fig. 109) in or out as required.

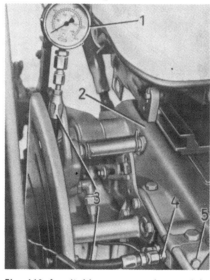

Fig. 110–A suitable gage can be used for checking pressure only. Refer to paragraph 116.

1. Gage	
2. Rockshaft housing	
3. Connecting hose	4. Pressure port
	5. Pressure port

Fig. 111–Negative stop screw (1) should be backed out ¼ turn from the point where it contacts load control arm. Tighten locknut securely when adjustment is correct.

ADJUSTMENT

All Models

117. **ROCKSHAFT LINKAGE.** The following external adjustments should be made on rockshaft linkage after first checking fluid level and bringing system to operating temperature. Adjustments should be checked or made, in the order given.

118. NEGATIVE STOP SCREW. Refer to Fig. 111. Hold stop screw (1) and back off locknut (2) at least two turns. With engine not running and no load on lift arms, turn stop screw (1) in until it just contacts load control arm, back screw out ¼-turn and tighten locknut.

NOTE: Contact between stop screw and load control arm can be determined by removing oil filler cap and placing a screwdriver or similar tool against upper end of arm. Refer to Fig. 114.

119. DEPTH CONTROL NEUTRAL RANGE. With engine running at high idle speed, no load on hydraulic system and selector lever (2—Fig. 112) in upper ("D") position, move rockshaft control lever to bottom of quadrant, then slowly upward until lift arms just start to raise. Mark this position on quadrant rim.

Push control lever forward until lift arms start to settle and again mark

Fig. 112–View of rockshaft control quadrant showing points of adjustment.

1. Quadrant slot	3. Clamp screw
2. Selector lever	4. Plug

Fig. 113–To adjust load control mechanism, move selector lever (1) to "L" position and move quadrant lever until distance (a) measures 92 MM (3⅝ inches), then refer to Fig. 114.

quadrant. The two marks should be 4-6MM (5/32-7/32 inch) apart; if they are not, remove plug (4) and reaching through plug hole turn adjusting screw (1—Fig. 123) clockwise to increase neutral range or counterclockwise to decrease range.

120. CONTROL LEVER ADJUSTMENT. With engine running at high idle speed, no load on hydraulic system and selector lever (2—Fig. 112) in upper ("D") position, move control lever downward until lever pin is 8MM (5/16-inch) from bottom of slot (1) as shown at (a). Loosen clamp nut (3) and move lever arm clockwise until lift arms are fully lowered, then counterclockwise until lift arms just start to move upward. Tighten clamp nut at this point.

Move control lever slowly rearward until lift arms are fully raised and measure clearance between lever pin and upper end of slot (1). Clearance should be approximately equal to lower clearance (a). Equalize the difference by loosening clamp screw (3) and making minor adjustments as required.

121. LOAD CONTROL ADJUSTMENT. With engine running at high idle speed, no load on hydraulic system and selector lever (1—Fig. 113) in lower ("L") position, move control lever downward until distance (a) measures 3⅝ inches.

Remove filler cap as shown in Fig. 114. Loosen locknut and turn adjusting screw (1) counter-clockwise until lift arms start to lower, then clockwise until lift arms just start to rise. Tighten jam nut at this point without moving adjusting screw.

122. RATE OF DROP. The rate-of-drop adjusting screw is located on top of

rockshaft housing underneath the seat as shown in Fig. 115. Turn screw clockwise to slow rate of drop, or counterclockwise to increase the rate.

123. SELECTIVE CONTROL VALVE. Valve levers can be equalized (dual valve models) by loosening locknut and turning valve end in connecting ball cap.

To adjust the lowering speed, adjust rate-of-drop screw (Arrow—Fig. 116) located at bottom of each valve housing as shown.

MAIN HYDRAULIC PUMP

All Models

All Models are equipped with a crankshaft driven gear-type main hydraulic pump which is mounted ahead of the radiator and driven by the crankshaft through a coupling and drive shaft attached to crankshaft pulley. Refer to Fig. 117 for exploded view of drive unit and to Fig. 118 for installed view. Refer to paragraph 14 for the combined pump used.

124. REMOVE AND REINSTALL. To remove the pump, first remove right grille and air cleaner. Remove cotter pin from pump drive shaft and push shaft rearward out of front coupling. Remove retaining clamps and disconnect inlet and pressure lines at pump. Remove cap screws (4—Fig. 118) and lift off pump and mounting plate as an assembly. When installing, tighten the screws attaching drive coupling to pulley and socket head screws at-

taching pump to mounting bracket to a torque of 4.84 kg/m (35 ft.-lbs.). Tighten screws securing line adapters to pump body to 2.77 kg/m (20 ft.-lbs.) and mounting bracket to front support to 11.76 kg/m (85 ft.-lbs.). Refer to paragraph 14 for models with power steering.

125. OVERHAUL. Shaft couplings are riveted assemblies and parts are available. The only pump parts available separately are shaft seal and pump sealing rings. If shafts (gears), bearings or housings are damaged, renew the pump. Mark the parts as pump is disassembled, to be sure parts are installed in same position as before removal. Refer to paragraph 15 for models with power steering.

RESERVOIR, FILTER AND VALVES

All Models

126. RESERVOIR, FILTER AND INLET SCREEN. The transmission housing is reservoir for hydraulic system. Fluid capacity is 37.89 liters (10 US gallons) and recommended fluid is John Deere Type 303 or equivalent. Oil level dipstick is located on right side of transmission case and filler plug is at rear of rockshaft housing.

The full flow oil filter is located underneath transmission housing as shown at (1—Fig. 119). Filter element

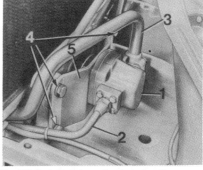

Fig. 118—Installed view of main hydraulic pump.

1. Pump	4. Mounting cap screws
2. Pressure line	5. Mounting plate
3. Inlet line	

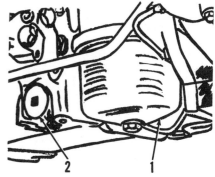

Fig. 119—Hydraulic oil filter is located beneath transmission housing as shown at (1). Plug (2) provides access to intake screen.

Fig. 114—With control levers positioned as shown in Fig. 113 turn adjusting screw (1) until lift arms just start to raise. Refer to paragraph 121.

Fig. 115—View of rockshaft housing showing Rate-Of-Drop adjusting screw.

Fig. 116—Arrow shows Rate-Of-Drop adjusting screw for selective control valves.

Fig. 117—Exploded view of front mounted main hydraulic pump and drive unit.

should be renewed every 500 hours, element can be removed without draining transmission. Every 1000 hours, drain and renew transmission fluid, remove and clean intake screen (2) and renew filter element.

127. FILTER BYPASS VALVE. The oil filter bypass valve is located in transmission housing directly above filter unit as shown in Fig. 120. To service the valve, remove plug (2) and withdraw valve cartridge (1) as an assembly.

The piston type valve should open at a pressure differential of 2.31 kg/cm² (28.5 psi). Approximately 4.54 kg (10 pounds) pressure should be required on end of valve to open bypass slot and valve should move smoothly. Drive out retaining spring (roll) pin and remove valve spring and piston if service is indicated.

128. LOW PRESSURE RELIEF VALVE. The low pressure relief valve is located in hydraulic pump inlet passage. Valve is accessible from outside by removing hex plug in clutch housing immediately forward of system filter as shown in Fig. 121.

Valve (1) maintains an inlet pres-

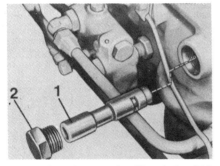

Fig. 120—Filter bypass valve (1) and plug (2) is located directly above filter unit (1–Fig. 119).

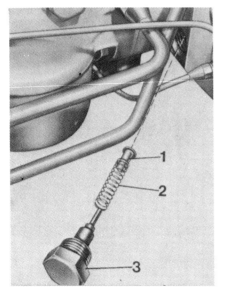

Fig. 121—Low pressure relief valve (1), spring (2) and plug (3) are located immediately forward of system filter as shown.

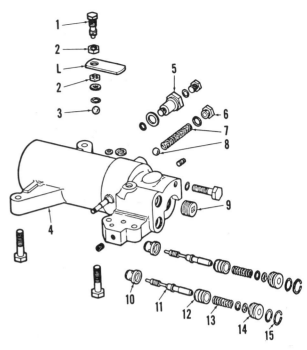

Fig. 122—Exploded view of rockshaft cylinder, valves and associated parts.

1. Throttle screw
2. Locknut
3. Steel ball
4. Cylinder hsg.
5. Outlet adapter
6. Plug
7. Spring
8. Steel ball
9. Plug
10. Valve seat
11. Valve
12. Sleeve
13. Spring
14. Plug
15. Snap ring
L. Drop rate lever

sure of approximately 4.20 kg/cm² (60 psi) to main hydraulic pump and normally bypasses a continuous flow of approximately 3.79 liters (1 gpm) at rated speed when system is in good condition. Spring (2) should have a free length of approximately 51MM (2 inches) and a pressure of 5.45 kg (12 lbs.) should be required to compress spring to a height of 19.05MM (¾-inch).

ROCKSHAFT HOUSING AND COMPONENTS

All Models

129. REMOVE AND REINSTALL. To remove the rockshaft housing, disconnect battery ground straps, remove transmission shield and disconnect wires from starter safety switch. Disconnect lift links from rockshaft arms and remove seat assembly. Disconnect oil return hose from manifold and rockshaft housing. Disconnect rear quick couplers if so equipped. Remove pilot

line between rockshaft housing and dump valve housing. Move selector lever to lower ("L") position and remove retaining cap screws. Attach a suitable hoist and lift rockshaft housing from transmission housing.

Install by reversing the removal procedure, making sure selector lever is in ("L") position and that roller link mates properly with load control arm cam follower. Tighten retaining cap screws to a torque of 11.76 kg/m (85 ft.-lbs.).

130. OVERHAUL. Remove rockshaft arms from rockshaft and selector lever from load control arm. Turn unit until bottom side is accessible. Remove control lever and quadrant. Remove remote cylinder adapter (5—Fig. 122) and unhook spring (3—Fig. 123) from pin on lift cylinder. Remove front, outside cylinder retaining cap screw and remaining cap screws, then lift cylinder from rockshaft housing disengaging selector arm from roller link as cylinder is removed.

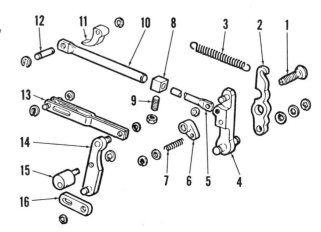

Fig. 123—Exploded view of control valve linkage.

1. Adjusting screw
2. Valve cam
3. Spring
4. Link
5. Control rod
6. Adjusting nut
7. Spring
8. Stop clamp
9. Set screw
10. Linkage tube
11. Cam
12. Pin
13. Selector link
14. Operating link
15. Pivot block
16. Link

NOTE: Do not lose throttle valve ball (3 —Fig. 122) which is free to fall out as cylinder is removed.

Remove cam and spacer from rockshaft then pull rockshaft from control arm and housing. Bushing and O-ring seal will be removed from one side of housing as rockshaft is withdrawn.

When assembling, remove throttle screw (1—Fig. 122) from housing. Install cylinder and the five retaining cap screws finger tight, then tighten outside (front) cap screw first, to the recommended 4.84 kg/m (35 ft.-lbs.). Tighten the other four cap screws after cylinder is properly positioned. Drop throttle valve ball (3) in threaded hole in housing where throttle screw (1) was removed. Reinstall throttle screw and turn down until it firmly contacts ball; tighten lower jam nut (2) until O-ring is seated in housing groove, turn down an additional ½ turn, point lever (L) to right and tighten upper jam nut firmly down on lever.

131. CYLINDER AND VALVE UNIT. To overhaul cylinder and valve unit, refer to Fig. 122. If piston will not slide from cylinder, remove closed-end plug (9) and push piston out with a convenient tool. Pressure and discharge valves (10 through 15) are identical, but worn-in parts should be kept together. Springs (14) have an approximate free length of 33.73MM (1-21/64 inches) and should test 3.76 kg (8.3 lbs.) when compressed to a height of 22MM (⅞ inch). Check valve spring (7) has an approximate free length of 94MM (3 11/16 inches) and should test 5.67 kg (12½ lbs.) when compressed to 64MM (2½ inches).

132. VALVE LINKAGE. Refer to Fig. 123. Linkage should move freely without binding or excessive looseness. Check actuating cam (2) for wear or scoring and other parts for wear, bending, binding or breakage. Use Fig. 123 as a guide for disassembly and reassembly. Adjust upper lift limit as outlined in paragraph 133.

133. ADJUSTMENT. With lift cover fully assembled, loosen set screw (9—Fig. 123) in adjusting clamp (8). Turn neutral range adjusting screw (1) until a minimum positive clearance exists between the two lobes of cam (2) and the two valves (11—Fig. 122). (Slight clearance between cam and one valve when cam touches the other valve.) Insert a ⅛-inch thick shim between rockshaft ram arm and rear housing wall and raise rockshaft arms until shim is securely held in place. Move rockshaft control lever if necessary until cam (2—Fig. 123) is in a neutral position (neither valve depressed). Slide clamp (8) against end of tube (10) and tighten set screw (9).

Remove the shim. Move selector

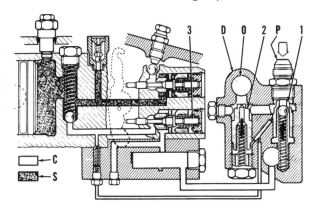

Fig. 124–Cross sectional view of dump valve, valve passages and rockshaft operating valve in neutral position. Main valve (1) is closed by spring pressure and outlet valve (2) opened by hydraulic pressure as shown by heavy arrows.

C. Circulating fluid
D. Dump valve
O. Outlet passage
P. Pump (pressure) passage
S. Static fluid

lever to lower ("L") position and rockshaft control lever to raised position, then check to be sure cam (2) starts to open upper (discharge) valve before rockshaft ram arm contacts housing wall.

DUMP VALVE

All Models

134. OPERATION. Refer to Fig. 124. The dump valve (D) contains two hydraulically actuated, spring loaded valves. When rockshaft control valve is in neutral position, main valve (1) is hydraulically balanced by the orifice passage which equalizes pressure on both sides of valve; and valve and ball are seated by spring pressure. The outlet control valve (2) is hydraulically unbalanced by the slightly higher pressure of circulating oil (C) and is held open against spring pressure, allowing fluid to flow from pump port (P) to outlet port (O) and on to sump.

When rockshaft control lever is moved to "Raise" position, pressure valve (3) is opened as shown in Fig. 125. Pressure is momentarily reduced at spring end of main valve (1) causing valve to move downward opening the passage to spring end of outlet control valve (2). The resultant flow restores hydraulic balance to outlet valve which then closes by spring pressure, shutting off the main hydraulic flow to outlet passage (O). Pressure fluid then flows through the opened check ball in main valve (1), through the opened pressure valve (3) to the rockshaft cylinder.

The flow continues until pressure valve (3) is closed; until rockshaft piston reaches the end of its stroke; or until system relief valve pressure is exceeded. When any of these conditions occur, hydraulic balance is restored to main valve (1) which closes by spring pressure unbalancing outlet valve (2) which then opens against spring pressure permitting resumed fluid flow through outlet passage (O) to the sump.

The orifice in closed end of outlet valve (2) provides lubrication flow for

transmission gears when outlet passage is closed to main hydraulic flow.

135. OVERHAUL. The dump valve is shown exploded in Fig. 126. Dump valve attaches to lower right side of transmission housing by connector (4), and can be removed after disconnecting pressure, return and pilot lines.

Valve spools (8 & 11) should slide smoothly in their bores without excessive looseness. Springs (7 & 9) are in-

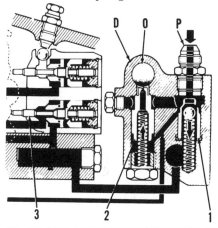

Fig. 125–Cross sectional view of dump valve in raising position. Main valve (1) is opened by hydraulic pressure and outlet valve (2) closed by spring pressure as shown by heavy arrows. Refer also to Fig. 124.

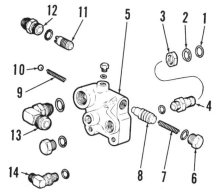

Fig. 126–Exploded view of dump valve showing component parts.

1. O-ring
2. Special washer
3. Jam nut
4. Fitting
5. Valve body
6. Plug
7. Spring
8. Outlet valve
9. Spring
10. Check ball
11. Main valve
12. Inlet fitting
13. Return fitting

terchangeable, have an approximate free length of 57MM (2¼ inches), and should test 5-5.9 kg (11-13 lbs.) when compressed to a height of 28.58MM (1⅛ inches).

Use Figs. 124, 125 and 126 as a guide when reassembling the valve and reinstall by reversing the removal procedure.

LOAD CONTROL (SENSING) SYSTEM

All Models

The load sensing mechanism is located in the rear of the transmission case. See Fig. 127 for an exploded view showing component parts.

The load control shaft (21) is mounted in tapered bushings and as load is applied to the shaft ends from the hitch draft links, the shaft flexes forward and actuates the load control arm (12) which pivots on shaft (10). Movement of the load control arm is transmitted to the rockshaft control valves via the roller link (13—Fig. 123) and control linkage and control valves are opened or closed permitting oil to flow to or from the rockshaft cylinder and piston.

136. **R&R AND OVERHAUL.** To remove the load sensing mechanism, remove the rockshaft housing assembly as outlined in paragraph 129, the left final drive assembly as outlined in paragraph 101, and the three-point hitch.

Remove cam follower spring (18—Fig. 127), then slide pivot shaft (10) to the left and lift out control arm (12) assembly. Removal of pivot shaft (10) can be completed if necessary by removing snap ring (11). Remove retainer ring (2) and retainer bushing (3) from right end of load control shaft and bump shaft from transmission case. The negative stop screw (7), located on rear of transmission case behind right final drive, can also be removed if necessary.

Inspect bushings (6) in transmission case and renew if necessary. Drive old bushings out by inserting driver through opposite bushing. New bushings are installed with chamfer toward inside. Check the special pin (20) for damage in area where it is contacted by load control shaft and renew if necessary. Also check contact areas of negative stop screw (7) and load control arm (12). Check load control shaft (21) to be sure it is not bent or otherwise damaged. Wear or damage to any other parts will be obvious.

Reassemble load sensing assembly by reversing the disassembly procedure and when installing hitch draft links, tighten retaining nuts until end play is removed between link collar and retaining ring, then tighten nut until next slot aligns with cotter pin hole and install cotter pin. After assembly is completed, adjust the negative stop screw as outlined in paragraph 118.

PRESSURE RELIEF VALVE

All Models

137. The pressure relief valve housing (1—Fig. 128) mounts on outside of selective control valves (2) on models so equipped; or directly on mounting bracket (3) on models without remote control.

Relief valve spring (6) should have a free length of approximately 44.45MM (1¾ inches) and should test approximately 54.43 kg (120 lbs.) when compressed to a height of 34.93MM (1⅜ inches). Adjust system pressure as outlined in paragraph 116 after valve is reassembled.

SELECTIVE CONTROL VALVES

All Models

138. Selective control valve linkage is shown in Fig. 128 and exploded view of valve is shown in Fig. 129.

Valve spool (11) and body (10) are only available in a valve assembly which includes items (1 through 9). All other components are available individually. Keep valve body and spool together as a matched set if more than one valve is disassembled at a time. Fill hollow of cover (15) with multipurpose grease when reassembling valve unit. Adjust the installed valve if necessary, as outlined in paragraph 123.

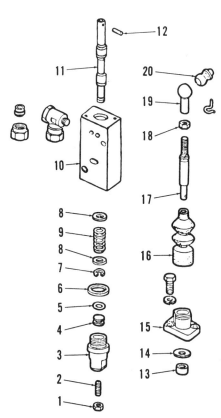

Fig. 129–Exploded view of selective control valve showing component parts.

1. Locknut		11. Valve spool	
2. Adjusting screw		12. Pin	
3. Retainer		13. Packing	
4. Spring cap		14. Washer	
5. O-ring		15. Cover	
6. Copper gasket		16. Boot	
7. Snap ring		17. Connector	
8. Washer		18. Locknut	
9. Centering spring		19. Ball socket	
10. Body		20. Ball	

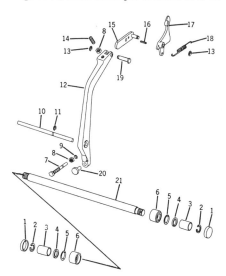

Fig. 127–Load control mechanism is located in rear of transmission case. Flexing of control shaft (21) actuates load control arm (12).

1. Plug (no rockshaft)	
2. Retaining ring	12. Load control arm
3. Retaining bushing	13. Retaining ring
4. Seal	14. Adjusting screw
5. "O" ring	15. Extension
6. Bushing	16. Spring pin
7. Negative stop screw	17. Cam follower
8. Jam nut	18. Spring
9. "O" ring	19. Pin
10. Pivot shaft	20. Special pin
11. Snap ring	21. Load control shaft

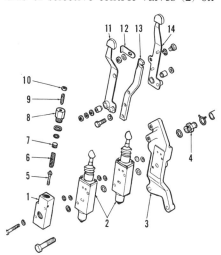

Fig. 128–Exploded view of mounting bracket and pressure relief valve showing selective control valve lever linkage.

1. Valve body	8. Plug
2. Selective control valves	9. Adjusting screw
3. Bracket	10. Locknut
4. Fitting	11. Outer lever
5. Valve plunger	12. Locking strap
6. Spring	13. Support
7. Spring cap	14. Inner lever

JOHN DEERE
(PREVIOUSLY JD-11)

Models ■ 720 Diesel ■ 730 Diesel

SHOP MANUAL

JOHN DEERE

720 DIESEL-730 DIESEL

(General Purpose, Hi-Crop & Standard)

IDENTIFICATION

Tractor serial number stamped on plate on right side of main case

INDEX (By Starting Paragraph)

CONDENSED SERVICE DATA
John Deere 720 Diesel and 730 Diesel

GENERAL

LIQUID CAPACITIES

Cooling System, Tractors With:
Electric Starting Motor...................... 6¼ Gals.
Gasoline Starting Engine...................... 7 Gals.
Crankcase (Diesel Engine)...................... 9 Qts.
(Gasoline Starting Engine)...................... 1½ Qts.
First Reduction Gear Cover...................... 1½ Qts.
Fuel Tank (Diesel Engine)...................... 20 Gals.
(Gasoline Starting Engine)...................... 1 Qt.
Hydraulic System ("Powr-Trol")...................... 13 Qts.
PTO Clutch .... 4½ Qts.
Remote Control Cylinder...................... 1 Qt.
Transmission (Diesel Engine)...................... 8 Gals.
(Gasoline Starting Engine)...................... ½ Pt.

MISCELLANEOUS

Front Wheels—Toe-in .... ⅛-⅜-inch
Speeds—Number Forward 6
Number Reverse 1
Electrical System (Tractors With Gasoline Starting Engine)
Voltage 6
Generator Make and Model...................... D-R 1100027
Regulator Make and Model...................... D-R 1118786
Starting Motor Make and Model...................... D-R 1107155
Electrical System (Tractors With Electrical Starting Motor)
Voltage .... Split Load 24-12
Generator Make and Model...................... D-R 1103021
Regulator Make and Model...................... D-R 1119219
Starting Motor Make and Model...................... D-R 1113801
Solenoid Make and Model...................... D-R 1119803

STARTING ENGINE

GENERAL

Make .... Own
No. Cylinders 4
Bore & Stroke—Inches .... 2 x 1½
Displacement—Cubic Inches 18.85
Cylinder Sleeves? .... Wet

TUNE-UP—SIZES—CLEARANCES

Tappet Gap (In. & Ex.)...................... 0.008-0.010
Valve Seat Angle (In. & Ex.)...................... 45°
Valve Face Angle (In. & Ex.)...................... 44½°
Valve Stem Diameter (In. & Ex.)...................... 0.2445-0.2455
Ignition Type .... Battery
Distributor Model .... Wico B4027
Firing Order 1-2-3-4
Breaker Gap 0.020
Timing Mark on Flywheel...................... "V" Mark, Spark No. 1
Plug Make .... Champion or AC
Champion Model J-8
AC Model 45M
Electrode Gap 0.025
Starting Motor—Make and Model...................... D-R 1107155
Carburetor .... Zenith TU
Engine Load—rpm 4500
Engine Slow Idle—rpm 4000

TUNE-UP—SIZES—CLEARANCES—(Continued)

Engine Fast Idle—rpm 5000
Piston Skirt Clearance...................... 0.002-0.004
Piston Pin Diameter...................... 0.4818-0.4822
Piston Removed From? .... Above
Camshaft End Play 0.003-0.009
Camshaft Front Journal Diameter...................... 1.248-1.249
Camshaft Center Journal Diameter...................... 1.060-1.061
Camshaft Rear Journal Diameter...................... 0.748-0.749
Camshaft Bearing Clearance...................... 0.002-0.005
Crankshaft End Play 0.005-0.008
Crankshaft Main Bearings, Number of?...................... 2
Crankshaft Main Journal Diameter...................... [1]1.124-1.125; [2]1.499-1.500
Crankshaft Main Bearing Clearance...................... [1]0.002-0.004; [2]0.001-0.003
Crankshaft Rod Journal Diameter (Crankpin)...................... 1.0615-1.0625
Crankshaft Rod Bearing Clearance...................... 0.001-0.003
Main & Rod Bearings Adjustable?...................... No

TIGHTENING TORQUES (Ft.-Lbs.)

Cylinder Head Nuts 67
Rod Bolts 7
Flywheel Nut 150
Spark Plugs 32
[1]Prior to tractor serial number 7214900
[2]After tractor serial number 7214899

DIESEL ENGINE

GENERAL

Make .... Own
No. Cylinders 2
Bore and Stroke—Inches...................... 6⅛ x 6⅜
Displacement—Cubic Inches 376
Compression Ratio .... 16 to 1
Cylinders Sleeved? .... No

TAPPETS & VALVES

Tappet Gap (In. & Ex.)...................... 0.020H
Valve Seat Angle (In. & Ex.)...................... 45°
Valve Face Angle (In. & Ex.)...................... 44½°
Valve Stem Diameter 0.496-0.497

DIESEL SYSTEM

Pumps and Nozzles—Make .... †
Injection Occurs .... 24° BTC
Injection Mark on Flywheel...................... "No. 1 INJ"

GOVERNED SPEED

Load—rpm 1125
Fast Idle—rpm 1250
Slow Idle—rpm 700

PISTONS—PINS—RINGS

Piston Skirt Clearance...................... See Paragraph 86
Piston Pin Diameter 2.3545-2.3550
Pistons Removed From? .... Front

CAMSHAFT & BEARINGS

End Play Adjustable .... No
Right Journal Diameter 1.499-1.500
Right Bearing Clearance 0.002-0.005
Center Journal Diameter 2.4975-2.4985
Center Bearing Clearance 0.002-0.005
Left Journal Diameter 1.4960-1.4970
Left Bearing Clearance 0.0025-0.0045

CRANKSHAFT & BEARINGS

End Play 0.005-0.010
Main Bearings, Number of?...................... 3
Main Journal Diameter 4.4975-4.4985
Right & Left Main Bearing Clearance...................... 0.0060-0.0080
Center Main Bearing Clearance 0.0021-0.0055
Main and Rod Bearings Adjustable?...................... No
Rod Journal Diameter (Crankpin)...................... 3.7480-3.7490
Rod Bearing Clearance 0.0015-0.0045

TIGHTENING TORQUES (Ft.-Lbs.)

Cylinder Head Nuts...................... 275
Rod Bolts...................... 200-220
Flywheel Clamp Bolts...................... 275

OIL PRESSURE

Adjustable? .... Yes
Recommended psi 25

†American-Bosch or Bendix-Scintilla

FRONT SYSTEM — TRICYCLE TYPE

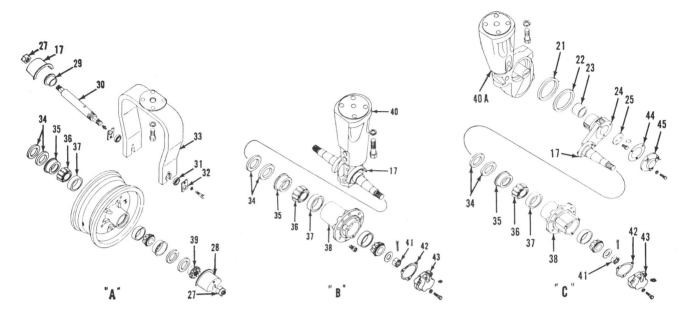

Fig. JD2001—Tricycle type front systems. A. Single front wheel. B. Conventional knuckle mounted dual front wheels.   C. "Roll-O-Matic".

17. Dust excluder
21. Felt washer
22. Retainer
23. Bushing
24. "Roll-O-Matic" knuckle
25. Thrust washer

27. Nut
28. Dust shield
29. Bearing spacer
30. Axle
31. Washer
32. Axle lock plate
33. Yoke

34. Felt washers
35. Retainer
36. Bearing cone
37. Bearing cup
38. Hub
39. Bearing adjusting nut

40. Pedestal extension (Conventional dual wheel)
40A. Pedestal extension ("Roll-O-Matic")
41. Nut
43. Cap
45. Cap

1. Tricycle type 720 and 730 diesel tractors are available with a convertible type, two-piece pedestal. The available convertible front systems include a fork mounted single front wheel, dual wheels of the conventional knuckle mounted type or dual wheels of the "Roll-O-Matic" type. Refer to Fig. JD2001. The procedure for disassembling and overhauling the fork mounted or the conventional knuckle mounted front wheels is evident.

"ROLL-O-MATIC"

4. OVERHAUL. The "Roll-O-Matic" unit can be overhauled without removing the unit from the tractor.

Support front of tractor and remove wheel and hub units. Remove knuckle caps (45—Fig. JD2001C). Unbolt and remove thrust washers (25). Pull knuckle and gear units from housing and remove felt washer (21). Check the removed parts against the values which follow:

Knuckle bushing inside diameter 1.8735-1.8755
Knuckle shaft outside diameter ......1.870 -1.872
Thickness of thrust washers (25)........0.156

Install knuckle bushings (23) with open end of oil groove toward gap between bushings as shown in Fig. JD2002. When the bushings are properly installed, there should be a gap of 1/32-1/16-inch between the bush-

Fig. JD2002 — "Roll-O-Matic" knuckle, showing the proper installation of bushings. Notice that open end of oil grooves in bushings is toward the 1/32-1/16-inch gap between the bushings.

Fig. JD2003 — When assembling the "Roll-O-Matic" unit, make certain that gears are meshed so that timing marks (M) are in register.

ings to allow grease from the fittings to enter grooves in bushings.

Soak felt washers (21—Fig. JD-2001C) in engine oil prior to installa-tion. Install one of the knuckles so that wheel spindle extends behind the vertical steering spindle. Pack the "Roll-O-Matic" unit with wheel bear-ing grease and install the other knuckle so that timing marks on gears are in register as shown in Fig. JD2003.

FRONT SYSTEM—AXLE TYPE

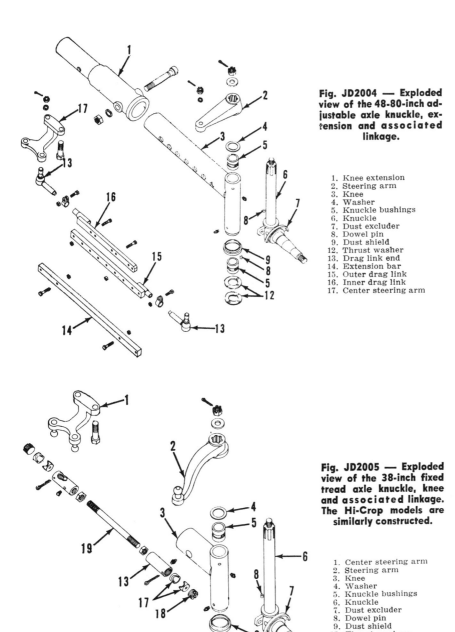

Fig. JD2004 — Exploded view of the 48-80-inch ad-justable axle knuckle, ex-tension and associated linkage.

1. Knee extension
2. Steering arm
3. Knee
4. Washer
5. Knuckle bushings
6. Knuckle
7. Dust excluder
8. Dowel pin
9. Dust shield
12. Thrust washer
13. Drag link end
14. Extension bar
15. Outer drag link
16. Inner drag link
17. Center steering arm

Fig. JD2005 — Exploded view of the 38-inch fixed tread axle knuckle, knee and associated linkage. The Hi-Crop models are similarly constructed.

1. Center steering arm
2. Steering arm
3. Knee
4. Washer
5. Knuckle bushings
6. Knuckle
7. Dust excluder
8. Dowel pin
9. Dust shield
12. Thrust washers
13. Drag link end
17. Stud bearings
18. Screw plug
19. Drag link rod

STEERING KNUCKLES

Except Standard

5. R&R AND OVERHAUL. To re-move either knuckle, remove wheel and hub assembly, disconnect drag link from steering arm and remove nut retaining steering arm to top of knuckle post. Using a knocker tool or brass drift, drive knuckle post free of steering arm. Inside diameter of new knuckle post bushings should be 1.504-1.506 and the kunckle post diameter should be 1.494-1.495.

When installing new bushings, press them in until flush with top and bot-tom of knee and make certain that oil groove in each bushing is in register with fittings in knee. Ream the bush-ings after installation, if necessary, to give the knuckle post the recom-mended clearance of 0.009-0.012 in the bushings.

When reassembling, install dust shield (9—Figs. JD2004 and 2005) on knee. Install thrust washers (12) on knuckle post and install knuckle and post assembly into knee, making cer-tain that dowel pin engages holes in thrust washers. Vary the number of 0.036 adjusting washers (4) until a maximum of 0.036 kunckle end play is obtained.

Standard

5A. R&R AND OVERHAUL. To re-move the steering knuckles (8—Fig. JD2006 or 2007), remove the front wheel and hub units, disconnect the steering arms from knuckles and re-move knuckle caps (4). Remove the taper bolt retaining nuts and drive the taper bolts forward and out of axle. Using a drift, bump spindle pins (5) out of axle and knuckle.

Spindle bushings (7) are pre-sized and if not distorted during installation, will require no final sizing. Recommended clearance between new spindle pins and knuckle bushings is 0.002-0.005. Install two thrust washers (9) between axle and lower fork of each knuckle.

DRAG LINKS AND TOE-IN

6. An adjustable ball socket is fitted to both ends of each drag link on some models and the link ends should be adjusted so they have no end play, yet do not bind. On other models, the tie-rod and drag link ends are of the non-adjustable automotive type.

With front wheels pointing straight ahead, toe-in should be ⅛-⅜-inch. If the adjustment is not as specified, adjust the length of each drag link or tie-rod an equal amount, until proper adjustment is obtained.

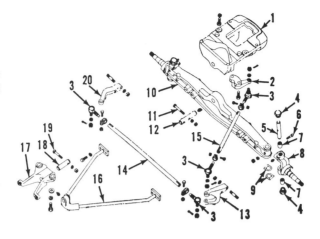

Fig. JD2006 — Exploded view of standard fixed tread front axle and associated parts. Refer to legend under Fig. JD2007.

AXLE PIVOT PINS AND BUSHINGS

Except Standard

7. To renew the axle pivot pins and bushings, support front of tractor and disconnect the center steering arm from the steering spindle. Unbolt axle pivot bracket from front end support and roll axle, pivot bracket and wheels assembly forward and away from tractor. Remove the retaining bolt and withdraw pivot bracket from axle. Inside diameter of new pivot pin bushings is 1.504-1.506 and diameter of new pivot pins is 1.494-1.495.

When installing a new bushing in axle or pivot bracket, press or drive bushing in until bushings are flush with castings and oil groove in each bushing is in register with grease fittings in axle and pivot bracket. Ream the bushings after installation, if necessary, to provide a clearance of 0.009-0.012 for the pivot pins. Pivot pins can be driven from axle and pivot bracket and new ones can be driven or pressed in. Be sure to measure pivot pin extension from axle or pivot bracket and install new ones to the same dimension.

Reassemble and reinstall the axle and pivot bracket assembly by reversing the removal procedure.

Standard

7A. Axle and radius rod are unbushed at pivot pin locations. The procedure for renewing the pins is evident.

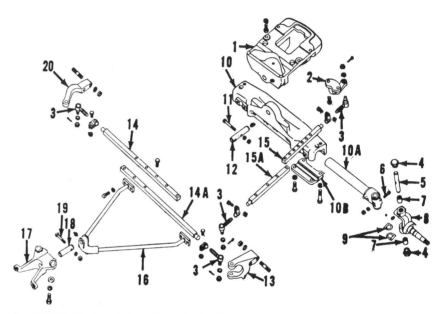

Fig. JD2007—Exploded view of standard adjustable tread front axle and associated parts.

1. Axle pivot bracket	9. Thrust washers	14A. Tie-rod extension
2. Center steering arm	10. Front axle	15. Drag link
3. Tie-rod and drag link ends	10A. Axle extension	15A. Drag link extension
4. **Knuckle cap**	10B. Extension clamp	16. Radius rod
5. Spindle pin	11. **Bolt**	17. Radius rod pivot bracket
6. Taper bolt	12. Axle pivot pin	18. Radius rod pivot pin
7. Spindle bushings	13. Right steering arm	19. Bolt
8. Steering knuckle	14. Tie rod	20. Left steering arm

MANUAL STEERING SYSTEM

STEERING SPINDLE AND PEDESTAL

8. **R&R AND OVERHAUL.** To remove the steering spindle (16—Fig. JD2010), first remove grille and disconnect the steering shaft coupling. Unbolt baffle plate from radiator top tank. Remove cap screws retaining the steering worm housing to the pedestal and bump the housings apart. Withdraw the steering worm and housing assembly from tractor. Support front of tractor. On axle type tractors, disconnect center steering arm from the steering spindle. Unbolt axle pivot

bracket from front end support and roll axle, pivot bracket and wheels assembly forward and away from tractor. On tricycle type tractors, unbolt and remove the lower spindle (or fork) and wheels assembly from the steering spindle. On all models, remove cup plug from top of pedestal and remove cap screw (1) retaining steering gear to spindle. Using a knocker tool or brass drift, bump spindle down and out of steering gear. Withdraw gear and save the adjusting washers (4) which are located under the gear.

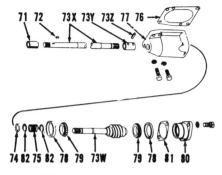

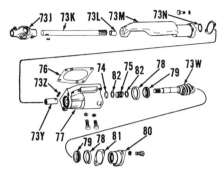

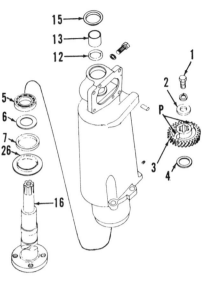

Fig. JD2008—Exploded view of 720 diesel manual steering worm shaft and associated parts. Shims (76) control backlash between worm and steering gear. Shims (81) control end play of steering worm.

Fig. JD2009—Exploded view of 730 diesel manual steering worm shaft and associated parts. Refer to legend for Fig. JD2008.

Fig. JD2010—Exploded view of convertible two-piece pedestal. Spindle end play is controlled by washers (4).

71. Bushing	73Z. Roll pin
72. Woodruff key	74. "O" ring packing
73J. Universal joint (730)	75. Spring
73K. Coupling shaft (730)	76. Shims
73L. Bushing (730)	77. Worm housing
73M. Coupling shaft	78. Bearing cup
support (730)	79. Bearing cone
73N. "O" ring (730)	80. Bearing housing
73W. Steering worm	81. Shims
73X. Steering shaft (720)	82. Washer
73Y. Coupling	

1. Cap screw	7. Cork washer
2. Washer	12. "O" ring
3. Steering gear	13. Bushing
4. Adjusting washers	15. Cup plug
5. Lower bearing	16. Spindle
6. Washer	26. Retainer

The spindle lower bearing (5) can be removed by using a suitable puller after removing retainer (26).

Inside diameter of a new spindle upper bushing (13) is 1.500-1.503 and the diameter of the steering spindle upper bearing surface should be 1.497-1.498. When installing the bushing, the beveled edge should be toward bottom of pedestal, the split should be toward the steering worm and top of bushing should be flush with top of bushing bore. Check the installed bushing to make certain that the steering spindle has the recommended clearance of 0.002-0.006 in the bushing. If there is evidence of oil leakage to the lower part of the pedestal, renew 'O' ring (12) which is located just below the bushing (13).

When installing the steering spindle, observe the following:

(a) Be careful not to damage the "O" ring (12) when installing the steering spindle.

(b) Vary the number of spacer washers (4) to give the spindle an end play of 0.004-0.040.

(c) With steering spindle positioned so that when the center steering arm is connected and the front wheels are pointing straight ahead, the hub projections (P) on steering gear should be parallel to center line of tractor.

Steering spindle to center steering arm screws should be tightened to a torque of 275 ft.-lbs.

STEERING MECHANISM

For the purposes of this discussion, the steering mechanism will include the steering worm and shaft and the steering gear. For R&R and overhaul of the steering spindle, refer to paragraph 8.

9. **ADJUSTMENT.** Three adjustments are provided on the steering mechanism: (1) steering spindle shaft end play; (2) steering (worm) shaft end play; (3) backlash between the worm and steering gear.

10. **STEERING SPINDLE SHAFT END PLAY.** The desired steering spindle shaft end play of 0.004-0.040 is controlled by varying the number of 0.036 adjusting washers (4— Fig. JD2010) which are located under the steering gear (3). To make the adjustment, it is necessary to remove the steering gear from tractor as outlined in paragraph 14.

11. **WORM SHAFT END PLAY.** The desired steering worm shaft end play of 0.001-0.004 is controlled by shims (81—Fig. JD2008 or 2009) which are located under the worm shaft front bearing housing. To check the end play, first remove grille, disconnect the steering shaft coupling and pull the steering wheel shaft rearward until the coupling is disengaged from the worm shaft. Attach a dial indicator in a suitable manner, move the worm shaft back and forth and check the end play. If the end play is not as specified, remove the front bearing housing and add or remove shims until the proper adjustment is obtained.

12. **BACKLASH.** The desired backlash between the steering gear and worm is ½-1 inch when measured at rim of steering wheel, and is controlled by shims (76—Fig. JD2008 or 2009) which are located between the pedestal and worm housing. If backlash is not as specified, remove grille and unbolt baffle plate from the radi-

ator top tank. Unbolt worm housing from pedestal, bump housings apart and add or remove shims (76) until backlash is as specified. Shims are available in thicknesses of 0.005 and 0.015. If excessive backlash cannot be eliminated, it will be necessary to renew the worm and steering gear; or, reposition the steering gear on the steering spindle so as to bring unworn teeth into mesh. Refer to paragraph 14. Note: Be careful not to confuse wear in steering shaft U-joints or couplings with gear unit backlash.

13. **OVERHAUL.** The steering mechanism can be overhauled without removing pedestal from tractor, as follows:

14. **STEERING GEAR.** To remove the steering gear (3—Fig. JD2010), first remove grille and disconnect the steering shaft coupling. Unbolt baffle plate from radiator top tank and remove cup plug from top of pedestal. Unbolt worm housing from pedestal and bump the housings apart. Withdraw wormshaft and housing assembly from tractor. Support front of tractor. On axle type tractors, disconnect center steering arm from steering spindle. Unbolt axle pivot bracket from front end support and roll axle and wheels assembly away from tractor. On all models, raise front of tractor, remove cap screw (1) retaining steering gear to spindle and using a knocker tool or brass drift, bump spindle down and out of steering gear.

Reinstall the steering gear by reversing the removal procedure and observe the following:
(a) If the same gear is being reinstalled, position the gear so that unworn teeth will mesh with the worm.
(b) Vary the number of spacer washers (4) to give the spindle an end play of 0.004-0.040.
(c) With front wheel (or wheels) pointing straight ahead, the hub projections (P) on steering gear should be parallel to center line of tractor.

15. STEERING WORM. Overhaul of the steering worm and components is accomplished as follows: Remove grille and disconnect the steering shaft coupling. Remove the worm shaft front bearing housing (80—Fig. JD2008 or 2009) and turn worm shaft forward and out of housing. The worm shaft housing can be removed from pedestal at this time. The need and procedure for further disassembly is evident after an examination of the unit. Inspect "O" ring seal (74) and renew if damaged.

When reassembling, adjust the worm shaft end play and the gear unit backlash as outlined in paragraphs 11 and 12.

POWER STEERING SYSTEM

Note: The maintenance of absolute cleanliness of all parts is of utmost importance in the operation and servicing of the hydraulic power steering system. Of equal importance is the avoidance of nicks or burrs on any of the working parts.

LUBRICATION

16. It is recommended that the power steering system be drained (but not flushed) once a year or every 1,000 hours. To drain the system, turn the steering wheel to the extreme left position and remove the grille. Remove the reservoir drain plug and allow system to drain. To remove the additional oil remaining in the steering cylinder, disconnect the front oil line from the control valve, pivot the line forward and turn the steering wheel to the extreme right position.

Refill the system with five U.S. quarts of John Deere power steering oil.

TROUBLE-SHOOTING

17. The following paragraphs outline the possible causes and remedies for troubles in the power steering system.

17A. DRIFTING TO EITHER SIDE could be caused by:
1. Valve housing not in correct relation to worm shaft housing. Center the valve housing as in paragraph 23.

17B. HARD STEERING could be caused by:
1. Insufficient volume of oil flowing to steering valve from flow control valve. Adjust the flow control valve as in paragraph 22.
2. Excessive tension on the centering cam spring. Adjust the spring as in paragraph 25C or 26C.
3. Insufficient end play or binding in the worm shaft bearings. See paragraph 25 or 26.
4. Insufficient backlash between the steering worm and gear. See paragraph 25B or 26B.

5. Excessive leakage past the steering vane seals in cylinder. To check, remove grille, disconnect the front oil line from steering valve, pivot the oil line forward and tighten the lower oil line connection. With the engine running at fast idle, hold the steering wheel to the extreme right turn position and measure the leakage from the oil line as shown in Fig. JD2015. Oil leakage should not exceed 1 quart in ½ minute. If leakage is excessive, renew the vane seals. Refer to paragraph 29B.
6. Binding in the rear steering shaft support bushing.

Fig. JD2015 — Checking for leakage past the power steering cylinder vane.

7. Pedestal improperly aligned. See paragraph 25A or 26A.
8. Insufficient pump pressure. Refer to paragraph 19.
9. Insufficient oil in system.
10. Foaming oil in system. Refer to paragraph 17D.

17C. EXCESSIVE INSTABILITY OF FRONT WHEELS. This condition is often referred to as shimmy or flutter. In some cases, flutter cannot be entirely eliminated, but the unit can be adjusted to the point where it is not objectionable. Possible causes of instability are:
1. Excessive volume of oil flowing from flow control valve to steering valve. Adjust the flow control valve as in paragraph 22.
2. Actuating sleeve set screw too loose in helix slot. Refer to paragraph 25.
3. Worn point on actuating sleeve set screw and/or damaged helix slot in steering worm shaft.
4. Insufficient tension on cam spring. Refer to paragraph 25C or 26C.
5. Excessive end play in the steering worm shaft, paragraph 25 or 26.
6. Excessive backlash between the steering worm and gear. Refer to paragraph 25B or 26B.
7. Unbalanced front wheels.
8. Loose or worn front wheel bearings.
9. Loose or worn "Roll-O-Matic" assembly.
10. Use of front wheel weights rather than front end weights.

17D. OVERFLOW OF FOAMING OF OIL could be caused by:
1. Air leak in system. To check, apply a light coat of oil to sealing surfaces and observe for leaks.
2. Wrong type oil. Use only John Deere power steering oil.

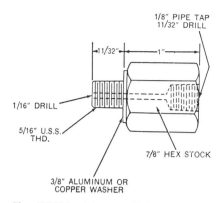

Fig. JD2016—Adapter which can be made to accommodate a pressure gage in the steering valve housing.

17E. LOCKING could be caused by:

1. Scored worm, or worm bearings adjusted too tight. Refer to paragraph 25 or 26.

2. Steering valve arm interference in groove of actuating sleeve.

3. Steering gear loose on vertical spindle. Tighten the cap screw to a torque of 190 ft.-lbs.

4. Bent steering spindle or scored steering spindle bearings. Refer to paragraph 29B.

5. Loose or broken movable vane retaining cap screws. Screws should be tightened to a torque of 208 ft.-lbs. Refer to paragraph 29B.

6. Insufficient clearance between actuating sleeve and the steering worm shaft or cam. Refer to paragraph 25C or 26C.

17F. VARIATION IN STEERING EFFORT WHEN TURNING IN ONE DIRECTION could be caused by:

1. A bent steering spindle. Refer to paragraph 29B.

SYSTEM OPERATING PRESSURE

19. The system relief valve is mounted in the steering valve housing as shown in Fig. JD2017. To check the system operating pressure, remove grille and using the adapter (Fig. JD-2016) install a suitable pressure gage (at least 1500 psi capacity) in flow control stop screw hole as shown in Fig. JD2017. With the engine running at fast idle rpm, cramp the front wheels to the extreme right or extreme left position, and observe the highest pressure reading which should be 1170-1210 psi.

If the operating pressure is not as specified, vary the number of washers (19—Fig. JD2024). If the addition of washers will not increase the pressure to within the specified limits, check for a faulty pump.

PUMP

20. **REMOVE AND REINSTALL.** To remove the power steering pump, first remove the grille and hood; then, proceed as follows: Drain the power steering reservoir and remove the pump pressure and suction oil lines. Loosen and disconnect fan belt. Unbolt fan shaft tube flange from pump and pump from mounting bracket.

Move the pump and fan assembly forward until the fan shaft coupling is just exposed. While holding the coupling rearward, move the pump assembly forward until the pump shaft is disengaged from the coupling.

Note: If coupling is not held rearward while pump is moved forward,

the fan shaft may come out of the rear coupling. If the shaft comes out of the rear coupling, it is difficult to re-install.

Move rear of pump toward right side of tractor, tip top of pump toward right side and withdraw the pump and fan assembly from left side of tractor.

Note: In some individual cases it is more convenient and often time will be saved by removing the fan blades before unbolting the pump from its mounting bracket.

Install the pump by reversing the removal procedure and fill the pump with oil before connecting the oil lines.

21. **OVERHAUL.** With the pump removed from tractor, thoroughly clean the exterior surfaces to remove any accumulation of dirt or other foreign material. Remove the pump housing retaining cap screws and using a plastic hammer, bump the pump housing from its locating dowels. Remove the pump body, follower gear and drive gear. Refer to Fig. JD2020. Extract the drive gear Woodruff key (2) from the fan shaft.

Place the pump cover and fan assembly on a press, depress the fan against spring pressure as shown in Fig. JD2018 and remove the split cone locks (9) and keeper (8). Remove assembly from press and withdraw the fan, pulley, spring and drive parts from the shaft. Press the pump drive shaft rearward out of pump cover, extract the bearing retaining snap ring and remove bearing from the cover. Inspect all parts and renew any which

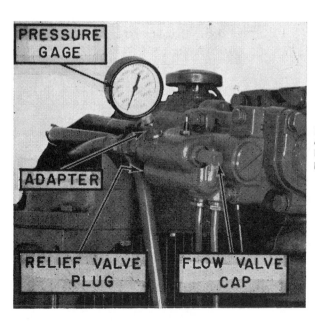

Fig. JD2017 — Adapter and pressure gage installation for checking the power steering system operating pressure.

Fig. JD2018 — Compressing the fan shaft spring (4) to permit removal of locks (9) and keeper (8).

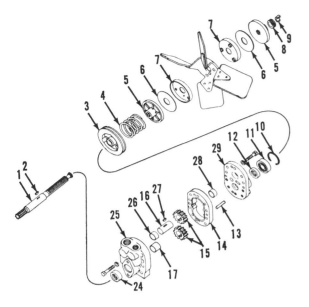

Fig. JD2019 — Exploded view of power steering pump. The system relief valve is located within the valve housing.

1. Fan shaft
2. Woodruff key
3. Pulley
4. Spring
5. Friction disc
6. Friction washer
7. Drive cup
8. Fan keeper
9. Locks
10. Snap ring
11. Bearing
12. Oil seal
13. Dowel pin
14. Pump body
15. Pumping gears
16. Idler gear shaft
17. Bushing
24. Oil seal
25. Pump housing
26. Bushing
27. Woodruff key
28. Bushing
29. Pump cover

To reassemble the pump, install the cover bearing and snap ring. Press the pump shaft into bearing from gear side of cover until shoulder on shaft seats against the bearing race. Install the two assembly cap screws in cover, then install the driving pulley, spring and fan parts in the sequence shown in Fig. JD2019. Install Woodruff key in the drive shaft and slide the drive gear into position. Drive gear must float on shaft and not bind on the Woodruff key. Install the follower gear, coat the mating surfaces of the pump cover and body with shellac and install the body so that edges of body are concentric with edges of cover. Coat the mating surfaces of the pump body and housing with shellac, install the housing and tighten the assembly cap screws securely.

are questionable. Bushings in pump body and cover should be pressed in with a piloted arbor. Oil seal in pump cover should be pressed in not more than $\frac{1}{16}$-inch below bottom of bearing bore.

Fan drive friction washers (6—Fig. JD2019) should be approximately $\frac{1}{16}$-inch thick, the friction spring (4) should exert 216-264 lbs. @ 1½ inches and the fan blade pitch should be 2 17/32 - 2 19/32 inches.

STEERING VALVES

22. ADJUST FLOW CONTROL VALVE. Turning the flow control valve adjusting screw inward will cause faster, easier steering action; but, the increase in steering speed can result in a decrease of front wheel stability (increased tendency of front wheels to flutter).

To make the adjustment, operate the steering system until the oil is at normal operating temperature; then, turn the adjusting screw in until the fastest turning screw is obtained without causing an objectionable amount of front wheel flutter. Refer to Fig. JD2021.

NOTE: Front wheel instability (flutter) may be due to one or more of the causes suggested in paragraph 17C.

23. CENTERING VALVE HOUSING. Housing (40—Figs. JD2021 and 2024) is properly centered when the effort required to turn the steering wheel in one direction is exactly the same as the effort required to turn the steering wheel in the other direction. The adjustment is made by loosening the valve housing retaining cap screws, tapping the housing either way as required and then tightening the cap screws.

If the wheel turns harder to the left, tap the housing rearward. Conversely, if the wheel turns harder to the right, tap the valve housing forward.

To accurately center the valve housing using a pressure gage, use the adapter (Fig. JD2016) and install a suitable pressure gage (at least 1500

Fig. JD2020 — Partially disassembled view of the power steering pump. Be sure to remove Woodruff key (2) before attempting to remove the fan shaft (1).

Fig. JD2021 — Adjusting the power steering flow control valve. To make the adjustment, turn the screw in to obtain fastest turning speed without causing an objectionable amount of front wheel flutter.

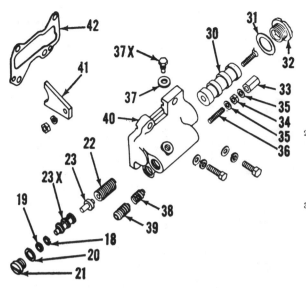

18. Washers
19. Shim washers
20. Gasket
21. Plug
22. Spring
23. Relief valve
23X. Relief valve sleeve
30. Steering valve
31. Gasket
32. Plug
33. Cap nut
34. Jam nut
35. Washer
36. Adjusting screw
37. Copper washer
37X. Stop screw
38. Flow control valve spring
39. Flow-control valve
41. Valve arm

With the valve housing installed, center the unit as in paragraph 23 and adjust the flow control valve as outlined in paragraph 22. Check the system operating pressure as outlined in paragraph 19.

STEERING WORM AND VALVE ACTUATING SLEEVE

Early 720 Diesel Models

25. **END PLAY ADJUSTMENT.** Two end play adjustments are provided in the steering worm and valve actuating sleeve assembly. The valve actuating sleeve set screw must be adjusted to provide an end play of 0.001-0.002 between the cone point of the set screw and the helix slot in the steering worm shaft. Then, the number of shims between the worm housing and the front bearing housing should be varied to obtain a worm shaft end play of 0.001-0.004. The two adjustments will therefore provide a total end play for the worm and actuating sleeve assembly of 0.002-0.006.

The two adjustments can be approximated, but since the end play should be held as near as possible to the lower limit, it is recommended that a dial indicator be used as follows:

Remove grille and mount a dial indicator with contact button resting on rear end of the valve actuating sleeve as shown in Fig. JD2026. Remove the worm shaft front bearing housing and all of the adjusting shims. Reinstall

psi) in flow control stop screw hole as shown in Fig. JD2017. With the engine running at fast idle speed and front wheels in the straight ahead position, loosen the steering valve attaching screws and tap the housing to the front or rear until the lowest pressure reading is obtained; then, tighten the housing retaining cap screws.

24. **REMOVE AND REINSTALL.** To remove the valve housing, remove grille, drain the power steering reservoir and disconnect the oil lines from valve housing. To facilitate reinstallation, mark the relative position of the valve housing with respect to the worm housing with a scribed line. Unbolt and remove the valve housing assembly from tractor.

24A. To install the valve housing, use a new gasket and tighten the assembly cap screws finger tight; also, be sure to align the previously affixed scribed lines. Reconnect the oil lines, fill the reservoir and start engine. Center the valve housing as in paragraph 23.

24B. **OVERHAUL.** With the valve housing removed, clean the unit and remove the flow control valve adjusting screw. Remove the union adapter from rear of housing and slide the flow control valve and spring out of the valve bore. Remove the end plug from housing and remove the steering valve arm by removing its retaining bolt as shown in Fig. JD2025. Remove the relief valve plug (21—Fig. JD2024) and withdraw the relief valve sleeve, valve and spring. Wash all parts in a suitable solvent and renew any that are nicked, grooved or worn.

Install the flow control valve spring, valve and adjusting screw, and initially adjust the valve as follows: Using a punch or similar tool, hold the valve against rear of flow valve passage and turn the adjusting screw in until it just contacts the spring; then, turn the screw in two additional turns. Install the union adapter, flow valve and arm and end plug. Assemble the relief valve, using the same number of adjusting washers as were removed.

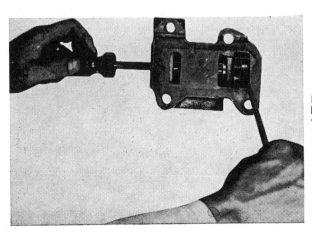

Fig. JD2025 — Removing bolt retaining the steering valve arm to the steering valve.

Fig. JD2026—Adjusting the early production 720 diesel steering valve actuating sleeve set screw. Adjustment is best accomplished by grinding a screwdriver tip as shown in Fig. JD2027.

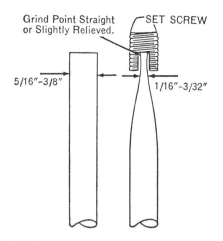

Fig. JD2027 — Dimensions for grinding a special screwdriver for adjusting the early production 720 valve actuating sleeve set screw.

43. Woodruff key
44. Steering shaft
45. Oil seal
46. Oil seal housing
47. Gasket
48. Valve actuating sleeve
49. Pins
50. Expansion plug
51. Cone point set screw
52. Bearing housing
53. "O" ring
54. Shims
55. Bearing cup
56. Bearing cone
57. Steering worm
58. Worm cam
59. Cam rod
60. Cotter pin
61. Spring
62. Washer
63. Nut
64. Worm housing
65. Shims
66. Pipe plug
67. Dowel pin

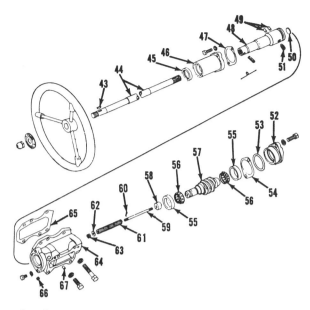

Fig. JD2029—Exploded view of typical early production 720 power steering worm, valve actuating sleeve and associated parts. Backlash between the worm and steering gear is controlled by shims (65).

the front bearing housing without the shims so that the worm shaft bearings have no end play. Remove the inspection hole plug and turn the steering wheel until the actuating sleeve set screw lines up with the inspection hole. Turn the set screw either way as required to obtain an end play of 0.001-0.002. It is desirable, however, to hold the end play adjustment as close to 0.001 as possible.

Note: Adjustment of the set screw will be made easier by grinding a screw driver tip to the dimensions shown in Fig. JD2027. After adjustment, lock the screw in position by staking.

With the set screw adjusted, remove the worm shaft front bearing housing, install the previously removed shims and check the end play which should now be 0.002-0.006. Note: This 0.002-0.006 end play includes the 0.001-0.002 end play previously obtained by the actuating sleeve set screw adjustment. The number of shims under the front

bearing housing should be varied to reduce the total end play as near to 0.002 as possible. Shims are available in thicknesses of 0.010 and 0.0025.

25A. STEERING SHAFT BIND. To obtain proper steering action, there should be a minimum of binding in the steering shaft. To check for binding with engine stopped, turn the steering wheel in either direction without turning the front wheels. When the steering wheel is released, spring pressure against the cam within the valve actuating sleeve should return the steering wheel to a neutral position.

If binding exists after the end play is adjusted as outlined in paragraph 25, check for misalignment of the steering shaft rear support. The support can be aligned after loosening the two retaining cap screws.

An improperly located pedestal can also cause binding. Beginning with the front center pedestal retaining cap screw, loosen every other cap screw and bump pedestal either way as shown in Fig. JD2028 until the steering shaft is free.

25B. BACKLASH. Backlash between the steering worm and gear is controlled by shims located between the worm shaft housing and the pedestal. To check the backlash, rotate the steering wheel back and forth without permitting the steering cams within the actuating sleeve to separate or the front wheels to turn. Backlash should be ⅜-¾-inch when measured at outer edge of steering wheel. Check the backlash with the front wheels in the extreme right, straight ahead and extreme left positions. Excessive variation in backlash between the three positions indicates a bent steering spindle. NOTE: Be careful not to con-

Fig. JD2028 — Centering the power steering pedestal to reduce binding in the system.

Fig. JD2030 — When removing the valve actuating sleeve on early production 720 diesel tractors, it is often necessary to use a punch and position the cam rod spring so it will clear the sleeve pins.

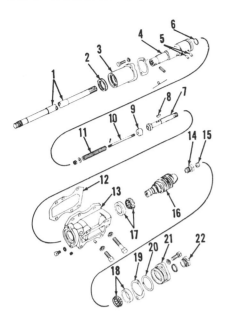

Fig. JD2031—Exploded view of late production 720 diesel steering worm, valve actuating sleeve and associated parts. Backlash between the worm and steering gear is controlled by shims (12). Series 730 diesel is similar.

1. Steering shaft	12. Adjusting shims
2. Oil seal	13. Worm housing
3. Oil seal housing	14. Adjusting bushing
4. Actuating sleeve	15. Lock screw
5. Dowels	16. Worm
6. Welch plug	17. Bearing cup and cone
7. Worm shaft	18. Bearing cup and cone
8. Woodruff key	19. Adjusting shim
9. Cam	20. "O" ring
10. Spring rod	21. Worm bearing housing
11. Cam spring	22. Screw plug

fuse wear in steering shaft couplings with gear unit backlash.

25C. R&R AND OVERHAUL. To remove the steering worm and valve actuating sleeve, first remove the

grille, disconnect the steering shaft coupling and drain the power steering oil reservoir. Disconnect the oil lines and remove the steering valve housing as outlined in paragraph 24. Unbolt and remove the worm shaft housing from pedestal, being careful not to damage or lose the backlash adjusting shims.

Remove oil seal housing (46—Fig. JD2029), actuating sleeve set screw (51) and actuating sleeve (48). When withdrawing the actuating sleeve, it may be necessary to insert a punch through the sleeve hole as shown in Fig. JD2030 to position the cam rod spring so it will clear the internal pins in the sleeve. Unbolt the worm shaft front bearing housing, save the adjusting shims and withdraw the worm. Remove cam spring (61—Fig. JD2029) and cam (58) from rod.

Clean and inspect all parts and renew any that are damaged and cannot be reconditioned. Lip of oil seal (45) goes toward worm. Renew the worm shaft if it is scored or worn at cam end, helix slot or worm. Burrs can be removed from the worm shaft keyway or the surface over which the actuating sleeve operates by using a fine stone. Check the valve actuating sleeve pins for damage, and bore of sleeve for being scored. Check fit of sleeve on worm shaft. If sleeve is tight on shaft or if bore is scored, remove the dowel pins and hone the sleeve to insure a free fit. Renew the pins if they are worn or damaged. Renew the actuating sleeve set screw if it has a loose fit or if cone point is worn.

When reassembling, use a valve spring tester and measure the length of spring (61) at 70-80 lbs. pressure.

Note: If spring will not exert 70-80 lbs. pressure at 2⅞ inches or more for a 3 inch free length spring or 1⅞ inches or more for a 2⅛-inch free length spring, obtain a new spring and measure its length at 70-80 lbs. pressure.

Then install the spring on the cam rod and tighten the adjusting nut until spring is compressed to the same length which yielded 70-80 lbs.

Reassemble the worm and actuating sleeve and adjust the end play as in paragraph 25. Install the unit and adjust the backlash as in paragraph 25B. Install the valve housing as outlined in paragraph 24A.

Late 720 Diesel-730 Diesel Models

26. END PLAY ADJUSTMENTS. Two end play adjustments are provided in the steering worm and valve actuating sleeve assembly. The number of shims between the worm housing and the front bearing housing should be varied to obtain a worm shaft end play of 0.001-0.004. Then, the worm helix width should be adjusted to provide the valve actuating sleeve with an end play of 0.0015-0.0025. To make the adjustments, proceed as follows:

Remove grille, disconnect steering shaft and pull the steering shaft rearward and free from the valve actuating sleeve. Remove the large plug (22—Fig. JD2031), install a cap screw in end of steering worm and mount a dial indicator as shown in Fig. JD2032. Use a suitable drift in actuating sleeve roll pin hole, rotate worm assembly in both directions and note the amount of end play in the worm bearings as registered on the dial indicator. If the indicator reading is not 0.001-0.004, add or deduct shims (19—Fig. JD2031) until proper end play is obtained. It is important that end play be adjusted as close to 0.001 as possible.

With the worm shaft end play properly adjusted, mount a dial indicator as shown in Fig. JD2033, reach into

SHIMS   0.001-0.004 END PLAY

Fig. JD2032 — Using dial indicator to check the steering worm end play on late 720 and all 730 diesel tractors.

0.0015-0.0025 END PLAY

Fig. JD2033 — Checking end play of late production steering valve actuating sleeve. Specified end play is 0.0015-0.0025.

the large screw plug opening in front bearing housing with your fingers and while pressing on the worm to keep it from moving, use a suitable drift in actuating sleeve roll pin hole, move the valve actuating sleeve back and forth to determine the amount of end play between the hardened dowel and the helix. If the indicator reading is not 0.0015-0.0025, adjust the helix width by loosening the special lock screw (15—Fig. JD2031), which is located inside bushing (14), and using a ⅞-inch socket, turn the bushing **in** to decrease or **out** to increase the end play clearance. Adjust the clearance as close to 0.0015 as possible. When adjustment is complete, hold the bushing from turning and tighten the lock screw.

Note: Never make the adjustment so tight that steering wheel will not return to neutral when released after being turned in either direction to operate the steering valve. Refer to paragraph 25A.

26A. **STEERING SHAFT BIND.** To check and/or adjust the steering shaft bind, refer to paragraph 25A.

26B. **BACKLASH.** To check and/or adjust the backlash between the steering worm and gear, refer to paragraph 25B.

26C. **R&R AND OVERHAUL.** To remove the steering worm and valve actuating sleeve, first remove the grille, disconnect the steering shaft coupling and drain the power steering oil reservoir and worm housing. Disconnect the oil lines and remove the steering valve housing as outlined in paragraph 24. Unbolt and remove the worm shaft housing from pedestal, being careful not to damage or lose the backlash adjusting shims. Remove the worm shaft front bearing housing (21—Fig. JD2031) and carefully measure the distance the adjusting bushing protrudes from the worm as shown in Fig. JD2034. Loosen the special lock screw, remove the adjusting bushing and withdraw the worm.

Remove oil seal housing (3—Fig. JD2031) and pull actuating sleeve and worm shaft assembly from housing. To remove the worm shaft from the actuating sleeve, use a pair of vise-grip pliers to turn the shaft about 90 degrees in a counter-clockwise direction so the cam and shaft keyways align.

Clean and inspect all parts and renew any that are damaged and can-

not be reconditioned. Lip of oil seal (2) goes toward worm. Renew worm if it is worn or if helix end is scored. Renew worm shaft (7) if it is worn or scored at cam or helix locations. Small burrs can be removed from wearing surfaces by using a fine stone. Check the valve actuating sleeve dowels (5) for damage, and bore of sleeve for being scored. Check fit of parts by inserting them in the sleeve. If binding exists, remove the dowel pins and hone the sleeve bore to insure a free fit. Renew the dowel pins if they are worn or damaged.

When reassembling, use a valve spring tester and measure the length of spring (11) at 70-80 lbs. pressure.

Note: If the spring will not exert 70-80 pounds pressure at 1⅞ inches or more, obtain a new spring and measure its length at 70-80 lbs. pressure. Then install spring on spring rod and tighten the adjusting nut until spring is compressed to the same length which yielded a pressure of 70-80 lbs.

Assemble worm shaft (7) and cam (9) with keyways aligned. Then insert the worm shaft assembly into the actuating sleeve so the keyways will slide under the dowels and when the cam is behind the front dowel, turn the worm shaft clockwise with a pair of vise-grip pliers until the cam seats. Install actuating sleeve and worm shaft assembly into worm housing and

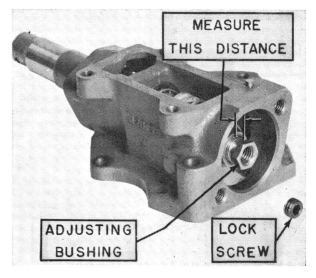

Fig. JD2034 — Checking protrusion of adjusting bushing from late production steering worm.

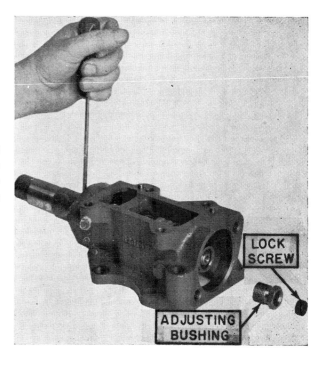

Fig. JD2035—Positioning the late production worm shaft with respect to the worm prior to installing adjusting bushing.

install bearing assembly (17). With Woodruff key properly positioned in worm shaft, slide worm into position.

Using a small screw driver as shown in Fig. JD2035, move the worm shaft until it lacks about 1/32-inch of being flush with end of worm; then, carefully install the adjusting bushing and screw it into the worm until all actuating sleeve end play is removed (dowel tight in helix). Measure the bushing protrusion as shown in Fig. JD2034. If the measured dimension is not almost identical to that measured before the bushing was removed, unscrew the bushing and start over.

When proper bushing protrusion is obtained, install lock screw, front bearing housing and oil seal housing. Check the end play adjustments as in paragraph 26.

Install the worm housing assembly and adjust the gear backlash as in paragraph 25B. Install the steering valve housing as in paragraph 24A.

SPINDLE, PEDESTAL AND CYLINDER

29B. R&R AND OVERHAUL. Remove the grille, disconnect the steering shaft coupling and drain the power steering reservoir. Disconnect the oil lines, unbolt worm housing from pedestal and remove the worm housing and valve housing assembly being careful not to lose the backlash adjusting shims located between the worm housing and pedestal. Using a feeler gage as shown in Fig. JD2036, measure the spindle end play clearance between the steering gear and

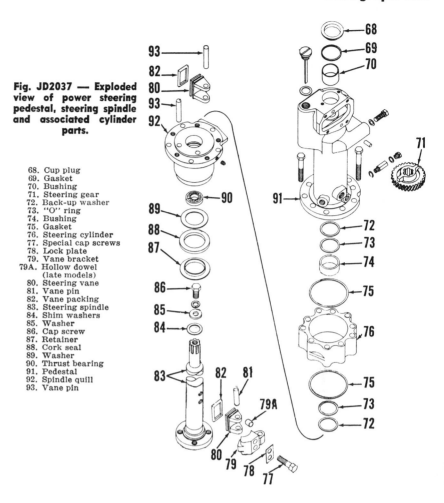

Fig. JD2037 — Exploded view of power steering pedestal, steering spindle and associated cylinder parts.

68. Cup plug
69. Gasket
70. Bushing
71. Steering gear
72. Back-up washer
73. "O" ring
74. Bushing
75. Gasket
76. Steering cylinder
77. Special cap screws
78. Lock plate
79. Vane bracket
79A. Hollow dowel (late models)
80. Steering vane
81. Vane pin
82. Vane packing
83. Steering spindle
84. Shim washers
85. Washer
86. Cap screw
87. Retainer
88. Cork seal
89. Washer
90. Thrust bearing
91. Pedestal
92. Spindle quill
93. Vane pin

Fig. JD2036—Checking the power steering spindle end play. Desired end play of 0.004-0.021 is controlled by shims under the gear.

the adjusting shims. If the clearance is not within the limits of 0.004-0.021, it will be necessary to add or remove the necessary amount of shims during reassembly. Remove the cup plug from top of pedestal and the cap screw and washer retaining the steering gear to vertical spindle. Using a wood block or brass drift and a heavy hammer, drive the steering gear upward until gear is loose on the spindle splines. Unbolt pedestal from front end support and lift pedestal, gear and adjusting shims from spindle.

Remove cylinder (76—Fig. JD2037) and unbolt wheel fork, lower spindle or center steering arm from the steering spindle (83). Unlock the cap screws (77) and unbolt the vane bracket from spindle. Lift the quill (92) and steering spindle assembly from tractor and bump spindle out of quill.

Lower surface of pedestal and upper surface of quill (92) form part of the cylinder sealing surface and must be smooth and free from nicks and scratches. If the attaching pin (81) for the movable steering vane has worn a groove in the upper surface

of the quill (92), renew quill and pin. The new type headed pin will eliminate grooving of the quill. Original inside diameter of the pedestal upper bushing (70) is 1.502-1.503. When installing the bushing, the beveled edge should be toward bottom of pedestal, the split should be toward the steering worm and top of bushing should be flush with top of bushing bore. New bushing is pre-sized and, if not damaged during installation, will require no final sizing. Original inside diameter of the pedestal lower bushing (74) is 2.502-2.503. To remove the bushing, extract the "O" ring located just above the bushing and use a suitable puller. Using a closely fitting mandrel, install bushing until it is flush with bottom of pedestal. Bushing is not pre-sized and must be honed to an inside diameter of 2.502-2.503. After honing, clean and lubricate the bushing. Install a new "O" ring and back-up washer above the lower bushing. Note: The upper end of the bushing forms the lower edge of the "O" ring groove. Install a new and well lubricated neoprene gasket in groove in lower flange of pedestal.

Bearing surface of spindle must be smooth and free from nicks or scratches. Side of thrust bearing (90) marked "THRUST HERE" must be installed toward bottom of quill (92). Install washer (89), new cork seal (88) and retainer (87). Install back-up washer (72) in quill, then install a new "O" ring on top of the washer. With "O" ring (73) well lubricated, slide spindle into quill, position the assembly on the front end support and bolt wheel fork, lower spindle or center steering arm in position and tighten the cap screws to a torque of 275 ft.-lbs.

Install the steering vane bracket (79), tighten the retaining cap screws to a torque of 208 ft.-lbs. and bend tab of lock plate over cap screw heads. Install vane seals, vanes and pins. Refer to Fig. JD2039. Install cylinder (76—Fig. JD2037) and position the cylinder so that all cap screw holes in cylinder and quill are aligned. Lubricate "O" ring (73) in lower part of pedestal and install pedestal. When lowering pedestal into position, install the end play adjusting shims (84) and steering gear. Tighten the gear retaining cap screw to a torque of 190 ft.-lb.

Fig. JD2039 — Correct installation of the steering vane bracket and vane. Screws (77) should be tightened to a torque of 208 Ft.-Lbs.

Note: The number of end play adjusting shims to be inserted between steering gear and pedestal were determined during disassembly and should provide the steering spindle with an end play of 0.004-0.021 when the pedestal retaining cap screws are securely tightened.

Beginning with the front center cap screw hole in pedestal, install a long cap screw in every other hole. Install the remaining shorter cap screws and tighten all of the pedestal cap screws to a torque of 150 ft.-lbs.

Assemble the remaining parts and adjust the steering gear backlash as in paragraph 25B. Remove any binding in the steering shaft as outlined in paragraph 25A.

STARTING ENGINE, COMPONENTS AND ACCESSORIES

John Deere 720 diesel tractors prior to serial number 7222600 are equipped with a four cylinder, V-type, valve-in-head, gasoline starting engine which has a bore of two inches and a stroke of one and one-half inches.

John Deere 720 diesel tractors after serial number 7222599 and all 730 diesel tractors are available with either the gasoline starting engine which is covered in the following paragraphs or with the 24-volt electric starting motor which is covered in paragraph 126A.

R&R ENGINE ASSEMBLY

30. To remove the starting engine assembly, first remove the tractor hood and on 730 diesel tractors the steering shaft. Drain the tractor cooling system and disconnect the battery cable. Disconnect the starting engine fuel line and remove the engine compartment door. Remove the diesel engine flywheel cover and the two cap screws retaining the flywheel cover back plate to the starting engine trans-

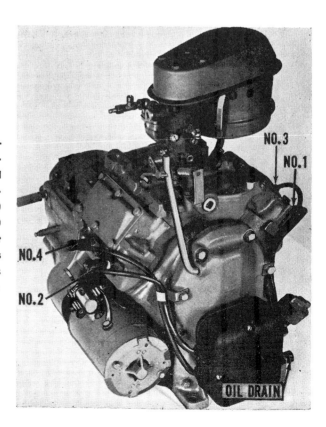

Fig. JD2045—Three-quarter view of the V-4 gasoline starting engine used on some 720 and 730 diesel tractors. On series 720 after 7222599 and all 730 diesel tractors a 24-volt electric starting motor is available in place of this gasoline starting engine.

Fig. JD2046—Starting engine cylinder head with valve cover removed. Tappet lever shaft is retained by Allen head set screw (4).

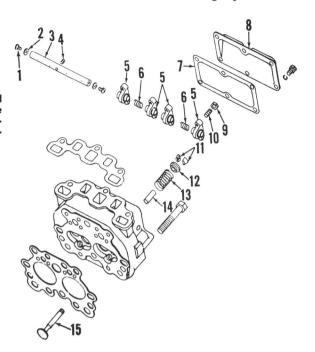

Fig. JD2048 — Exploded view of starting engine cylinder head, tappet levers and valves. Valve tappet gap is 0.008-0.010.

1. Machine screw
2. Packing washer
3. Tappet lever shaft
4. Set screw
5. Tappet levers
6. Spacer spring
7. Gasket
8. Valve cover
9. Jam nut
10. Adjusting screw
11. Keepers
12. Valve spring cap
13. Valve spring
14. Valve guide
15. Valve

mission. Remove the starting engine air cleaner, water intake pipe, exhaust pipe and water return pipe. Disconnect the choke rod, throttle rod and governor arm, and remove the throttle and choke levers from side of steering shaft support. Disconnect the starting engine fuel shut-off rod at coupling, disconnect the speed-hour meter cable at the diesel engine governor, disconnect the heat indicator sending unit from the diesel engine water outlet casting and remove the diesel engine oil pressure gage line. Disconnect wires from distributor and starting motor switch, and disconnect linkage from starting motor lever and the diesel engine speed control rod. Disconnect the starting control lever rod. Remove the four cap screws retaining the starting engine to the main case and lift starting engine from tractor, being careful not to lose the small shim washers which are located between the starting engine and tractor main case.

Fig. JD2047—Sequence for tightening the starting engine cylinder head cap screws. Specified torque value is 67 Ft.-Lbs.

When reinstalling the starting engine, reverse the removal procedure and be sure to install the same number of shim washers between the starting engine and the main case as were originally removed. Also, be sure the starting engine oil drain hole gasket is in position in the right hand diesel engine crankcase cover before lowering the starting engine into position.

With the starting engine lowered to its normal position, check the backlash between the starting engine pinion and the diesel engine flywheel ring gear at several (at least 3) locations around the flywheel. If the backlash at any point is less than 0.010, shims should be added between the main case and the four bolting flanges of the starting engine.

CYLINDER HEADS

31. It is possible to remove the starting engine cylinder heads without completely removing the starting engine from tractor. This procedure, however, is impractical and often very time consuming.

With the starting engine removed as outlined in paragraph 30, proceed to remove either cylinder head as follows: Remove manifold and valve covers. Remove the machine screw (1—Fig. JD2046) from each end of the tappet lever shaft and remove the Allen head set screw (4) which positions the tappet lever shaft in the cylinder head. With the valves closed, slide the tappet lever shaft out of head and remove the tappet levers,

spacer springs and push rods. Unbolt and remove cylinder head from engine.

NOTE: Do not turn the engine crankshaft with cylinder heads off unless a 7/16-inch cap screw and washer are installed as shown in Fig. JD2053 to hold the cylinder sleeves in position.

When reassembling, install head gasket on block with flange side up, then install head and tighten the cap screws finger tight. Install the manifold and tighten the retaining cap screws securely; then, tighten the cylinder head retaining cap screws in the sequence shown in Fig. JD2047 and to a torque of 67 Ft.-Lbs. Install the push rods, tappet levers and spacer springs. Slide the tappet lever shaft into position, making certain that the set screw hole aligns with the Allen head set screw. Install and tighten the set screw. Install a new packing washer (2—Fig. JD2048) at each end of tappet lever shaft and install and tighten the machine screw (1) in each end of shaft.

VALVES AND SEATS

32. To remove the valves, remove the cylinder heads as outlined in paragraph 31, compress the valve springs and remove the split cone keepers.

Intake and exhaust valves are not interchangeable and seat directly in the cylinder head with a face angle of 44½ degrees and a seat angle in the cylinder head of 45 degrees. Seats should be concentric with guides within 0.005. Valve seat width is 0.052-0.072 for the intake, 0.030-0.040

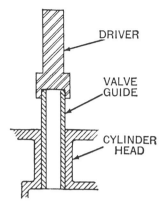

Fig. JD2049—Starting engine valve guides should be installed with a cup shaped driver as shown.

for the exhaust. Seats can be narrowed using 20 and 70 degree stones. Valve stem diameter is 0.2445-0.2455 for both the intake and the exhaust.

Valve tappet gap is 0.008-0.010 for both the intake and the exhaust. Adjust the clearance for each valve when the other valve of the same cylinder is in the wide open position.

VALVE GUIDES

33. Intake and exhaust valve guides are interchangeable and can be driven from the cylinder head if renewal is required. When installing new guides, coat the outer surface of the guides

with sealing compound and, using a cup shaped driver as shown in Fig. JD2049, drive the guides in until distance from end of guide to top of valve seat is $\frac{27}{32}$-inch.

Ream the guides after installation to provide a valve stem clearance of 0.0005-0.0025. Valve stem diameter is 0.2445-0.2455 for both the intake and the exhaust.

VALVE SPRINGS AND TAPPETS

34. Intake and exhaust valve springs are interchangeable and can be removed after removing the cylinder heads as outlined in paragraph 31. Springs should test 37-45 pounds when compressed to a height of $1\frac{13}{16}$-inch. Renew any spring which is rusted, discolored, or does not meet the foregoing pressure specifications.

35. The barrel type tappets (cam followers) ride directly in the cylinder block bores and are available in standard size only. Any tappet can be removed after removing the cylinder heads as outlined in paragraph 31.

VALVE TIMING

36. To check valve timing, first remove timing gear cover as outlined in paragraph 40. Valves are properly timed when "V" mark on the camshaft gear is in register with "V" mark on crankshaft gear as shown in Fig. JD2050.

VALVE TAPPET LEVERS

39. The tappet levers (rocker arms) and shaft can be removed from either cylinder head after performing the preliminary work of removing the grille, steering shaft and hood; however, most mechanics prefer to perform the additional work of removing the complete engine, which requires very little additional time due to the increased accessibility of the engine components when engine is removed.

Remove the valve covers and the machine screw (1—Fig. JD2046) from each end of the tappet lever shaft. Remove the Allen head set screw (4) which positions the tappet lever shaft in the cylinder head, and with the valves closed, slide the tappet lever shaft out of head and remove the tappet levers and spacer springs. Intake and exhaust valve tappet levers are interchangeable.

Check all parts and renew any which are excessively worn.

Shaft diameter .... 0.310 -0.311
Tappet lever bore..... 0.3125-0.3145

Thoroughly clean the oil holes in the tappet levers and shaft before reassembly. Install the tappet levers and spacer springs and slide the tappet lever shaft into position, making certain that the set screw hole aligns with the Allen head set screw. Install and tighten the set screw. Install a new packing washer (2—Fig. JD2048) at each end of the tappet lever shaft and install and tighten the machine screw in each end of the shaft.

TIMING GEARS AND COVER

40. **TIMING GEAR COVER.** To remove the timing gear cover, first drain the starting engine crankcase. Remove the ignition distributor cover, disconnect the spark plug wires and remove the four machine screws retaining the distributor to the timing gear cover, then remove the distrib-

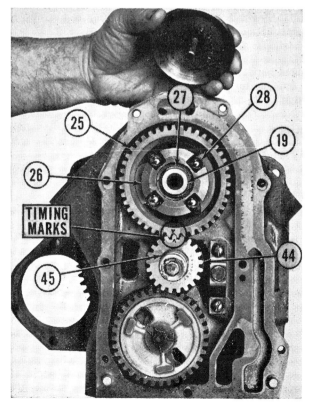

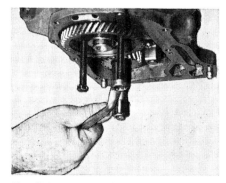

Fig. JD2050—The starting engine valves are properly timed when "V" marks on camshaft gear and crankshaft gear are in register.

Fig. JD2051—Using jack screws to pull the starting engine camshaft gear from the shaft. Refer to text.

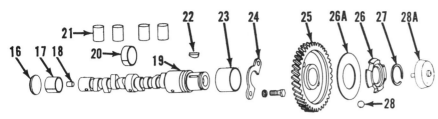

Fig. JD2052—Exploded view of the starting engine camshaft, bearings and cam followers. Shaft end play is controlled by thrust plate (24).

16. Expansion plug
17. Rear bushing
18. Dowel pin
19. Camshaft
20. Center bushing
21. Cam followers

22. Woodruff key
23. Front bushing
24. Thrust plate
25. Camshaft gear
26. Governor ball retainer

26A. Thrust washer (after 7214899)
27. Snap ring
28. Governor flyball
28A. Governor ball race

utor. Remove the diesel (main) engine decompression control rod and the clutch control rod. Disconnect the starting engine governor linkage from throttle linkage, then unbolt and remove the timing gear cover. The crankshaft front oil seal can be renewed at this time and should be installed with lip facing the flywheel end of engine.

When reassembling, check and adjust the ignition timing as outlined in paragraph 59.

41. **TIMING GEARS.** To remove the timing gears, first remove the timing gear cover as in preceding paragraph. To remove the camshaft gear, withdraw the governor ball race and balls. Remove snap ring (27—Fig. JD2050) and ball retainer (26). Turn the camshaft gear until the two puller holes are directly over the two camshaft thrust plate retaining cap screws; then, using two ⅜-inch puller cap screws as shown in Fig. JD2051, pull the camshaft gear from shaft. The crankshaft gear can be pulled from shaft after removing its retaining snap ring.

When reassembling, mesh the "V" mark on the camshaft gear with the "V" mark on the crankshaft gear as shown in Fig. JD2050.

Fig. JD2053 — Cap screw and washer installed to hold the starting engine cylinder sleeves in position when cylinder heads are off.

CAMSHAFT AND BUSHINGS

42. To remove the camshaft, first remove the starting engine as outlined in paragraph 30 and the cylinder heads as in paragraph 31. Withdraw the cam followers from the cylinder block bores. Remove the distributor cover, distributor and timing gear cover. Withdraw the governor ball race and balls, remove snap ring (27 —Fig. JD2050) and remove the ball retainer (26). Turn the camshaft gear until the two puller holes are directly over the two camshaft thrust plate retaining cap screws; then, using two ⅜-inch puller screws as shown in Fig. JD2051, pull the camshaft gear from shaft. Unbolt and remove the camshaft thrust plate (24—Fig. JD2052) and withdraw the camshaft from the cylinder block.

Check the camshaft and bushings against the values which follow:

Camshaft Journal Diameter
Front 1.248-1.249
Center 1.060-1.061
Rear 0.748-0.749

Bushing Inside Diameter
Front 1.251-1.253
Center 1.063-1.065
Rear 0.751-0.753

Journal Running
Clearance 0.002-0.005

Camshaft End Play.......0.003-0.009

If the camshaft end play is not as specified, renew the thrust plate (24).

To renew the camshaft bushings after camshaft is out, unbolt clutch housing from cylinder block and remove clutch, flywheel and expansion plug behind the camshaft rear bushing. Drive out the old bushings and install the new ones using a piloted drift. Ream the bushings after installation, if necessary, to provide a journal running clearance of 0.002-0.005.

When installing the camshaft gear, mesh the valve timing marks as shown in Fig. JD2050. Retime the ignition as outlined in paragraph 59.

ROD AND PISTON UNITS

43. Connecting rod and piston units are removed from above as follows: Remove the starting engine as outlined in paragraph 30 and the cylinder heads as in paragraph 31. Install a $\frac{7}{16}$-inch cap screw and washer as shown in Fig. JD2053 to hold the cylinder sleeves in position when removing the rod and piston units. Drain the crankcase and remove the crankcase cover. Remove the water pump.

Pistons and connecting rods are not marked with cylinder numbers and they should be identified before removal. Bend the lock plate tabs away from connecting rod bearing cap screws, remove the connecting rod bearing caps and push the connecting rod and piston units out of cylinder sleeves.

44. When reassembling, install the rod and piston units in their respective cylinders so that depressions in top of pistons are toward top of engine as shown in Fig. JD2053. Install the connecting rod bearing caps so that raised projections on rod and cap are in register as shown at X in Fig. JD2054 and tighten the cap screws to a torque of 7 Ft.-Lbs.

PISTONS, RINGS AND SLEEVES

45. The aluminum alloy pistons have a skirt diameter of 1.996-1.997 and are equipped with two compression rings and one oil control ring. Pistons are available in standard size only and have a skirt clearance of 0.002-0.004 in the 1.999-2.000 diameter cylinder sleeves. Pistons and/or sleeves should be renewed if the skirt clearance exceeds 0.007.

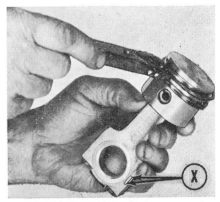

Fig. JD2054 — Installing rings on starting engine piston. When installing rod caps, make certain that raised projections (X) are in register.

Desired piston ring side clearance is 0.0015-0.0031 for the top compression ring, 0.0010-0.0026 for the second compression ring and the oil control ring. The oil ring can be installed with either side up. White dot on compression rings must face toward top of piston.

If the cylinder sleeves are worn, push them out of the cylinder block and remove any accumulation of cooling system sediment from engine water jacket and sleeve seating surfaces. Remove any foreign material or burrs which may prevent proper sleeve seating. Install a new "O" ring seal in cylinder block groove, lubricate the lower seating surface of the sleeve and push the sleeve into position.

PISTON PINS

46. The 0.4818-0.4822 diameter floating type piston pins are retained in the piston pin bosses by snap rings and are available in standard size only. Piston pins should have 0.0000-0.0008 clearance in piston and 0.0002-0.0016 clearance in the connecting rod. Piston can be installed either way on connecting rod, but rod and piston units must be installed as outlined in paragraph 44.

CONNECTING ROD

47. The aluminum alloy connecting rods ride directly on the 1.0615-1.0625 diameter crankshaft crankpins with a running clearance of 0.001-0.003. If clearance exceeds 0.005, renew the worn part; do not file rod or cap. Connecting rod bearing bore is 1.0635-1.0645.

Connecting rod can be installed either way in piston, but rod and piston units must be installed as outlined in paragraph 44.

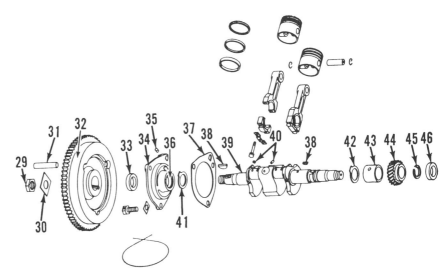

Fig. JD2056 — Exploded view of the starting engine pistons, rods, crankshaft and associated parts. Shaft end play of 0.005-0.008 is controlled by shims (37).

29. Nut
30. Lock plate
31. Clutch drive pin
32. Flywheel
33. Oil seal
34. Main bushing housing
35. Dowel pin
36. Rear bushing
37. Shims
38. Woodruff key
39. Crankshaft
40. Pins
41. Thrust washer
42. Thrust washer
43. Front bushing
44. Crankshaft gear
45. Snap ring
46. Oil seal

CRANKSHAFT AND MAIN BEARING BUSHINGS

48. The crankshaft is supported in two precision type main bearing bushings which can be renewed after removing the crankshaft. To remove the crankshaft, first remove the connecting rod and piston units as outlined in paragraph 43 and proceed as follows:

Remove the distributor cover, distributor and timing gear cover. Remove the gear retaining snap ring and pull the crankshaft gear from shaft. Unbolt clutch housing from cylinder block and remove the clutch and flywheel. Remove the crankshaft rear bearing housing, being careful not to lose the shims located between the bearing housing and crankshaft and withdraw the crankshaft from the cylinder block as shown in Fig. JD2055.

Check the crankshaft and main bearings against the values which follow:

Main Journal Diameter
Series 720 prior 7214900.1.124-1.125
Series 720 after 7214899
  and Series 730........1.499-1.500

Bushing Inside Diameter
Series 720 prior 7214900.1.127-1.128
Series 720 after 7214899
  and Series 730........1.501-1.502

Shaft Clearance in Bushings
Series 720 prior 7214900.0.002-0.004
Series 720 after 7214899
  and Series 730........0.001-0.003

49. Main bearing bushings are available in standard size only. When installing new bushings, use a piloted driver and make sure the bushing oil holes are toward bottom of engine.

Install the crankshaft by reversing the removal procedure and be sure that tab on thrust washer (42—Fig. JD2056) fits into slot in cylinder block. The shaft rear oil seal (33) located in the bearing housing should be installed with lip facing inward. Install rear bearing housing and vary the number of shims (37) to obtain a shaft end play of 0.005-0.008. Securely tighten and safety wire the housing retaining cap screws.

Fig. JD2055 — Removing crankshaft from starting engine cylinder block. Crankshaft end play is controlled by shims (37).

Fig. JD2057 — The oil pump driving gear can be removed after removing its retaining hairpin snap ring (59).

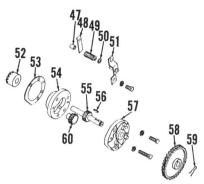

Fig. JD2058—Exploded view of the starting engine oil pump and relief valve. Shims (53) control backlash between the oil pump and water pump bevel gears.

47. Relief valve seat	53. Shims
48. Leaf spring	54. Pump body
49. Relief valve spring	55. Driver gear
50. Shim washers	56. Woodruff key
51. Relief valve bracket	57. Pump cover
52. Bevel gear	58. Driving gear
	59. Snap ring
	60. Idler gear

Install the remaining parts by reversing the removal procedure. The crankshaft front oil seal (46) located in timing gear housing should be installed with lip facing the flywheel end of engine.

CRANKSHAFT REAR OIL SEAL

50. The crankshaft rear oil seal (33—Fig. JD2056) is located in the rear main bearing housing and can be renewed after removing the flywheel as outlined in paragraph 51. Seal should be installed with lip facing timing gear end of engine.

FLYWHEEL

51. To remove the flywheel, first remove the starting engine as outlined in paragraph 30, then unbolt and separate the clutch housing from the engine. Remove the clutch cover assem-

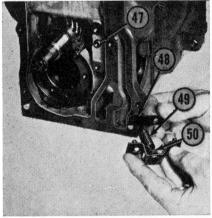

Fig. JD2059 — Normal starting engine oil pressure of 25-30 psi at high idle rpm is regulated by the leaf-type spring (48). Pressure is increased by the addition of washers (50)

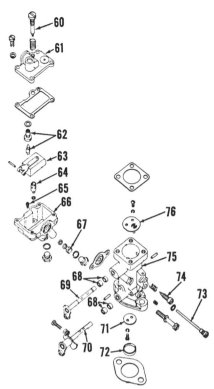

Fig. JD2060—Exploded view of the Zenith carburetor used on the gasoline starting engine. The unit is provided with adjustments for both idling and power range.

60. Load screw	69. Choke shaft
61. Fuel bowl cover	70. Throttle shaft
62. Float valve & seat	71. Throttle disc
63. Float	72. Deflector ring
64. Main jet	73. Idle jet
65. Compensator jet	74. Idle adjusting
66. Fuel bowl	needle
67. Metering well	75. Carburetor body
68. Felt washer & retainer	76. Choke disc

bly and lined plate. Remove the flywheel retaining nut. To loosen flywheel from crankshaft, pry on flywheel from behind while bumping crankshaft with a soft hammer. To install starter ring gear, heat same to 550 deg. F. and install gear on flywheel so that beveled end of teeth will face toward starting motor pinion.

When installing the flywheel, tighten the retaining nut to a torque of 150 Ft.-Lbs. and bend tab of lock plate against flat of nut.

OIL PUMP AND RELIEF VALVE

52. **OIL PUMP.** To remove the oil pump, first remove the timing gear cover as outlined in paragraph 40. Remove the hairpin snap ring and withdraw the pump drive gear as shown in Fig. JD2057. Unbolt and withdraw pump from cylinder block, being careful not to lose or damage the shims located between oil pump and block.

Unbolt and remove the pump cover from the pump body and disassemble the remaining parts. The pump driv-

ing bevel gear (52—Fig. JD2058) can be pressed from shaft if renewal is required. New bevel gear should be installed flush with end of shaft. When reassembling, coat mating surfaces of pump body and cover with shellac and tighten the cover retaining cap screws finger tight. Shift the cover until pumping gears are free, then tighten the cap screws.

52A. Install pump by reversing the removal procedure and use the same number of shim gaskets (53) between pump body and cylinder block as were removed. Installing the original number of gaskets should provide a backlash of 0.004-0.006 between the water pump and oil pump bevel gears. If there is any reason to suspect that the bevel gear backlash is not as specified, remove the pump and apply a 0.003 thick piece of paper directly on the bevel gear teeth. Then install the pump and turn the pump drive gear, thereby passing the 0.003 piece of paper between the bevel gears. If the piece of paper shows a heavy impression but is not punctured, the backlash is within the specified limits.

53. **RELIEF VALVE.** The recommended starting engine oil pressure of 25-30 psi at high idle rpm is regulated by a spring loaded leaf type relief valve located within the timing gear cover. Refer to Fig. JD2059. Oil pressure can be checked on early 720 models by removing the pipe plug, shown in Fig. JD2047, from side of cylinder block and installing a master gage. On later models, it is necessary to connect a gage in series with the oil line connected to the automatic fuel shut-off unit.

If the pressure is less than 20 psi, remove the timing gear cover as outlined in paragraph 40, and add adjusting washers (50—Fig. JD2059). The addition of one washer will increase the oil pressure approximately 5 psi.

The relief valve spring (49—Fig. JD2058) should have a free length of approximately $1\frac{1}{16}$-inches and should require 22-25 ounces to compress it to a height of ¾-inch. If the relief valve seat (47) is renewed, press it in until outer surface of seat is level with the relief valve bracket pads.

CARBURETOR

54. The starting engine on 720 tractors prior to tractor serial number 7214900 is originally equipped with a Zenith 0-11681, model TU3½ x 1C down draft type carburetor shown exploded in Fig. JD2060. On series 720

tractors after tractor serial number 7214899 and series 730 tractors, the starting engine is equipped with a Zenith D-12172 model TU3 x 1C carburetor. Distance from furthest face of float to machined surface of float bowl cover is $1\frac{3}{16}$-inches.

To adjust the carburetor, first start engine and let it run until warm. With speed control in the slow idle position, back the idle adjusting screw (74—Fig. JD2061) out until engine slows down or falters, then turn the screw in until engine runs smoothly.

With the starting engine motoring the diesel engine on full compression, turn the load adjusting screw (60) in until engine slows down or falters and back the screw out until engine runs smoothly; then, back the screw out one or two more notches. Normal load screw adjustment is $\frac{5}{8}$-turn open.

GOVERNOR

55. The flyball type governor weight unit is mounted on the front of the camshaft as shown in Fig. JD2050 and can be overhauled after removing the timing gear cover as outlined in paragraph 40. When reassembling, make sure that slot in ball retainer (26) engages the Woodruff key in camshaft.

56. **ADJUSTMENT.** Before attempting to adjust the governed speed, free up all linkage and renew any parts causing lost motion. Disconnect the throttle rod (90—Fig. JD2061) from governor arm (80) and with throttle butterfly and governor arm in the wide open position, adjust the length of the throttle rod so it is short one full turn of the adjusting nut (N).

Reconnect the throttle rod, start engine and turn the throttle stop screw either way as required to obtain a slow idle speed of 4000 rpm. The high idle, no load speed of 5000 rpm is ad-

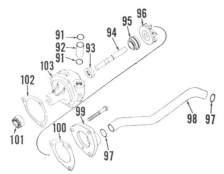

Fig. JD2062—Exploded view of the starting engine water pump. Shims (102) control backlash between water pump and oil pump bevel gears.

91. "O" ring	97. "O" ring
92. Short water pipe	99. Pump cover
93. Oil seal	100. Gasket
94. Pump shaft	101. Bevel gear
95. Pump seal	102. Shims
96. Impeller	103. Pump housing

justed by loosening the two cap screws retaining the governor spring bracket (79) to manifold and moving it away from carburetor to increase speed or toward carburetor to decrease speed. With diesel engine at operating temperature, the starting engine should motor the diesel engine under full compression at 200 rpm.

Note: The starting engine speed can be checked at right end of crankshaft after removing the distributor cover.

COOLING SYSTEM

The starting engine cooling system is connected to the diesel engine cooling system. Water is circulated through the starting engine by an impeller type water pump mounted on the starting engine cylinder block.

57. **R&R AND OVERHAUL WATER PUMP.** To remove the water pump, drain tractor cooling system and starting engine crankcase. Remove the starting engine water inlet pipe by unbolting the bracket from right side

of diesel engine cylinder block and pulling inlet pipe out of water pump. Remove the pump retaining cap screws. Remove the pump cover and the by-pass pipe, turn pump counterclockwise to clear the short water pipe (92—Fig. JD2062) and withdraw the pump from cylinder block, being careful not to lose the gasket shims (102) located between pump and block.

To disassemble the pump, press the pump shaft out of the bevel drive gear (101).

Install seal (93) with lip facing impeller. Press bevel drive gear on shaft until hub of gear is flush with end of shaft. Impeller should be just flush with the machined surface of the pump body as shown in Fig. JD2063.

Install pump by reversing the removal procedure, use the original number of shims between water pump and cylinder block and renew all "O" rings. Backlash between the water pump and oil pump bevel gears should be a recommended 0.004-0.006. If there is any reason to suspect the backlash is not as specified, check the adjustment with paper as in paragraph 52A.

IGNITION SYSTEM

58. Wico distributor model B4027 is used. The unit is mounted on the timing gear cover and actuated by a cam which is retained to the end of the crankshaft by a cap screw. There is a coil, condenser and set of contacts for each two cylinders. The unit does not incorporate an automatic spark advance mechanism.

Refer to Fig. JD2064. The gap for each set of breaker contacts is 0.020. Breaker contact spring tension is 18-26 oz. Firing order is 1-2-3-4. When

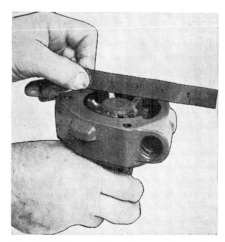

Fig. JD2061 — Starting engine carburetor installation, showing mixture and linkage points of adjustments.

60. Load adjusting screw
74. Idle adjusting needle
78. Governor spring
79. Spring bracket
80. Governor arm
90. Throttle rod
N. Jam nut

Fig. JD2063 — When the starting engine water pump is properly assembled, the impeller vanes should be just flush with the machined surface of the pump housing.

Fig. JD2064 — The starting engine distributor is equipped with two sets of contacts, two condensers and two coils. Contact gap is 0.020.

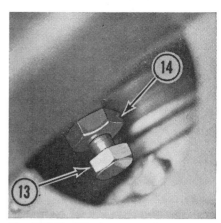

Fig. JD2068—Starting engine clutch can be adjusted after removing the clutch housing inspection hole cover as shown.

CLUTCH

62. ADJUSTMENTS. If the starting engine clutch slips when motoring the diesel engine under full compression, adjust the clutch and linkage as follows:

With the starting engine control lever pulled all the way to the rear (engaged position), the transmission shift lever stop pin should project 7/32-inch beyond the shift lever stop plate as shown in Fig. JD2066. Note: Measurement is taken between the cotter pin and the stop plate. If the measurement is less than 7/32-inch, remove the inspection hole cover from clutch housing as shown in Fig. JD2068, loosen the three jam nuts (14) and turn each of the clutch adjusting screws (13) in an equal amount until the shift lever stop pin protrudes the specified 7/32-inch. Move the control lever forward to the neutral position and check to make certain that the

viewed from flywheel end of engine, cylinders Nos. 1 & 3 are on left hand side (front of tractor) and cylinders No. 2 & 4 are on right hand side (rear of tractor). See Fig. JD2045.

59. TIMING. To time the distributor, remove the inspection cover from clutch housing and using a screw driver or similar tool, turn the flywheel in a counter-clockwise direction (viewed from flywheel end) until number one cylinder is coming up on compression stroke and continue turning flywheel until "V" marks on flywheel and clutch housing are aligned as shown in Fig. JD2065. Loosen the four distributor mounting screws and rotate the distributor in a clockwise direction (viewed from distributor

end of engine) as far as possible. Turn on the ignition switch, and while holding the number one spark plug wire about ¼-inch from engine block, slowly rotate distributor in a counter-clockwise direction until a spark occurs at the end of the spark plug wire; then, tighten the mounting screws securely.

STARTING MOTOR

60. The starting engine is equipped with a 6-volt, automotive type Delco-Remy 1107155 starting motor. Refer to paragraph 126 for test specifications.

Fig. JD2065—Starting engine ignition timing marks on flywheel and clutch housing.

Fig. JD2066—The starting engine transmission shift lever stop pin protrusion should be 7/32-inch (between cotter pin and stop plate). The measurement should be taken when the starting engine control lever is pulled all the way to the rear (engaged position). See text.

Fig. JD2068A—To align the starting engine clutch operating lever with the decompression lever, loosen the jam nut (34) and turn the connector coupling either way as required to align the levers.

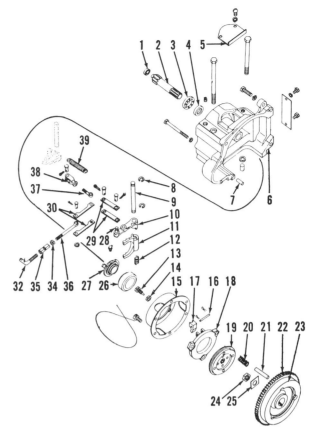

Fig. JD2070—Exploded view of starting engine clutch, clutch housing and associated linkage.

1. Clutch shaft outer bearing	14. Jam nut	28. Pin
2. Clutch shaft	15. Clutch cover	29. Fork arm links
3. Clutch shaft inner bearing	16. Toggle pin	30. Operating links
4. Oil seal	17. Toggle	32. Starting control lever link
5. Inspection cover	18. Pressure plate	34. Jam nut
6. Clutch housing	19. Lined disc	35. Coupling
7. Dowel pin	20. Spring	36. Clutch and transmission operating link
8. Retainer ring	21. Clutch drive pin	37. Transmission operating link
9. Fork shaft	22. Ring gear	38. Operating link yoke
10. Fork arm	23. Flywheel	39. Spring
11. Fork	24. Nut	
12. Spring	25. Lock plate	
13. Clutch adjusting screw	26. Release bearing	
	27. Bearing carrier	

clutch completely releases. If the clutch does not release when the control lever is moved to neutral position, back-out the three adjusting screws an equal amount until a clean release is obtained; then, obtain the specified stop pin protrusion as follows: Disconnect the transmission operating link yoke (38—Fig. JD2070) and unhook spring (39) from the shifter lever. Remove roll pin shifter spring (46Y—Fig. JD2074) and shifter arm (46Z). Turn the operating link yoke (38—Fig. JD2070) out to provide 7/32-inch stop pin protrusion.

After the afore mentioned adjustments are completed vary the length of the operating link by turning the coupling (35—Fig. JD2068A) either way as required to align the clutch operating lever with the decompression lever.

63. R & R AND OVERHAUL CLUTCH. To remove the starting engine clutch, it will be necessary to first remove the starting engine as outlined in paragraph 30 or, raise the

Fig. JD2069—The starting engine clutch release bearing (26) and carrier can be withdrawn after disconnecting spring (39).

starting engine enough to permit removal of clutch housing and transmission assembly. Remove the cap screws retaining the clutch housing to the engine and separate the units.

The release bearing can be removed after disconnecting spring (39—Fig. JD2069).

Fig. JD2071 — After removing bearing (61) and snap ring (60), the overrunning clutch parts can be disassembled from the transmission drive shaft.

Unbolt and remove the clutch cover and lined disc from flywheel. The need and procedure for further disassembly is evident after an examination of the unit and reference to Fig. JD2070.

When reassembling, install release springs (20) on the drive pins (21) and install the lined disc (19) with

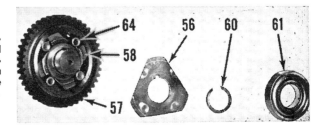

Fig. JD2072—Use a suitable puller to remove the bearing from the transmission shaft.

bearing. Remove the five cap screws retaining the transmission housing to clutch housing and separate the units. Withdraw clutch shaft and drive shaft from transmission housing and using a suitable puller as shown in JD2072, remove bearing from drive shaft. Remove snap ring (60—Fig. JD2071) and slide the over-running clutch plate from drive shaft. Remove the transmission drive gear (57) and disassemble the over-running clutch parts. The need for further disassembly is evident.

Thoroughly clean and check all parts for damage or wear. Renew the brake plate lining (52—Fig. JD2074) if it is worn.

Use Figs. JD2070 and 2074 as a guide during reassembly and adjust the linkage as outlined in paragraph 62.

Fig. JD2073—Starting engine clutch adjustment. Screw (13) should be adjusted to obtain ⅝-inch clearance between the bearing contacting surface of each toggle and the rear face of the cover.

rate the units. Disconnect the transmission operating link yoke (38—Fig. JD2070) and spring (39) from the shifter lever. Remove roll pin and withdraw the shifter lever shaft (48—Fig. JD2074) and shifter lever (46X), shifter spring (46Y) and the shifter arm (46Z). Remove the clutch release

hub facing away from flywheel. Install pressure plate (18) and bolt cover assembly to flywheel. Be sure to safety wire the cover retaining cap screws. With the actuating toggles depressed until the pressure plate contacts the clutch lined disc, there should be a clearance of ⅝-inch between the release bearing contacting surface of each toggle and the rear surface of the clutch cover as shown in Fig. JD2073. If adjustment is not as specified, loosen the jam nut and turn each of the toggle adjusting screws (13) **in** to decrease or **out** to increase clearance.

Install the remaining parts by reversing the removal procedure.

TRANSMISSION

64. R&R AND OVERHAUL. To remove the starting engine transmission, it will be necessary to first remove the starting engine as outlined in paragraph 30 or, raise the starting engine enough to permit removal of clutch housing and transmission assembly. Remove the cap screws retaining the clutch housing to the engine and sepa-

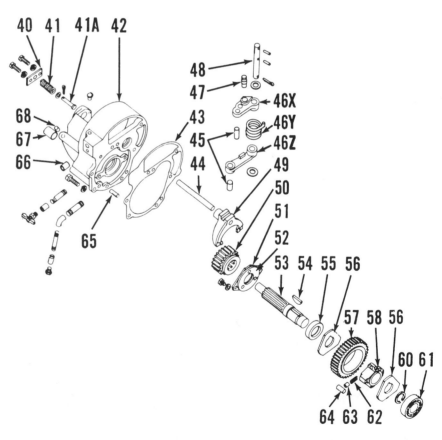

Fig. JD2074—Exploded view of starting engine transmission. Refer also to Fig. JD2070.

40. Shift lever stop
41. Spring
41A. Shifter lever stop pin
42. Transmission housing
43. Gasket
44. Shifter shaft
45. Dowel pins
46X. Shifter lever
46Y. Shifter arm spring
46Z. Shifter arm
47. Pin

48. Shifter lever shaft
49. Transmission shifter
50. Sliding pinion
51. Brake plate
52. Brake lining
53. Transmission drive shaft
54. Woodruff key
55. Oil seal
56. Over-running clutch plate
57. Transmission drive gear

58. Over-running clutch hub
60. Snap ring
61. Bearing
62. Spring
63. Plunger
64. Clutch roller
65. Dowel pin
66. Bushing
67. Bushing
68. Dowel pin

DIESEL ENGINE AND COMPONENTS

CYLINDER HEAD

The cylinder head on 720 tractors prior to serial number 7214900 is fitted with hard drawn copper tubes to receive the diesel injection nozzles. Refer to paragraph 113 for removal and reinstallation of nozzle tubes. Series 720 tractors after serial number 7214899 and 730 diesel tractors are not equipped with renewable type nozzle sleeves. The injection nozzle tips extend through small holes drilled into the cylinder head between the inlet and exhaust valves. A heat exchanger, which acts as an air inlet manifold for the diesel engine, is mounted on the cylinder head. Exhaust gasses from the starting engine flow into the heat exchanger and warm the initial incoming air for the diesel engine. Water outlet pipes and an exhaust manifold are also mounted on the head.

70. REMOVE AND REINSTALL. To remove the diesel engine cylinder head, first drain the cooling system and proceed as follows: Remove the heat exchanger outlet pipe with muffler and the tappet lever cover. Disconnect the tappet lever oil line and remove the tappet levers assembly and push rods. Remove the injection pump compartment cover and remove both of the high pressure fuel lines connecting the injection pumps to the injection nozzles.

CAUTION: When disconnecting fuel lines from pumps, do not permit delivery valve holders (Fig. JD2081) to turn. When fuel lines are removed, immediately cap the connectors on pumps and nozzles to eliminate the entrance of dirt or other foreign material.

Note: Some mechanics prefer to remove the injection nozzles to prevent any possible damage to same when removing the cylinder head.

Remove exhaust pipe, exhaust manifold, air cleaner cup, fuel filters assembly and the lower water pipe. Unbolt the inlet manifold and the water outlet casting from the cylinder head. Remove the cylinder head retaining stud nuts, including the four which are located in the injection pump compartment, slide the head forward on studs and lift cylinder head from tractor.

Before installing cylinder head, clean carbon and all gasket surfaces, paying particular attention to the small holes through which the injection nozzle tips project. Check to make certain that the crankcase ventilating hole and the oil return holes are open.

Coat the steel side of the cylinder head gasket with No. 3 Permatex or equivalent and install with this side toward block. Slide the cylinder head into position. Note: Lead washers are used under all cylinder head stud nuts except the four nuts which are located in the injection pump compartment. Spring steel lock washers are used under the four nuts in the pump compartment. Tighten the cylinder head stud nuts in the sequence shown in Fig. JD2082 and to a torque of 275 Ft.-Lbs. A torque wrench cannot be used on the four nuts in the pump compartment, but they must be tightened securely. Be sure to remove the surplus lead which oozes from under the stud nuts when they are tightened.

If the injection nozzles were removed from cylinder head, the copper washers which are located between the nozzles and the head must be renewed; or, the old washers may be reannealed by heating to a cherry red and quenching in water before installation.

Adjust the tappet gap as outlined in paragraph 72A. Run the engine until it reaches operating temperature; then, re-torque the cylinder head stud nuts, readjust the tappet gap and adjust the decompression control as outlined in paragraph 71.

71. ADJUST DECOMPRESSION CONTROL. With flywheel cover and tappet lever cover removed, turn flywheel forward until the inlet valve for number one (left hand) cylinder opens and closes and continue turning flywheel until the "No. 1 TDC" mark on flywheel rim aligns with pointer on timing gear housing. Mount a dial indicator so that indicator contact button is depressed to near the extreme inward position and contacting the exhaust valve spring cap. Zero the indicator dial. Note: Make certain that indicator body does not touch tappet lever. Pull the starting control lever back to locked position. At this time, the indicator should show that the decompression linkage has opened the

Fig. JD2081 — When disconnecting the high pressure fuel lines from the fuel injection pumps, make certain that the delivery valve holders do not turn.

Fig. JD2082 — Diesel engine cylinder head nut tightening sequence. Nuts should be tightened to a torque of 275 Ft.-Lbs.

exhaust valve 0.055-0.080. Turn the flywheel forward 180 degrees (½ turn) and make the same check for No. 2 cylinder. If indicator readings are not as specified, adjust length of the decompression control rod at coupling (Fig. JD2083) until both exhaust valve openings are within specifications.

VALVES AND SEATS

72. Inlet and exhaust valves are not interchangeable, and seat directly in cylinder head with a face angle of 44½ degrees and a seat angle of 45 degrees. Desired valve seat width is 0.161-0.175 for the inlet, 0.126-0.140 for the exhaust. Seats can be narrowed, using 20 and 70 degree stones. Valve stem diameter is 0.496-0.497 for both the inlet and the exhaust.

Adjust both the inlet and exhaust tappet gap to 0.020 hot as outlined in the following paragraph.

72A. Remove flywheel cover and turn the flywheel forward until the inlet valve for number one (left hand) cylinder opens and closes and continue turning flywheel until the "No. 1 TDC" mark on flywheel rim aligns with pointer on timing gear housing as shown in Fig. JD2084. Make sure the

Fig. JD2085 — Exploded view of the diesel engine cylinder head and associated parts. Series 720 diesel after serial number 7214899 and 730 diesel tractors aren't equipped with renewable nozzle sleeves (18).

1. Tappet lever shaft
2. Washer
3. L. H. exhaust tappet lever
4. Bracket
5. Spring
6. Inlet tappet lever
7. Jam nut
8. R. H. exhaust tappet lever
9. Tappet adjusting screw
10. Valve keepers
11. Spring cap
12. Valve spring
13. Valve guide
14. Lead washers
15. Valve
16. Head gasket
17. "O" ring
18. Nozzle sleeve
19. Washer
20. Machine screw
21. Cover support
22. Coupling nut
23. Dowel pin
24. Cotter pin

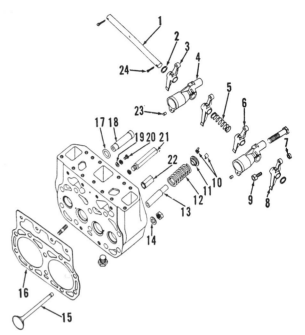

decompression lever is not holding the exhaust valves open and adjust the inlet and exhaust tappets for number one cylinder to 0.020 (hot) which equals ⅓ turn of the adjusting screws.

Fig. JD2083—The decompression control rod can be adjusted at coupling so that both exhaust valves are open 0.055-0.080 when the decompression control lever is pulled back to the locked position. Although a 70 diesel is shown, 720 and 730 diesel tractors are similar.

Fig. JD2084 — Inlet and exhaust tappet gap should be adjusted to 0.020 hot when pistons are at TDC on compression stroke. Although a 70 diesel is shown, 720 and 730 diesel tractors are similar.

Turn the flywheel forward 180 degrees (½ turn) and adjust the tappets for the No. 2 cylinder in the same manner.

VALVE GUIDES AND SPRINGS

73. Inlet and exhaust valve guides are interchangeable and can be pressed from cylinder head if renewal is required. Press new guides into cylinder head so that smaller O.D. of guides will be toward valve springs. Distance from port end of guides to gasket surface of cylinder head is 2⅛ inches.

Ream new guides after installation to an inside diameter of 0.5011-0.5025. The 0.496-0.497 diameter valve stems have a clearance of 0.0041-0.0065 in the guides.

74. Inlet and exhaust valve springs are interchangeable and have a free length of approximately 4 inches. Springs should require 41-51 lbs. to compress them to a height of 3¼ inches and 68-84 lbs. to compress them to a height of 2¾ inches. Renew any spring which is rusted, discolored or does not meet the foregoing pressure test specifications.

VALVE TAPPET LEVERS
(Rocker Arms)

75. The tappet levers assembly can be removed after removing the heat exchanger outlet pipe and tappet lever cover and disconnecting the tappet lever oil line.

The 0.858-0.859 diameter tappet lever shaft should have a clearance of 0.002-0.005 in the 0.861-0.863 diameter tappet lever bores.

Excessive wear of any of the components of the tappet lever assembly is corrected by renewing the parts. Intake and exhaust tappet levers are not interchangeable. When reinstalling the assembly, make certain that supports (4—Fig. JD2085) engage the positioning dowels (23) in cylinder head. Adjust tappet gap as in paragraph 72A and the decompression control as in paragraph 71.

VALVE TIMING

76. Inlet valve opens 12 degrees BTDC and closes 28 degrees ABDC, giving an inlet period of 220 degrees. Exhaust valve opens 48 degrees BBDC and closes 12 degrees ATDC, giving an exhaust period of 240 degrees. Valves are properly timed when the "V" marks on crankshaft gear, camshaft gear and idler gear are in register as shown in Fig. JD2090.

TIMING GEAR HOUSING

Removal of the timing gear top cover is not necessary for removal of timing gear housing; however, in the cases where it will be necessary, special attention will be called to the removal of the cover.

Series 720 Diesel Prior to Serial Number 7214900

77. **REMOVE AND REINSTALL.** To remove the timing gear housing, first remove the flywheel cover and before removing the flywheel, use a dial indicator as shown in Fig. JD2086 and check the crankshaft end play by engaging and disengaging the clutch with force. If the end play is not between the limits of 0.005-0.010, remember the measurement so the end play can be adjusted to the specified value during reassembly.

Fig. JD2086 — Crankshaft end play should be 0.005-0.010. Although a 70 diesel is shown early 720 tractors are similar.

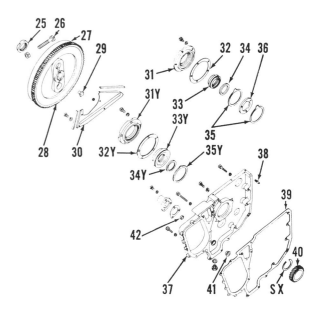

Fig. JD2087 — Diesel engine timing gear housing, flywheel and associated parts. Crankshaft end play should be 0.005-0.010. Parts (31, 32, 33, 34, 35 and 36) are used only on 720 tractors prior to tractor serial number 7214900.

25. Flywheel lock nut
26. Flywheel clamp bolt
27. Flywheel
28. Ring gear
29. Spacer drive pin
30. Back plate
31. Thrust cover
31Y. Oil slinger housing
32. Shims
32Y. Gasket
33. Spacer
33Y. Spacer
34. Seal
34Y. Seal
35. Thrust washers
35Y. Thrust washer
36. Thrust plate
37. Timing gear housing
38. Dowel pin
39. Gasket
40. Crankshaft gear
41. Bushing
42. Plug
SX. Snap ring

Unstake and remove the flywheel lock nut (25—Fig. JD2087) and loosen the two flywheel clamp bolts (26). Using a suitable puller, move flywheel out toward end of crankshaft and using a hoist, remove flywheel from crankshaft. Remove the flywheel cover backplate.

Unbolt and remove the crankshaft thrust cover (31), being careful not to damage or lose the paper or steel shims located between the thrust cover and the timing gear housing. Pull flywheel spacer (33) from crankshaft and withdraw the thrust plate and the two fiber thrust washers. Note: The neoprene oil seal (34) located in the flywheel spacer can be renewed at this time.

Install nuts on the two housing locating dowels and tighten the nuts until dowels are free. Remove the housing retaining cap screws and remove timing gear housing from tractor.

When reassembling, make certain that the "Powr-Trol" pump drive gear thrust washer is in position on pump drive shaft, install timing gear cover, bump the locating dowels into main case and tighten the cover retaining cap screws securely. Using heavy grease as an adhesive, install one of the fiber thrust washers (35) in recess of timing gear housing; then, install the metal thrust plate (36), making certain that keyway in plate engages the protruding end of the crankshaft gear key. Remove the belt pulley dust cover and while having another man buck-up the right end of the crank-

shaft with a heavy bar, align the driving slot in flywheel spacer with "V" mark on left end of crankshaft as shown in Fig. JD2088, install the flywheel spacer and bump it tightly against the steel thrust plate (36—Fig. JD2087).

NOTE: If right end of crankshaft is not bucked-up, it is possible that crankshaft will move toward right enough to allow the steel thrust plate to drop out of position.

Again, using heavy grease as an adhesive, install the other fiber thrust washer (35) in recess of thrust cover and install the thrust cover with word "TOP" toward top of tractor. Use the original shim pack between cover and timing gear housing, adding or deducting shims as necesary to allow 0.005-0.010 end play as determined be-

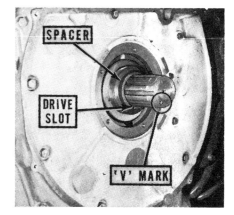

Fig. JD2088—When installing the flywheel spacer on early 720 diesel tractors, make certain that the driving slot in spacer is aligned with the "V" mark on end of crankshaft.

fore flywheel was removed. Install the thrust cover cap screws finger tight and using a 0.003 thick strips of shim stock as shown in Fig. JD2089, center the thrust cover about the flywheel spacer and tighten the thrust cover retaining cap screws to a torque of 35 Ft.-Lbs.

Now, while having another man buck-up the right end of the crankshaft as before, align the "V" mark on flywheel with "V" mark on left end of the crankshaft, install flywheel and drive flywheel solidly against flywheel spacer. Note: If the "V" marks on flywheel spacer, crankshaft and flywheel are properly aligned, drive pin (29—Fig. JD2087) will engage drive slot in spacer (33). Install the flywheel nut, tighten it securely and lock it in position by peening a portion of the nut into one of the crankshaft splines. Tighten the flywheel clamp bolts to a torque of 275 Ft.-Lbs.

Series 720 Diesel After Serial Number 7214899-730 Diesel

77A. **REMOVE AND INSTALL.** To remove the timing gear housing, first remove the flywheel cover. Unstake and remove the flywheel lock nut (25 —Fig. JD2087) and loosen the two flywheel clamp bolts (26). Using a suitable puller, move flywheel out toward end of crankshaft and using a hoist, remove flywheel from crankshaft. Remove the flywheel cover backplate.

Unbolt and remove the crankshaft oil slinger housing (31Y), pull the flywheel spacer (33Y) from crankshaft and withdraw the thrust washer (35Y).

Note: The neoprene oil seal (34Y) located in the flywheel spacer can be renewed at this time.

Install nuts on the two housing locating dowels and tighten the nuts until dowels are free. Remove the housing retaining cap screws and remove timing gear housing from tractor.

When reassembling, make certain that the "Powr-Trol" pump drive gear thrust washer is in position on pump drive shaft, install timing gear cover, bump the locating dowels into main case and tighten the cover retaining cap screws securely. Using heavy grease as an adhesive, install the fiber thrust washer (35Y) on the crankshaft; then, locate the flywheel spacer on the crankshaft so that the flywheel drive pin will engage the slot in the spacer when the flywheel is installed. Install and bolt the oil slinger housing to the timing gear housing without any gasket between the slinger housing and the gear housing; then, remove the oil slinger housing and reinstall same with the gasket in position and 0.003 shim stock inserted as shown in Fig. JD2089 which will serve to center the slinger housing. Reinstall flywheel making certain that the drive pin engages the slot in the flywheel spacer, then adjust the crankshaft end play as outlined in paragraph 92A.

TIMING GEARS

78. As shown in Fig. JD2090, the timing gear train consists of the camshaft gear, crankshaft gear, one single idler gear, one double idler gear, the oil pump drive gear and the hydraulic "Powr-Trol" pump drive gear. The larger portion of the double idler gear meshes with a spur gear on the governor shaft.

79. **REMOVE AND REINSTALL.** To remove the timing gears, first remove the timing gear housing as outlined in paragraph 77 or 77A; then, proceed as outlined in the appropriate following paragraph.

80. **CRANKSHAFT GEAR.** On 720 diesels after serial number 7214899, series 730 diesel and early 720 diesel which have a later crankshaft installed, the crankshaft gear is positioned on the shaft by a snap ring. With the timing gear housing and the snap ring removed, the crankshaft gear can be pulled from crankshaft by using a suitable puller. To facilitate installation of the gear, heat same in oil and install gear so the "V" mark is in register with "V" mark on idler gear as shown in Fig. JD2090. While another man is bucking up right end

Fig. JD2089 — Install thrust cover or oil slinger housing with word "TOP" toward top of tractor. Center the cover (or slinger housing) about the flywheel spacer using 0.003 thick strips of shim stock as shown.

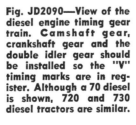

Fig. JD2090—View of the diesel engine timing gear train. Camshaft gear, crankshaft gear and the double idler gear should be installed so the "V" timing marks are in register. Although a 70 diesel is shown, 720 and 730 diesel tractors are similar.

of crankshaft, seat the gear firmly against shoulder on crankshaft by using a brass drift and hammer.

81. CAMSHAFT GEAR. With the timing gear housing removed, proceed to remove the camshaft gear as follows: Remove the starting engine as outlined in paragraph 30. Remove the cap screws retaining the fan shaft front bearing housing (or power steering pump housing) to its support and unbolt the fan shaft rear bearing housing from the governor housing. Remove the governor lever housing from left side of cylinder block. Unbolt governor housing from tractor main case, raise left side of governor housing enough to clear camshaft gear and block the governor housing in this position. Remove the gear retaining nut and lock washer and pull gear from camshaft.

When reinstalling the gear, mesh "V" mark on camshaft gear with "V" mark on idler gear as shown in Fig. JD2090.

82. DOUBLE IDLER GEAR. With the timing gear housing removed, proceed to remove the double idler gear as follows: Remove the governor assembly as outlined in paragraph 118. Remove the camshaft gear retaining nut and lock washer and pull camshaft gear from shaft. Remove snap ring retaining idler gear to shaft and remove gear and bearing assembly. Bearing can be removed from gear after removing one of its retaining snap rings. If idler gear shaft is damaged, it can be pulled from main case. The procedure on series 720 diesel is shown in Fig. JD2091.

When reinstalling the double idler gear, mesh "V" mark on larger gear

with "V" mark on crankshaft gear and "V" mark on smaller gear with "V" mark on camshaft gear as shown in Fig. JD2090.

83. OIL PUMP DRIVE GEAR, SINGLE IDLER GEAR AND "POWR-TROL" PUMP DRIVE GEAR. With the timing gear housing removed, the procedure for removing the gears is evident. The "Powr-Trol" pump drive gear, however, can be removed without removing the timing gear housing by removing the pump as outlined in paragraph 191.

CAMSHAFT

The diesel engine camshaft has six lobes, two of which actuate the injection pump cam followers. The camshaft is drilled for pressure lubrication and is carried in three bearings. The right hand and center bearings are pre-sized, steel-backed, babbitt lined bushings, whereas the left hand bearing is cast iron.

84. To remove the camshaft, first remove the governor and housing assembly as outlined in paragraph 118.

Remove the timing gear housing as outlined in paragraph 77 or 77A. Remove the heat exchanger outlet pipe and tappet lever cover and back-off the tappet adjusting screws. Remove the injection pump compartment cover and back-off the injection pump push rod adjusting screws. Remove the injection pump cam followers (Fig. JD2092) and loosen the set screw (46) which retains the shaft left bearing in the main case. Remove the camshaft gear from shaft, remove the double idler gear retaining snap ring and remove the double idler gear Withdraw the camshaft and left bearing assembly from the main case.

If the right and center camshaft bushings are to be inspected and/or renewed, it will be necessary to remove the reduction gear cover as in paragraph 136.

Check the shaft journals and bearings against the following values.

Camshaft journal diameter
Right	1.4990-1.5000
Center	2.4975-2.4985
Left	1.4960-1.4970

Camshaft bearing inside diameter
Right	1.5020-1.5040
Center	2.5005-2.5025
Left	1.4995-1.5005

Camshaft journal clearance in bearings
Right	0.0020-0.0050
Center	0.0020-0.0050
Left	0.0025-0.0045

Inspect camshaft lobes for wear or scoring and make certain that the drilled oil passages are open and clean. When installing the right and center bushings, press them in with a piloted arbor and make certain that oil holes in bushings are in register with oil holes in main case.

When reassembling, install the cast iron left bearing on shaft, then install the assembly in the main case bore, making certain that set screw (46—Fig. JD2092) will engage the hole in the left bearing. Tighten the set screw to a torque of 8 Ft.-Lbs. and lock same by tightening the jam nut. Install the double idler gear and camshaft gear. Note: There is no adjustment for camshaft end play.

Assemble the remaining parts, time the injection pumps as outlined in paragraph 105, adjust the tappets as outlined in paragraph 72A, adjust the decompression control as in paragraph 71 and bleed the fuel system as in paragraph 102.

CAM FOLLOWERS

84A. The mushroom type cam followers operate directly in the main case bores and are available in standard size only. As shown in Fig. JD2093, the cam followers are grooved to accommodate the decompression shaft which raises the exhaust valve cam followers and decompresses the engine.

To remove the cam followers, first remove the camshaft as outlined in paragraph 84. Disconnect the control rod from the decompression shaft arm (Fig. JD2094), loosen the locking screw and pull the decompression shaft from the main case. Remove the diesel engine crankcase covers, and withdraw the cam followers through the crankcase cover openings.

Fig. JD2091—Using suitable puller to remove idler gear shaft on series 720 diesel.

Fig. JD2092 — The diesel engine camshaft left bearing is positioned in the main case by set screw (46).

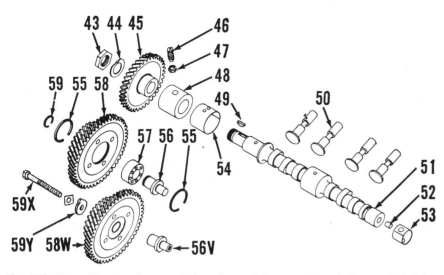

Fig. JD2093 — Diesel engine camshaft and cam followers. Two of the camshaft lobes actuate the diesel injection pumps. Parts (56V, 58W, 59X and 59Y) are used on 730 diesel models in place of parts (55, 56, 57, 58 and 59) used on 720 diesel tractors.

43. Nut	50. Cam followers	56. Idler gear shaft
44. Lock plate	51. Camshaft	56V. Idler gear shaft
45. Camshaft gear	52. Dowel pin	57. Idler gear bearing
46. Set screw	53. Camshaft right	58. Double idler gear
47. Jam nut	bushing	58W. Double idler gear
48. Camshaft left	54. Camshaft center	and bearing
bearing	bushing	59. Snap ring
49. Woodruff key	55. Snap ring	59X. Cap screw
		59Y. Retaining washer

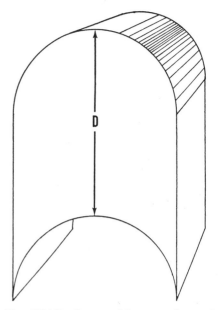

Fig. JD2095—Suggested home made wood block which can be used to push the connecting rod and piston units forward, out of cylinder block. It is more convenient to use three blocks. Dimension (D) should be approximately as follows: first block, 2⅞ inches; second block, 4 inches; third block, 5⅝ inches. Refer to text.

84B. The roller type injection pump cam followers (Fig. JD2092) can be removed after first removing the governor housing as per paragraph 118. Recheck the injection pump timing as outlined in paragraph 105, after reinstallation is complete.

CONNECTING ROD AND PISTON UNITS

85. To remove the connecting rod and piston units, first drain cooling system and remove the hood. Remove the diesel engine cylinder head as in paragraph 70 and the crankcase covers as in paragraph 93. Check the connecting rods and bearing caps for assembly marks as follows: No. 1 cylinder is on left (flywheel) side of tractor and the connecting rod and cap has one assembly punch mark. No. 2 connecting rod and cap has two assembly punch marks.

Before removing the units, check the bearing running clearance using plastic bearing fitting gage or equivalent. When checking the bearing clearance, rod bolts should be tightened to a torque of 200-220 ft.-lbs. Normal operating clearance is 0.0015-0.0045. If the clearance exceeds 0.007, the bearing inserts and/or crankshaft crankpins are excessively worn.

Remove the bearing caps and bearing inserts. Also, remove carbon accumulation and ridge from unworn portion of cylinders to prevent dam-

age to ring lands and to facilitate piston removal. Using a piece of 2x4 wood, make a set of blocks shown in Fig. JD2095. Using the smallest block between connecting rod and crankshaft crankpins, turn the crankshaft and push the rod and piston unit forward. Continue this process with the next larger block, and so on, until assemblies can be withdrawn from cylinder block.

Fig. JD2094—The diesel engine decompression shaft can be withdrawn after first disconnecting the control rod from the decompression shaft arm and loosening the locking screw.

When reassembling, punch marks on rod and cap must be in register and face toward top of engine. Number one cylinder is on left side of tractor. When installing No. 2 rod and piston unit, the No. 2 crankshaft crankpin must be in the forward position so that large end of connecting rod will clear the crankshaft counterweight. Tighten the self-locking connecting rod nuts to a torque of 200-220 ft.-lbs.

PISTONS AND RINGS

86. Pistons and rings are available in standard, as well as 0.045 oversize. Each of the aluminum alloy pistons are tapered, cam ground and fitted with six rings. The top compression ring is chrome plated, the next three rings are unplated compression rings, the fifth is a ventilated oil ring and the sixth is a cast iron wiper ring. Desired side clearance of new rings in piston grooves is as follows:

No. 1 (top) 0.006-0.007
Nos. 2, 3 & 4 ....0.0025-0.0045
No. 5 ....0.0015-0.0035
No. 6 ....0.001-0.004

Desired piston to cylinder clearance is as follows:
Top of skirt (just above piston pin)
  Parallel to piston pin.0.0218-0.0240
  At right angles to
    piston pin 0.0153-0.0177
Bottom of skirt
  Parallel to piston pin.0.0151-0.0175
  At right angles to
    piston pin 0.0086-0.0110

Piston skirt clearance is best determined by measuring the cylinder bore and the pistons as follows:

Standard cylinder bore..6.1246-6.1260

Rebore size (0.045
oversize) 6.1696-6.1710

Piston diameter at top of skirt
(just above piston pin)
Parallel to piston pin..6.1018-6.1028
Right angles to piston
pin 6.1083-6.1093

Piston diameter at bottom of skirt
Parallel to piston pin..6.1085-6.1095
Right angles to piston
pin 6.1150-6.1160

Install piston rings as follows: The ventilated oil control ring and the top (chrome plated) compression ring are not tapered and can be installed with either side facing the closed end of piston. The three unplated compression rings and the cast iron skirt wiper ring are tapered and must be installed with "dot" toward closed end of piston.

PISTON PINS AND BUSHING

87. The 2.3545-2.3550 diameter floating type piston pins are retained in the piston pin bosses by snap rings and are available in standard size as well as oversizes 0.003 (marked yellow) and 0.005 (marked red). The piston pin end of connecting rod is fitted with a renewable cast iron bushing. Piston pin should have 0.0000-0.0011 clearance in piston and 0.0010-0.0025 clearance in the connecting rod bushing.

Assemble piston to connecting rod with the word "Top" on piston head facing same side as the cylinder identifying punch marks on the rod.

CONNECTING RODS AND BEARINGS

88. Connecting rod bearings are of the steel-backed, copper-lead-lined, slip-in, precision type which can be renewed after removing the crankcase covers and the connecting rod bearing caps. To remove the crankcase covers, follow the procedure outlined in paragraph 93.

Normal connecting rod bearing running clearance is 0.0015-0.0045. If the clearance exceeds 0.007, the bearing inserts and/or crankshaft crankpins are excessively worn. Crankshaft crankpin diameter is 3.7480-3.7490. Connecting rod bearing inserts are available in standard size as well as undersizes of 0.002, 0.004, 0.020, 0.022, 0.030 and 0.032.

When installing new bearing inserts make certain that the bearing shell projections engage the milled slot in

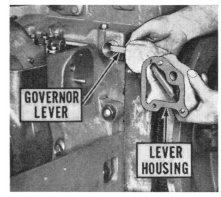

Fig. JD2096—Installing the governor lever housing on cylinder block. Make certain that the governor lever is properly inserted in the slotted pivot in governor case.

connecting rod and bearing cap and that cylinder identifying punch marks on rod and cap are in register and face toward top of engine. No. 1 cylinder is on left (flywheel) side of tractor and the connecting rod and cap has one identifying punch mark. No. 2 connecting rod and cap has two identifying punch marks. Tighten the self-locking connecting rod nuts to a torque of 200-220 Ft.-Lbs.

CYLINDER BLOCK

89. To remove the diesel engine cylinder block, first remove the cylinder head as outlined in paragraph 70 and the connecting rod and piston units as in paragraph 85. Remove the governor lever housing from left side of cylinder block and disconnect the fuel line from right side on block. Remove the injection pumps assembly and withdraw the injection pump push rods. Remove the cap screws retaining the cylinder block to the front end

support and stud nuts retaining block to main case. Back-out the implement mounting screws located in the front end support, pull the block forward and withdraw the unit, front end first, over left side of front end support.

Before reinstalling the cylinder block, make certain that oil return hole at bottom center of block is open and clean. When installing the block, tighten the stud nuts to a torque of 275 Ft.-Lbs. and be sure to install the two cap screws retaining the cylinder block to the front end support. Install and time the injection pumps as in paragraph 105. When installing the governor lever housing, make certain that governor lever (Fig. JD2096) is properly inserted in the slotted pivot in governor case. To check this, hold the governor lever housing in position, push the speed control lever back and forth and watch or feel to see that governor lever moves sideways in both directions. If the governor lever is not properly inserted in the slotted pivot, there will be no positive connection between the injection pump racks and speed control lever or governor. Bleed the fuel system as in paragraph 102.

CRANKSHAFT, MAIN BEARINGS AND SEALS

The crankshaft is supported in three main bearings. The right and left main bearings are of the aluminum alloy sleeve type which are pressed into main bearing housings. The sleeve type main bearing bushings are available either as individual parts or as a unit with the main bearing housings, on a factory exchange basis. The exchange main bearing and housing units

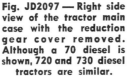

Fig. JD2097 — Right side view of the tractor main case with the reduction gear cover removed. Although a 70 diesel is shown, 720 and 730 diesel tractors are similar.

are pre-sized; but if a replacement bushing is pressed into the old housing, it will require final sizing to provide a running clearance of 0.0060-0.0080. The center main bearing is of the steel-backed, copper-lead, precision insert type retained in a two piece center bearing housing.

The right main bearing housing contains two single oil seals, installed back to back, to prevent mixing of transmission oil with the engine lubricating oil. The seals are furnished already installed in an exchange main bearing housing; or, the seals are available as individual replacement parts. Oil leakage at left end of crankshaft is prevented by the flywheel spacer acting as an oil slinger, and by a neoprene oil seal installed in the spacer.

90. **MAIN BEARINGS.** Although most repair jobs associated with the crankshaft and main bearings will require removal of all three main bearings, there are infrequent instances where the failed or worn part is so located that repair can be accomplished safely by removing only the right or the left bearing housing (center main bearing is removed with crankshaft). In effecting such localized repairs, time will be saved by observing the following paragraphs as a general guide.

90A. RIGHT MAIN BEARING HOUSING AND SEALS. To remove the right main bearing housing, remove the clutch, belt pulley and reduction gear cover. Withdraw spacer (S—Fig. JD2097) and remove the first reduction gear (RG) and the power shaft idler gear (IG). Remove snap ring (SR) and using a suitable puller, remove the power shaft drive gear (DG) from crankshaft. Unlock and remove the main bearing housing retaining cap screws and, using three of the screws as jack screws in the tapped holes provided in the bearing housing, remove the housing as shown in Fig. JD2098. The two opposed oil seals can be removed from the main bearing housing at this time, but the new seals should not be installed until after the bearing housing is reinstalled and permanently bolted to the main case.

Bearing housing has a 0.000-0.005 tight fit in the main case bore; therefore, when installing the housing, pull it evenly into place with the attaching cap screws. Tighten the retaining caps screws to a torque of 75 Ft.-Lbs. and lock the screws with tab washers. Install the inner oil seal with lip fac-

Fig. JD2098—Using jack screws to remove the right hand main bearing housing. All three main bearing housings have 0.000-0.005 tight fit in the main case bores.

ing inward, fill the space between the seals with gun grease to provide initial lubrication and install the outer seal with lip facing outward. To facilitate installation of the power shaft drive gear, heat gear in oil and drive the gear into position with a brass drift.

90B. LEFT MAIN BEARING HOUSING. To remove the left main bearing housing, remove the timing gear housing as in paragraph 77 or 77A and governor housing as outlined in paragraph 118. Remove the camshaft gear retaining nut and lock washer and pull gear (45—Fig. JD2099) from camshaft. Extract the retaining snap rings and remove the double and single idler gears (58 and 123). Using a suitable puller, remove the crankshaft gear from shaft. Remove the main bearing housing oil line and unlock and remove the main bearing housing retaining cap screws. Using three of the cap screws as jack screws in the tapped holes provided in the bearing housing, remove the housing by tightening the screws evenly against the main case.

The bearing housing has a 0.000-0.005 tight fit in the main case bore; therefore, when installing the housing, pull it evenly into place with the attaching cap screws. Tighten the retaining cap screws to a torque of 75 Ft.-Lbs. and lock the screws with tab washers. To facilitate installation of the crankshaft gear, heat same in oil and while another man is bucking up right end of crankshaft, drive the gear

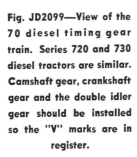 

Fig. JD2099—View of the 70 diesel timing gear train. Series 720 and 730 diesel tractors are similar. Camshaft gear, crankshaft gear and the double idler gear should be installed so the "V" marks are in register.

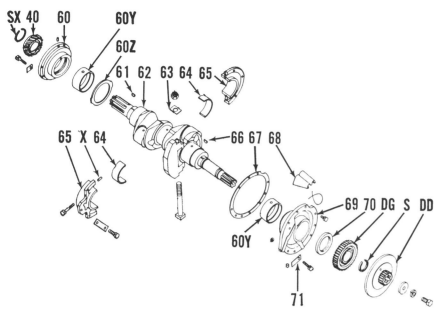

Fig. JD2100—Diesel engine crankshaft and main bearing housings. The center main bearing and housing are removed with crankshaft. Snap ring (SX) and thrust washer (60Z) are used in production beginning with 720 diesel serial number 7214900.

DD. Clutch drive disc	60Z. Thrust washer	65. Center main bearing
DG. Powershaft drive gear	61. Woodruff key	housing
S. Snap ring	62. Crankshaft	66. Woodruff key
SX. Snap ring	63. Lock plate	67. Gasket
40. Crankshaft gear	64. Center main bearing	68. Oil baffle
60. Left main bearing	inserts	69. Right main bearing
housing		housing
60Y. Left and right main		70. Oil seals (2 used)
bearing bushings		71. Lock plate

into position with a brass drift. Mesh "V" marks on crankshaft gear, double idler gear and camshaft gear as shown in Fig. JD2099.

91. **CRANKSHAFT.** To remove the crankshaft, first remove the right main bearing housing as in paragraph 90A, the left main bearing housing as in paragraph 90B and proceed as follows: Remove the crankcase covers as in paragraph 93 and disconnect the connecting rods from crankshaft crankpins. Remove the counterweight retaining bolt and tap counterweight from crankshaft. Working through right opening in main case, unwire and remove the cap screws retaining the center main bearing housing to main case web. To break the center bearing housing loose from the main case web, tap on left end of shaft with a heavy, soft hammer. Withdraw the crankshaft through right opening in main case, being careful not to nick or damage the bearing journals. Using an Allen wrench, separate the center main bearing halves and remove the bearing inserts.

Check the crankshaft and main bearings against the values which follow:

Main journal diameter 4.4975-4.4985

Crankpin diameter 3.7480-3.7490

Pulley journal
diameter 2.4320-2.4340
Right and left main
bearing bore 4.5045-4.5055
Center main
bearing bore 4.5006-4.5030
Pulley bushing bore
(installed) ....2.4375 min.

Right and left main bearing oil
clearance (new).......0.0060-0.0080
Center main bearing oil
clearance (new) 0.0021-0.0055
Pulley bushing oil
clearance (new) 0.0035-0.0055

Main bearings and/or crankshaft should be renewed when oil clearance exceeds 0.008 for the center main bearing, 0.015 for the right and left main bearings. Main bearings are available in standard size as well as undersizes of 0.002, 0.004, 0.020, 0.022, 0.030 and 0.032.

Assemble the center main bearing inserts and housing to crankshaft and tighten the Allen screws securely. Install crankshaft and center main bearing assembly, being careful not to nick or damage the bearing journals. Word "Front" on center bearing housing must face toward front of tractor. The center bearing housing has a 0.000-0.005 tight fit in the main case bore; therefore, when installing the housing,

pull it evenly into place with the attaching cap screws. Tighten the center main bearing cap screws to a torque of 50 Ft.-Lbs. and the left and right bearing housing cap screws to 75 Ft.-Lbs. Lock the screws with safety wire or lock plates.

FLYWHEEL AND CRANKSHAFT END PLAY

Series 720 Diesel Prior to Serial Number 7214900

92. To remove the diesel engine flywheel on these early models, first remove the flywheel cover and using a dial indicator as shown in Fig. JD2101, check the crankshaft end play by engaging and disengaging the clutch with force. If the end play is not between the limits of 0.005-0.010, remember the measurement so the end play can be adjusted to the specified value during reassembly.

Unstake and remove the flywheel lock nut and loosen the two flywheel clamp bolts. Using a suitable puller, move flywheel out toward end of crankshaft and using a hoist, remove flywheel from tractor.

If the previously measured crankshaft end play was not within the limits of 0.005-0.010, remove the thrust cover (Fig. JD2102) and add or remove shims, located between thrust cover and timing gear housing, as necessary to provide the correct amount of end play. Reinstall the thrust cover with word "TOP" toward top of tractor and tighten the screws finger tight. Using 0.003 thick strips of shim stock as shown, center the thrust cover about the flywheel spacer and tighten the thrust cover retaining cap screws to a torque of 35 Ft.-Lbs.

Fig. JD2101 — Using a dial indicator to measure the crankshaft end play. Desired end play is 0.005-0.010.

To install the flywheel ring gear, heat same uniformly to about 550 degrees F. and place gear on flywheel so that beveled end of ring gear teeth face toward the diesel engine crankcase.

Remove the belt pulley dust cover and while having another man buckup the right end of the crankshaft with a heavy bar, align the "V" mark on flywheel with "V" mark on left end of crankshaft, install flywheel and drive flywheel solidly against the flywheel spacer. Install the flywheel nut, tighten it securely and lock it in position by peening a portion of the nut into one of the crankshaft splines. Tighten the flywheel clamp bolts to a torque of 275 Ft.-Lbs.

Series 720 Diesel After Serial Number 7214899-All Series 730 Diesel

92A. To remove the diesel engine flywheel on these models proceed as follows: Remove flywheel cover, flywheel lock nut and loosen both fly-

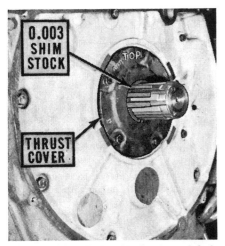

Fig. JD2102 — Early 720 diesel crankshaft end play 0.005-0.010 is controlled by shims located behind the thrust cover.

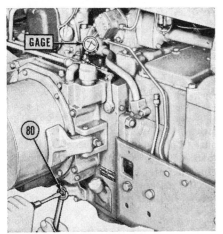

Fig. JD2103—Using a master gage to check the diesel engine oil pressure which should be 25 psi with engine at operating temperature and running at high idle rpm.

Fig. JD2104 — Diesel engine oil filter, filter head, pressure regulator and associated tubing and connections. Plate (85A) is used on tractors after 720 serial number 7225046.

72. Tappet lever oil pipe
73. Cylinder to tappet lever oil pipe
74. By-pass valve cap
75. Camshaft oil pipe
76. Pressure relief valve bushing
77. Relief valve spring
78. Coil spring pilot
79. Pressure adjusting spring
80. Pressure adjusting screw
81. Gasket
82. Relief valve housing
83. Jam nut
84. Cap nut
85. Oil filter body
85A. Plate
86. Gasket
87. Rivet
88. Right main bearing pipe
89. Filter relief pipe
90. Oil filter head
91. Case to left main bearing housing pipe
92. Oil filter outlet
93. Oil filter element
94. Snap ring
95. Spacer
96. End plate
97. Spring
98. Washer
99. Gasket
100. Cover
101. Copper washer
102. Nut
103. Filter head to main case pipe
103X. Center main bearing oil pipe
104. Copper gasket
105. By-pass valve spring
106. By-pass valve ball

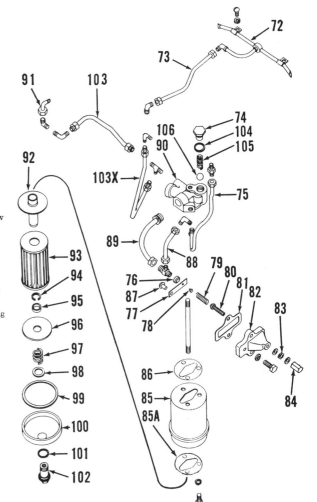

wheel retaining bolts. Swing flywheel in a hoist, attach a suitable puller and pull flywheel from crankshaft.

To install flywheel starter ring gear, heat same evenly to 550 degrees F. and place gear on flywheel so that the beveled ends of the teeth face toward the diesel engine crankcase. Install flywheel on crankshaft so that "V" mark on flywheel is in register with "V" mark on crankshaft.

The diesel engine crankshaft end play is determined by the position of the flywheel on the crankshaft. To adjust crankshaft end play, proceed as follows: Drive flywheel on crankshaft and mount a dial indicator so that the indicator contact button is resting on the flywheel. Engage and disengage clutch with force several times and observe crankshaft end play as shown on the dial indicator. Continue driving flywheel on the crankshaft until the proper end play of 0.005-0.010 is obtained. Tighten both flywheel retaining bolts to a torque of 275 Ft.-Lbs. Install flywheel lock nut and secure same by staking.

CRANKCASE COVERS

93. On tractors equipped with the gasoline starting engine, procedure for removing either cover is evident after first removing the starting engine as outlined in paragraph 30.

On tractors equipped with the 24-volt electric starting motor, procedure for removing the right cover is evident after removing the hood. To remove the left crankcase cover, first remove the starting motor.

ENGINE OILING SYSTEM

94. **OIL PRESSURE-ADJUST.** Desired diesel engine oil pressure is 25 psi, with oil warm and engine operating at the high idle rpm. To check and/or adjust the oil pressure, disconnect the oil pressure gage line at transmission case and install a master pressure gage as shown in Fig. JD2103. With engine at normal operating temperature and running at the high idle rpm, remove cap nut and turn the adjusting screw (80) **in** to increase or **out** to decrease the oil pressure.

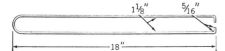

Fig. JD2105 — Home made tool which can be used for removing the oil filter outlet as shown in Fig. JD2106. Tool can be made from 3/16-inch cold rolled steel rod.

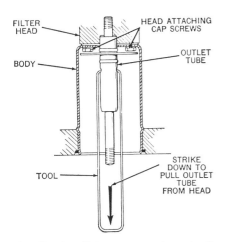

Fig. JD2106—Using the tool shown in Fig. JD2105 to remove the oil filter outlet.

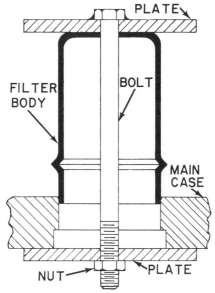

Fig. JD2107—Suggested home made tool for installing the oil filter body. The long bolt can be welded to the upper plate.

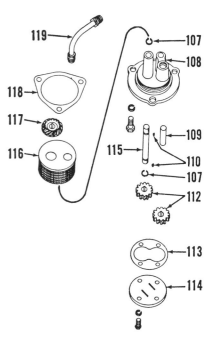

Fig. JD2108—Exploded view of the diesel engine oil pump. The unit is driven, via the drive unit shown in Fig. JD2109, from the timing gear train.

107. Snap ring	114. Oil pump cover
108. Oil pump body	115. Oil pump drive shaft
109. Idler gear shaft	116. Oil pump screen
110. Woodruff keys	117. Pump drive gear
112. Pumping gears	118. Shim gaskets
113. Gasket	119. Oil pump discharge pipe

95. OIL PRESSURE REGULATOR — R&R AND OVERHAUL. The oil pressure regulator assembly can be removed from right side of tractor main case after removing the retaining cap screws. The oil pressure relief coil spring (79—Fig. JD2104) should have a free length of 1 25/32 inches and should require 22-25 ounces to compress it to a height of ¾-inch.

96. FILTER HEAD AND BY-PASS VALVE — R&R AND OVERHAUL. The oil filter by-pass valve ball (106—Fig. JD2104) and related parts can be removed from oil filter head (90) after removing the crankcase covers as outlined in paragraph 93.

To remove the oil filter head, first drain the crankcase and remove the crankcase covers. Disconnect oil lines from filter head.

NOTE: If special tools are available, it is possible to disconnect these oil lines by working through the crankcase opening. Due to space limitations, it is often very time consuming, and if nipples connecting oil lines to filter head are not tight in filter head, it is oftentimes impossible for the average man to disconnect the oil lines by working through the crankcase opening; in which case, the following procedure is used.

96A. Disconnect connecting rods from crankshaft and remove the tappet lever cover, tappet levers assembly and push rods. Remove the injection pump compartment cover, injection pumps and pump push rods. Remove the exhaust pipe, exhaust manifold, water outlet casting and inlet manifold (or heat exchanger.) Remove the lower water pipe. Remove governor lever housing from left side of block and disconnect fuel line from right side of block. Unbolt cylinder block from main case and front end support. Slide cylinder head and cylinder block assembly forward as far as possible. Working through front opening in main case, disconnect the oil lines from filter head.

96B. After oil lines are disconnected, remove the filter element. Using 3/16-inch round, cold rolled steel rod, make-up a tool similar to that shown in Fig. JD2105. Hook jaws of puller tool into holes of the filter outlet as shown in Fig. JD2106, then strike the bottom of the tool in a downward motion to remove the oil filter outlet. Working through bottom of filter body, remove three cap screws retaining filter head to filter body. Withdraw filter head from crankcase.

Normal servicing of the filter head includes cleaning the oil passages and overhauling the by-pass valve assembly. The by-pass valve spring (105—Fig. JD2104) has a free length of 1¾ inches and should require 6-8 ounces to compress it to a height of 1 21/64 inches.

When reassembling, use a suitable piece of pipe, to drive the filter outlet tube into position. When installing oil filter element, tighten the cover nut only enough to eliminate oil leakage. Install the governor lever housing as outlined in paragraph 118, time the injection pumps as in paragraph 105 and bleed the fuel system as in paragraph 102.

97. OIL FILTER BODY — RENEW. Zero or low oil pressure can be caused by a distorted oil filter body (85—Fig. JD2104). Distortion is usually caused by over tightening the filter bottom retaining nut. Nut (102) should be tightened only enough to eliminate oil leakage at this point. If leakage occurs, with normal tightening, renew gasket (99).

98. To renew the oil filter body, remove the oil filter head as outlined in paragraph 96 and proceed as follows: Place a cylindrical wooden block in filter body and jack-up block, forcing filter body out of crankcase recess. Withdraw body from crankcase.

To install oil filter body, coat the portion below the "bead" with white lead or equivalent sealing compound to prevent leaks and facilitate draw-

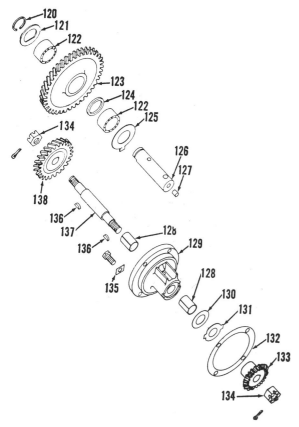

Fig. JD2109 — Diesel engine oil pump drive unit. Idler gear (123) meshes with the crankshaft gear as shown in Fig. JD2099.

120. Snap ring
121. Thrust washer
122. Bearing
123. Idler gear
124. Spacer
125. Thrust washer
126. Idler gear shaft
127. Dowel pin
128. Bushing
129. Drive housing
130. Shim washer
131. Thrust washer
132. Shims
133. Bevel gear
134. Nut
135. Lock plate
136. Woodruff key
137. Pump drive shaft
138. Drive gear

ber of shims (118) to provide a backlash of 0.004-0.006 between the bevel gears. Backlash can be measured directly at the gears by working through the crankcase cover openings.

The act of setting the backlash of the bevel gears will generally serve to place them in the correct mesh position. However, if the gear shafts or the main case are to be renewed, it is advisable to check the mesh position as in paragraph 100A.

100. **OIL PUMP DRIVE UNIT— R&R AND OVERHAUL.** To remove the oil pump drive unit, first remove the timing gear housing as outlined in paragraph 77 or 77A. Extract snap ring (120—Fig. JD2109) and remove idler gear (123). Unbolt and withdraw the drive unit from the main case.

The procedure for disassembling the unit is evident. Bushings (128) are not pre-sized and must be reamed after installation to provide a clearance of 0.004-0.0064 for the 0.999-1.000 diameter drive shaft. When reassembling, adjust the drive shaft end play to 0.0025-0.005 by varying the number of shims (130). Shims are 0.0025 and 0.010 thick.

After end play of drive shaft has been adjusted, the unit should be reinstalled using the same steel shims and same number of new shim gaskets (132) as were originally installed. Desired backlash of teeth of bevel drive gear (133) with teeth of bevel gear on oil pump is 0.004-0.006. This can be checked when the drive unit is installed by mounting a dial gauge as shown in Fig. JD2110. Because the radius of the helical gear (138) is larger than the radius of the bevel gear (133—Fig. JD2109), backlash will be 0.0055-0.008 when checked at the helical gear. If backlash is not as specified, vary the number of shims (132).

The act of setting the backlash of the bevel gears will generally serve to place them in the correct mesh position; however, if the gear shafts or the main case are to be renewed, it is advisable to check the mesh position as outlined in paragraph 100A.

100A. **MESH POSITION OF BEVEL GEARS.** Bevel gears (133 — Fig. JD2109) and (117 — Fig. JD2108) should be meshed so that heels are in register and backlash is from 0.004-0.006. To observe the mesh position it is necessary to remove the crankcase covers as in paragraph 93. The

ing body into recess of crankcase. Place filter head on the body to make certain that cap screw holes in body and head are aligned and that oil lines will align with filter head connections; then, remove filter head. To draw body into crankcase, use a long bolt and two steel plates as shown in Fig. JD2107. Tighten the nut until bead of body seats against crankcase. Install remainder of parts by reversing the removal procedure.

99. **OIL PUMP—R&R AND OVERHAUL.** To remove the diesel engine oil pump, first remove the crankcase covers as in paragraph 93. Drain crankcase and remove oil line (119—Fig. JD2108) which runs from the oil pump to the filter head. Remove the oil pump retaining cap screws from underneath the main case and withdraw the pump unit from below. Disassemble pump and check body, gears and shafts against the values which follow:

Drive shaft bore in
  pump body 0.627-0.629
Diameter of drive shaft..0.624-0.625
Clearance between gear
  teeth and body 0.006-0.010
Gear bore in pump body 2.090-2.092
Gear diameter 2.082-2.084
Gear clearance in bore..0.006-0.010

Diameter of idler
  gear shaft 0.624-0.625
Idler gear shaft bore
  in pump body 0.627-0.629
Clearance between gears
  and pump cover 0.018-0.022
Assemble the pump by reversing the disassembly procedure and when installing the pump, vary the num-

Fig. JD2110 — Dial indicator mounted for checking backlash between the diesel engine oil pump drive bevel gears. Idler gear shaft is shown at (126) and the oil pump drive gear at (138).

desired mesh position and backlash of the bevel gears is obtained by using the proper combination of shims (132 —Fig. JD2109) and (118—Fig. JD2108) between the respective housings and the main case. Mount a dial indicator as shown in Fig. JD2110 for checking the bevel gear backlash. Because the radius of the helical gear is larger than the radius of the bevel gear, the bevel gear backlash will be 0.0055-0.008 when checked at the helical gear. Backlash can also be checked at the bevel gears by working through the crankcase cover openings.

DIESEL FUEL SYSTEM

The diesel engine fuel system consists of four basic units; the fuel transfer pump, fuel filters, injection pumps and injection nozzles. When servicing any unit associated with the fuel system, the maintenance of absolute cleanliness is of utmost importance.

101. **QUICK CHECKS—UNITS ON TRACTOR.** If diesel engine does not start or does not run properly, and the diesel fuel system is suspected as the source of trouble, refer to the Diesel System Trouble-Shooting Chart and locate points which require further checking. Many of the chart items are self-explanatory; however, if the difficulty points to the fuel filters, fuel transfer pump, injection nozzles and/or injection pumps, refer to the appropriate section which follows.

FILTERS AND BLEEDING

102. The fuel filtering system consists of a glass sediment bowl and strainer, and two stages of renewable element type filters. The fuel system should be purged of air whenever the filters have been removed or when the fuel lines have been disconnected. Under normal operating conditions, the sediment bowl should be cleaned every 10 hours of operation, the first stage filter element should be renewed every 450 hours of operation and the second stage filter should be renewed at major overhauls.

The general procedure for testing the condition of the filters and bleeding the system is as follows:

Shut off the fuel, remove and thoroughly clean the glass sediment bowl which is located under the fuel tank.

Reinstall the bowl loosely, turn on the fuel and allow it to flow until bowl is full, then tighten the bowl securely. Remove the first bleed plug, allow fuel to flow until free of air, then reinstall the plug. Remove the second bleed plug, allow fuel to flow until free of air, then reinstall the plug. On early models so equipped, remove the third bleed plug, allow fuel to flow until free of air, then reinstall the plug. If fuel does not flow freely at a bleed point, it indicates a clogged filter.

TRANSFER PUMP

The gear type fuel transfer pump is located on the right side of governor housing and is driven by the slotted end of the governor shaft. A spring and ball type relief valve which is lo-

DIESEL SYSTEM TROUBLE-SHOOTING CHART

	1. Engine Hard to Start	2. Lack of Power	3. Engine Surges (Idle Speeds)	4. Engine Smokes or Knocks	5. Excessive Fuel Consumption	6. Fuel in Pump Compartment	7. Sticking Pump Plunger
Lack of fuel	★						★
Water or dirt in fuel	★	★					★
Clogged fuel system	★	★					
Air in fuel and injection system	★	★					
Inferior fuel	★	★					
Cratered injection nozzle tips	★	★	★	★	★		
Clogged nozzle spray tip holes	★	★					
Sticky nozzle valves	★	★	★	★	★		
Faulty injection pump timing	★	★		★			
Worn or stuck pump plungers	★	★					
Improperly calibrated or balanced pumps	★	★			★		
Faulty transfer pump	★	★					
Engine speed too low		★					
Faulty or incorrectly adjusted governor			★	★			
Worn or sticky pump delivery valves			★	★			
Excessively worn pump rack and control sleeve			★				
Dirty nozzle spray tip		★					
System leaks					★		
Loose pump delivery valve holder		★			★	★	
Damaged delivery valve holder gasket		★			★	★	
Loose or split fuel lines		★				★	
Leaking pump-bracket or bracket-cylinder gasket						★	
Delivery valve holder too tight							★
Bent racks or linkage			★				★

All troubles listed in columns 1, 2, 4 and 5 could be caused by derangement of parts other than the diesel pump and injector system.

cated in the fuel pump housing maintains a normal delivery pressure of 18-26 psi. Seals (2—Fig. JD2113) prevent fuel from entering the diesel engine crankcase, and crankcase oil from entering the fuel pump.

103. **TESTING.** Remove first bleed plug and install a suitable pressure gage as shown in Fig. JD2112. Start the diesel engine and observe pressure reading which should be 18-26 psi. Add washers (11 — Fig. JD2113) between the by-pass valve spring (10) and housing to increase the pressure.

NOTE: Early pumps, identified by a plugged hole just above the pump casting number, require a cup shaped adjusting washer of $\frac{1}{2}$-inch diameter. Later pumps don't have the identifying plugged hole and should be adjusted by adding or deducting flat $\frac{3}{8}$-inch adjusting washers. Refer to Fig. JD2113A. Early tractors may have the later pump installed.

104. **OVERHAUL.** The procedure for removing and disassembling the pump is evident after an examination of the unit and reference to Fig. JD2113.

Failure of pump to deliver fuel at an adequate pressure may be caused by a faulty by-pass valve. Free length of by-pass valve spring is $1\frac{1}{4}$-inches and the spring should require 5.5-7.5 ounces to compress it to a height of $\frac{7}{8}$-inch. Check pump body, gears and shafts against the values which follow:

Drive shaft diameter . . 0.484 -0.485
Radial clearance between
    gear teeth and body . 0.0015-0.0030
Gear bore in pump
    body 1.1130-1.1150
Diameter of gears 1.1090-1.1100
Diameter of idler
    gear shaft 0.484-0.485
Clearance between gears
    and cover 0.003-0.004
Gear thickness 0.2805

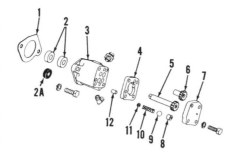

Fig. JD2113 — **Exploded view of the fuel transfer pump. The pump is driven from the governor shaft. Early models which don't have the felt washer (2A) between the seals (2) should have one installed when the pump is serviced.**

1. Gasket
2. Oil seals
2A. Felt washer
3. Pump housing
4. Fuel pump body
5. Pump drive gear
6. Pump idler gear
7. Pump cover
8. Relief valve piston
9. By-pass ball
10. By-pass spring
11. Adjusting washers
12. Dowel pin

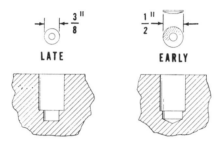

LATE   EARLY

Fig. JD2113A — **Early transfer pumps, at right, require a cup shaped adjusting washer of $\frac{1}{2}$-inch diameter. Late pumps, at left, should be adjusted by adding or deducting flat $\frac{3}{8}$-inch washers.**

When reassembling, use a thin coat of shellac on both sealing surfaces of pump body. Use a very small amount of No. 3 Permatex on O.D. of inner seal (2) and press the seal in until it bottoms. Install felt washer (2A). Outer seal should be installed flush with pump surface. Make certain that seal lips are opposed as shown in Fig. JD2114.

NOTE: Early production tractors were not fitted with a felt washer between the seals. When working on these early models, obtain and install the felt washer (John Deere part No. R201-39R).

INJECTION PUMPS

The diesel injection pumps are located in a compartment in the Diesel engine cylinder block. Fuel enters the injection pumps, from the filters, under a normal pressure of 18-26 psi and leaves the pumps under a pressure of 2400-2600 psi. The reciprocating pump plungers are actuated by cam followers from the Diesel engine camshaft.

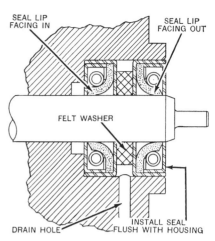

Fig. JD2114—**Partial sectional view of the fuel transfer pump showing the proper installation of the seals and felt washer.**

Injection pumps may be either Bendix-Scintilla or American Bosch; and, although both makes operate in the same manner and are similarly constructed, two pumps of different makes should never be installed in any one tractor.

105. **TIMING.** Remove flywheel cover, tappet lever cover and injection pump compartment cover. Turn flywheel in running direction until the inlet valve for No. 1 (left hand) cylinder opens and closes and continue turning flywheel until the "No. 1 INJ" mark on flywheel is in register with mark on timing gear housing as shown in Fig. JD2115. At this time, the scribe mark on number 1 injection pump plunger follower should align with an index in the window of the pump body. On Bosch pumps, the in-

Fig. JD2112—**Gage installation for checking the fuel transfer pump pressure. Pressure gage should read 18-26 psi.**

Fig. JD2115 — **"No. 1 INJ" mark on flywheel in register with mark on the timing gear housing.**

dex is a scribe line on the sides of the window as shown at right in Fig. JD2116. On Scintilla pumps, the index is a sheet metal pointer as shown at the left. If index marks do not align, adjust the length of the injection pump push rod with the screw located just behind pump, until proper register is obtained. Turn flywheel one-half turn (180 degrees) and align "No. 2 INJ" mark on flywheel with mark on timing gear housing. Check and time number 2 injection pump in the same manner.

CAUTION: Make certain that scribe line does not disappear from window in pump housing when flywheel is rotated ½ turn (180 degrees) from injection timing mark. If scribe line does disappear, the pump is not properly timed and internal pump parts may be destroyed.

106. **TESTING.** If the diesel engine does not run properly, or not at all, and the quick checks outlined in paragraph 101 point to faulty injection pumps, the pumps must be renewed, cleaned or overhauled.

In order to obtain balanced fuel delivery to each cylinder, the pumps are synchronized by adjusting the connecting linkage between the pump racks. Since elaborate testing equipment (equipment which was especially designed for use in testing and adjusting the John Deere injection system) is required to make the adjustment, the linkage should NEVER be disturbed unless such equipment is available.

Note: It is recommended that any work which must be performed on the injection pumps be done by a John Deere factory approved service station.

107. **EXCHANGE UNITS.** An exchange unit consisting of a pair of pumps, interconnecting linkage and mounting bracket is available through the John Deere parts department.

108. **INJECTION PUMPS—R&R.** To remove the injection pumps and mounting bracket unit, proceed as follows: Remove the tappet lever cover and injection pump compartment cover. On tractors equipped with the gasoline starting engine, remove the heat exchanger outlet pipe. On all models, remove the fuel lines connecting injection pumps to injection nozzles and immediately cap the four connectors. Completely back-off both injection pump push rod adjusting screws. Remove the rack spring plug from right side of cylinder block and extract the rack spring, Unbolt and remove the pump assembly from cylinder block.

CAUTION: Do not pry on the racks or any connecting linkage. If linkage is bent or distorted, it will be unsuitable for further use.

After assembly is removed from tractor, tape or cork the fuel inlet openings which are located in the underside of the pump mounting bracket. Note: Injection pump push rod units can be removed at this time.

CAUTION: In removing the injection pumps from mounting bracket, do not disturb the pump connecting linkage adjustment as in doing so, the pumps will be thrown out of synchronization.

Mark position of pumps on the mounting bracket. Pumps are not interchangeable from side to side and must be installed in their original position. Unbolt and remove pumps from bracket and don't disassemble unless a test stand is available.

INJECTION NOZZLES

Injection nozzles may be either Bendix-Scintilla or American Bosch; and, although both makes operate in the same manner and are similarly constructed, two nozzles of different makes should never be installed in any one tractor.

109. **R&R AND TEST.** If the engine does not run properly, or not at all, and the quick checks outlined in paragraph 101 point to faulty nozzles, remove the injection pump compartment cover and tappet lever cover and proceed as follows: Disconnect fuel lines at nozzles and remove the nozzle retaining nuts and clamps. If a nozzle tester is available, check the

nozzle as in paragraph 111. If a special tester is not available, mount nozzles as shown in Fig. JD2124, making certain that fuel lines are tight.

WARNING: Fuel leaves the injection nozzle tips with sufficient force to penetrate the skin. When testing, keep your person clear of the nozzle spray.

Place diesel engine speed control lever in wide-open position, hold injection pump racks open, motor the diesel engine with the starting engine, and observe the nozzle spray pattern.

In a good nozzle, the fuel comes out in a finely-atomized mist. A lopsided spray pattern indicates that holes in nozzle tip are partially clogged, and a stream of fuel or a poorly-atomized spray indicates that nozzle tip holes may be worn oversize. If cleaning and/or nozzle tip renewal does not restore unit and a special nozzle tester is not available for further checking, send the complete nozzle assembly to an official diesel service station for overhaul or, install an exchange unit.

110. **EXCHANGE UNITS.** Exchange nozzles are available through the John Deere parts department.

111. **NOZZLE TESTER.** A complete job of testing and adjusting the nozzle requires the use of a special nozzle tester. The nozzle should be tested

Fig. JD2124—Diesel engine injection nozzle mounted in position for observing spray pattern and checking discharge.

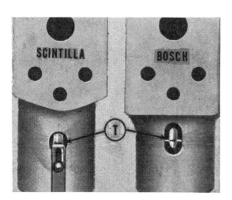

Fig. JD2116 — Either Bendix-Scintilla or American Bosch injection pumps are used. Notice differences in construction at timing windows (T).

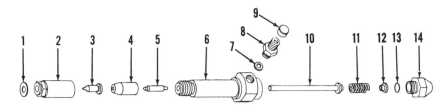

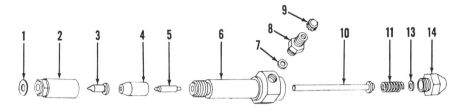

BENDIX SCINTILLA

AMERICAN BOSCH

Fig. JD2125 — Exploded views of the fuel injection nozzles. Pressure spring cap (14) should never be removed unless suitable equipment is available for checking the nozzle cracking pressure.

1. Copper washer
2. Nozzle tip retaining nut
3. Nozzle tip
4. Needle valve seat
5. Needle valve
6. Nozzle holder body
7. Gasket
8. Inlet nipple
10. Pressure spring pin
11. Pressure spring
12. Pressure spring seat
13. Adjusting shims
14. Pressure spring cap

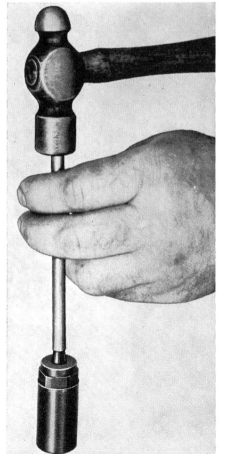

Fig. JD2126—If the injection nozzle tip is frozen in the retaining nut, a steel tube and hammer will facilitate removal.

for leakage, spray pattern and opening (or cracking) pressure. Operate the tester lever until oil flows and attach the nozzle and holder assembly.

Note: Only clean, approved testing oil should be used in the tester tank.

Close the tester valve and apply a few quick strokes to the lever. If undue pressure is required to operate the lever, the nozzle valve is plugged and same should be serviced as in paragraph 112.

CHATTERING AND POPPING. When operating the tester handle at a steady rate of 25-35 strokes per minute, the needle valve should open and close with a distinct, audible chattering or popping sound. If the chattering or popping sound is not heard, the needle valve and seat are dirty, broken or excessively worn. Service the valves as in paragraph 112.

SPRAY PATTERN. Operate the tester handle and observe the spray pattern which should be a finely-atomized mist. A lopsided spray pattern indicates that holes in nozzle tip are partially clogged, and a stream of fuel or a poorly-atomized spray indicates that nozzle tip holes may be worn oversize. The nozzle tip should be cleaned or renewed.

OPENING PRESSURE. While operating the tester handle, observe the gage pressure at which the spray occurs. The gage pressure should be at

least 2400 psi. If the pressure is not as specified, remove nozzle spring cap and add or deduct shims (13—Fig. JD2125) until the opening pressure is 2400 psi. Note: If a new pressure spring has been installed in the nozzle holder, adjust the opening pressure to 2600 psi.

LEAKAGE. The nozzle valve should not leak at a pressure of 100-300 pounds less than the opening pressure of 2400-2600 psi. To check for leakage, actuate the tester handle slowly and as the gage needle approaches the specified pressure, observe the nozzle tip for drops of fuel. If the nozzle valve and seat are in good condition, the nozzle tip will be reasonably dry. Also, observe the nozzle leak-off pipe where a small amount of leakage is normal. If the leakage through the leak-off pipe exceeds 1 cc. per minute or if the nozzle tip is wet, the valve and seat should be reconditioned.

DRIBBLE. When operating the tester handle with quick strokes, the fuel injection should start and stop without producing any dribble at the spray tip. Dribble could be caused by the method of handling the tester. Therefore, make the check several times and wipe the tip after each check. A persistent dribble indicates worn or dirty parts.

112. NOZZLES—R&R AND OVERHAUL. To remove nozzles from tractor, remove the tappet lever cover and injection pump compartment cover. Remove fuel lines connecting injection pumps to injection nozzles and immediately cap the four connectors. Remove both nozzle retaining nuts and clamps and pull nozzles from cylinder head, taking care not to damage the fuel leak-off pipes.

Before disassembling, wash complete nozzle in clean diesel fuel and blow off with clean, dry, compressed air. Carbon accumulation from exterior of nozzle spray tip can be removed with a wire brush. Set up a pan with a clean cloth in the bottom and fill pan with clean diesel fuel. As parts are removed from nozzles, place same on cloth in pan, but do not allow parts to touch one another. Clamp nozzle in a soft jawed vise, and remove nozzle tip retaining nut (2—Fig. JD2125). Nozzle valve assembly (4 & 5) and spray tip (3) will be removed with the nut. Push spray tip out of nut. If spray tip is frozen to the retaining nut, use a piece of steel tubing which will fit about half-way down on the tip and tap tip out of the nut as shown in Fig. JD2126.

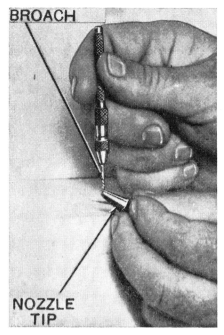

Fig. JD2127—The six tip orifices in the injection nozzle tip can be cleaned with a 0.0098 broach.

CAUTION: Do not remove cap (14—Fig. JD2125) unless equipment is available for setting the nozzle cracking pressure. Be careful to keep each matching needle valve and seat together.

NOZZLE TIP. Clean the six tip orifices with a 0.0098 broach (John Deere part No. AF2291R) mounted in a pin vise, as shown in Fig. JD2127.

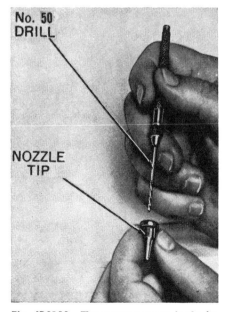

Fig. JD2128—The center passage in the injection nozzle tip can be cleaned with a No. 50 drill. A wire brush will facilitate removal of carbon deposits from outside of tip.

Fig. JD2129 — Needle valve and seat should be renewed if seating surfaces are worn or if the cylindrical surfaces are scratched or have a dull gray appearance.

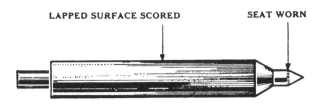

Clean center passage in tip with a No. 50 drill. See Fig. JD2128. Test the size of the tip orifices with a 0.012 piano wire. If wire will **not** go in the orifices, the tip is all right. If wire **does** go in, orifices are excessively worn and nozzle tip should be renewed.

NEEDLE VALVE AND SEAT. Inspect needle valve and seat with a magnifying glass. Renew valve and seat if seating surfaces are scored or if cylindrical surfaces are scratched or have a dull gray appearance. Both of these conditions are illustrated in Fig. JD2129. The needle valve and seat are available only as a matched unit. If valve and seat are in good condition, examine flat surface of valve seat where it contacts the nozzle holder. Discoloration and small pits can be removed by lapping with a coarse rouge until a smooth surface is obtained, then finishing with a polishing rouge—using the technique and equipment shown in Fig. JD2130.

Test fit of needle valve in its seat as follows: Hold seat in a vertical position and start needle valve in its seat. Valve should slide slowly into the seat under its own weight. Note: Dirt particles, too small to be seen by the naked eye, will restrict the valve action. If needle valve sticks, and it is known to be clean, free-up valve in seat, using a mixture of mutton tallow and clean diesel fuel and rotating the needle valve in its seat.

NOZZLE HOLDER BODY. Examine flat surface where it contacts the needle valve seat. Discoloration and small pits can be removed by lapping

with a coarse rouge until a smooth surface is obtained, then finishing with a polishing rouge.

REASSEMBLE. Before reassembling any parts, rinse same in clean fuel. Reassemble parts by reversing the disassembly procedure, using Fig. JD2125 as a general guide. Test nozzles as outlined in paragraph 111.

Reinstall nozzle in cylinder head by reversing the removal procedure, using a new gasket (1) between the nozzle and the head. If new copper gasket is not available, old gasket can be reclaimed by annealing, by heating same to a cherry red and quenching in cold water.

113. NOZZLE SLEEVE — RENEW. On 720 diesel tractors prior to serial number 7214900 the diesel engine cyl-

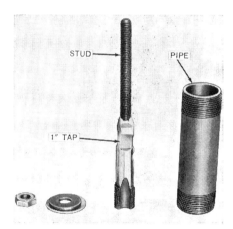

Fig. JD2130A — A tool for removing the early 720 diesel nozzle sleeves, similar to the one shown, can be made by welding a stud to a 1-inch tap. See Fig. JD2130B.

Fig. JD2130 — When lapping flat surfaces, use the "figure eight" motion as shown.

Fig. JD2130B — Removing an injection nozzle sleeve using the puller tool shown in Fig. JD2130A.

inder head was equipped with renewable type sleeves to receive the injection nozzles. To remove the nozzle sleeves, a tool similar to the one shown in Fig. JD2130A can be used. Remove the injection nozzles, then turn the tap into the sleeve before removing the sleeve retaining machine screw and washer. (This will keep the sleeve from turning in cylinder head.) After the tap is in place, remove the machine screw, install a length of pipe over the tap stud and using a washer and nut on the stud, pull the sleeve out of the head (Fig. JD2130B). Make

certain that all old material is removed from the head.

When installing the new sleeve, first make certain that opening is free from burrs and new sleeve "O" ring is greased and properly installed. Coat sleeve with light film of Permatex, then slip the sleeve into the cylinder head and while rotating the sleeve slightly, press the sleeve in place. When the sleeve is properly aligned, use a suitable piloted arbor and drive the sleeve in until the sleeve bottoms in the head, then install the sleeve retaining screw and washer.

GOVERNOR

114. **GOVERNOR LEVER ADJUSTMENT.** With the diesel engine stopped, remove the injection pump compartment cover and move the speed control lever to the fast idle position. At this time the injection pump racks should read thirteen. If they do not, remove the expansion plug from the governor lever housing and turn the adjusting screw (Fig. JD2131) until the racks are in the number thirteen position.

If the thirteen rack position cannot be obtained, remove the rack adjusting screw, loosen lock nut (N) and turn the eccentric screw (S) clockwise (viewed from top) until a rack reading of at least sixteen is obtained.

NOTE: If the racks will not open to sixteen, the governor housing should be removed and the flat end of the injection pump control rod ground off until the sixteen rack position can be obtained by turning the pivot pin.

Now, continue turning the pivot pin in a clockwise direction until a rack reading of fourteen is obtained and tighten the jam nut (N—Fig. JD2131). Reinstall the rack adjusting screw and

turn same in until the rack position is thirteen, then tighten the jam nut.

114A. **SPEED AND STOP ADJUSTMENTS.** Run engine until it reaches operating temperature. With the speed control lever in the fast idle position, turn the fast idle stop screw (W—Fig. JD2132) until the high idle no load speed is 1250 rpm when checked at belt pulley dust cover.

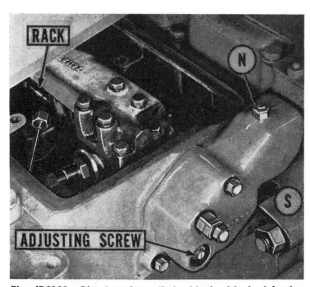

Fig. JD2131—Diesel engine cylinder block with the injection pump compartment cover removed.

Fig. JD2132 — Diesel engine speed and stop adjustments. Although a 70 diesel is shown series 720 and 730 diesel tractors are similar.

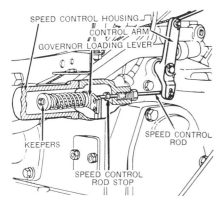

Fig. JD2132A—Cut-away drawing of speed adjusting parts on John Deere 720 diesel tractors prior to serial number 7201174.

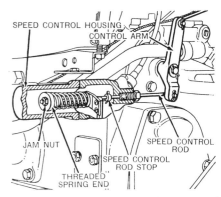

Fig. JD2132B—Cut-away drawing of speed adjusting parts on John Deere 720 diesel after serial number 7201173 and 730 diesel tractors.

Move the speed control lever to the slow idle position, at which time, the engine speed should be 700 rpm. If the speed is not as specified, loosen the cap screw (Y) and move the slow idle adjusting arm (Z) as required.

Pull out the speed control lever stop button on dash and pull the speed control lever all the way back. If the engine does not stop, loosen the diesel engine stop screw (X) and carefully pull the speed control lever back until the engine stops. Then, turn the stop screw in until it just contacts the stop arm of the control mechanism. If the engine fails to stop refer to paragraph 114B or 114C.

114B. On 720 diesel tractors prior to serial number 7201174, stop the engine and refer to Fig. JD2132A. Remove the speed control housing and the two screws attaching the governor loading lever to lever shaft. Disconnect the speed control rod from the control arm and slide the control rod and connected parts forward until the speed control rod stop is exposed. Loosen the Allen head set screw, slide the stop forward until the spring is compressed 0-1/32-inch and tighten the set screw. Reassemble and adjust the engine speeds and stop as in paragraph 114A.

114C. On 720 diesel after serial number 7201173 and 730 diesel tractors proceed as follows: With the engine running, move the speed control lever to the stop position, loosen the cap screw (Y—Fig. JD2132), move the adjusting arm (Z) until the engine stops and retighten the cap screw. Remove the speed control housing (Fig. JD2132B) and loosen the jam nut. Start engine and move control lever to the slow idle position; then, turn the threaded spring end as re-

quired to obtain a slow idle speed of 700 rpm. Tighten the jam nut and reinstall the speed control housing. Readjust the engine high idle speed as outlined in paragraph 114A.

OVERHAUL

115. Normal overhaul of the governor consists of removing and overhauling the shaft and weights assembly only; and can be accomplished without removing governor housing from tractor. If, however, the fan drive bevel gear is damaged, it will be necessary to remove the complete governor assembly as well as the fan shaft assembly in order to renew and adjust the matched set of bevel gears. Refer to paragraph 123 for adjustment of the bevel gears.

116. **SHAFT AND WEIGHTS.** To remove the governor shaft and weights assembly, first remove the flywheel cover, flywheel cover back plate and the inside governor lever access plate (36—Fig. JD2133). Working through the opening exposed by the removal of the access plate, remove the two screws retaining the inside governor lever (11) to the lever shaft (12). Remove the speed-hour meter bearing quill (59) and withdraw the drive shaft (56). Unbolt and remove the governor shaft bearing housing (1), being careful not to damage or lose the bevel gear adjusting shims (3). Withdraw the governor shaft and weights assembly from governor housing.

Thoroughly clean the removed unit and carefully examine the component parts for damage or excessive wear. Sleeve (9) must turn freely on the governor shaft without any binding tendency. Bevel gear (43) and drive gear (7) can be removed from governor shaft by using a suitable puller or press. When reassembling, press the bevel gear on the shaft until the shoulder on the gear seats against snap ring on shaft. Note: If this bevel gear requires renewal, the mating bevel gear on fan shaft must also be renewed (refer to paragraph 121B or 122) and the bevel gear mesh and backlash must be adjusted as outlined in paragraph 123. Bearing cone on right end of shaft must be pressed tightly against the bevel gear.

Drive gear (7) must be pressed on the shaft until gear seats against the snap ring; then, the bearing cone must be pressed tightly against the drive gear.

Install the governor shaft assembly by reversing the removal procedure, making certain that the spring (42) and the washer (41) are properly located in right side of governor housing as the shaft goes into place. Also, make certain that the slot in right end of governor shaft aligns with the fuel transfer pump drive shaft. Install the shaft bearing housing (1), using the same number of shims (3) as were removed and make certain that oil hole gasket (2) is in position.

When installing the speed-hour meter shaft and quill, be careful not to drop shaft into governor housing. Shaft can be guided into position by working through the access opening in front of housing. When installing the inside governor lever (11), make certain that lever does not ride on the governor sleeve.

After reassembly is complete adjust the governor lever as in paragraph 114 and the engine stop and speeds as in paragraph 114B or 114C.

REMOVE AND REINSTALL

118. To remove the complete governor housing on models with the gasoline starting engine, first remove the starting engine as in paragraph 30. On all models equiped with the 24-volt electric starting motor, remove the hood and flywheel cover. On series 730 tractors equipped with the 24-volt electric starting motor, disconnect one end of the center (long) steering shaft and fold same to one side out of the way. On all models, shut fuel off at tank, disconnect the fuel lines from sediment bowl and remove fuel tank including the tank brackets. Disconnect the speed-hour meter cable and the throttle linkage. Remove the governor lever housing

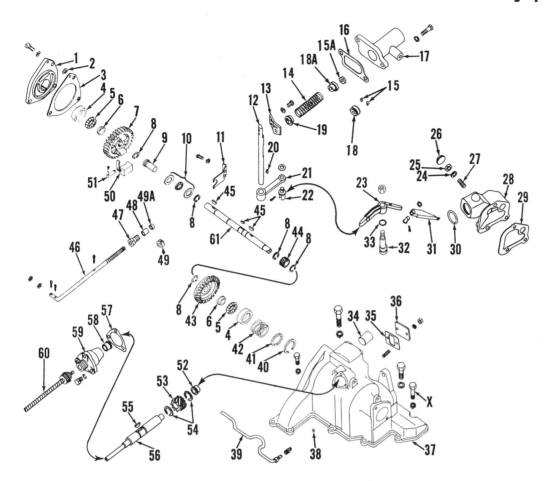

Fig. JD2133 — Exploded view of the diesel engine governor. The fan shaft is driven by bevel gear (43), the speed-hour meter is driven by spiral worm gear (44) and fuel transfer pump is driven by the slotted right end of governor shaft (61). On 720 diesel tractors prior to tractor serial number 7201175, unthreaded end (18) was used and the end of the speed control rod (46) was notched to receive the keepers (15).

1. Bearing housing
2. Gasket
3. Shims
4. Bearing cup
5. Balls and retainer
6. Bearing cone
7. Drive gear
8. Snap ring
9. Sleeve
10. Thrust bearing
11. Inside governor lever
12. Lever shaft
13. Loading lever
14. Speed control spring
15. Keepers
15A. Nut
16. Gasket
17. Speed control housing
18. Spring end
18A. Threaded spring end
19. Spring fulcrum
20. Woodruff key
21. Governor lever shaft arm
22. Pivot
23. Governor lever
24. Lead washer
25. Jam nut
26. Expansion plug
27. Adjusting set screw
28. Governor lever housing
29. Gasket
30. "O" ring
31. Governor lever stop
32. Lever pivot pin
33. "O" ring
34. Tube
35. Gasket
36. Access plate
37. Housing
38. Pipe plug
39. Oil pipe
40. Snap ring
41. Bearing washer
42. Spring
43. Bevel gear
44. Worm gear
45. Woodruff keys
46. Speed control rod
47. Packing gland
48. Packing
49. Rod stop
49A. Threaded stop
50. Weight
51. Weight pin
52. Bushing
53. Worm drive gear
54. Snap rings
55. Woodruff key
56. Speed-hour meter drive shaft
57. Gasket
58. Bushing
59. Quill
61. Governor shaft

Fig. JD2134—Installing the governor lever housing on cylinder block. Make certain that the governor lever is properly inserted in the slotted pivot in the governor case.

(28—Fig. JD2133) from left side of cylinder block and loosen the cap screws retaining the fan shaft front bearing housing (or power steering pump) to its support. Unbolt the fan shaft rear bearing housing from governor housing and governor housing from main case. Lift governor assembly from tractor being careful not to damage or lose the bevel gear adjusting shims which are located between the fan shaft rear bearing housing and the governor housing.

Install the governor, reversing the removal procedure and be sure to install the same number of shims between the fan shaft rear bearing housing and governor housing as were originally removed. The special shoulder type cap screws go in the right front and left rear holes of governor case. When installing the governor lever housing, make certain that governor lever (23) is properly inserted in the slotted pivot (22) in governor case. Refer to Fig. JD2134. To check this, hold the governor lever housing in position, push the speed control lever back and forth and watch or feel to see that governor lever moves sideways in both directions. If the governor lever is not properly inserted in the slotted pivot, there will be no positive connection between the injection pump racks and speed control lever or governor. Bleed the fuel system as outlined in paragraph 102 and adjust the governor lever as in paragraph 114 and the engine stop and speeds as in paragraph 114B or 114C.

COOLING SYSTEM

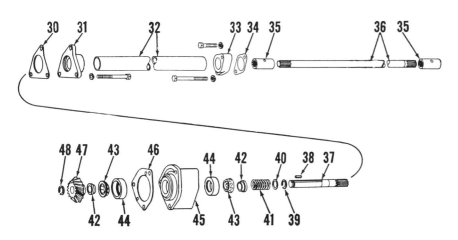

Fig. JD2138 — Exploded view of the fan shaft drive used on tractors equipped with power steering.

30. Gasket
31. Bearing housing cover
32. Fan drive tube
33. Flange
34. Gasket
35. Couplings
36. Coupling shaft
37. Drive shaft
38. Key
39. Snap ring
40. Washer
41. Spring
42. Bearing cones
43. Bearing balls
44. Bearing cups
45. Bearing housing
46. Shims
47. Bevel gear
48. Snap ring

RADIATOR

120. REMOVE AND REINSTALL. To remove the radiator, first drain cooling system and remove hood, air cleaner and upper water pipe. On models with power steering, drain the power steering reservoir and remove the power steering pump pressure and return lines. On all models, loosen the fan belt and unbolt water pump from the radiator lower tank. Unbolt and remove the steering worm housing from pedestal. Remove the radiator retaining screws and using a hoist, lift radiator from tractor.

FAN SHAFT AND DRIVE

Models With Power Steering

121. FAN AND FAN SHAFT. On tractors equipped with power steering the fan shaft is also the power steering pump drive shaft. To remove the fan shaft and overhaul the drive parts, refer to paragraphs 20 and 21.

121B. FAN DRIVE. An exploded view of the fan drive is shown in Fig. JD2138. With the fan shaft and power steering pump assembly removed as outlined in paragraph 20, unbolt the bearing housing cover (31) from bearing housing and withdraw the coupling shaft (36) and tube. Remove the bearing housing and drive shaft assembly from governor housing being careful not to damage or lose the bevel gear adjusting shims (46).

To disassemble the drive unit, press bevel gear (47) farther onto the drive shaft and remove snap ring (48); then, press drive shaft out of bevel gear. When reassembling, press bevel gear on the drive shaft far enough to install snap ring (48); then, press shaft in the opposite direction until the gear is in firm contact with the snap ring.

Note: If the bevel gear (47) is renewed, the mating bevel gear on governor shaft must also be renewed (refer to paragraph 116) and the bevel gear mesh and backlash must be adjusted as outlined in paragraph 123.

When installing the drive unit, reverse the removal procedure and be sure to install the same number of shims (46) as were removed.

Models With Manual Steering

122. FAN, FAN DRIVE AND FAN DRIVE SHAFT. To remove the fan shaft and governor assembly on tractors equipped with the gasoline starting engine, first remove starting engine as outlined in paragraph 30. On all tractors equipped with the 24-volt electric starting motor, remove the hood and flywheel cover. On series 730 tractors equipped with the 24-volt electric starting motor, disconnect one end of the center (long) steering shaft and fold same to one side out of the way. On all models, shut off the fuel, disconnect the fuel lines from fuel tank and transfer pump and remove fuel tank, including tank bracket. Drain the cooling system and remove the upper water pipe and the cylinder head water outlet casting. Remove the exhaust pipe, air cleaner body and generator. Disconnect the speed-hour meter cable and throttle linkage. Remove the governor lever housing from left side of cylinder block and unbolt the fan shaft front bearing housing from its support. Remove the cap screws retaining governor housing to main case and lift governor and fan shaft assembly from tractor.

Separate the fan shaft assembly from governor case, being careful not to damage or lose the bevel gear adjusting shims (46—Fig. JD2139).

If the bevel gear (47) or rear bearing are to be renewed, press the bevel gear farther onto the drive shaft and remove snap ring (48); then, press drive shaft out of bevel gear and remove the bearing. When reassembling, press the bevel gear on the drive shaft far enough to install snap ring (48); then, press the fan shaft in opposite direction until the gear is in firm contact with the snap ring.

Note: If the bevel gear (47) is renewed the mating bevel gear on governor shaft must also be renewed (refer to paragraph 116) and the bevel gear mesh and backlash must be adjusted as outlined in paragraph 123.

To disassemble the removed fan shaft, place the assembly in a press and remove the half-moon locks (9) and keeper (8) as shown in Fig. JD-2140. Remove the assembly from press, disassemble and check the parts against the values which follow:

Fan blade pitch (Inches)......$2\frac{17}{32}$-$2\frac{19}{32}$
Thickness of friction washers..$\frac{1}{16}$ inch
Friction spring
strength...216-264 lbs. @ $1\frac{1}{2}$ inches
Bearing take-up spring
strength...171-209 lbs. @ $1\frac{7}{16}$ inches

Use Fig. JD2139 as a general guide during reassembly and be sure to renew all "O" ring packings and felt washers.

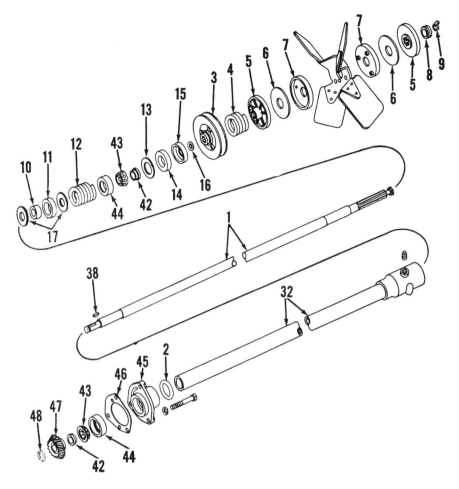

Fig. JD2139—Exploded view of the fan shaft, fan and associated parts used on tractors equipped with manual steering.

1. Fan shaft	10. Packing	38. Key
2. "O" ring	11. Retainer	42. Bearing cones
3. Pulley	12. Spring	43. Bearing balls
4. Spring	13. Washer	44. Bearing cups
5. Friction disc	14. Felt washer	45. Rear bearing
6. Friction washer	15. Felt retainer	housing
7. Fan drive cup	16. "O" ring	46. Shims
8. Fan keeper	47. Washers	47. Bevel gear
9. Locks	32. Tube and front	48. Snap ring
	bearing housing	

FAN DRIVE BEVEL GEARS

The fan drive bevel gears (43—Fig. JD2133 and 47—Fig. JD2138 or Fig. JD2139) are available in matched pairs only. Therefore, if either gear is damaged, it will be necessary to renew both gears and adjust them for the proper mesh and backlash.

123. To renew the bevel gears on models with manual steering, first remove the governor and fan shaft assembly from tractor and separate the fan shaft assembly from governor. On models with power steering, remove the power steering pump, separate the fan drive assembly from governor and remove governor. On all models, press bevel gear (47—Fig. JD2138 or 2139) farther onto the drive shaft and remove snap ring (48); then, press shaft out of bevel gear. When reassembling, press the bevel gear on the shaft far enough to install snap ring (48); then, press the fan shaft in opposite direction until gear is in firm contact with snap ring. Remove the governor shaft assembly from governor housing (refer to paragraph 116) and using a suitable puller, remove the bevel pinion from governor shaft. Press new bevel pinion on shaft until pinion seats against snap ring; then, press the bearing cone against the bevel pinion (43—Fig. JD2133). Reinstall governor shaft in housing.

Before installing fan shaft or governor assembly on tractor, temporarily bolt fan shaft to governor case, observe mesh position of bevel gears and add or remove shim gaskets (46—Fig. JD2138 or 2139) until heels of gears are in register. Mount a dial indicator in a suitable manner and check the bevel gear backlash. Add or remove shim gaskets (3—Fig. JD2133) until backlash is 0.004-0.006.

After proper backlash is obtained, recheck the mesh position.

THERMOSTAT

124. The thermostat is located in the cylinder head water outlet casting. To remove the thermostat, drain the cooling system, remove the upper water pipe and the thermostat cover.

WATER PUMP

125. **R&R AND OVERHAUL.** To remove the water pump, drain cooling system, disconnect fan belt from pump and remove the lower water pipe. Disconnect the by-pass line. Unbolt water pump from radiator bottom tank and withdraw from tractor.

Fig. JD2140 — When disassembling the fan shaft, compress the assembly enough to remove the half moon locks (9) and keeper (8). After releasing the pressure, the remaining parts can be removed.

Use a suitable puller and remove the drive pulley. Remove retainer ring (90—Fig. JD2141) and press shaft and bearing assembly out of pump

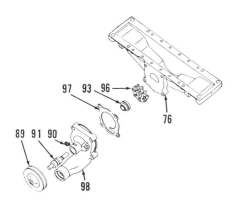

Fig. JD2141—Exploded view of the diesel engine water pump and radiator bottom tank.

76. Radiator bottom tank	93. Seal
89. Pulley	96. Impeller
90. Retainer ring	97. Gasket
91. Shaft and bearing	98. Pump housing

housing. Seal (93) can be renewed at this time. The shaft and bearing (91) are available as an assembled unit only.

When reassembling, coat outer surface of seal with a thin coat of shellac and press seal into housing. Install the shaft and bearing unit in pump housing and install retainer ring (90) so that end of ring locks over lug on the housing. Support pump shaft from underneath side and press pulley on shaft until pulley is flush with end of shaft. Place impeller on flat surface and press body and shaft assembly onto impeller until highest vane on impeller is flush with mounting surface of pump body.

NOTE: If any vane protrudes beyond the mounting face of pump housing, the protruding vane will strike the radiator bottom tank when pump is installed.

Install pump on tractor by reversing the removal procedure.

STARTING MOTOR

6-Volt System

126. On 720 and 730 diesel tractors equipped with the gasoline starting engine and a 6-volt electrical system, a Delco-Remy No. 1107155, 6-volt electric starting motor is mounted on the starting engine clutch housing and is accessible after removing the tractor hood. Test specifications are as follows:

Brush spring tension (min.)..24 oz.
No load test:
    Volts 5.65
    Amperes 70
    RPM 5500
Lock test:
    Volts 3.25
    Amperes 550
    Torque (Ft.-Lbs.) 11

24-Volt System

126A. On 720 and 730 diesel tractors equipped with the 24-volt electrical system a Delco-Remy No. 1113801, 24-volt electric starting motor is mounted on a bracket located on the top of the tractor main case and is accessible after removing the tractor hood. The Delco-Remy 1119803 solenoid switch is mounted on the starting motor. If the starting motor solenoid switch is removed, or if for any other reason the starting motor pinion does not correctly engage the flywheel ring gear, adjust the solenoid and pinion as in the following paragraph.

The starting motor solenoid and pinion can be adjusted with the starter and solenoid unit installed on the tractor; however, it is recommended that the unit be removed from the tractor,

ELECTRICAL SYSTEM

John Deere 720 diesel tractors after serial number 7222599 and all 730 diesel tractors are available with either a gasoline starting engine or a 24-volt electric starting motor. Series 720 tractors prior to serial number 7222600 and later tractors equipped with the gasoline starting engine have an elec-

trical system composed of 6-volt equipment. Series 720 and 730 tractors equipped with the electric starter have a 24-volt split-load system using 12-volts for the lights and accessories and 24-volts for cranking and charging. A wiring diagram of the 24-volt electrical system is shown in Fig. JD2142.

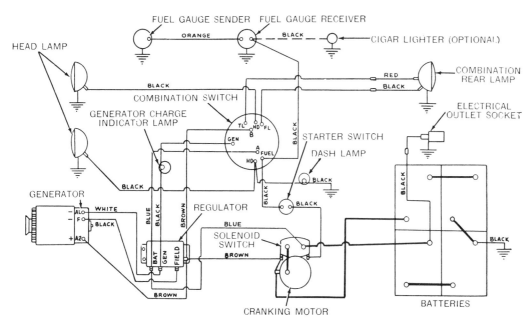

Fig. JD2142 — Wiring diagram of series 720 and 730 diesel tractors equipped with a 24-volt electrical system.

and mounted in a vise or holding fixture. Connect one wire from a 6-volt battery to one of the solenoid control winding terminals, engage the solenoid manually and hold the solenoid in the engaged position by connecting another wire from the battery to the remaining solenoid control winding terminal as shown in Fig. JD2143. Hold the pinion (Fig. JD2143A) toward the starting motor to remove any play in the linkage and measure the clearance between the forward edge of the pinion teeth and the inside edge of the motor drive housing. If the clearance is not 21/64-25/64-inch, disconnect the plunger stud from the starting motor shift lever and turn the stud in or out as required until the recommended clearance is obtained.

To renew the solenoid contact disc (Fig. JD2143B), first remove the solenoid assembly from the starting motor. Remove the compression spring retaining pin, retainer and compression spring from one end of the solenoid and the terminal plate from the other end. Remove the cotter pin, the castellated nut and the contact disc. Install the new contact disc with the heavy rolled edge of the center bushing toward the plunger end. Install the castellated nut and turn same on the plunger until the distance from the contact disc to the edge of the housing is $1\frac{1}{32}$-inches (Fig. JD2143B). Note: During this measurement it is important that the compression spring be removed. Reassembly of the solenoid is reverse of the disassembly procedure. Check and adjust the starting motor pinion setting as in the preceeding paragraph.

Specification data for both the starting motor and the solenoid are as follows:

Starting Motor 1113801
    Brush spring tension........35 oz.

No load test:
    Volts 23.0
    Amperes 100
    RPM 8000
Lock test:
    Volts 3.0
    Amperes 500
    Torque (Ft.-Lbs.) 28
Solenoid 1119803
At 80° F., consumption should be:
Both windings
    Amperes 41.6-45.8
    Volts 24
Hold-in windings
    Amperes 7.3-8.3
    Volts 24

GENERATOR AND REGULATOR

6-Volt System

126B. A Delco-Remy No. 1100027, 6-volt generator is used on tractors equipped with a gasoline starting engine. Generator output is controlled by a Delco-Remy No. 1118786, two unit regulator. Specification data for both units are as follows:

Generator 1100027
    Brush spring tension........28 oz.
    Field draw
        Volts 6.0
        Amperes 1.85-2.03
    Output (cold)
        Volts 8.0
        Amperes 35.0
        RPM 2650
Voltage Regulator 1118786
    Ground polarity ....Positive
    Cut-out relay:
        Air gap 0.020
        Point gap 0.020
        Closing voltage range 5.9-6.7
            Adjust to 6.3
    Voltage regulator:
        Air gap 0.075
        Setting voltage range.....6.8-7.4
            Adjust to 7.1

24-Volt System

126C. A Delco-Remy No. 1103021 24-volt generator is used on tractors equipped with the 24-volt electrical system. Generator output is controlled by a Delco-Remy No. 1119219, three unit voltage regulator. Specification data for both units are as follows:

Generator 1103021
    Brush spring tension ....28 oz.
    Field draw
        Volts 24.0
        Amperes 75-.85
    Output (cold)
        Volts 28.5
        Amperes 10.0
        RPM 2500
Voltage regulator 1119219
    Ground polarity ....Negative
    Cut-out relay:
        Air gap, inches 0.017
        Point gap, inches.........0.032
        Closing voltage range.....24-27
    Voltage regulator:
        Air gap, inches 0.075
        Voltage setting range....27.5-29.5
    Current regulator:
        Air gap, inches 0.075
        Regulator setting range...8.5-11.5

Fig. JD2143A—With the solenoid engaged as in Fig. JD2143, dimension between front face of teeth and inside edge of motor housing should be 21/64-25/64-inch.

Fig. JD2143 — Jumper wires from a 6-volt battery to the two small posts on the solenoid will hold the solenoid in the engaged position for checking the pinion setting. Refer to text.

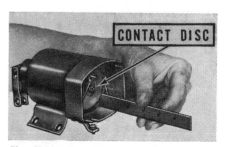

Fig. JD2143B — The solenoid contact disc should be 1 1/32 inches from the edge of housing when measured as described in text.

CLUTCH, BELT PULLEY AND PULLEY BRAKE

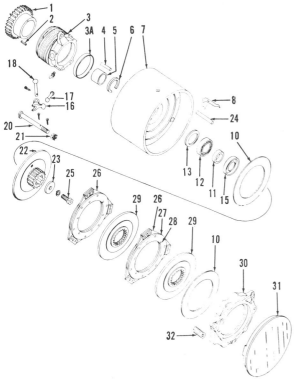

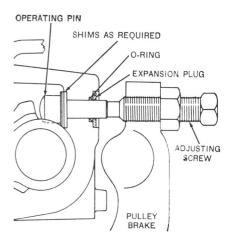

Fig. JD2144 — Exploded view of belt pulley and clutch assembly. Clutch is adjusted with nuts (21).

1. Drive gear
2. Key
3. Operating sleeve
3A. Grease retainer
4. Drive pin
5. Bushing
6. Snap ring
7. Pulley
8. Spring
10. Clutch facing
11. Bearing inner race
12. Bearing
13. Bearing washer
15. Bearing retainer
16. Clutch dog
17. Dog toggle
18. Dog pin
20. Operating·bolt
21. Adjusting nut
22. Clutch drive disc
23. Washer
26. Facing disc
27. Facing
28. Rivet
29. Sliding drive disc
30. Adjusting disc
31. Pulley dust cover
32. Spring

Fig. JD2147 — Top sectional view of very early production series 720 diesel pulley brake operating pin and related parts. Notice the position of the adjusting shims.

CLUTCH ADJUSTMENT

127. To adjust the clutch, remove the belt pulley dust cover and the cotter pin from each of the three clutch operating bolts. Place the clutch operating lever in the engaged position (lever fully forward) and tighten each adjusting nut (21—Fig. JD2145) a little at a time and to the same tension. Check tightness of clutch after each adjustment by disengaging and re-engaging clutch. When the adjustment is correct, a distinct snap will occur

when the clutch is engaged and 40-80 lbs. will be required at the end of the operating lever to lock the clutch in the engaged position with engine running at idle speed.

PULLEY BRAKE ADJUSTMENT
Series 720 Prior Ser. No. 7214900

127A. Loosen jam nut (10—Fig. JD2146) and back out the adjusting

screw (9) until the screw does not contact the pulley brake operating pin when clutch lever is pulled back. Then, with the clutch lever in the released position, the belt pulley should rotate freely without any bind.

If binding exists, remove the belt pulley as outlined in paragraph 129, clutch fork shaft (6—Fig. JD2158) and clutch fork (31). Withdraw the headed pulley brake operating pin (21), and add one adjusting washer (20) and recheck. When properly adjusted and with the clutch lever in the disengaged position, the pulley should rotate freely and have an end play of 1/16-1/8-inch on the crankshaft. Refer also to Fig. JD2147.

To adjust the pulley brake, disengage the tractor clutch and pull forward on the brake shoe to insure that the operating pin is contacting the clutch fork. Then while holding the shoe firmly against the pulley, turn the adjusting screw (9—Fig. JD2146) in until it just contacts the operating pin. Turn the screw in ¼-turn more and tighten the jam nut (10).

Series 720 Ser. No. 7214900-7218699

127B. Loosen the jam nut (10—Fig. JD2146) and back-out the adjusting screw (9) until the screw does not contact the pulley brake operating pin when clutch lever is pulled back. Then, with the clutch lever in the released position, the belt pulley should rotate freely without any bind.

Fig. JD2145—Clutch is adjusted by tightening nuts (21) evenly.

Fig. JD2146—Pulley brake adjusting screw and jam nut used on very early production 720 diesel tractors. Refer also to Fig. JD2147.

Fig. JD2148—On some 720 diesel tractors the pulley brake operating pin adjusting shims are externally located as shown.

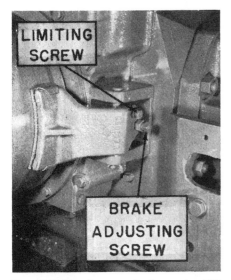

Fig. JD2149 — Late production series 720 and series 730 diesel pulley brake adjusting screws.

Fig. JD2150—Using a screw driver to hold the belt pulley outward.

If binding exists, remove the pulley brake assembly and shim retainer (14 —Fig. JD2158) and remove one adjusting shim (15) and recheck. When properly adjusted and with the clutch lever in the disengaged position, the pulley should rotate freely and have an end play of 1/16-1/8-inch on the crankshaft. Refer also to Fig. JD2148.

To adjust the pulley brake, disengage the tractor clutch and pull forward on the brake shoe to insure that the operating pin is contacting the clutch fork. Then while holding the shoe firmly against the pulley, turn the adjusting screw (9—Fig. JD2146) in until it just contacts the operating pin. Turn the screw in ¼-turn more and tighten the jam nut (10).

Series 720 After Ser. No. 7218699-Series 730

127C. These late production tractors can be identified in that they have two adjusting screws in the pulley brake as shown in Fig. JD2149. This late construction differs from both early constructions as follows:

The fork stop has been removed from the pulley brake operating pin and adjustable stop screws are located between clutch collar and housing. Pivot pin between the clutch collar and the clutch fork is located in a slot and spring loaded. The clutch lever now moves into a neutral position when the clutch is disengaged and further rearward movement of the clutch lever compresses the springs and moves the pivot pin in the slot to engage the pulley brake. A limit screw has been added to the brake and serves as a stop to prevent the pulley brake spring from holding the clutch in partial engagement when the lever is in neutral.

The adjustable stop screws are located in the clutch collar and the stop pads are machined on the main bearing housing. Access to the screws is obtained through rectangular openings in top and bottom of reduction gear cover. Proceed to the following paragraph.

127D. Before making the pulley brake adjustment, make certain that the crankshaft end play is properly adjusted as in paragraph 92A and the clutch is properly adjusted as in paragraph 127.

With the clutch disengaged, use a large screw driver or pry bar as shown in Fig. JD2150 and hold the pulley out

Fig. JD2151 — Checking the clutch fork stop screw adjustment on late models. The belt pulley is removed for illustrative purposes only.

against the fixed clutch drive disc. Working through the previously mentioned access openings, adjust the clearance between the stop screws and the machined stop bosses or pads to 0.060. The procedure for checking the adjustment is shown in Fig. JD2151 where the pulley is removed for illustrative purposes only. Reinstall the access opening covers.

With the clutch lever in neutral position, turn the brake adjusting screw (Fig. JD2149) inward until the pulley brake just touches the pulley and a slight drag is felt when pulley is rotated. Tighten the jam nut. Engage the clutch and adjust the limiting screw until there is 1/32-inch clearance between the pulley and the pulley brake facing. Tighten the jam nut.

CLUTCH

128. **R&R AND OVERHAUL.** To remove the clutch discs and facings, remove the pulley dust cover and adjusting disc (30—Fig. JD2144). Withdraw the lined and unlined discs. Remove cap screw (25) retaining the clutch drive disc (22) to crankshaft and using jack screws as shown in Fig. JD2157, remove the clutch drive disc and inner facing.

Worn, badly glazed or oil soaked facings should be renewed. A facing that is in usable condition, is quite rigid. Any facing that bends easily should be renewed. Renew release springs (32—Fig. JD2144) if they are rusted, distorted or do not exert 45-55 lbs. pressure when compressed to a length of 1⅝-inches.

Use Fig. JD2144 as a guide during reassembly and install the clutch drive disc so that "V" mark on drive disc is in register with flat spot at end of one of the crankshaft splines. Adjust the clutch as outlined in paragraph 127.

Fig. JD2157—Using ½-inch jack screws to remove the clutch drive disc.

Fig. JD2158 — Series 720 diesel (prior to 7218700) clutch and belt pulley operating parts exploded from the first reduction gear cover. Parts (7, 8, 20 and 21) are used on tractors prior to 7214900. Parts (14, 15, 16 and 17) are used on tractors 7214899 through 7218699.

1. Nut, special
2. Nut
3. Bushing
4. Operating rod
5. Yoke
6. Clutch fork shaft
7. "O" ring
8. Expansion plug
9. Operating pin adjusting screw
10. Jam nut
11. Pulley brake
12. Brake lining
13. Spring
14. Shim retainer
15. Steel shims
16. Operating pin
17. "O" ring
20. Shim washers
21. Operating pin
22. Fork spring
23. Clutch collar
25. Gasket
26. Dust cover
27. Cover
28. Pulley guard
29. First reduction gear cover
30. Plug
31. Clutch fork
31W. Spring
31X. Reinforcing spring
31Y. Plate
31Z. Rivet
32. Fork pivot
33. Lock plate
34. Limit screw
34Z. Jam nut
35. Pivot bolt
92. Limit screw

Fig. JD2159—Clutch and pulley brake operating parts used on 720 diesel after serial number 7218699 and all 730 diesel tractors. Refer to legend under Fig. JD2158.

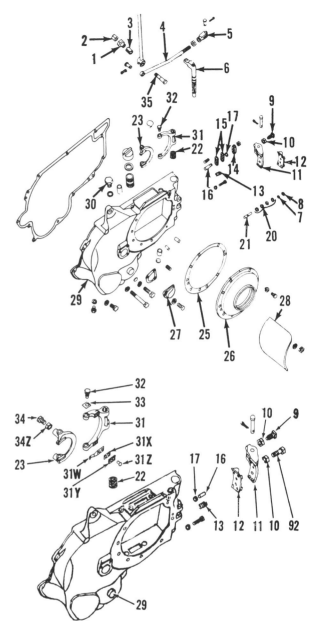

BELT PULLEY

129. **REMOVE AND REINSTALL.** To remove the belt pulley, first remove the clutch facings as in paragraph 128 and disconnect the clutch operating rod from the clutch fork shaft. Remove pivot pin and withdraw pulley brake (11 — Fig. JD2158 or 2159). Unbolt cover (26—Fig. JD2158) from reduction gear case and withdraw pulley assembly from tractor. Reinstallation is reverse of the removal procedure.

130. **OVERHAUL.** To disassemble the removed pulley, use a punch and hammer as shown in Fig. JD2160 and remove the bearing retainer and bearing (12—Fig. JD2144). Remove the operating bolts (20), pins (18), dogs (16) and toggles (17). Using a suitable puller, remove drive gear (1) and slide operating sleeve from pulley.

Check pulley bushing (5) and the engine crankshaft against the values which follow:

Crankshaft diameter at pulley
bushing 2.432-2.434
Pulley bushing inside
diameter (Min.) 2.4375
Suggested clearance between pulley
bushing & crankshaft . . 0.0035-0.0055

If clearance between pulley bushing (5) and crankshaft is excessive, renew the bushing. Using a piloted drift, press bushing into pulley until bushing seats against snap ring (6). If bushing is carefully installed, no final sizing will be required.

When reassembling, heat gear (1) to approximately 300 deg. F., install gear with long hub toward pulley and lubricate toggles (17) as follows:

Any one of the following special lubricants are recommended by John Deere for lubricating the clutch toggles:

Calumet Viscous Lubricant, 10X, manufactured by Standard Oil Company of Indiana.

No. 102 Cosmulube, manufactured By E. F. Houghton and Co., 303 West Lehigh Ave., Philadelphia 33, Penn.

On early production 720 diesel tractors pack the recess in end of each clutch toggle with lubricant and install lubricated end of toggle into sockets in the clutch operating sleeve. Lubricate opposite end of each toggle with the same special lubricant.

On late production 720 diesel tractors and all series 730 diesel tractors, lubricate end of toggle which goes into the clutch dog cup with one of the special lubricants. After pulley is assembled, place the operating sleeve in the engaged position (away from pulley gear) and force the special grease into the grease fitting until grease appears around toggle ends. This procedure fills the grease reservoir in the operating sleeve and provides toggle lubrication via small holes to the toggle sockets.

CLUTCH CONTROLS

131. **R&R AND OVERHAUL.** The clutch fork (31—Fig. JD2158 or 2159) and the clutch collar (23) can be renewed after removing the belt pulley as outlined in paragraph 129. The procedure for subsequent disassembly is evident.

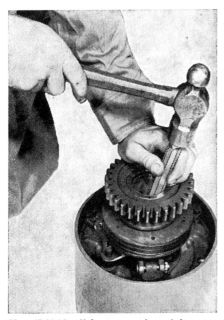

Fig. JD2160—Using a punch and hammer to remove the pulley bearing.

Fig. JD2161—Series 720 diesel prior to serial number 7218700. Clutch fork (31) and collar (23) installation as viewed after pulley has been removed.

TRANSMISSION

OVERHAUL

136. **FIRST REDUCTION GEAR COVER.** To remove the first reduction gear cover, disconnect the clutch operating rod and drain oil from cover. Remove belt pulley as outlined in paragraph 129. Move the right wheel out, unbolt and remove the right brake assembly. Remove the transmission drive shaft right bearing cover, extract cotter pin and remove nut from end of transmission drive shaft. Remove cap screws retaining the reduction gear cover to main case and bump end of drive shaft with a soft hammer to loosen the reduction gear cover. Pry cover from its locating dowels and withdraw the first reduction gear cover from tractor.

When reassembling, soak new reduction gear cover gasket until gasket is pliable, shellac gasket to main case and install reduction gear cover by reversing the removal procedure. Make certain that oiler gear (47—Fig. JD2183) meshes properly with the first reduction gear before tightening the cover cap screws. Pour about 1½-quarts of transmission oil into reduction gear cover.

137. **TRANSMISSION TOP COVER.** To remove the transmission top cover and shifter quadrant assembly on tractors equipped with the gasoline starting engine, disconnect the wires from the electric starting motor and remove the motor. Disconnect the

Fig. JD2183—Inside view of the reduction gear cover, showing the installation of the oil slinger gear.

44. Transmission drive shaft right bearing	46. Oil slinger gear pin
	47. Oil slinger gear

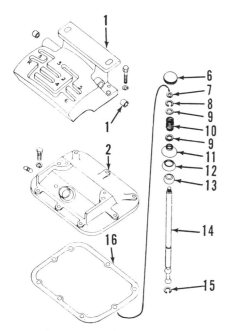

Fig. JD2184—Transmission cover and shifter quadrant assembly.

1. Quadrant	11. Fulcrum ball
2. Transmission cover	socket cover
7. Lock washer	12. Ball socket seal
8. Snap ring	13. Fulcrum ball
9. Washer	14. Shift lever
10. Spring	15. Snap ring
	16. Gasket

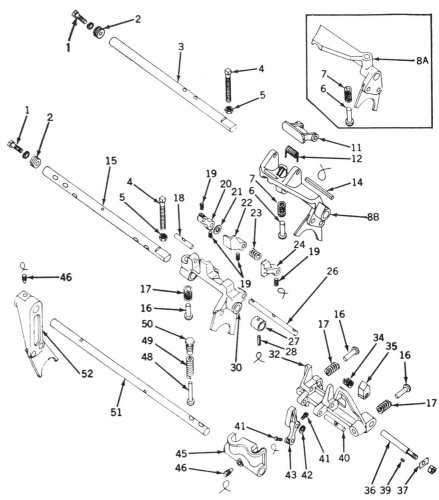

Fig. JD2185 — Exploded view of the transmission shifter shafts and shifters. Overdrive shifter (8A) is used on 720 diesel tractors prior to serial number 7202781. On 720 diesel after 7202780 and all 730 diesel tractors, parts (8B, 11, 12 and 14) are used in place of shifter (8A).

1. Cap screw	20. Fourth and sixth	36. Second and fifth
2. Adjusting screws	speed shifter link	speed shifter arm
3. Overdrive shifter	21. Washer	shaft
shaft	22. Sixth speed shifter	37. Lock plate
4. Set screw	arm	39. Woodruff key
5. Jam nut	23. Spring	40. Shifter pawl lock
6. Shifter pawl	24. Fourth speed	41. Set screw
7. Shifter pawl spring	shifter arm	42. Washer
8A. Overdrive shifter	26. Fourth and sixth	43. Fifth speed shifter
8B. Overdrive shifter	speed shifter arm	arm
11. Overdrive shifter	**shaft**	45. First and third
lock	27. Fourth and sixth	speed **shifter**
12. Spring	speed shifter spacer	46. Set screw
14. Roll pin	28. Roll pin	48. Shifter pawl
15. Shifter shaft	30. Fourth and sixth	49. Spring
16. Shifter pawl	speed shifter	50. Pawl retainer
17. Shifter pawl	32. Second, fifth and	51. First and third
spring	reverse speed shifter	speed shifter shaft
18. Shifter pawl lock	34. Spring	52. First and third
19. Set screws	35. Second speed	speed shifter arm
	shifter arm	

starting engine clutch linkage and the decompression control rod. On tractors equipped with the 24-volt electric starting motor, disconnect the decompression control rod. On all models, remove the grille and loosen the cap screws retaining front of hood to the radiator. Remove the two cap screws retaining the steering shaft and hood support to the gear shift quadrant. Remove the seat and disconnect the battery cable and the wires which go through the gear shifter quadrant from instrument panel. Raise and block-up the steering shaft support and rear of hood approximately two inches. Remove the cap screws retaining the shifter quadrant and transmission cover assembly to main case and with the transmission shifted to the neutral position, withdraw the assembly from tractor.

When installing the cover, make certain that end of gearshift lever enters the shifter gates at left side of shifters.

138. SHIFTER SHAFTS AND SHIFTERS. To remove the transmission shifter shafts and shifters, first remove the engine flywheel as outlined in paragraph 92 or 92A and transmission top cover as in the preceding paragraph. Move the left wheel out and remove the timing gear cover as outlined in paragraph 77 or 77A.

On 720 models prior to Ser. No. 7202781 slide overdrive shifter (8A—Fig. JD2185) along shaft (3) until the detent pawl rises and hold the pawl in the raised position by inserting a cotter pin in pawl hole. On 720 models after Ser. No. 7202780 and all 730 models, raise shifter lock (11), slide overdrive shifter (8B), along shaft (3) until the detent pawl rises and hold the pawl in the raised position by inserting a cotter pin in the pawl hole. On all models, loosen both jam nuts

(5) and back-off both set screws (4). Remove cap screw (1) and adjusting screw (2) from left end of shaft (3), then turn shaft (3) until detents are out of alignment, withdraw shaft (3) from left side of main case and remove the overdrive shifter (8A or 8B) from above.

Release the pawl locks, slide the fourth and sixth speed shifter (30) and the second, fifth and reverse shifter (32) along shaft (15) until the detent pawls rise and hold the pawls in the raised position by inserting a cotter pin in the pawl holes. Remove cap screw (1) and adjusting screw (2) from left end of shaft (15). Unwire and remove roll pin (28) which positions spacer (27) on the shaft. Turn shaft (15) until detents are out of alignment, withdraw shaft (15) from left side of main case and remove shifter assemblies from above.

Remove the threaded plug (50), then unwire and remove set screw (46) from shifter (45). Withdraw shifter fork (52) and shaft (51) from left and remove shifter (45) from above.

New shifter yokes can be riveted to shifters if old yokes are worn or bent. Renew any shifter shaft that is worn around the detents. Renew any pawl that is worn out-of-round at ball end.

When reinstalling the shifter shafts and shifters, refer to Fig. JD2185 as a guide and reverse the removal procedure. Position all shifters in neutral and turn adjusting screw (2) at left end of center shaft until the gates in the shifters align with the gate in the first and third speed shifter as shown in Fig. JD2185A. But be sure that the center shifter shaft is held to the left against the adjusting screw when making the adjustment. Tighten set screw (4) and jam nut (5), at right end of shaft, securely; then, install and tighten cap screw (1—Fig. JD2185) in left end of shaft. Turn adjusting screw (2) at left end of front shaft until the left edge of the overdrive shifter aligns with left edge of lug on second and fifth speed shifter as shown in Fig. JD2185A. But be sure that the front shifter shaft is held to the left against the adjusting screw when making the adjustment. Tighten set screw (4) and jam nut (5), at right end of shaft securely; then, install and tighten cap screw (1—Fig. JD2185) in left end of shaft.

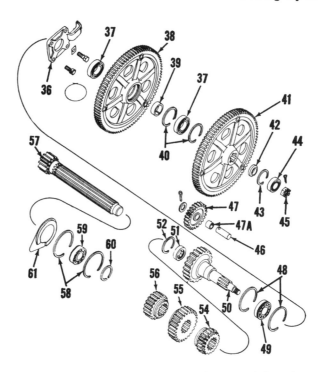

Fig. JD2186—Exploded view of the transmission sliding gear shaft and associated parts. Oil slinger gear (47) is mounted in the first reduction gear cover as shown in Fig. JD2183.

36. Bearing cover	46. Oil slinger gear pin	55. Fourth and sixth sliding
37. Bearing	47. Oil slinger gear	pinion
38. Powershaft idler	47A. Bushing	56. Second, fifth and re-
39. Spacer	48. Snap rings	verse sliding pinion
40. Snap rings	49. Bearing	57. Sliding gear shaft
41. First reduction gear	50. Transmission drive shaft	58. Snap rings
42. Spacer	51. Pilot bearing	59. Bearing
43. Snap ring	52. Snap ring	60. Snap ring
44. Bearing	54. Sliding gear shaft drive	61. Oil retainer
45. Nut	gear	

139. SLIDING GEAR SHAFT. To remove the sliding gear shaft (57—Fig. JD2186) and gears, remove the transmission case top cover as outlined in paragraph 137 and the shifter shafts and shifters as in the preceding paragraph 138. Remove the sheet metal oil retainer (61) and extract outer snap ring (58) from main case. Withdraw the sliding gear shaft from left side of main case and remove gears from above. Pilot bearing (51)

can be removed from the transmission drive shaft (50) after removing snap ring (52).

Using Fig. JD2186 as a guide, reinstall the sliding gear shaft by reversing the removal procedure. Install outer snap ring (58) with gap in snap ring spanning the oil passage in main case. Install the sheet metal oil retainer (61) with flat spot adjacent to oil passage in main case.

140. TRANSMISSION DRIVE SHAFT. To remove the transmission drive shaft (50—Fig. JD2186), remove the clutch, belt pulley and the first reduction gear cover as outlined in paragraph 136. Remove the sliding gear shaft as in the preceding paragraph 139. Withdraw the first reduction gear (41) and idler gear (38). Remove the countershaft right bearing housing (71—Fig. JD2187) so the countershaft assembly will drop down enough to clear the gear on inner end of the drive shaft. Remove bearing cover (36—Fig. JD2186) and withdraw drive shaft from the transmission top opening.

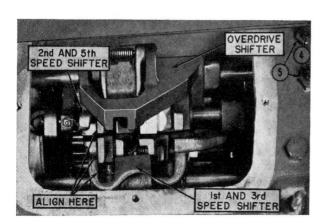

Fig. JD2185A — Shifter installation showing proper alignment of shifter gates.

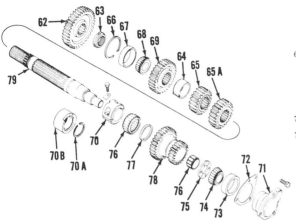

62. First and third speed
    sliding gear
63. Nut
64. Spacer
65. Differential drive pinion
65A. Fourth and sixth
    speed gear
66. Snap ring
67. Bearing cup
68. Bearing cone
69. Second and fifth speed
    gear
70. Spacer (prior 7215213)
70A. Snap ring (after
    7215212)
70B. Spacer (after 7215212)
71. Bearing housing
72. Shims and gaskets
73. Bearing cup
74. Bearing cone
75. Thrust washer
76. Roller bearing
77. Spacer
78. Countershaft idler gear
79. Countershaft

Fig. JD2187—Exploded view of the transmission countershaft. Shaft end play of 0.001-0.004 is controlled by shims and gaskets (72).

When reinstalling the shaft, refer to Fig. JD2186 as a reference and reverse the removal procedure.

141. **COUNTERSHAFT.** To remove the transmission countershaft (79—Fig. JD2187), remove the sliding gear shaft as in paragraph 139 and the transmission drive shaft as in paragraph 140. Unstake and remove nut (63) from left end of countershaft. On 720 models prior to Ser. No. 7215213 equipped with the two-piece spacer (70), unbolt and remove the spacer halves. On late 720 and all 730 models, bump countershaft toward right and spacer toward left until snap ring (70A) comes out from under spacer (70B); then, disengage the snap ring from the shaft groove. On all models, bump countershaft out right side of case and remove gears from above.

On early 720 models with the two-piece spacer (70), it is recommended that the two-piece spacer be discarded and the later production one-piece spacer (70B) and snap ring (70A) be installed.

When reinstalling the countershaft, use the same number of shims (72—Fig. JD2187) as were originally removed and tighten nut (63) securely. Mount a dial indicator with contact button resting on left end of countershaft and check the countershaft end play which should be 0.001-0.004. If end play is not as specified, remove the countershaft right bearing housing and add or remove the necessary amount of shims (72). After the countershaft end play is properly adjusted, remove bearing housing (71); then, reinstall same after the transmission drive shaft is in place.

DIFFERENTIAL, FINAL DRIVE AND REAR AXLE

The differential unit is mounted on the spider of a spur (ring) gear which meshes with a driving pinion on the transmission countershaft. Pressed through the spider is the differential cross shaft which forms the journals for the integral differential side gears and spur (bull) pinions. The outer ends of the differential cross shaft carry taper roller bearings which support the differential unit. The remainder of the final drive includes the final drive (bull) gears, located in the rear axle housing, and the wheel axle shafts to which the bull gears are splined.

143. **REMOVE AND REINSTALL.** To remove the differential, first drain the hydraulic system and main case. Remove the platform and disconnect the hydraulic "Powr-Trol" lines. Remove seat and disconnect battery cable and wiring harness at rear of tractor. On "hi-crop" models, remove the drive chains as in paragraph 148A. Support tractor under main case and unbolt rear axle or drive housing from main case. With the rear axle or drive housing assembly supported so that it will not tip, roll the unit rearward and away from tractor. On all except "hi-crop" models, remove the left brake assembly and on "hi-crop" models, disengage inner snap rings (S—Fig. JD2194) and slide the snap rings and drive sprockets in toward differential. On all models, remove the differential left bearing quill (1—Fig. JD2192 or JD2194) and withdraw the differential assembly, right end first, as shown in Fig. JD2193.

When reinstalling, be sure to renew packing (6—Fig. JD2192 or JD2194), use the same thickness of shims (2) as were originally removed, mount a dial indicator so that contact button is resting on side of the ring gear and check the differential end play which should be 0.001-0.004. If end play is not as specified, remove the left bearing quill and add or remove the required amount of shims (2).

144. **OVERHAUL.** With the differential removed as outlined in the preceding paragraph, proceed to disassemble the unit as follows: Remove snap rings (4—Fig. JD2192 or 2194).

Fig. JD2191—When pressing the differential cross shaft in differential, be sure to support the spider with a piece of pipe as shown. The differential shown is a model 50, but the same procedure applies to series 720 and 730.

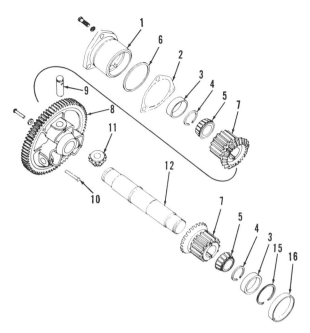

Fig. JD2192 — Exploded view of the differential assembly used on all models except "Hi-Crop". Differential end play is adjusted with shims (2).

1. Left bearing quill	8. Ring gear and spider
2. Shims	9. Pinion shaft
3. Bearing cup	10. Rivet
4. Snap rings	11. Bevel pinion
5. Bearing cone	12. Differential cross shaft
7. Differential side gear	15. Snap ring
and bull pinion	16. Bearing cover

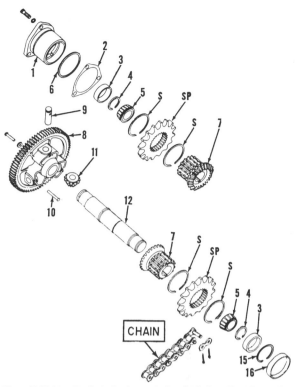

Fig. JD2194—Exploded view of the "Hi-Crop" differential, drive sprockets and chains. Sprockets (SP) are located on the differential side gears by snap rings (S). Refer to legend under Fig. JD2192.

Use a suitable puller and remove the bearing cones (5). Withdraw the combination differential side gears and spur (bull) pinions (7). Remove rivets (10), extract pinion shafts (9) and remove bevel pinions (11).

Fig. JD2193—When removing the differential, withdraw the right end first. If the differential only is to be removed, it is not necessary to remove both brake assemblies. Refer to text.

Neither the differential spider nor spur ring gear (8) is available separately. If either part is damaged, renew the entire assembly. Check the disassembled parts against the values which follow:

I. D. of pinions (7)......2.597-2.599
Diameter of differential
   shaft (12) 2.592-2.593

NOTE: If shaft (12) is worn, press the old shaft out, support differential spider on a piece of pipe and press the new shaft in place in a manner similar to that shown in Fig. JD2191.

I. D. of bevel pinions (11—
   Fig. JD2192 or 2194)..1.114-1.116
Diameter of pinion
   shafts (9) 1.1085-1.1100

When reassembling, reverse the disassembly procedure and press bearing cones (5) on the differential shaft until they seat. Install new snap rings (4).

FINAL DRIVE
Except "Hi-Crop"

145. **R&R REAR WHEEL.** Support rear of tractor, turn wheel until rack in axle shaft is in the up position and unscrew the three cap screws (33—Fig. JD2195) approximately $\frac{5}{16}$-inch. Turn the two jack screws (32) clockwise until outer groove in each screw

is flush with outer surface of wheel hub. Turn pinion shaft screw (34) and remove wheel.

When reinstalling, make certain both rear wheels are set the same distance from the centerline of the tractor. A slotted hole in back of battery box indicates tractor centerline.

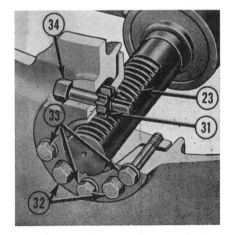

Fig. JD2195—Rear wheels can be removed by loosening cap screws (33), turning jack screws (32) in and turning pinion screws (34).

146. AXLE SHAFT OUTER FELT SEAL RENEW. To renew the rear wheel axle shaft outer felt seal (21—Fig. JD2197), remove wheel as in the preceding paragraph and using a cold chisel and hammer as shown in Fig. JD2196, drive the felt seal retainer (22—Fig. JD2197) from the axle housing. Withdraw the seal retainer and felt seal.

When reassembling, renew the seal retainer, and using a brass drift and hammer, drive the retainer in until it seats against recess in housing.

147. FINAL DRIVE (BULL) GEAR, WHEEL AXLE SHAFT & BEARINGS AND/OR AXLE SHAFT INNER OIL SEAL. Drain hydraulic system, transmission and final drive housing. Remove tractor seat and if three-point hitch is installed, remove the upper links, draft links and lift links. Disconnect the draft link support from the drawbar and loosen the supports attached to rockshaft (basic) housing. Remove the platform and disconnect the hydraulic lines. Disconnect battery cable and wiring harness at rear of tractor. Support the complete basic housing assembly and attached units in a chain hoist so arranged that the complete assembly will not tip. Unbolt basic housing from rear axle housing and move the complete assembly away from tractor.

CAUTION: This complete assembly is heavy and due to the weight concentration at the top, extra care should be exercised when swinging the assembly in a hoist.

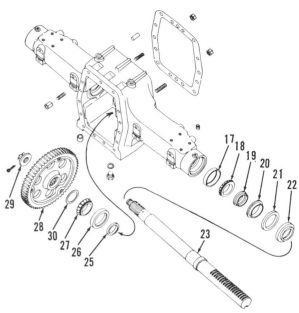

Fig. JD2197 — Exploded view of the rear wheel axle shaft and housing assembly. Refer to Fig. JD2200 for "Hi-Crop" models.

17. Bearing cup
18. Bearing cone
19. Bearing spacer
20. Inner retainer for felt seal
21. Felt seal
22. Outer retainer for felt seal
23. Wheel axle shaft
25. Inner oil seal
26. Bearing cup
27. Bearing cone
28. Final drive (bull) gear
29. Adjusting nut

Remove cotter pin and loosen, but do not remove the adjusting nut (29—Fig. JD2197). Using a hammer and a long taper wedge as shown in Fig. JD2198, force axle shaft loose from the bull gear. Remove nut (29) and withdraw gear.

Withdraw axle shaft and inspect housing between inner and outer seals for presence of transmission oil. If oil is found, the inner seal (25—Fig. JD2197) should be renewed. The need for further disassembly is evident.

When reassembling, pack the axle shaft outer bearing with wheel bearing grease. Alternately tighten nut (29) and bump outer end of axle shaft to assure proper seating of the taper roller bearings. The proper adjustment is when the axle shaft has an end play of 0.001-0.004; then, tighten to the nearest castellation and install the cotter pin. When installing the basic housing, turn the power (PTO) shaft to engage the coupling splines.

FINAL DRIVE "Hi-Crop"

148. BULL GEAR, WHEEL AXLE SHAFT, BEARINGS AND/OR SEALS. Support rear of tractor and remove wheel, inner bearing cover (1—Fig. JD2199) and housing lower cover (18). Remove nuts (3), bearing cone (4) and snap ring (21) from inner end of wheel axle shaft (15). Bump the axle shaft out of housing and withdraw the bull gear. The need and procedure for further disassembly is evident after an examination of the unit.

When reassembling, tighten the bearing adjusting nut (3) enough to provide 0.002 deflection of the housing when measured at the inner bearing boss.

148A. R&R CHAIN AND SPROCKET. To remove the drive chains and sprockets, first remove the basic housing or rear cover from the final drive shaft housing. Support rear of tractor and turn the rear wheels until the chain master link is accessible. Re-

Fig. JD2196 — Removing outer retainer for axle shaft outer felt seal.

Fig. JD2198—The axle shaft can be forced out of the final drive (bull) gears by loosening the adjusting nut and driving a long tapered wedge between the nuts.

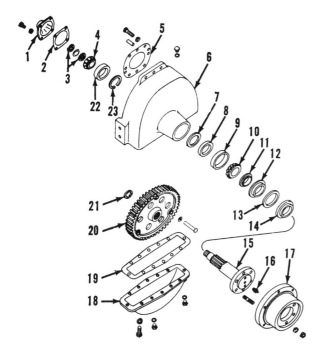

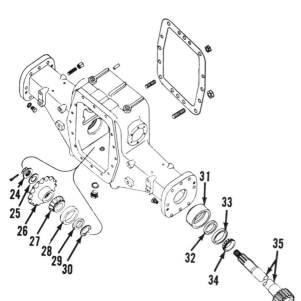

Fig. JD2199 — Exploded view of "Hi-Crop" final drive housing, bull gear and wheel axle shaft.

1. Bearing cover
2. Gasket
3. Nuts
4. Bearing cone
5. Gasket
6. Housing
7. Washer
8. Oil seal
9. Bearing cup
10. Bearing cone
11. Spacer
12. Inner retainer
13. Felt seal
14. Outer retainer
15. Axle shaft
16. Grease fitting
17. Wheel hub
18. Housing cover
19. Gasket
20. Bull gear
21. Snap ring
22. Bearing cup
23. Snap ring

Fig. JD2200 — Exploded view of series 720 and 730 diesel "Hi-Crop" final drive shaft and housing. Nut (24) should be tightened enough to provide shaft (35) with an end play of 0.001-0.004.

24. Nut
25. Washer
26. Driven sprocket
27. Bearing cone
28. Bearing cup
29. Oil seal
30. Washer
31. Adapter
32. Oil seal
33. Bearing cup
34. Bearing cone
35. Drive shaft

149A. R&R DRIVE SHAFT AND PINION. To remove the drive shaft (35—Fig. JD2200) support rear of tractor, remove the driven sprocket (26) as outlined in paragraph 148A and remove the final drive (bull) gear housing from the drive shaft housing. Withdraw the drive shaft from the housing.

When reassembling, tighten nut (24) enough to provide the drive shaft with an end play of 0.001-0.004. Tighten the drop housing to drive shaft housing screws to a torque of 275 Ft.-Lbs.

BRAKES

Except "Hi-Crop"

150. ADJUSTMENT. To adjust the brakes, tighten the adjusting screw as shown in Fig. JD2201 to reduce the pedal free travel to approximately three inches.

151. R&R BRAKE SHOES. To remove the brake shoes for lining replacement, loosen adjusting screw (41

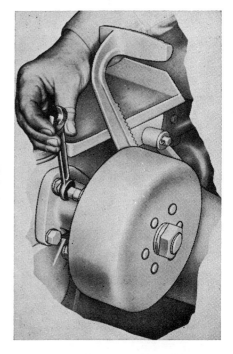

Fig. JD2201—Brakes should be adjusted to provide a pedal free travel of approximately three inches.

move the master link and withdraw the chains. Remove the nut and washer from inner end of shaft (35—Fig. JD2200) and remove the driven sprocket.

Reinstall in reverse order and tighten nut (24) to obtain a drive shaft end play of 0.001-0.004. Make certain that the drive chain master link is installed with open end toward center of tractor, and check and adjust the chain slack as in the following paragraph.

NOTE: To remove the drive sprockets which are mounted on the differential side gears, it is necessary to remove the differential as outlined in paragraph 143.

149. ADJUST CHAINS. To check and/or adjust the slack in the drive chains, remove the inspection covers from side of main case. The chains should have ½ to 1¾-inch slack. To adjust the slack, remove the brake drum and the brake housing retaining bolts. Turn the brake housing one hole at a time, until the desired chain slack is obtained and reinstall the housing retaining bolts.

After adjusting the chain slack, remove the brake shoe carrier from the brake housing and re-position the carrier so that brake pedal is in the proper position.

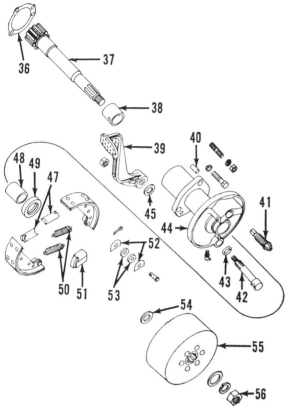

Fig. JD2202 — Exploded view of typical brake used on all except "Hi-Crop" models. Adjustment is accomplished by turning screw (41).

36. Gasket	43. "O" ring	50. Springs
37. Brake shaft	44. Housing	51. Cam
38. Inner bushing	45. Washer	52. Washers
39. Pedal	47. Adjusting pins	53. Rollers
40. Dowel pin	48. Outer bushing	54. Thrust washer
41. Adjusting screw	49. Oil seal	55. Brake drum
42. Pedal shaft		56. Nut

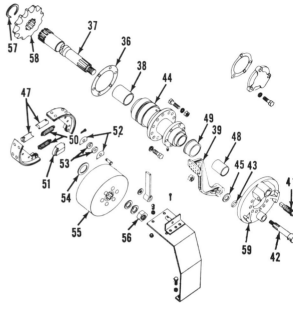

Fig. JD2204 — Exploded view of brakes used on "Hi-Crop" models.

36. Gasket	44. Housing	53. Rollers
37. Brake shaft	45. Washer	54. Washer
38. Inner bushing	47. Adjusting pins	55. Drum
39. Pedal	48. Outer bushing	56. Nut
41. Adjusting screw	49. Dust guard	57. Snap ring
42. Brake lever shaft	50. Springs	58. Sprocket
43. "O" ring	51. Brake cam	59. Brake shoe carrier
	52. Washers	

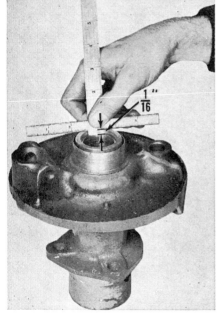

Fig. JD2203 — The brake shaft bushings should be installed to a depth of 1/16-inch below end of housing as shown.

—Fig. JD2202), remove nut (56) and using a rawhide hammer, bump brake shaft (37) inward to free drum from shaft. Withdraw the brake drum, pry shoes away from adjusting pins to release spring tension and remove shoes.

Install brake shoes by reversing the removal procedure and adjust the brakes as outlined in paragraph 150.

152. R&R AND OVERHAUL BRAKE ASSEMBLY. The procedure for removing and disassembling the brakes is evident after an examination of the unit and reference to Fig. JD2202. Be sure to mark the relative position of pedal shaft (42) with respect to pedal (39) and housing (44). Check the disassembled parts against the values which follow:

I. D. of inner bushing
(38) ....2.0030 Min.

Diameter of brake shaft
at inner bushing......1.999-2.000

I. D. of outer bushing
(48) ....1.5000 Min.

Diameter of brake shaft at
outer bushing 1.494-1.495

Defective bushings can be removed from the brake housing by sawing through the bushing wall with a hack saw and driving bushing out, but be careful not to damage the brake housing during the sawing operation.

Install new bushings by using a suitable piloted drift and to a depth of $\frac{1}{16}$-inch as shown in Fig. JD2203.

When reassembling, vary the number of thrust washers (54—Fig. JD2202) to provide a brake shaft end play of 0.004-0.044.

"Hi-Crop"

153. ADJUSTMENT. The procedure for adjusting the brakes is the same as outlined in paragraph 150.

154. R&R BRAKE SHOES. To remove the brake shoes for lining replacement, loosen adjusting screw (41 —Fig. JD2204), remove the brake drum guard, remove nut (56) and remove drum. Pry the shoes away from the adjusting pins to release spring tension and remove shoes.

Install shoes by reversing the removal procedure and adjust the brakes as in paragraph 150.

155. R&R AND OVERHAUL BRAKE ASSEMBLY. To remove the brake assembly, first remove the transmission rear cover or basic housing, extract snap ring (57—Fig. JD2204) and remove sprocket (58). Remove the brake drum guard, then unbolt and remove assembly from tractor. Before disassembling the unit, mark the relative position of pedal shaft (42) with respect to pedal (39) and carrier (59).

The remainder of the overhaul procedure is similar to the brakes used on non-"hi-crop" models. Refer to paragraph 152. When reinstalling the brakes, adjust the drive chain slack as in paragraph 149.

POWER TAKE-OFF SYSTEM

CLUTCH, OUTPUT SHAFT, DRUM AND OIL PUMP

165. ADJUST CLUTCH. To adjust the clutch on all series 730, series 720 tractors after Ser. No. 7210335, and early 720 models on which the improved powershaft clutch has been installed, proceed as follows: Remove the cap nut, exposing the adjustment indicator rod (21—Fig. JD2205) and guide (20). Engage the clutch, then pull back on the clutch pedal to take up all slack in the linkage but do not disengage the clutch. Hold the adjustment indicator rod in and note the position of the rod with respect to end of rod guide (20). Disengage clutch and be sure pedal is latched, then again note position of indicator rod (21) in relation to end of guide. If clutch is properly adjusted, the distance the rod has moved will be equal to the length of one land and one groove (0.090) on the rod as shown. If adjustment is not as specified, proceed as follows:

If tractor is equipped with a universal three-point hitch, remove the right hand draft link and draft link support. Remove the clutch adjusting hole cover from side of clutch housing, and with the clutch pedal in neutral (disengaged but not latched), turn the power (output) shaft until the adjusting cam locking screw is visible through opening in clutch housing as shown in Fig. JD2206. Now, latch the pedal in the disengaged position and turn the locking screw in until head of same clears slot in the adjusting cam. To tighten the clutch, turn the adjusting cam counter-clockwise (viewed from rear), one notch at a time and recheck the adjustment. Continue this procedure until adjustment is as specified, then turn the locking screw outward into one of the notches of the adjusting cam. Reinstall the adjusting hole cover and indicator rod cap nut.

If the powershaft fails to stop when clutch is disengaged and pedal latched, adjust the powershaft brake as follows: Disconnect yoke (Fig. JD2207) at rear end of operating rod and shorten the rod ½-turn at a time until proper adjustment is obtained.

The procedure for adjusting the powershaft clutch on early 720 models on which the improved powershaft

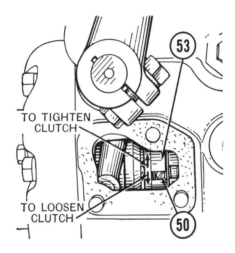

Fig. JD2206 — Side view of powershaft clutch housing with adjusting hole cover removed. Adjusting cam (53) is locked in position with screw (50).

Fig. JD2205—Checking powershaft clutch adjustment. Refer to text for proper procedure.

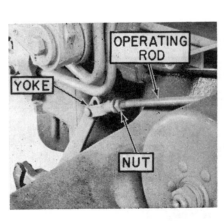

Fig. JD2207—The length of the powershaft clutch operating rod can be varied to obtain proper adjustment of the powershaft brake.

Fig. JD2208 — Using a thin ruler to check the powershaft clutch adjustment on early 720 models. Refer to text for proper procedure.

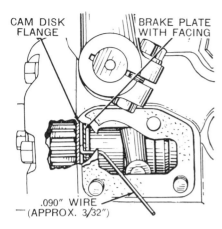

CAM DISK FLANGE

BRAKE PLATE WITH FACING

.090" WIRE (APPROX. 3/32")

Fig. JD2209 — Using a 3/32-inch rod to check powershaft clutch adjustment on early models.

clutch has not been installed, is similar to the above procedure except there is no indicator rod for checking the adjustment. Proper adjustment is determined by measuring how far the clutch cam disc moves rearward from the brakeplate as the clutch is engaged. The desired movement of 0.090 can be checked by using a thin ruler through plug hole in clutch housing as shown in Fig. JD2208 or, by using a $\frac{3}{32}$-inch wire through opening in side of clutch housing as shown in Fig. JD2209. Keep in mind, however, that power (output) shaft on these early models can be turned only when the clutch is engaged.

167. OVERHAUL CLUTCH AND OUTPUT SHAFT. To remove the clutch and/or power (output) shaft, first drain oil from clutch housing and engage the clutch. Remove the powershaft guard and disconnect the clutch operating linkage. Remove the cap screws and nuts retaining the

clutch housing cover to the clutch housing and withdraw clutch, cover and powershaft assembly. It may be necessary to pry cover from its locating dowels.

NOTE: The clutch drum cannot be serviced at this time. If the drum requires renewal, refer to paragraph 168.

Inspect the clutch fork (25—Fig. JD2210) and fork shoes (23). To remove the fork, punch assembly marks on the fork and shaft, remove ring (26) and withdraw fork shaft (27).

167A. Support the removed clutch and housing cover assembly in a soft jawed vise and using two large screwdrivers, disengage the clutch as shown in Fig. JD2211. Remove snap ring (43) retaining clutch plates to clutch shaft and if there are shim washers under the snap ring, save them for reinstallation. Withdraw clutch plates, discs, release springs, adjusting cam, balls and clutch collar. On early 720 models so equipped, unlock the brake plate retaining cap screws, unscrew the screws evenly to avoid damaging the brake plate and withdraw cam with disc, springs and washer. If there are shims under the brake plate, save them for reinstallation. On later 720 models and all 730 models, the brake plate is retained by snap rings on the support pins.

Remove bearing cover (34—Fig. JD2212), and oil seal housing (55). Extract snap ring (9) from rear end of clutch shaft and bump or press clutch shaft (11) out of bearing (10). Lift powershaft and gear assembly from cover.

To disassemble the clutch shaft on late models, press the clutch cam down far enough to remove snap ring (61). Lift off the hardened washer (63) and adjusting washer (62).

The nine outer clutch pressure springs (14) should require 83-93 pounds to compress them to a height of 1⅜ inches.

The clutch release springs (48) should require 12½-15½ pounds to compress them to a height of $2\frac{9}{16}$ inches.

Inspect clutch shaft and pilot bushing (41) which is located in clutch drum. If new bushing is carefully installed with a piloted arbor, it will not require final sizing.

Inspect clutch facings to make certain they are in good condition and be sure plates and discs are not worn or out-of-flat more than 0.008. Inspect all other parts and renew any which are questionable.

On late models, assemble the nine large springs (14) in cam (15), then install a small spring (13) into each third large spring. Lay washer (12) over the springs and install shaft (11). Place the assembly on a press as shown in Fig. JD2213 and press the cam down until springs are compressed solid. Then release pressure on the cam and allow it to move upward $\frac{5}{32}$-inch only. Install sufficient adjusting washers (62) under hardened cam washer (63) so snap ring (61 —Fig. JD2212) can just be inserted. Install snap ring, but make certain that the hardened washer is next to the snap ring. Install clutch shaft bearing into cover so that shielded side of bearing is toward inside, then place powershaft assembly in cover. Install bearing washer (W), press clutch shaft assembly into position

F2500R

26 25

27

PUNCH MARKS

8   23   41

Fig. JD2210 — With the powershaft clutch removed, the clutch fork (25) can be removed after removing the retainer ring.

43

Fig. JD2211—Disengaging the pto clutch, using two large screw drivers. Removal of snap ring (43) will permit disassembly of the clutch.

housing, install the same quantity as were removed. Install brake plate (19) with facing side next to cam and turn the cap screws (SS) down evenly, a little at a time, to avoid bending or breaking the brake plate. When the cap screws are tight, lock them in position with the tab washers. Install clutch shaft bearing cover (34).

On all models, install clutch collar (22), balls (17) and adjusting cam assembly (49, 50, 51 and 52). Turn the adjusting cam until hub (49) is about $\frac{1}{16}$-inch out of the cam. Install plate (45) with facing up, followed by a steel drive disc (46). Install a faced disc (47) with opening for release spring in line with oil hole in shaft, then install another steel drive disc (46). Continue alternating the discs in this order until all lined and unlined plates are installed. Install springs (48). Install the thick driven plate (45) with facing down and install snap ring (43). NOTE: If washers (44) were originally installed under the snap ring, be sure the same number are in place. Install oil seal (54) in oil seal housing so that lip of seal will face front of tractor then install the oil seal housing and tighten the screws finger tight.

To adjust the clutch on all 730 series, 720 after serial number 7210335, and earlier models on which the improved powershaft clutch has been installed, proceed as follows: Remove the cap nut, exposing the adjustment indicator rod (21—Fig. JD2205) and guide (20) and note the position of the rod with respect to end of rod guide (20) when clutch is disengaged.

Fig. JD2212—Exploded view of typical powershaft clutch assembly. Shims (S) and items (SS) are used only on early 720 models which have not been converted to the later construction. Washer (W), pin and snap rings (SR) and parts (20, 21, 30, 61, 62 and 63) are used only on late models and early models which have been converted to the late construction.

S. Shims	14. Spring	30. Cap nut	47. Clutch disc with facing
SR. Pins and snap rings	15. Clutch cam disc	31. Gasket	48. Spring
SS. Brake plate retaining cap screw	16. Clutch cam	32. Bushing	49. Clutch adjusting hub
W. Washer	17. Ball	33. Oil seal	50. Set screw
1. Power (output) shaft	18. Clutch cover	34. Bearing cover	51. Spring
3. Bearing cup	19. Brake plate with facing	35. Pump housing	52. Rivet
4. Bearing cone	20. Adjustment indicator guide	36. Dowel pin	53. Clutch adjusting cam
5. Powershaft gear	21. Adjustment indicator	37. Idler gear	54. Oil seal
6. Snap ring	22. Clutch collar	38. Idler gear shaft	55. Oil seal housing
7. Bearing cone	23. Clutch fork shoe	39. Bearing	56. Shim
8. Bearing cup	24. Washer	40. Clutch drum and shaft	57. Oil filler plug
9. Snap ring	25. Clutch fork	41. Bushing	58. Dowel pin
10. Bearing	26. Snap ring	42. Thrust washer	59. Gasket
11. Clutch shaft	27. Clutch fork shaft	43. Snap ring	61. Snap ring
12. Washer	28. Woodruff key	44. Washer	62. Adjusting washer
13. Spring	29. Oil seal	45. Clutch plate with facing	63. Hardened washer
		46. Clutch drive disc	

and install snap ring (9). Install brake plate (19) with facing side next to flange on cam then install retainer rings on the support pins and clamp the rings tightly into the pin grooves. Install clutch shaft bearing cover (34).

On early models which do not have the improved clutch, install the clutch shaft bearing (10) into cover, press

clutch shaft into position and install snap ring (9). Place powershaft assembly into cover. Install washer (12) and place the nine large springs in position on washer (12), then install a small spring (13) into each third large spring. Place cam (15) over the springs. If there were shims (S) installed between the brake plate and

Fig. JD2213—Compressing cam for the purpose of installing adjusting washers (62), hardened washer (63) and snap ring.

Engage clutch and again note the position of indicator rod (21) in relation to end of guide. If clutch is properly adjusted, the distance the rod has moved will be equal to the length of one land and one groove on the rod as shown in Fig. JD2205. If indicator rod movement is insufficient, disengage clutch, using two screwdrivers as in Fig. JD2211, turn the cam locking screw inward until head of same clears slot in adjusting cam and turn the adjusting cam counter-clockwise (viewed from rear) to increase the indicator rod movement, then recheck the adjustment. If indicator rod moves an excessive amount, turn the adjusting cam in the opposite direction.

When the clutch is properly adjusted, turn the locking screw outward into one of the notches in the adjusting cam.

After the clutch is properly adjusted, disengage it and check the clearance between adjusting cam and rear heavy clutch plate as shown in Fig. JD2214. The adjusting cam should not contact the clutch plate and at the same time, the clearance should not exceed $\frac{1}{16}$-inch. If a $\frac{1}{16}$-inch washer can be inserted, loosen the clutch adjustment and add an adjusting washer (44—Fig. JD2212). If the adjusting cam contacts the clutch plate, remove an adjusting washer (44) and recheck the clutch adjustment.

The procedure for adjusting the powershaft clutch on early models on which the improved clutch has not been installed, is similar to the above procedure except there is no indicator rod for checking the adjustment. Proper adjustment is determined by measuring how far the clutch cam disc moves rearward from the brake plate as the clutch is engaged. The desired movement of 0.090 can be checked with a feeler gage as shown in Fig. JD2215.

Before installing the clutch, use a spare clutch drum or straight edge, align clutch plates and engage clutch. Install a new thrust washer (42—Fig. JD2212) with smooth side toward shaft, install the complete assembly, making certain that clutch fork shoes engage grooves in clutch collar. Tighten the clutch cover retaining screws to a torque of 56 Ft.-Lbs.

Mount a dial indicator in a suitable manner and check the power (output) shaft end play which should be 0.001-

0.004. If end play is not as specified, vary the number of gasket shims (56) and recheck.

168. OVERHAUL CLUTCH DRUM AND OIL PUMP. To remove the clutch drum, first drain hydraulic system, clutch housing and main case. Remove clutch, clutch housing cover assembly and clutch fork as outlined in paragraph 167. Remove platform, disconnect hydraulic lines and remove seat. Disconnect battery cable and wiring harness at rear of tractor. Support the complete basic housing assembly and attached units in a chain hoist so arranged that the complete assembly will not tip. Unbolt basic housing from rear axle housing and move the complete assembly away from tractor.

CAUTION: This complete assembly is heavy and due to the weight concentration at the top, extra care should be exercised when swinging the assembly in a hoist.

Remove the cap screws retaining the clutch oil pump body to front side of basic housing and remove pump body and idler gear as shown in Fig. JD2216. Pull clutch drum (40) from basic housing.

Fig. JD2216—Power shaft clutch oil pump body and idler gear removed from front face of basic (rockshaft) housing. The pump drive gear and shaft is also the clutch drum shaft.

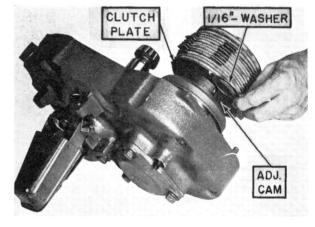

Fig. JD2214—Using 1/16-inch washer to check adjusting cam clearance.

Fig. JD2215—Using feeler gage to check early model pto clutch adjustment prior to final assembly on tractor.

The drum shaft is supported in an anti-friction bearing (39—Fig. JD-2212) which should be renewed if its condition is questionable.

The front end of the clutch drum shaft is supported by bushing (32). Bushing (32) is pre-sized and will not require reaming if carefully installed.

Check the clutch oil pump parts against the values which follow:

O.D. of idler gear......2.115-2.117
I.D. of idler gear bore
  in housing 2.121-2.123
O.D. of idler gear shaft.0.9994-1.0000
I.D. of idler gear
  bushing 1.002-1.003
Thickness of idler gear..0.997-0.998
Depth of idler gear bore
  in housing 1.000-1.005

When reassembling, coat mating surfaces of pump body and basic housing with shellac and use shim stock around splines of drum shaft to avoid damaging the pump housing oil seal. Refer to Fig. JD2217.

Fig. JD2222 — Exploded view of all components of power take-off drive. Idler gear (6) rotates on the transmission drive shaft.

1. Woodruff key
2. Drive gear
3. Snap ring
4. Bearings
5. Snap rings
6. Idler gear
7. Spacer
9. Spring
10. Cotter pin
11. Shifter lever
12. Snap ring
13. Woodruff key
14. Shifter arm
15. Shifter shaft spool
16. Shifter shaft bearing
17. Shifter shaft and yoke
18. Rivet
19. Coupling
20. Coupling shaft
21. Coupling
22. Bearing cover
23. Shims
26. Bevel driven gear
27. Bearings
28. Locking nut
29. Snap ring
32. Bevel drive gear & shaft
34. Shims
36. Sliding gear
37. Spring
38. Ball
45. Bearing cone
46. Bearing cup
47. Drive shaft quill
48. Adjusting nuts
49. Locking washer

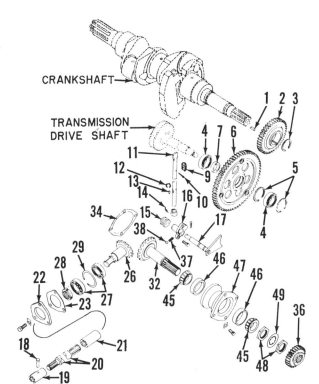

DRIVING GEARS, BEVEL GEARS AND SHIFTER

169. RENEW DRIVE AND IDLER GEARS. To renew the powershaft drive gear (2—Fig. JD2222) and/or idler gear (6), first remove engine clutch, belt pulley and first reduction gear cover as outlined in paragraph 136. Withdraw spacer, first reduction gear and powershaft idler gear. Remove snap ring (3) and using a suitable puller, remove the drive gear from crankshaft.

The idler gear is carried on two anti-friction type bearings (4) which should be inspected and renewed if their condition is questionable.

When reassembling, heat the powershaft drive gear in hot oil and drive it in place with a brass drift.

171. OVERHAUL BEVEL GEARS AND SHIFTER. To overhaul the power take-off driving bevel gears and associated parts, first drain the hydraulic system, main case and first reduction gear cover. Remove the differential as outlined in paragraph 143 and the first reduction gear cover as in paragraph 136. Working through rear opening in main case, remove cotter pin (10—Fig. JD2222) from shifter lever and pull the lever part way out of the main case. Bump shift-

er arm (14) down and off the shaft. Remove Woodruff key (13), snap ring (12), spring (9) and withdraw shifter lever from above. Remove cotter pin from inner end of shifter shaft (17), slide spool (15) from shifter shaft and unbolt shifter shaft bearing from main case. Remove shifter shaft (17), bearing (16) and sliding gear (36).

Remove cap screws retaining quill (47) to main case and withdraw quill

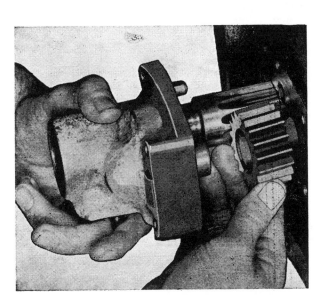

Fig. JD2217—Assembling the power take-off clutch oil pump. Mating surfaces of pump housing and basic housing should be coated with shellac.

Fig. JD2223—Method of removing the pto shifter lever (11) and shifter arm (14) from main case.

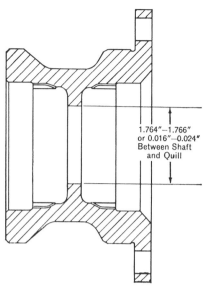

Fig. JD2224—The inside diameter of the power shaft drive shaft quill should be 1.764-1.766 as shown.

and drive shaft assembly. To disassemble the unit, remove nuts (48). Working through rear opening in main case, remove locking nut (28) and push the driven shaft forward and out of the case boss. Remove bearing retainer (22). The need and procedure for further disassembly is evident.

Refer to Fig. JD2224 and check I.D. of drive shaft quill which should be 1.764-1.766 as shown. This should provide a shaft diametral clearance of 0.016-0.024. If I.D. of quill is too small, ream out the quill to provide the specified clearance.

When installing drive shaft (32—Fig. JD2222) into quill (47), tighten inner nut (48) enough to provide the

shaft with an end play of 0.001-0.005. Install locking washer and outer nut (48). Install the front driven shaft bearing (27) and pull it into case boss until it is against the snap ring. Install rear bearing (27), shims (23) and retainer (22) and tighten the retainer screws finger tight. Install the assembled drive shaft and quill assembly, using the original number of shims (34). When the bevel gears are properly installed, the heels will be in register and the backlash will be 0.012-0.018. If mesh and backlash are not as specified, vary the number of shims (23 and 34) until the proper conditions are obtained.

Install the remaining parts by reversing the removal procedure.

POWR-TROL (HYDRAULIC LIFT)

175. Custom "Powr-Trol" is available on 720 and 730 diesel series general purpose tractors in the following six ways:

1. Rear rockshaft only.
2. Rear rockshaft, load and depth control and three-point hitch.
3. Rear rockshaft and single remote cylinder (with or without load and depth control and three-point hitch).
4. Rear rockshaft and two remote cylinders (with or without load and depth control and three-point hitch).
5. Rear rockshaft, single remote cylinder and single control front mounted rockshaft (with or without load and depth control and three-point hitch).
6. Rear rockshaft, two remote cylinders and double control front mounted rockshaft (with or without load and depth control and three-point hitch).

Custom "Powr-Trol" is available on series 720 and 730 diesel standard tractors in the following six ways:

1. Rear rockshaft only.
2. Single remote cylinder only.
3. Two remote cylinders only.

4. Rear rockshaft, load and depth control and three-point hitch.
5. Rear rockshaft and single remote cylinder (with or without load and depth control and three-point hitch).
6. Rear rockshaft and two remote cylinders (with or without load and depth control and three-point hitch).

Note: Some models which have the rear rockshaft only, may have a dash pot in the rockshaft housing so they can be converted to load and depth control.

Tractors equipped with a single remote cylinder valve housing will accommodate either a single acting or a double acting remote cylinder; whereas, on tractors equipped with a dual remote cylinder valve housing, the No. 1 circuit will accommodate either a single acting or a double acting remote cylinder, but the No. 2 circuit will accommodate only double acting cylinders.

Note: The maintenance of absolute cleanliness of all parts is of utmost importance in the operation and servicing of the hydraulic system. Of equal importance is the avoidance of nicks or burrs on any of the working parts.

LUBRICATION AND BLEEDING

176. It is recommended that the "Powr-Trol" working fluid be changed twice a year. After the system is completely drained, refill the reservoir, operate the system several times to bleed out any trapped air and refill the reservoir to the full mark on dip stick.

Note: If other than John Deere cylinders are used, always retard the piston before checking the fluid level.

Capacity is 13 U.S. qts., plus one qt. for each remote cylinder.

For temperatures above 90 degrees F., use SAE 30 single viscosity oil or SAE 10W-30 multi-viscosity oil; for 32 degrees F. to 90 degrees F., use SAE 20-20W single viscosity oil or SAE 10W-30 multi-viscosity oil; for temperatures below 32 degrees F., use SAE 10W single viscosity oil or SAE 5W-20 multi-viscosity oil.

TROUBLE-SHOOTING

177. The following paragraphs should facilitate locating troubles in the hydraulic system.

177A. **INSUFFICIENT LOAD RESPONSE.** Could be caused by:

a. Upper link in low sensitivity hole of load control yoke, paragraph 181.

b. Implement is not level.

c. Implement does not penetrate due to angle of ground engaging elements, check implement operator's manual.

d. Implement not trailing properly, check implement adjustment and alignment of ground engaging elements.

e. Dash pot orifice plugged, paragraph 202.

f. Load control arm to shaft key sheared.

g. Excessive initial loading of load control spring, paragraph 181.

h. Rockshaft control valve opening too wide, paragraph 184.

i. Improper front end weighting, see tractor operator's manual.

177B. EXCESSIVE LOAD RESPONSE. Could be caused by:

a. Upper link in high sensitivity hole of load control yoke, paragraph 181.

b. Broken load control arm.

c. Dash pot inlet valve leaking, paragraph 202.

d. Faulty "O" ring on dash pot piston or spring retainer, paragraph 202.

e. Dash pot relief valve leaking, paragraph 187.

f. Dash pot relief valve opening pressure too low, paragraph 187.

g. Low initial loading of load control spring, paragraph 181.

h. Excessive "suck" in implement, see implement operator's manual.

i. Implement ground-engaging elements in poor condition.

j. Implement not trailing properly, check adjustment of implement and alignment of ground engaging elements.

177C. INSUFFICIENT TRANSPORT CLEARANCE. Could be caused by:

a. Lift links adjusted too long.

b. Upper link too long.

c. Less than full rockshaft rotation, paragraph 185.

d. Incorrect implement mast angle, see implement operation manual.

e. Implement mast braces too long.

f. Flexible elements of implement not properly supported, see implement operator's manual.

g. Wear or looseness in implement.

h. Implement not level.

177D. HITCH FAILS TO LIFT. Could be caused by:

a. Excessive load.

b. Pump disengaged.

c. Insufficient fluid in reservoir.

d. Faulty pressure or return line.

e. By-pass valve in rockshaft control valve sticking.

f. Faulty pump, paragraph 191.

g. Faulty rockshaft control linkage, paragraph 196.

h. Remote cylinder control levers not in neutral position.

i. Rockshaft control valve gasket or "O" ring has failed, paragraph 200.

j. Restricted oil return passages.

177E. HITCH SETTLES UNDER LOAD OR "HUNTS" IN TRANSPORT POSITION. To determine whether the trouble is in the control valve housing or in the rockshaft piston, cylinder or associated parts refer to paragraph 181A. The following are possible causes:

a. Rockshaft cylinder seal has failed, paragraph 199.

b. Rockshaft piston seal has failed, paragraph 199.

c. Check valve in rockshaft control valve housing is leaking, paragraph 184.

d. Check valve plug gasket leaking.

e. Rockshaft control valve gasket or "O" ring has failed, paragraph 200.

f. Rockshaft overload relief valve is leaking, paragraph 183.

g. Throttle valve gaskets leaking.

h. Porous or failed rockshaft cylinder or rockshaft housing.

i. Check valve improperly adjusted, paragraph 184.

177F. LESS THAN FULL ROCKSHAFT ROTATION. Could be caused by:

a. Control lever linkage too short, adjust linkage for 75 degree rotation.

b. Twisted load control yoke shaft.

c. Sheared load control arm key.

d. Load control springs are short, paragraph 188.

e. Loose or bent linkage.

f. Control valve spring unhooked or failed.

g. Insufficient oil in rockshaft housing.

177G. WORKING DEPTH CHANGES OR CONTROL LEVER MOVES DURING OPERATION. Could be caused by:

a. Friction brake adjustment loose on control shaft, paragraph 186.

b. Retaining ring upset from groove in control shaft, paragraph 196.

177H. HITCH FAILS TO LOWER OR LOWERS SLOWLY. Could be caused by:

a. Throttle valve closed, paragraph 179.

b. Weak or failed rockshaft return spring.

c. Failure in linkage.

d. Oil viscosity too high, paragraph 176.

e. Check valve stem lock nut loose, paragraph 184.

177I. RELIEF VALVE OPERATES CONTINUOUSLY. Could be caused by:

a. Control lever linkage too long, adjust rockshaft travel to 75 degrees.

b. Remote cylinder control lever not in neutral position.

c. Load too heavy.

d. Relief valve opening pressure too low, paragraph 182, 189 or 190.

177J. OIL OVERHEATS (LOAD AND DEPTH CONTROL SYSTEM). Could be caused by:

a. Relief valve operates continuously, paragraph 177R.

b. Insufficient oil in reservoir.

c. Excessive load response, paragraph 177B.

d. Excessive load.

e. Continued use of lift, stop once in a while to allow oil to cool.

f. Worn pump, paragraph 191.

177K. REMOTE CYLINDER WILL NOT LIFT LOAD OR WILL NOT FUNCTION WHEN NOT LOADED. Could be caused by:

a. Air in remote cylinder, paragraph 176.

b. Pump disengaged.

c. Faulty pump, paragraph 191.

d. System overloaded.

e. Relief valve opening pressure too low, paragraph 189 or 190.

f. Gasket or "O" ring failed between remote cylinder control valve housing and rockshaft housing.

g. Faulty oil lines, packings or couplings.

h. Porous control valve housing.

i. Control shaft Woodruff key failed or missing.

j. Missing inner check valve ball on return side, paragraph 193 or 194.

k. Single valve housing, or No. 1 circuit of dual valve housing set for single-acting cylinder operation, and hoses crossed, paragraph 178.

l. Thermal relief valve missing.

177L. REMOTE CYLINDER WILL NOT LOWER. Could be caused by:

a. Air in remote cylinder, paragraph 176.

b. Missing inner check valve ball on return side, paragraph 193 or 194.

c. Control shaft Woodruff key failed or missing.

d. Faulty oil line coupler or coupler not completely engaged.

177M. REMOTE CYLINDER CONTROL LEVER DOES NOT RETURN TO NEUTRAL. Could be caused by:

a. Failed centering spring, paragraph 193 or 194.

b. Dual valve housing assembled without secondary relief valve balls.

c. Valve housing porous between relief valve and detent passages.

d. Control valve snap ring failed or disengaged, paragraph 193 or 194.

e. Control valve sticking, paragraph 193 or 194.

f. Detent sticking, paragraph 193 or 194.

g. Single valve housing, or No. 1 circuit of dual valve housing set for single-acting cylinder operation, paragraph 178.

h. Single-acting cylinder by-pass screw not seated in housing. Screw must be firmly seated. If early models have thick bodied by-pass screw, discard it and install the late thin-bodied screw.

177N. REMOTE CYLINDER CONTROL LEVER WILL NOT LATCH IN FAST OPERATING POSITION. Could be caused by:

a. Remote cylinder overloaded.

b. Relief valve pressure set too low, paragraph 189 or 190.

c. Detent stuck in the disengaged position, paragraph 193 or 194.

d. Detent spring has failed, paragraph 193 or 194.

e. Detent has failed, paragraph 193 or 194.

f. Defective oil line coupler or coupler not engaged.

177P. REMOTE CYLINDER SETTLES UNDER LOAD. Could be caused by:

a. Dirty check valve, paragraph 193 or 194.

b. Remote cylinder adapter packing has failed.

c. Single-acting cylinder by-pass screw open while using double acting cylinder, paragraph 178.

d. Valve housing assembled without outer check valve, paragraph 193 or 194.

e. Thermal or overload relief valve leaking, paragraph 193 or 194.

f. Remote cylinder rod seal leaking, paragraph 203.

g. Remote cylinder piston ring failure, paragraph 203.

h. Remote cylinder casting failure, paragraph 203.

i. Outer check valve ball leaking or missing, paragraph 193 or 194.

177Q. "POWR-TROL" OIL OVERHEATS (REMOTE CYLINDER SYSTEM). Could be caused by:

a. Control lever being held in engaged position after remote cylinder reaches end of stroke.

b. Low oil supply, paragraph 176.

c. Relief valve pressure set too high, paragraph 189 or 190.

d. Control lever does not return to neutral, paragraph 177M.

e. Rockshaft against stop at full raise. Adjust rockshaft travel to 75 degrees from lowered position.

177R. RELIEF VALVE OPERATES CONTINUOUSLY. Could be caused by:

a. Control lever not returning to neutral, paragraph 177M.

177S. NOISY PUMP. Could be caused by:

a. Oil viscosity too high, paragraph 176.

b. Insufficient oil supply in reservoir, paragraph 176.

c. Air leak in oil return line.

d. Pump drive shaft oil seal leaking.

OPERATING ADJUSTMENTS

178. SINGLE OR DOUBLE ACTING CYLINDER ADJUSTMENT. To operate a single-acting cylinder on tractors equipped with a single cylinder valve housing, remove cap nut (Fig. JD2225) and slotted adjusting screw located under the nut. Thread the adjusting screw into the cap nut as far as it will go, then reinstall the cap nut on the valve housing. To operate a double-acting cylinder on tractors equipped with a single cylinder valve housing, remove the cap nut (Fig. JD2225) and turn the slotted adjusting screw into the valve housing until it seats; then, reinstall the cap nut.

Fig. JD2225—Single or double acting remote cylinder adjustment for models with a single cylinder valve housing.

Fig. JD2226—Single or double acting remote cylinder adjustment for models with a dual cylinder valve housing.

Fig. JD2227 — Location of throttle valve screw cap nut. The throttle valve controls the speed of rockshaft drop.

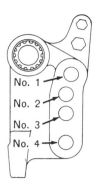

Fig. JD2228—Identification of holes in the load control yoke.

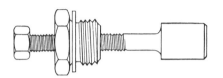

Fig. JD2228A—The test fixture shown can be used to determine the cause of the lift arms falling under a load. See text.

178A. On tractors equipped with a dual cylinder valve housing, adjustment of the No. 1 circuit for use with a single or double-acting remote cylinder is made the same as outlined in paragraph 178 except the cap nut and the slotted screw are located on top of the adapter which attaches to right front of valve housing as shown in Fig. JD2226.

179. **SPEED OF ROCKSHAFT DROP.** The speed at which the rockshaft (and attached implement) drops is regulated by a throttle valve on side of rockshaft housing. To make the adjustment, remove cap nut (Fig. JD2227), loosen the jam nut and turn the slotted screw **in** to decrease or **out** to increase speed of rockshaft drop.

Adjustment should provide a smooth drop of the implement. After adjustment is complete, tighten the jam nut and install cap nut firmly on gasket seat.

180. **ADJUST HYDRAULIC STOP REMOTE CYLINDER.** Any length of stroke within an 8-inch range may be selected. To adjust the stroke, lift the piston rod stop lever, slide the adjustable stop along piston rod to the desired position and press the stop lever down. If clamp does not hold securely, lift stop lever, rotate it ½-turn clockwise and press in place. Make certain, however, that the adjustable stop is located on the piston rod so that stop arm contacts one of the flanges on stop.

181. **LOAD CONTROL ADJUSTMENT.** As shown in Fig. JD2228, the load control yoke has four attaching holes for the upper link. These holes determine the sensitivity of the load and depth control system. The bottom hole (No. 4) is the most sensitive and the top hole (No. 1) the least sensitive. The proper hole to use is usually specified in the John Deere implement operators' manuals. Also, markings under a pointer attached to the control yoke make it possible to see that the proper hole is being used.

The pointer must be adjusted properly, as follows: When there is no load on the control yoke, the rear edge of the pointer should align with the front edge of the front indicating mark. If it does not align, loosen the two nuts and reposition the pointer until proper alignment is obtained.

When tractor and implement are in operation, the rear edge of the pointer should float over the center mark below the pointer. If pointer is over front mark, there is not enough compression on the load control spring and the link should be moved to a lower hole. If the rear edge of the pointer is over the rear mark, there is too much compression on the load control spring and the link should be moved to a higher hole.

Note: On self-gauging implements for which load and depth control is not required, install the upper link in top (No. 1) hole in control yoke. When using self-gauging implements, the control lever should be pushed to the forward end of the quadrant.

SERVICE TESTS AND ADJUSTMENTS

181A. If the rockshaft will not stay up under load, the difficulty is in either the control valve housing or the rockshaft piston, cylinder or associated parts. To determine which section is the cause, a test fixture (shown in Fig. JD2228A) is available from R.B. Precision Tools, 313 Morse St., Ionia, Michigan. Method of using the test fixture is as follows: With engine not running and implement lowered to the ground, install the test fixture as shown in Fig. JD2228B in place of the regular throttle valve. Make certain that the plunger is turned out partially. Start engine and raise implement all the way up, turn the test fixture screw all the way in until the valve seats on rockshaft housing, then stop engine. If the implement lowers, the trouble is in the rockshaft piston, cylinder or associated parts. If the implement stays up, the trouble is in the control valve housing.

182. **ROCKSHAFT RELIEF VALVE.** When tractor is not equipped with remote cylinder valve housings, the rockshaft relief valve is located in a housing which is bolted to rear face of rockshaft housing.

To check and/or adjust the rockshaft relief valve opening pressure, remove the rockshaft throttle valve assembly from the right side of rockshaft housing and connect a suitable pressure gage (at least 1800 psi capacity) to the opening from which the throttle valve was removed. Refer to Fig. JD2229.

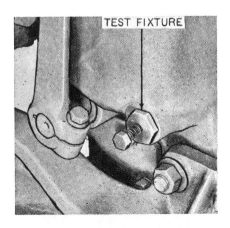

Fig. JD2228B — The test fixture shown in Fig. JD2228A should be installed as shown in place of the regular throttle valve. See text.

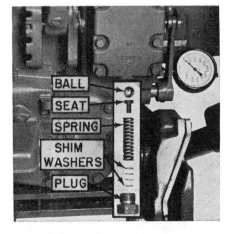

Fig. JD2229 — Checking and adjusting the rockshaft relief valve opening pressure. Specified pressure is 1300-1400 psi.

Disconnect rod from the rockshaft control lever and with engine running and pump engaged, pull back on the rockshaft control arm to raise the rockshaft. The rockshaft will travel beyond its normal range and the rockshaft arm will strike the housing cover, preventing further travel of the rockshaft and causing the relief valve to open. The relief valve opening pressure as shown on the gage should be 1300-1400 psi. CAUTION: Operate system with relief valve open only long enough to observe the opening pressure.

If the pressure is not as specified, remove the relief valve plug (Fig. JD2229) and add adjusting washers to increase the pressure or remove washers to decrease pressure. One washer changes the pressure approximately 40 psi.

When adjustment is complete, connect the rockshaft control rod, install the throttle valve assembly and adjust the rockshaft dropping speed as in paragraph 179.

When tractors are equipped with remote cylinder valve housing, the remote cylinder relief valves also protect the rockshaft circuit. Refer to paragraph 189 or 190.

183. ROCKSHAFT CYLINDER OVERLOAD RELIEF VALVE. The rockshaft cylinder overload relief valve is located in the rockshaft control valve housing which is bolted to front face of the rockshaft housing. Whenever the control valve is in neutral, the circuit to the rockshaft cylinder is sealed off from the rockshaft relief valve by the check valve. The overload relief valve then protects the cylinder circuit under these conditions. For example, the need for this protection can be appreciated because there could be considerable shock loads transmitted to the rockshaft

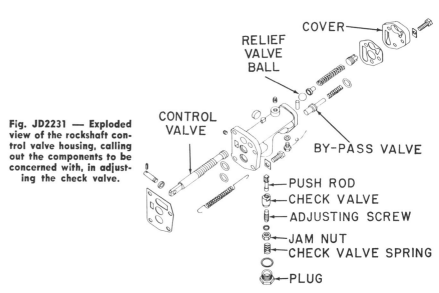

Fig. JD2231 — Exploded view of the rockshaft control valve housing, calling out the components to be concerned with, in adjusting the check valve.

cylinder and piston when transporting heavy implements over rough ground.

The overload relief valve is adjusted to open at 1450-1550 psi and should not normally require field adjustment. To check and/or adjust the overload relief valve opening pressure, proceed as follows:

Remove the rockshaft throttle valve assembly from right side of rockshaft housing and connect a suitable pressure gage (at least 1800 psi capacity) to the opening from which the throttle valve was removed. Refer to Fig. JD2229. Remove the rockshaft relief valve plug, adjusting washers and spring as shown. Replace the spring with a spacer made from a short piece of shaft, then screw the plug tightly against the spacer to lock the relief valve ball against its seat.

Disconnect rod from the rockshaft control lever and with engine running and pump engaged, pull back on the rockshaft control arm to raise the rockshaft. The rockshaft will travel

beyond its normal range and the rockshaft arm will strike the housing cover, preventing further travel of the rockshaft and causing the overload relief valve to open. The overload relief valve opening pressure as shown on the gage should be 1450-1550 psi.

CAUTION: Operate system with relief valve open only long enough to observe the opening pressure.

If the pressure is not as specified, remove the rockshatt (basic) housing assembly as outlined in paragraph 195. Remove the rockshaft control valve cover to expose the adjusting screw as shown in Fig. JD2230. Remove the lock screw and turn the adjusting screw **in** to increase or **out** to decrease the opening pressure. Each ½-turn of the adjusting screw will change the opening pressure approximately 50 psi. When adjustment is complete, install the lock screw and control valve cover.

Note: While the rockshaft housing assembly is removed, it is good practice to check

Fig. JD2230 — Rockshaft cylinder overload relief valve adjusting screw. Specified relief valve opening pressure is 1450-1550 psi.

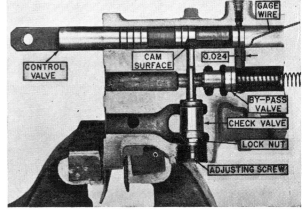

Fig. JD2232 — Cut-away view of the rockshaft control valve housing showing the points for checking the check valve adjustment.

the adjustment of the rockshaft check valve as in paragraph 184.

Install rockshaft housing assembly on tractor and connect the rockshaft control rod. Remove the relief valve plug (Fig. JD2229) and the home made spacer. Reassemble the relief valve as shown and be sure to install the same number of adjusting washers as were removed. Install the throttle valve assembly and adjust the rockshaft dropping speed as in paragraph 179.

184. ROCKSHAFT CHECK VALVE. To check and/or adjust the rockshaft check valve, remove the rockshaft control valve housing as outlined in paragraph 200. Remove cover (Fig. JD-2231) from control valve housing. Also refer to Fig. JD2232 as well as JD-2231. Remove check valve plug, check valve spring and check valve assembly. Position the control valve so the forward section covers about half of the passage which leads down to the by-pass valve. Reinstall the check valve and insert a home made spacer instead of the check valve spring. Then tighten the plug nut against the spacer so the check valve is firmly seated.

It should now be possible to easily move the control valve back and forth, a short distance, in the housing. If binding exists, the check valve is not seating because the check valve push rod is adjusted too long; in which case, remove the plug and spacer, loosen jam nut and back-off the adjusting screw enough to permit free movement of the control valve when the check valve is seated.

With the check valve seated, pull out on the control valve until the cam surface on the control valve contacts the check valve push rod; at which time, it should be possible to just insert a 0.024 diameter gage wire into the passage which leads down to the by-pass valve as shown in Fig. JD2232. If wire does not enter passage, remove the check valve and back-off the adjusting screw as necessary. If wire enters too freely, turn the adjusting screw in to decrease the clearance. Tighten the jam nut and recheck the adjustment as follows:

Remove the check valve plug and spacer. Insert gage wire in by-pass passage and push the control valve against the wire. Hold the check valve on its seat with your fingers. If the valve is properly adjusted, the slightest pulling of control valve out of the housing will start to open the check valve.

Install the check valve spring, retainer plug and housing cover. Install control valve housing assembly and rockshaft housing.

185. ROCKSHAFT ROTATION. The total rotation of the rockshaft when hydraulically operated should be 75 degrees. To adjust the rotation, lower the rockshaft to the lowest position and with engine running and pump engaged, adjust the length of the control rod (Fig. JD2233) at yoke so that when control lever is moved to the extreme rear position on the quadrant, the rockshaft will rotate 75 degrees up from its lowest position.

186. CONTROL LEVER FRICTION BRAKE. The control lever is held at selected positions by a friction brake on the control shaft in rockshaft hous-

ing. The brake should be adjusted so that a force of 12-15 lbs. applied at end of lever, is required to move the lever. To adjust the brake, remove plug (Fig. JD2234), exposing a self-locking nut (720 series) or cap screw (730 series) on the control shaft. Turn the nut or screw clockwise to increase and counter-clockwise to decrease the force required to move the lever.

187. DASH POT RELIEF VALVE. The dash pot relief valve is located within the rockshaft housing and can be removed as shown in Fig. JD2235, after removing the housing cover. The opening pressure of the valve is factory set to open at 225-250 psi and should not normally be adjusted. If the operation of the relief valve is questionable, it is recommended that a complete new valve unit be installed. If, however, a new valve is not available, remove the relief valve and attach it to a suitable hydraulic pump with a pressure gage. If the opening pressure is not as specified, turn the adjusting screw in to increase or out to decrease the pressure.

One method of checking the relief valve, using the hydraulic system of a tractor equipped for remote cylinder operation is shown in Fig. JD2236. Use a small steel ball and C-clamp to close off the bleed hole in the valve body as shown and be sure the shut-off valve is open. With engine running and hydraulic pump engaged, slowly move the control lever to the slow raise position and observe the pressure reading as the relief valve opens.

NOTE: Be sure to hold a can over the valve to prevent undesirable spraying of oil.

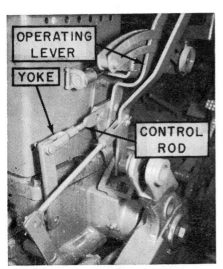

Fig. JD2233 — Rockshaft operating lever, control rod and adjusting yoke. Full rotation of rockshaft should be 75 degrees.

Fig. JD2234—Removal of plug will expose nut or screw for adjusting the control lever friction brake.

Fig. JD2235—Using bent wrench to remove the dash pot relief valve. The factory set valve should open at 225-250 psi.

188. LOAD CONTROL SPRINGS.
The rear end of the load control springs should be carefully located with respect to the machined valve housing mounting surface on the rockshaft housing as shown in Fig. JD-2237. Dimension (D), which should be $1\frac{7}{16}$-inches, can be checked after removing the load control yoke as in paragraph 197.

The position of the springs is varied by adding or deducting shim washers in front of the springs.

189. SINGLE REMOTE CYLINDER VALVE HOUSING RELIEF VALVE.
The relief valve in the single remote cylinder valve housing should open at 1230-1300 psi. To check the pressure, connect a pressure gage (at least 1800 psi capacity) and shut-off valve to the remote cylinder couplings as shown in Fig. JD2238. With the hydraulic pump engaged, start engine and allow to run for a few minutes before starting the test. With the shut-off valve open, move the control lever to the fast raise position. The control lever will remain in this position. Slowly close the shut-off valve and note the pressure gage reading the instant the control lever returns to neutral.

If the relief valve opening pressure is not 1230-1300 psi, remove the relief valve plug (Fig. JD2239) and add or deduct adjusting washers as required. Each washer will change the opening pressure approximately 40 psi.

190. DUAL REMOTE CYLINDER VALVE HOUSING RELIEF VALVE.
The dual remote cylinder valve housing has two separate circuits, each having its own relief valve. The specified opening pressure is 1430-1500 psi for the number one circuit and 1230-1300 psi for the number two circuit.

To check the pressure for the No. 1 circuit, connect a pressure gage (at least 1800 psi capacity) and shut-off valve to the remote cylinder couplings as shown in Fig. JD2238. With the hydraulic pump engaged, start engine and allow to run for a few minutes before starting the test. With the shut-off valve open, move the control lever to the fast raise position. The control lever will remain in this position. Slowly close the shut-off valve and note the pressure gage reading the instant the control lever returns to neutral.

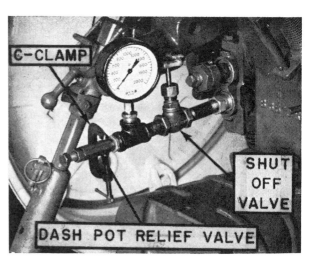
Fig. JD2236—One method of testing the dash pot relief valve, using the hydraulic system of a tractor equipped for remote cylinder operation.

Fig. JD2237 — Using straight edge and scale to check position of the load control springs. Dimension (D) should be 1 7/16 inches.

Fig. JD2238 — Pressure gage and shut-off valve hook-up for checking the single remote cylinder valve housing relief valve. This same hook-up can be used to check the No. 1 circuit on dual valve housings.

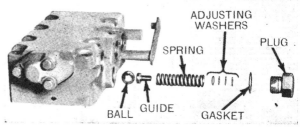
Fig. JD2239—Pressure relief valve exploded from the single remote cylinder valve housing.

If the relief valve opening pressure is not 1430-1500 psi, remove the relief valve cap (Fig. JD2240) and add or deduct adjusting washers as required. Each washer will change the opening pressure approximately 40 psi.

To check the relief valve opening pressure for the No. 2 circuit, make certain that the control lever for the No. 1 circuit is in neutral position and connect a pressure gage (at least 1800 psi capacity) and shut-off valve to the remote cylinder outlets in the housing as shown in Fig. JD2240. With the hydraulic pump engaged, start engine and with the shut-off valve open, move the control lever to the fast raise position. The control lever will remain in this position. Slowly close the shut-off valve and note the pressure gage reading the instant the control lever returns to neutral.

If the relief valve opening pressure is not 1230-1300 psi, remove the relief valve cap (Fig. JD2240) and add or deduct adjusting washers as required. Each washer will change the opening pressure approximately 40 psi.

PUMP

191. R&R AND OVERHAUL. The hydraulic "Powr-Trol" pump is mounted under the front part of the tractor main case and is driven by the engine timing gear train. To remove the pump, drain the hydraulic system and disconnect the hydraulic lines. Place the pump shifter handle in the "ON" position, unbolt and withdraw pump from tractor, being careful not to lose thrust washer (3—Fig. JD-2241).

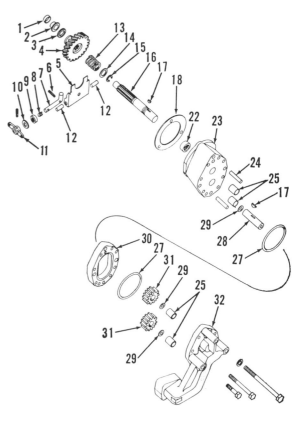

Fig. JD2241 — Exploded view of the hydraulic "Powr-Trol" pump.

1. Plug
2. Bushing
3. Thrust washer
4. Pump drive gear
5. Shifter yoke
6. Roll pin
7. Shifter
8. Spring
9. Oil seal
10. Washer
11. Shifter handle
12. Dowel pins
13. Spring
14. Washer
15. Snap ring
16. Drive shaft
17. Woodruff keys
18. Gasket
22. Oil seal
23. Pump housing
24. Dowel pins
25. Bushings
27. "O" rings
28. Idler shaft
29. Thrust washers
30. Pump body
31. Pumping gears
32. Pump cover

The procedure for disassembling the pump is evident after an examination of the unit and reference to Fig. JD-2241. After pump is disassembled, wash all parts in a solvent and examine them for damage or wear. Pump specifications are as follows:

O. D. of pumping
　gears 2.4485-2.4495
I. D. of pump body......2.452-2.454
Radial clearance between
　gears and body....0.00125-0.00275
Pumping gear
　thickness 0.9044-0.9050
Pump body thickness..0.9075-0.9085
Clearance between gears
　and cover 0.0025-0.0041
I. D. of pump shaft
　bushings (25) 1.0025-1.0035
O. D. of pump shaft...0.9994-1.0000
Clearance between bushings
　and shaft 0.0025-0.0041
I. D. of bushing (2)....0.8109-0.8115
O. D. of pump shaft at
　bushing (2) 0.8082-0.8092
Clearance between bushing
　(2) and shaft 0.0017-0.0033

Bushings (2 & 25) are pre-sized and if carefully installed should need no final sizing. When reassembling, renew all "O" rings and seals and fill the space between the lips of seal (22) with gun grease. Tighten the cover retaining screws to a torque of 32 Ft.-Lbs. Check to be sure that pump drive shaft turns freely. If it doesn't, loosen the pump assembly screws and center the cover and body by tapping; then, retighten the screws.

Fig. JD2240—One of the pressure relief valves exploded from dual remote cylinder valve housing.

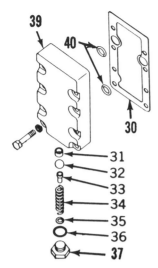

Fig. JD2242 — Exploded view of the rock-shaft relief valve and housing.

30. Gasket	35. Adjusting washers
31. Relief valve seat	36. Washer
32. Relief valve ball	37. Plug
33. Spring guide	39. Housing
34. Relief valve spring	40. "O" rings

ROCKSHAFT RELIEF VALVE HOUSING

When tractor is not equipped with remote cylinder valve housings, the rockshaft relief valve is located in a housing which is bolted to rear face of rockshaft housing.

192. R&R AND OVERHAUL. To remove the rockshaft relief valve housing, thoroughly clean the housing and surrounding area and drain the oil from rockshaft housing. Unbolt and remove the relief valve housing.

Remove plug (37—Fig. JD2242) and extract the remaining parts. Thoroughly clean all parts and examine them for damage or wear. If ball (32) or seat (31) are renewed, seat the new ball by tapping it with a brass drift and hammer.

When reassembling, be sure to install the same number of adjusting washers (35) as were removed and after the unit is installed on tractor, check the opening pressure as outlined in paragraph 182.

SINGLE REMOTE CYLINDER VALVE HOUSING

193. R&R AND OVERHAUL. To remove the single remote cylinder valve housing, thoroughly clean the housing and surrounding area and drain the oil from rockshaft housing. Disconnect linkage from the control shaft arm, disconnect the oil lines from adapter, then unbolt and remove the housing.

To overhaul the removed unit, refer to Fig. JD2243 and proceed as follows: Remove the oil line adapter, nut (28),

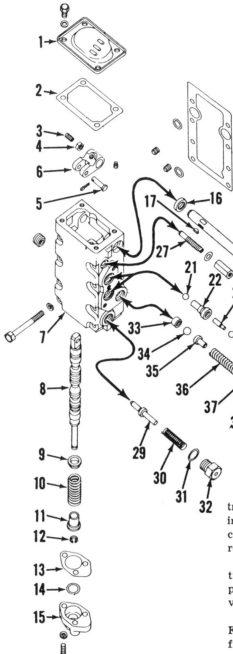

Fig. JD2243 — Exploded view of single remote cylinder valve housing.

1. Housing cover
2. Gasket
3. Set screw
4. Jam nut
5. Pin
6. Control valve arm
7. Valve housing
8. Control valve
9. Valve spring upper retainer
10. Control valve spring
11. Valve spring lower retainer
12. Snap ring
13. Gasket
14. Shim
15. Control valve cover
16. Oil seal
17. Woodruff key
18. Control arm and shaft
21. Check valve ball (2 used)
22. Check valve (2 used)
23. Check valve metering shaft (2 used)
24. Spring (2 used)
25. Retainer (2 used)
26. Check valve ball (2 used)
27. By-pass screw
28. Cap nut
29. Control valve detent
30. Detent spring
31. Washer
32. Plug
33. Relief valve seat
34. Relief valve ball
35. Relief valve spring guide
36. Relief valve spring
37. Washer
38. Washer
39. Plug

by-pass screw (27) and the detent assembly (29, 30, 31 and 32). Remove the relief valve assembly (34, 35, 36, 37, 38 and 39) making certain none of the adjusting washers (37) are lost or damaged. Remove the two check valve retainers (25), then remove both check valve assemblies (21, 22, 23, 24 and 26) making certain all valve parts are removed and none are lost. Remove cover (1), then loosen jam nut (4) and set screw (3). Control shaft (18) can then be moved far enough to remove Woodruff key (17), after which shaft can be removed from valve housing. Unbolt and remove control valve cover (15) making certain washers (14) are not lost or damaged. Move con-

trol valve arm (6) to position shown in Fig. JD2244 and remove pin. The control valve assembly can now be removed from housing.

Inspect for cracks, porous conditions, excessive wear, missing or failed parts, dirt or metal particles and valves which do not seat properly.

The seat on outer check valves (22—Fig. JD2243) can be lapped in, using fine lapping compound. If ball seating surfaces on parts (22 and 33) or if steel balls (21, 26 and 34) are renewed, the ball can be seated by tapping it lightly against its seat with a brass drift and hammer. If oil seal (16) is renewed, install it with lip facing inward.

All parts should be lubricated with "Powr-Trol" oil before reassembly. Reinstall control valve assembly, installing pin (5) through control valve and control valve arm. Slide control shaft (18) through oil seal (16) and position Woodruff key, then, move the control shaft into the control arm and secure with set screw (3) and jam nut (4). Install control valve cover (15) using the same number of washers (14) as were removed. Check free end play of control valve with dial in-

Fig. JD2244 — Removing the control valve arm on single remote cylinder valve housing.

dicator as shown in Fig. JD2245 and add or deduct washers (14—Fig. JD-2243) until the recommended free end play of 0.010 is obtained without compressing the neutralizing spring, then install cover (1).

Install both check valve assemblies (21, 22, 23, 24 and 26) and be sure spring retainers (25) are secure. Install detent assembly (29, 30, 31 and 32) and by-pass screw (27). Install relief valve assembly (33, 34, 35, 36, 37, 38 and 39) using same number of adjusting washers (37) as were removed. Valve housing retaining bolts should be tightened to a torque of 36 Ft.-Lbs. After unit is reinstalled on tractor, check relief valve opening pressure as outlined in paragraph 189.

DUAL REMOTE CYLINDER VALVE HOUSING

194. R&R AND OVERHAUL. To remove the dual remote cylinder valve housing, thoroughly clean the housing and surrounding area and drain the oil from rockshaft housing. Disconnect linkage from the control shaft arm, disconnect the oil lines from adapter, then unbolt and remove the housing.

To overhaul the removed unit, remove the adapter casting, refer to Fig. JD2246 and proceed as follows:

Fig. JD2245—Checking control valve end play on single remote cylinder valve housing. Desired end play, without compressing the neutralizing spring, is 0.010.

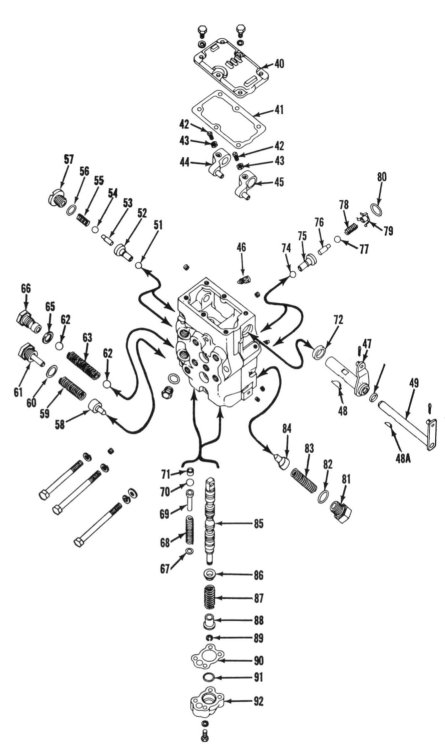

Fig. JD2246—Exploded view of the dual remote cylinder valve housing.

42. Set screw
43. Jam nut
44. L. H. control arm
45. R. H. control arm
46. Thermal relief valves (2 used)
47. Control arm & sleeve
49. Control arm and shaft
51. Check valve ball (2 used)
52. Check valve (2 used)
53. Check valve metering shaft (2 used)
54. Check valve ball (2 used)
55. Check valve spring (2 used)
56. "O" ring
57. Plug (2 used)
58. Control valve detent
59. Detent spring
60. Washer
61. Plug
62. Detent relief valve ball
63. Detent relief valve spring
65. Gasket
66. Detent relief valve seat
67. Shim washers
68. Relief valve spring (2 used)
69. Spring guide (2 used)
70. Relief valve ball (2 used)
71. Relief valve seat (2 used)
72. Oil seal
74. Check valve ball (2 used)
75. Check valve (2 used)
76. Check valve metering shaft (2 used)
77. Check valve ball (2 used)
78. Check valve spring (2 used)
79. Retainers (2 used)
80. "O" ring
82. Washer
83. Detent spring
84. Control valve detent
85. Control valve (2 used)
87. Control valve spring (2 used)
88. Spring lower retainer (2 used)
89. Snap ring (2 used)
91. Shims

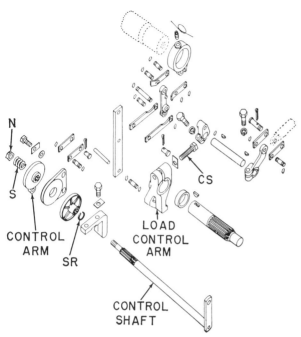

N

S

CONTROL ARM

SR

CS

LOAD CONTROL ARM

CONTROL SHAFT

Fig. JD2249—Exploded view of series 720 diesel rockshaft control linkage. On series 730 diesel tractors a cap screw is used in place of the nut (N).

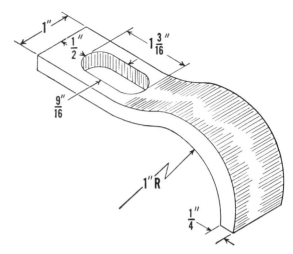

Fig. JD2251—Special tool which can be made and used to clamp the dash pot piston forward.

tapping it lightly against its seat with a brass drift and hammer. If oil seal (72) is renewed, install with lip facing inward. Both thermal relief valves (46) should be renewed if their condition is questionable.

All parts should be lubricated with "Powr-Trol" oil before reassembly. Reinstall both control valve assemblies (85, 86, 87, 88 and 89) and both primary relief valves (68, 69 and 70). Install the same number of washers (67 and 91) and install control valve covers (92). NOTE: Make certain none of the washers (91 and/or 67) are lost, damaged or mixed with those of the other circuit. Reinstall control shafts and control arms (44 and 45), then tighten set screws (42) and jam nuts (43). Check free end play of each control valve with a dial indicator in a manner similar to that shown in Fig. JD2245 and add or deduct washers—(91—Fig. JD2246) until the recommended free end play of 0.010 is obtained without compressing the neutralizing springs, then install cover (40). Reinstall all valve and detent

Remove retainers (79) and withdraw both of the No. 1 circuit check valve assemblies (74, 75, 76, 77 and 78). Unscrew plugs (57) and extract both of the No. 2 circuit check valves assemblies (51, 52, 53, 54, 55 and 56). Remove plug (81) and the No. 1 circuit detent assembly (82, 83 and 84). Remove plug (61) and the No. 2 circuit detent assembly (58, 59 and 60). Unscrew plug (66) and the secondary relief valve assembly (62, 63 and 65). NOTE: Make certain none of the parts from any of these assemblies are mixed, lost or damaged. Remove cover (40), loosen nuts (43) and set screws

(42), then slide control arm (45) enough to remove key (48). Withdraw both, control shafts and control arms. Remove Woodruff key (48A) and separate the control arms. Unbolt and remove both control valve covers (92), then withdraw both primary relief valve assemblies (67, 68, 69 and 70). NOTE: Washers (67 and 91) must not be lost, damaged or mixed with washers from the other circuit. Withdraw control valves from bottom of housing.

Inspect for cracks, porous conditions, excessive wear, missing or failed parts, dirt or metal particles and valves which do not seat properly.

The seat on outer check valves (75 and 52) can be lapped in, using fine lapping compound. If ball seating surfaces on parts (52, 71 and 75) or if steel balls (51, 54, 70, 74 and 77) are renewed, the ball can be seated by

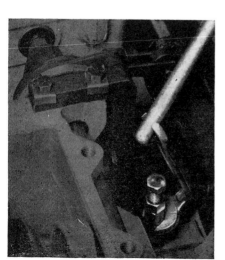

Fig. JD2250—Using bent wrench to remove the dash pot relief valve. The factory set valve should open at 225-250 psi.

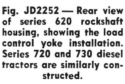

Fig. JD2252 — Rear view of series 620 rockshaft housing, showing the load control yoke installation. Series 720 and 730 diesel tractors are similarly constructed.

assemblies in the sequence shown in Fig. JD2246. Valve housing retaining bolts should be tightened to a torque of 36 Ft.-Lbs. After unit is reinstalled on tractor, check each relief valve opening pressure as outlined in paragraph 190.

ROCKSHAFT HOUSING AND COMPONENTS

195. R&R ASSEMBLY. To remove the complete rockshaft (basic) housing assembly, first drain the hydraulic system and main case. Remove tractor seat and if three-point hitch is installed, remove the upper links, draft links and lift links. Disconnect the draft link supports from the drawbar and loosen the supports attached to rockshaft housing. Remove the platform and disconnect the hydraulic lines. Disconnect battery cable and wiring harness at rear of tractor. Support the complete housing assembly and attached units in a chain hoist so arranged that the complete assembly will not tip. Unbolt housing assembly from rear axle housing and move the complete assembly away from tractor.

CAUTION: This complete assembly is heavy and due to the weight concentration at the top, extra care should be exercised when swinging the assembly in a hoist.

Reinstall the assembly by reversing the removal procedure and make certain that powershaft splines engage properly.

196. CONTROL SHAFT AND LINKAGE. To remove the control shaft and linkage, remove seat, batteries, battery box and rockshaft housing cover. Disconnect control rods. Remove access plug from left side of rockshaft housing and remove nut (N—Fig. JD-2249) and spring (S). Pull control shaft toward right, disengage snap ring (SR) from shaft, then withdraw the shaft and snap ring. The need and procedure for further disassembly is evident.

The control plate lining must be in good condition and the aluminum control discs must not be scored, grooved or cracked.

When reassembling, be sure to securely seat snap ring (SR) in groove of control shaft. If early production tractors have a spring steel snap ring, discard it and install the later production soft steel ring. Be sure the short spline on the control shaft matches the corresponding groove in the control arm.

Adjust the control lever friction brake as in paragraph 186.

Fig. JD2254—Using the special tool shown in Fig. JD2251 to clamp the dash pot piston forward.

197. LOAD CONTROL YOKE. To remove the load control yoke, remove seat, batteries, battery box and rockshaft housing cover. Bend a ¾-inch wrench as shown in Fig. JD2250 and remove the dash pot relief valve assembly, but be careful not to disturb the adjusting screw which controls the relief valve opening pressure. Make up a tool, using the dimensions shown in Fig. JD2251, and using a ½ x 1-inch cap screw and washer, install the tool as shown in Fig. JD2254 and leave the screw slightly loose. Using a large screw driver or pry bar, pry the dash pot piston forward and tighten the clamp retaining screw securely. Loosen the load control arm retaining cap screw (CS—Fig. JD2249) and drive the roll pin (Fig. JD2252) out of the yoke. Remove the coupler bracket (or cover), screw a ⅜ x 4½-inch cap screw and washer into the load control shaft and using a pry bar as shown in Fig. JD2253, remove the load control shaft. Remove the load control yoke and withdraw the yoke roll pin from pocket in housing.

The need and procedure for further disassembly is evident. Before installing the load control yoke, check the control spring adjustment as outlined in paragraph 188.

Install the remaining parts by reversing the removal procedure.

198. ROCKSHAFT. To remove the rockshaft, remove seat, batteries, battery box and rockshaft housing cover. Remove lift arm from left end of rockshaft and disconnect the return spring. Remove lift arm from right end of rockshaft, then unwire and remove the set screw retaining the operating ring to the shaft. Refer to Fig. JD2255. Disengage the snap ring,

Fig. JD2253—Using cap screw, washer and pry bar to remove load control shaft from the rockshaft housing.

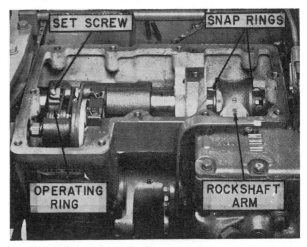

Fig. JD2255—Top view of rockshaft housing with top cover removed.

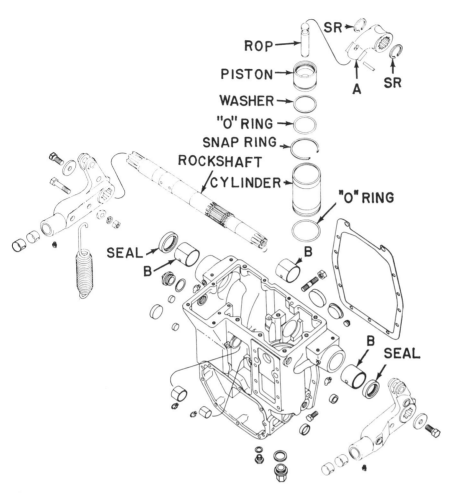

Fig. JD2256—Typical rockshaft housing showing an exploded view of the rockshaft, rockshaft operating cylinder and associated parts. Bushings (B) are pre-sized.

through the center bearing and install one of the snap rings (SR) over end of shaft with long edge of ends of snap ring toward left. Position the rockshaft arm and piston rod assembly in housing and slide shaft just through the arm but make certain that "V" mark on shaft is in register with "V" mark on arm as shown in Fig. JD2257. Position right snap ring on shaft so that long edge of ends of snap ring is toward right. Push the rockshaft into position and engage the snap rings in the shaft grooves. Install the operating ring set screw and safety wire. Install the oil seals with lips facing inward. Install lift arms so that "V" mark on arms is in register with "V" mark on ends of shaft.

199. **ROCKSHAFT OPERATING CYLINDER.** To remove the rockshaft cylinder piston, first remove the rockshaft and rockshaft arm as outlined in paragraph 198. Refer to Fig. JD-2258, remove cap nut, loosen the jam nut and remove the throttle valve adjusting screw. Lift the rockshaft piston from the cylinder.

To remove the cylinder after piston is out, first remove the control shaft as outlined in paragraph 196; then, remove the cylinder retaining snap ring and lift cylinder from rockshaft housing. An L-shaped bar will be helpful in lifting out the cylinder.

Examine all parts and renew any which are damaged or worn. Renew piston rod if it is worn or scored at piston contacting end.

When reassembling, install "O" ring on cylinder, lubricate the "O" ring with gun grease and push cylinder

(SR—Fig. JD2256) located just to the right of rockshaft arm (A), from its groove in the rockshaft and bump the rockshaft toward left until the snap ring (SR), located just to the left of the rockshaft arm, can be disengaged from its groove. Withdraw rockshaft from left and remove the rockshaft arm and piston rod assembly from above. Remove and discard the shaft oil seals.

Inspect the rockshaft bearing journals and bushings located in rockshaft housing. Replacement bushings are pre-sized and if not distorted during installation will require no final sizing. When installing the bushings, be sure to align oil hole in bushing with oil passages in the housing.

When reassembling, slide rockshaft through left side of housing, position the operating ring. Push shaft just

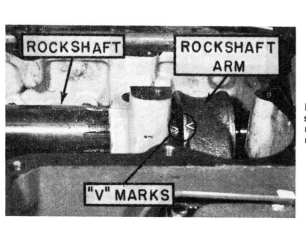

Fig. JD2257 — When assembling the rockshaft and rockshaft arm, the "V" marks must be in register.

Fig. JD2258—Throttle valve screw cap nut and jam nut.

into housing. Install the cylinder retaining snap ring with open end toward front of tractor. Install the piston "O" ring and backing washer with
backing washer toward open end of
piston. Lubricate piston with "Powr-
Trol" oil and push piston into cylinder.
Install rockshaft as in paragraph
198 and control shaft as in paragraph
196. Adjust the control lever friction
brake as in paragraph 186 and the
throttle valve as outlined in paragraph 179.

**200. ROCKSHAFT CONTROL
VALVE.** To remove the rockshaft control valve, first remove the complete
rockshaft (basic) housing assembly
as outlined in paragraph 195.

Note: Be sure to check for the possibility
of leakage at control valve cover, check
valve plug and between control valve housing and rockshaft housing. System malfunc-
tions could be caused by leakage at these
points and further disassembly would be unnecessary.

Remove cover from top of rockshaft
housing, disconnect the control valve
from the operating linkage by unhooking the return spring, pulling the
cotter pin and removing the connecting pin. Unbolt and remove the control valve housing from rockshaft
housing.

Remove cover from housing and
remove pin (6—Fig. JD2260), return
spring (26), control valve (1) and
by-pass valve (13, 14 and 15).

Remove plug (22), spring (23) and
check valve (18, 19, 20, 24 and 25).
Remove cap screw (17), unscrew plug
(10) counting the number of turns
required to remove it so it can be reinstalled to the same position. Remove the overload relief valve (7, 8
and 9).

Check all parts for damage or wear
and renew any which are questionable. Control valve (1) and by-pass
valve (15) must slide freely in the
housing bores. Examine the check
valve and its seat in housing and
check the seal with Prussian blue.
Valve can be reseated by using fine
lapping compound. The overload relief valve ball seat can be reconditioned by lapping with fine compound.
To do so, make up a lapping tool by
welding a 10-inch rod to a new relief
valve ball (Part No. F 2772R). Be sure
to use a new ball when reassembling.
Install spring (9) and turn the nut
(10) **in** the same number of turns that
were required to remove the nut. Install cap screw (17). Examine finished

1. Control valve
2. Pipe plug
3. Valve housing
4. Plug
5. Pipe plug
6. Dowel pin for return spring
7. Relief valve ball
8. Spring guide
9. Relief valve spring
10. Relief valve adjusting screw
11. Cover
12. Gasket
13. "O" ring
14. By-pass valve spring
15. By-pass valve
16. Washer
17. Cap screw
18. Check valve push rod
19. Check valve adjusting screw
20. Jam nut
21. Washer
22. Plug
23. Check valve spring
24. Washer
25. Check valve
26. Return spring
27. Gasket
28. "O" ring

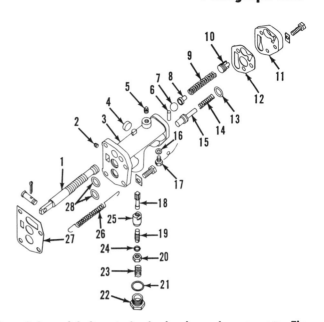

Fig. JD2260—Exploded view of the rockshaft control valve housing and components. The overload relief valve (7, 8, 9 and 10) should be open at 1450-1550 psi.

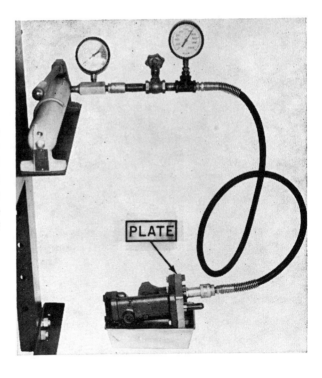

Fig. JD2261—Using pump of hydraulic press, shut-off valve, gage and adapter plate shown in Fig. JD2262 to test the rockshaft control valve housing when unit is off tractor.

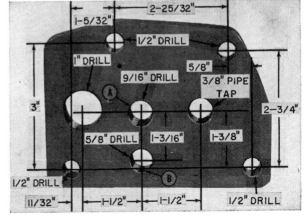

Fig. JD2262 — Home-made adapter plate used with the test layout shown in Fig. JD2261.

surface near rear end of control valve (1) where the letter "T" is stamped. Install valve with letter "T" toward top of housing. Install by-pass valve (15) and spring (14). Install check valve and carefully adjust its position as outlined in paragraph 184. Install return spring and its pin (6). Position new "O" ring and gasket and install cover (11). Test the control valve housing assembly as in the following paragraph 201, then reinstall the control valve by reversing the removal procedure.

201. TESTING HOUSING — OFF TRACTOR. The overload relief valve can be tested, as outlined in paragraph 183, by installing the unit on the tractor; but, this entails considerable assembly and disassembly.

A means of testing these valves off the tractor is shown in Fig. JD2261, wherein the use of the pump on a hydraulic press is shown, but any hydraulic pump with a capacity of about 1800 psi. can be used.

An adapter plate (Fig. JD2262) must be made and assembled to the control valve housing with "O" ring seals and a gasket.

Pump the pressure up to near the overload relief valve opening pressure of 1450-1550 psi. and close the shut-off valve. The pressure may drop with the valve closed if the overload relief valve ball is not entirely seated. The ball will not seat completely until the film of oil is displaced from the seat. If this occurs, pump up the pressure two or three times to give the ball a chance to seat. If the seal is good, the pressure should then stay up. If it doesn't, either the overload relief valve or check valve is leaking.

Leakage past the rockshaft check valve will be indicated by the appearance of oil at hole (A). If the overload relief valve leaks, oil will appear at opening (B).

Be careful when building up pressure. If the overload relief valve opens (moves away from its seat), oil that escapes past the valve will also appear at opening (B). As oil is pumped into the control valve housing, the pressure increase will be indicated on the gage and when the opening pressure of the overload relief valve is reached (1450-1550 psi.), the opening of the valve will be indicated by a slight drop in pressure.

If the overload relief valve does not open at the specified pressure, adjust the pressure by means of the adjusting nut (10—Fig. JD2260). Lock the adjustment by replacing the lock screw (17) so it engages a castellation on the nut.

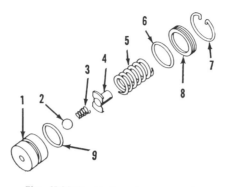

Fig. JD2263—Dash pot components.

1. Piston	6. "O" ring
2. Check valve ball	7. Snap ring
3. Check valve spring	8. Spring retainer
4. Check valve cup	9. "O" ring
5. Piston spring	

Fig. JD2264—Using a suitable home-made tool to compress the dash pot piston spring and retainer.

If the valves leak, further lapping is required as outlined in paragraph 200.

202. DASH POT PISTON. To remove the dash pot piston, first remove the complete rockshaft (basic) housing assembly as outlined in paragraph 195. Compress the spring retainer with a tool similar to that shown in Fig. JD2264, remove snap ring (7—Fig. JD2263) and extract the remaining parts.

Renew any damaged or worn parts and be sure the opening in the dash pot piston is open and clean. Lubricate all parts and reassemble.

REMOTE CYLINDER (HYDRAULIC STOP TYPE)

203. OVERHAUL. To disassemble the unit, remove oil lines and end cap (11—Fig. JD2265). Remove stop valve (7) and bleed valve (5) by pushing stop rod (1) completely into cylinder. Withdraw stop valve from bleed valve, being careful not to lose the small ball (3). Remove nut from piston rod, being careful not to distort the rod and remove piston and rod. Push stop rod (1) all the way into cylinder and drift out Groov pin (19). Remove piston rod guide (27).

Examine all parts for being excessively worn and renew all seals. Wiper seal (30) should be installed with sealing lip toward outer end of bore. Install stop rod "V" seal assembly (20, 21, 22 and 26) with sealing edge toward cylinder. Complete the assembly by reversing the disassembly procedure and install the piston rod stop.

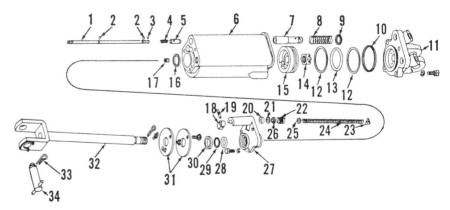

Fig. JD2265—Exploded view of the hydraulic stop type remote control cylinder.

1. Stop rod	12. Packing washer	24. Stop rod spring
2. Snap rings	13. "O" ring	25. Stop rod washer
3. Ball	14. Nut	26. Packing adapter male
4. Bleed valve spring	15. Piston	27. Piston rod guide
5. Bleed valve	16. Piston rod guide gasket	28. "O" ring
6. Cylinder	17. Pipe plug	29. Packing washer
7. Stop valve	18. Stop rod arm	30. Wiper seal
8. Stop valve spring	19. Groov pin	31. Rod stops
9. Gasket	20. Packing adapter female	32. Piston rod
10. Gasket	21. "V" packing	33. Locking pin
11. End cap	22. Packing spring	34. Attaching pin
	23. Stop rod washer	

JOHN DEERE
(PREVIOUSLY JD-17)

Models ■ 80 ■ 820 ■ 830 (2 Cyl. Diesel Models)

SHOP MANUAL

JOHN DEERE

MODELS 80-820-830

IDENTIFICATION

Tractor serial number is stamped on plate on right side of main case.
MODEL 80 tractors have serial numbers from 8000000 to 8199999.
MODEL 820 tractors have serial numbers from 8200000 to 8299999.
MODEL 830 tractors have serial numbers beginning with 8300000.
All models produced in non-adjustable axle version only.

INDEX (By Starting Paragraph)

CONDENSED SERVICE DATA

GENERAL

LIQUID CAPACITIES

Cooling System
80 and 820............................8¾ Gals.
830 with Gasoline Starting Engine...............8 Gals.
830 with Electric Starting Motor..................7½ Gals.

Crankcase (Diesel Engine)
Model 80...........................15 Qts.
Model 820 and 830........................14 Qts.

Crankcase (Starting Engine)
Model 80..........................1 Qt.
Model 820 Prior 8203100...................1 Qt.
Model 820 After 8203099 and 830............1½ Qts.

Power Steering.........................5½ Qts.
P. T. O. Clutch.........................3¾ Qts.
Transmission (Diesel Engine)...............3¼ Gals.
Transmission (Starting Engine)...............½ Pt.

MISCELLANEOUS

Electrical System (Tractors with Gasoline Starting Engine)
Voltage...........................6
Generator Make and Model................D-R1100027
Regulator Make and Model................D-R1118786
Starting Motor Make and Model............D-R1107155

Electrical System (Tractors with Electric Starting Motor)
Voltage.....................Split Load 24-12
Generator Make and Model................D-R1103021
Regulator Make and Model................D-R1119219
Starting Motor Make and Model.......D-R1113801 or 1113830
Solenoid Make and Model.......D-R1119803 or 1119832

Toe-In..........................⅛-⅜ inch
Transmission Speeds Forward................6
Transmission Speeds Reverse................1
Belt Pulley rpm.................Same as Engine
P. T. O. rpm—Load.....................536
P. T. O. rpm—No Load..................596

STARTING ENGINE

GENERAL

Make...........................Own
No. Cylinders........................4
Bore & Stroke—Inches.................2 x 1½
Displacement—Cubic Inches...............18.85
Cylinder Sleeves?.....................Wet

TUNE-UP—SIZES—CLEARANCES

Tappet Gap (In. & Ex.)...............0.008-0.010
Valve Seat Angle (In. & Ex.)................45°
Valve Face Angle (In. & Ex.).................44½°
Valve Stem Diameter (In. & Ex.)........0.2445-0.2455
Ignition Type.......................Battery
Distributor Model...................Wico B4027
Firing Order......................1-2-3-4
Breaker Gap........................0.020
Timing Mark on Flywheel..........."V" Mark, Spark No. 1
Plug Make.................Champion or AC
Champion Model......................J-8
AC Model.........................45M
Electrode Gap......................0.025
Starting Motor...................D-R 1107155
Carburetor.....................Zenith TU

TUNE-UP—SIZES—CLEARANCES—(Continued)

Engine Load rpm......................4500
Engine Slow Idle rpm..................4000
Engine Fast Idle rpm..................5000
Piston Skirt Clearance................0.002-0.004
Piston Pin Diameter.................0.4818-0.4822
Piston Removed From?..................Above
Camshaft End Play..................0.003-0.009
Camshaft Front Journal Diameter.........1.248-1.249
Camshaft Center Journal Diameter.........1.060-1.061
Camshaft Rear Journal Diameter..........0.748-0.749
Camshaft Bearing Clearance.............0.002-0.005
Crankshaft End Play.................0.005-0.008
Crankshaft Main Bearings, Number of?.........2
Crankshaft Main Journal Diameter.... [1]1.124-1.125; [2]1.499-1.500
Crankshaft Main Bearing Clearance...[1]0.002-0.004; [2]0.001-0.003
Crankshaft Rod Journal Diameter (Crankpin).....1.0615-1.0625
Crankshaft Rod Bearing Clearance.........0.001-0.003
Main & Rod Bearings Adjustable?..................No

TIGHTENING TORQUES (Ft.-Lbs.)

Cylinder Head Nuts........................67
Rod Bolts...........................7
Flywheel Nut........................150
Spark Plugs..........................32

[1]Prior to tractor serial number.................8203100
[2]After tractor serial number.................8203099

DIESEL ENGINE

GENERAL

Make..........................Own
No. Cylinders........................2
Bore—Inches........................6⅛
Stroke—Inches........................8
Displacement—Cubic Inches...............471½
Compression Ratio....................16 to 1
Cylinders Sleeved?.....................No

TUNE-UP—SIZES—CLEARANCES

Compression Pressure (Cranking Speed)............467 psi
Tappet Gap (In.)...................0.015H
(Ex.)...................0.020H
Valve Seat Angle (In. & Ex.)................45°
Valve Face Angle (In. & Ex.)................44½
Valve Stem Diameter..................0.496-0.497
Diesel Pumps Make........American-Bosch or Bendix-Scintilla
Diesel Nozzles Make........American-Bosch or Bendix-Scintilla
Injection Occurs.....................25° BTDC
Injection Mark on Flywheel.............."No. 1 INJ"
Engine Load rpm......................1125
Engine Fast Idle...................1250 max.
Engine Slow Idle (Recommended)...............750
Piston Skirt Clearance.............See Paragraph 85

TUNE-UP—SIZES—CLEARANCES—(Continued)

Piston Pin Diameter.................2.3545-2.3550
Camshaft End Play..................0.005-0.009
Camshaft Right Journal Diameter..............1.499-1.500
Camshaft Right Bearing Clearance............0.002-0.005
Camshaft Center Journal Diameter............2.4975-2.4985
Camshaft Center Bearing Clearance............0.002-0.005
Camshaft Left Journal Diameter............1.4985-1.4995
Camshaft Left Bearing Clearance............0.002-0.004
Crankshaft End Play..................0.005-0.010
Crankshaft Main Bearings, Number of..............3
Crankshaft Main Journal Diameter.........4.4980-4.4990
Left and Right Main Bearing Clearance..........0.0055-0.0075
Center Main Bearing Clearance.............0.0016-0.0050
Rod Journal Diameter (Crankpin)...........3.9991-4.0005
Rod Bearing Clearance...............0.0015-0.0049
Main & Rod Bearings Adjustable?..................No
Oil Pressure Adjustable?..................Yes
Oil Pressure Recommended psi................25

TIGHTENING TORQUES (Ft.-Lbs.)

Cylinder Head Nuts.....................275
Rod Nuts........................200-220
Flywheel Bolts.......................334

FRONT AXLE SYSTEM

STEERING KNUCKLES AND ARMS

1. To remove the steering knuckles (9—Fig. JD1500), remove front wheel and hub units, disconnect steering arms from knuckles and remove knuckle caps (11). Remove taper bolt retaining nuts and drive taper bolts (6) forward and out of axle main member. Using a drift, bump spindle pins (7) out of axle and knuckles.

Knuckle bushings (8) are pre-sized and if not distorted during installation will need no final sizing. Install two thrust washers (12) between axle and lower fork of steering knuckle.

TIE ROD AND DRAG LINK
Model 80

2. To adjust toe-in of the front wheels, it is necessary to disconnect tie rod ends from steering arms. Tie rod ends should be adjusted to provide the recommended toe-in of ⅛-⅜ inch. Tighten jam nuts (25—Fig. JD 1501) when adjustment is complete. Adjust tie rod and drag link ends (ball joints) so they have no end play, yet do not bind.

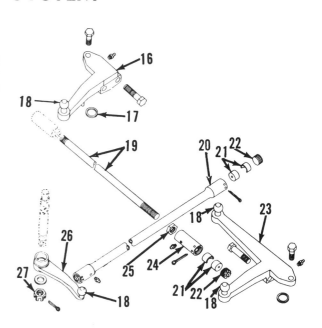

Fig. JD1501 — Exploded view of Model 80 steering arms, tie rod, drag link and associated parts.

16. Left hand steering arm
17. Rubber gasket
18. Steering socket ball
19. Tie rod
20. Drag link
21. Ball stud bearing
22. Screw plug
23. Right hand steering arm
24. Tie rod end
25. Jam nut
26. Steering gear arm
27. Steering gear arm retaining nut

Models 820-830

3. To adjust the toe-in of the front wheels, loosen clamps (29—Fig. JD 1502) and turn tie rod to provide the proper toe-in of ⅛-⅜-inch. Tighten clamps when adjustment is complete. Tie rod or drag link ends are not adjustable to compensate for wear.

The length of the drag link should be adjusted so the steering arm does not hit its stop when steering wheel is turned to both extreme positions. If proper length cannot be obtained, it is likely that the steering gear arm (26) is on the wrong spline of vertical steering shaft.

AXLE AND RADIUS ROD
PIVOT PINS

4. The non-adjustable front axle pivots on a pin (3—Fig. JD1500) which is retained in the front end support by a bolt (2). The rear end of the unbushed radius rod pivots on a pin (15) which is retained in the front end support by a rivet (14). A 1⅜-inch diameter axle pivot pin is used on 80 and 820 and regular duty 830 front axles and must have a free fit in axle. On 830 heavy duty front axles, a 1.498-1.500 diameter axle pivot pin is used and should have a diametral clearance of 0.006-0.022 in the pre-sized axle bushing. The radius rod pivot pin on all models is 1⅜-inch diameter and should have a diametral clearance of 0.005-0.010 in the radius rod. The method of removal of either pin is obvious and is accomplished without disturbing any other parts.

RADIUS ROD

5. **REMOVE AND REINSTALL.** To remove radius rod (13—Fig. JD1500), jack up front end of tractor and disconnect drag link from right hand steering arm. Remove pivot pins (3 & 15) and roll axle, tie rod, wheels and radius rod assembly away from tractor. Remove radius rod to axle retaining nuts (5) and bump radius rod out of the axle main member.

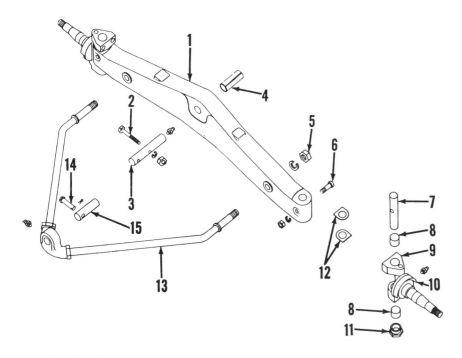

Fig. JD1500—Exploded view of the radius rod, knuckle and front axle assembly.

1. Axle main member
2. Axle pivot pin retaining bolt
3. Axle pivot pin
4. Axle pivot pin bushing (820 & 830)
5. Radius rod retaining nut
6. Taper bolt
7. Spindle pin
8. Knuckle bushing
9. Knuckle
10. Dust excluder
11. Knuckle cap
12. Thrust washers
13. Radius rod
14. Radius rod pivot pin retaining rivet
15. Radius rod pivot pin

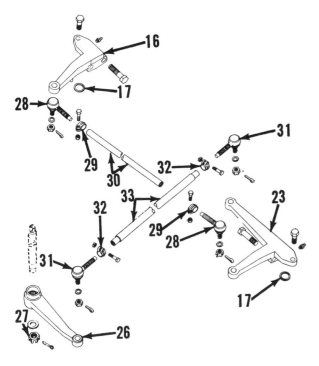

Fig. JD1502 — Exploded view of models 820 and 830 steering arms, tie rod, drag link and components.

16. Left hand steering arm
17. Rubber gasket
23. Right hand steering arm
26. Steering gear arm
27. Steering gear arm retaining nut
28. Tie rod end
29. Clamp
30. Tie rod
31. Drag link end
32. Clamp
33. Drag link

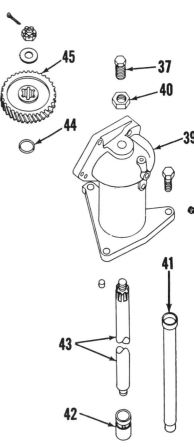

Fig. JD1503A—Exploded view of the manual steering gear and associated parts.

37. Vertical steering shaft adjusting screw	41. Steering gear shaft tube
39. Steering gear and vertical steering shaft housing	42. Steering gear shaft bushing
40. Adjusting screw lock nut	43. Steering gear shaft
	44. "O" ring
	45. Steering gear

MANUAL STEERING SYSTEM

7. **ADJUSTMENT.** Three adjustments can be made on the steering gear unit: (1) the steering worm shaft end play; (2) the vertical shaft end play; and (3) the backlash between the steering worm and the steering gear. Before adjusting the steering gear unit, it is recommended that the front end of the tractor be raised so

that all unnecessary load is removed from the steering mechanism.

8. STEERING WORM SHAFT END PLAY. Remove hood and cowl. Remove cap screws retaining cowl side support to wormshaft housing. Remove cap screws retaining the steering worm shaft housing to the steering gear housing and remove worm shaft housing from the tractor. Remove steering worm bearing housing and add or remove shims (34— Fig. JD1503) until the worm shaft has 0.001-0.004 end play.

9. VERTICAL STEERING SHAFT END PLAY. Remove hood from tractor, loosen jam nut and turn adjusting screw (37—Fig. JD1503) **down** until it is tight against the top of vertical steering shaft. Then back-off the adjusting screw ⅛ turn and secure it by tightening jam nut (40).

10. BACKLASH ADJUSTMENT. Remove hood and cowl from tractor. Remove cap screws retaining cowl side support to worm shaft housing. Remove cap screws retaining the steering worm shaft housing to the steering gear housing and separate the housings. Add or remove shims (38— Fig. JD1503), which are located between the two housings, until there is ½-1 inch free movement measured at rim of steering wheel.

Check for binding through entire range of steering wheel travel. If the steering gear binds, or has an exces-

sive amount of backlash in any position, it will be necessary to renew the worm and steering (sector) gear; or, re-position same to bring unworn teeth into mesh.

11. **GEAR ASSEMBLY — R&R.** To remove the complete steering gear assembly, proceed as follows: Remove hood and disconnect diesel engine speed control rod at top of transmission case cover. Disconnect gas line, wires and controls from the gasoline starting engine. Disconnect the head light wires and both of the shutter control rods. Unbolt instrument panel and pull same rearward. Disconnect dash light wire. Unbolt and raise cowl enough to unclip headlight wires from rear under side of cowl; then, remove cowl from tractor. Detach clip from top of steering gear housing. On tractors equipped with the gasoline starting engine, disconnect wire from starting engine distributor. Unclip generator wires and heat indicator tube from starting engine timing gear housing. On all models, move wires and lines

Fig. JD1503 — **The manual steering gear unit. Screw (37) controls the vertical steering shaft end play; shims (38) control backlash between the worm and steering gear; and shims (34) control end play of worm.**

toward right side of tractor as far as possible. Remove cap screws retaining gear housing to main case, disconnect steering gear arm from vertical steering shaft and lift steering gear assembly from tractor.

12. **OVERHAUL.** To overhaul the steering gear, it is first necessary to remove the unit from tractor as in paragraph 11. Steering gear shaft bushing (42—Fig. JD1503A) which is driven into the bottom portion of the

tractor main case, can be renewed at this time. The bushing is pre-sized and if carefully installed, will need no reaming.

Procedure for disassembly of the removed gear unit is obvious. Worm-shaft bushing which is located at rear of steering column is pre-sized and if not distorted during installation will require no final sizing. When reassembling, adjust the unit as outlined in paragraphs 8, 9 and 10.

13. **STEERING SHAFT TUBE—RENEW.** To renew the vertical steering shaft tube (41—Fig. JD1503A), first remove steering gear unit as in paragraph 11, drive bushing (42) out through bottom of main case and tube out through the top. Install new tube from above and using a piloted driver, bump the tube down until lower shoulder of tube bottoms in the main case. Install new, pre-sized steering gear shaft bushing (42).

POWER STEERING SYSTEM

NOTE: The maintenance of absolute cleanliness of all parts is of utmost importance in the operation and servicing of the hydraulic power steering system. Of equal importance is the avoidance of nicks or burrs on any of the working parts.

LUBRICATION

14. It is recommended that the power steering system be drained (but not flushed) once a year or every 1,000 hours. To drain the system, jack up front end of tractor enough to relieve weight from the front wheels. Remove the drain plug and allow the system to drain, then turn the steering wheel from one extreme to the other to remove trapped oil from the cylinder.

Refill the system with approximately 5½ U. S. quarts of John Deere power steering oil.

TROUBLE-SHOOTING

15. **DRIFTING TO EITHER SIDE** could be caused by:

1. Valve housing not in correct relation to worm shaft housing. Center the valve housing as in paragraph 20.

15A. **HARD STEERING** could be caused by:

1. Insufficient volume of oil flowing to steering valve from flow control valve. Adjust the flow control valve as in paragraph 19.

2. Excessive tension on the centering cam spring. Adjust the spring as in paragraph 27A or 27D.

3. Insufficient end play or binding in the worm shaft bearings. See paragraph 25 or 27B.

4. Insufficient backlash between the steering worm and gear. See paragraph 26.

5. Excessive leakage past the steering vane seals in cylinder. Renew the vane seals as in paragraph 28.

6. Binding in the steering shaft bushing.

7. Insufficient pump pressure. Refer to paragraph 16.

8. Insufficient oil in system.

9. Foaming oil in system. Refer to paragraph 15C.

10. Seal between reservoir and cylinder leaking. Refer to paragraph 28.

15B. **EXCESSIVE INSTABILITY OF FRONT WHEELS.** This condition is often referred to as shimmy or flutter. Possible causes of instability are:

1. Excessive volume of oil flowing from flow control valve to steering valve. Adjust the flow control valve as in paragraph 19.

2. Actuating sleeve set screw too loose in helix slot. Refer to paragraph 24.

3. Worn point on actuating sleeve set screw and/or damaged helix slot in steering worm shaft.

4. Insufficient tension on cam spring. Refer to paragraph 27A or 27D.

5. Excessive end play in the steering worm shaft. See paragraph 25 or 27B.

6. Excessive backlash between the steering worm and gear. Refer to paragraph 26.

7. Unbalanced front wheels.

8. Loose or worn front wheel bearings.

9. Loose or worn tie rod ends.

10. Excessively worn spindle pins.

15C. **OVERFLOW OR FOAMING OF OIL** could be caused by:

1. Air leak in system. To check, apply a light coat of oil to sealing surfaces and observe for leaks.

2. Wrong type of oil. Use only John Deere power steering oil.

15D. **LOCKING** could be caused by:

1. Scored worm, or worm bearings adjusted too tight. Refer to paragraph 25 or 27B.

2. Steering valve arm interference in groove of actuating sleeve.

3. Steering gear loose on steering gear shaft. Tighten the retaining nut securely.

4. Bent steering gear shaft or scored steering gear shaft bearings. Refer to paragraph 28.

5. Loose or broken movable vane retaining cap screws. Screws should be tightened to a torque of 208 ft.-lbs. Refer to paragraph 28.

6. Insufficient clearance between actuating sleeve and the steering worm shaft or cam. Refer to paragraph 27A or 27D.

15E. **VARIATION IN STEERING EFFORT WHEN TURNING IN ONE DIRECTION** could be caused by a bent steering gear shaft. Refer to paragraph 28.

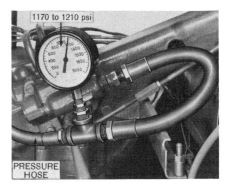

Fig. JD1506—Pressure gage installation for checking the power steering relief valve opening pressure. Opening pressure should be 1170-1210 psi.

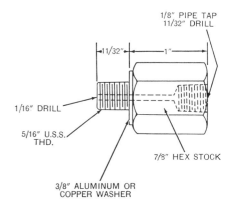

Fig. JD1506A — Adapter which can be made to accommodate a pressure gage in the steering valve housing.

SYSTEM OPERATING PRESSURE

16. The system relief valve is located in the steering valve housing as shown in Fig. JD1509. To check the system operating pressure, install pressure checking gage as illustrated in Fig. JD1506, start engine and slowly turn the steering wheel to the extreme right or left. Note the highest pressure reading on the gage, which should be 1170-1210 psi.

The relief valve operating pressure can also be checked as follows: Use the adapter (Fig. JD1506A) and install a suitable pressure gage (at least 1500 psi) in flow control stop screw hole in valve housing. With the engine running at fast idle rpm, turn the steering wheel to the extreme right or left and note the highest pressure reading on the gage, which should be 1170-1210 psi.

If the operating pressure is not as specified, vary the number of washers (3—Fig. JD1509) which are located under plug (1). If the addition of washers will not increase the pressure to within the specified limits, look for a faulty pump.

PUMP

17. **REMOVE AND REINSTALL.** To remove the power steering pump, first drain power steering system and remove fan and water pump drive belts. Unbolt coupling shaft tube from rear of pump, disconnect the pressure hose and the suction tube and remove the pump retaining cap screws.

Move the pump assembly forward until the fan shaft coupling is exposed; then, while holding the coupling rearward, remove pump from tractor.

NOTE: If the coupling is not held rearward while pump is moved forward, the fan shaft may come out of the rear coupling.

Install the pump by reversing removal procedure and fill pump with power steering oil before connecting oil lines.

18. **OVERHAUL.** With the pump removed from the tractor, thoroughly clean all exterior surfaces to remove any accumulation of dirt or other foreign material. Remove the pulley retaining cap screw (25—Fig. JD1507) and with suitable puller, remove pulley from pump shaft. Unbolt and separate the pump housing (40) and body (34) from the pump cover (31). Remove the drive gear from shaft (46) and withdraw follower gear and shaft from the pump.

Note: There is a Woodruff key (37) in the follower shaft as well as the pump drive shaft.

Remove the snap ring (28) and using a soft hammer, bump the shaft

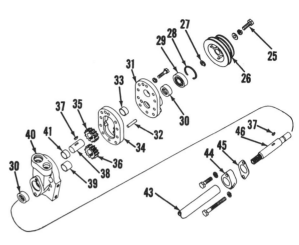

Fig. JD1507 — Exploded view of the power steering pump which is located on the forward end of the fan drive shaft.

26. Pump pulley
27. Snap ring
28. Snap ring
29. Bearing
30. Oil seal
31. Pump cover
32. Dowel pin
33. Bushing
34. Pump body
35. Idler gear
36. Drive gear
38. Idler gear shaft
39. Bushing
40. Pump housing
41. Bushing
43. Fan drive shaft tube
44. Pump flange
45. Gasket
46. Pump drive shaft

(46) and bearing (29) from the cover. Remove the snap ring (27) and press the shaft out of the bearing. Thoroughly clean all parts except the sealed bearing (29) in a suitable solvent.

Pump specifications are as follows:
Gear diameter 2.0825-2.0835
Thickness of
pump body 0.6270-0.6280
Gear bore in pump
body 2.0860-2.0880
I.D. of installed
bushings 1.0020-1.0030
Shaft diameter 0.9994-1.0000
Gear thickness 0.6250-0.6256

If bushings are scored or excessively worn they should be renewed. When installing oil seals (30) they should be pressed in until they bottom in their respective bores. Renew any other parts which are questionable.

When reassembling, press the pump drive shaft into the bearing and install the snap ring (27). Install the shaft and bearing in the pump cover and secure with snap ring (28). Install the Woodruff keys in the gear shafts and slide the gears into position. Gears must not bind on the Woodruff key or

shaft. Be sure that both cover and body mating surfaces are clean and free from burrs. Apply a thin coat of shellac on body surface and install the cover. Follow the same procedure when installing the housing. Tighten the assembly cap screws to a torque of 36 Ft.-Lbs., then check to be sure the pump shaft turns freely.

After pump is installed on tractor, check the system operating pressure as outlined in paragraph 16.

STEERING VALVES

19. **ADJUST FLOW CONTROL VALVE.** Turning the flow control valve adjusting screw inward will cause faster, easier steering action; but the increase in steering speed can result in a decrease of front wheel stability (increased tendency of front wheels to flutter).

To make the adjustment, operate the steering system until the oil is at normal operating temperature and remove cap nut; then, turn the adjusting screw (17—Fig. JD1508) in until the fastest turning speed is obtained without causing an objectionable amount of front wheel flutter.

Fig. JD1508 — Adjusting the power steering flow control valve. To make the adjustment, turn screw (17) in to obtain the fastest turning speed without causing an objectionable amount of front wheel flutter.

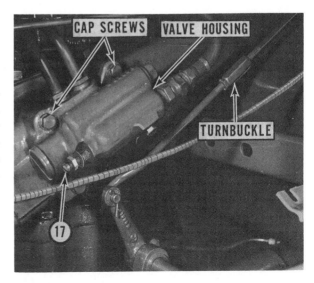

20. CENTERING VALVE HOUSING. The steering valve housing (Fig. JD1508) is properly centered when the effort required to turn the steering wheel in one direction is exactly the same as the effort required to turn the steering wheel in the other direction. The adjustment is made by loosening the four valve housing retaining cap screws, tapping the housing either way as required and then tightening the cap screws.

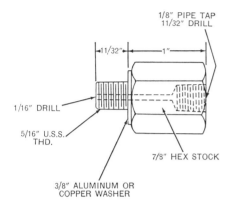

Fig. JD1508A — Adapter which can be made to accommodate a pressure gage in the steering valve housing.

To accurately center the valve housing using a pressure gage, use the adapter (Fig. JD1508A) and install a suitable pressure gage (at least 1500 psi) in flow control stop screw hole in valve housing. With the engine running at fast idle rpm and the front wheels in the straight ahead position, loosen the steering valve attaching screws and tap the housing to the front or rear until the lowest pressure reading is obtained; then, tighten the housing retaining cap screws.

21. REMOVE AND REINSTALL. To remove the valve housing, remove the hood, drain the power steering system and disconnect the oil lines from valve housing. To facilitate reinstallation, mark the relative position of the valve housing with respect to the worm housing with a scribed line. Unbolt and remove the valve housing assembly from tractor.

22. To install the valve housing, use a new gasket and tighten the assembly cap screws finger tight; also, be sure to align the previously affixed scribed lines. Reconnect the oil lines, fill the reservoir and start the engine. Center the valve housing as in paragraph 20.

23. OVERHAUL. With the valve housing removed, clean the unit and

remove the flow control valve adjusting screw (17—Fig. JD1509). Remove the union adapter from rear of housing and slide the flow control valve and spring (20 & 21) out of the valve bore. Remove the end plug (13) from housing and remove the steering valve arm (8) by removing its retaining bolt as shown in Fig. JD1509A. Remove the relief valve plug (1—Fig. JD1509) and withdraw the relief valve sleeve, valve and spring. Wash all parts in a suitable solvent and renew any that are nicked, grooved or worn.

Install the flow control valve spring, valve and adjusting screw, and initially adjust the valve as follows: Using a punch or similar tool, hold the valve against rear of flow valve passage and turn the adjusting screw in until it just contacts the spring; then turn the screw in two additional turns. Install the union adaptor, control valve and arm and end plug. Assemble the relief valve, using the same number of adjusting washers as were removed.

With the valve housing installed, center the unit as outlined in paragraph 20 and adjust the flow control valve as outlined in paragraph 19. Check the system operating pressure as outlined in paragraph 16.

STEERING WORM AND VALVE ACTUATING SLEEVE
Models 80-Early Production 820

24. VALVE ACTUATING SLEEVE SET SCREW ADJUSTMENT. To accurately adjust the valve actuating sleeve set screw to the recommended end play of 0.001-0.002 (always try to adjust this end play to the minimum specification), the steering shaft and worm housing must be removed from

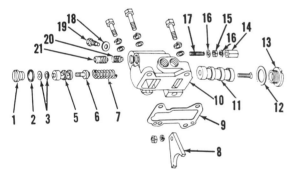

Fig. JD1510 — Adjustment of model 80 and early production 820 power steering valve actuating sleeve set screw is made easier if a screw driver is ground to the dimensions shown in Fig. JD1510B.

Fig. JD1509 — Exploded view of the power steering valve housing, pressure relief valve, flow control valve and components.

1. Plug
2. Gasket
3. Shim washers
5. Relief valve sleeve
6. Relief valve
7. Spring
8. Valve arm
9. Gasket
10. Steering valve housing
11. Steering valve
12. Gasket
13. Plug
14. Cap nut
15. Jam nut
16. Washer
17. Adjusting screw
18. Copper washer
19. Stop screw
20. Flow control valve spring
21. Flow control valve

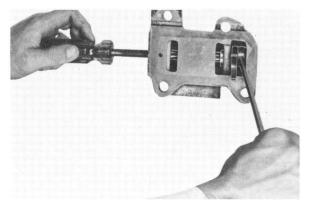

Fig. JD1509A—Removing the screw retaining the steering valve arm to the steering valve.

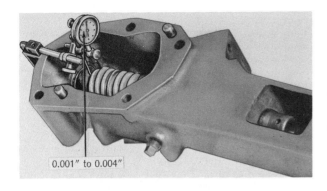

0.001" to 0.004"

Fig. JD1510A—The steering worm end play should be adjusted by varying the shims under bearing housing to provide an end play of 0.001-0.004.

the tractor. The following, however, is an approximate adjustment which can be made while the steering worm shaft housing is on the tractor. Remove pipe plug and with a screw driver (ground to the dimensions shown in Fig. JD1510B) inserted as shown in Fig. JD1510, screw the self-locking adjusting screw down snug into the helix slot in the steering worm shaft; then, back off the screw ⅛ turn. NOTE: If old screw is loose in actuating sleeve, use a new screw. If the new screw is loose, the actuating sleeve should be renewed. Refer to paragraph 27. After adjustment, lock the screw in position by staking.

25. STEERING WORM END PLAY ADJUSTMENT. To make the adjustment, first remove the worm housing assembly as outlined in paragraph 27 and proceed as follows: Mount a dial indicator as shown in Fig. JD1510A and check the worm shaft end play which should be 0.001-0.004. If the end play is not as specified, vary the number of shims (73—Fig. JD1511), which are located under bearing housing (75). Always try to adjust this end play to the minimum specification of 0.001.

26. BACKLASH. Backlash between the steering worm and gear is controlled by shims (54—Fig. JD1511) located between the worm shaft housing and the gear housing. To check the backlash, rotate the steering wheel back and forth without permitting the steering cams within the actuating sleeve to separate or the front wheels to turn. Backlash should be ⅜-¾ inch when measured at outer edge of the steering wheel. To vary the number of shims, it is necessary to remove the worm housing assembly as outlined in paragraph 27.

27. R&R AND OVERHAUL. To remove the steering worm, worm housing and valve actuating sleeve assembly, first remove the tractor hood and drain the power steering unit. Discon-

nect the diesel engine speed control rod at top of transmission case cover. Disconnect the gas line, wires and controls from the gasoline starting engine. Disconnect the headlight wires and both of the shutter control rods. Unbolt the instrument panel and pull same rearward. Disconnect dash light wire. Unbolt and raise cowl enough to unclip headlight wires from rear underside of cowl; then, remove cowl from tractor. Disconnect the power steering pressure hose from the steering valve, remove the two steering valve-to-cylinder oil tubes, then unbolt and remove the valve assembly. Remove the five cap screws which attach the worm shaft housing to the steering gear housing and remove the worm shaft housing assembly from the tractor, being careful not to damage or lose any of the shims located between the two housings.

27A. Remove the roll pin (58—Fig. JD1511), then unbolt and remove the bearing housing (75) being careful not to damage or lose shims (73). Remove the actuating sleeve set screw (59) and withdraw the worm gear from the actuating sleeve and housing. Extract the worm gear rear bearing cup and withdraw the actuating sleeve. Remove cam spring (65) and cam (67) from rod (66).

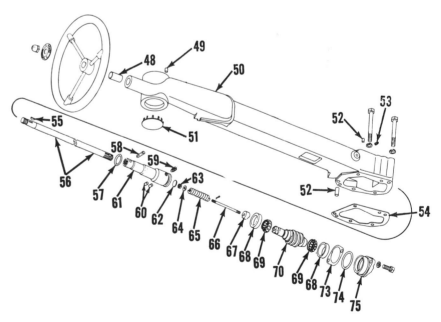

Fig. JD1511 — Exploded view of model 80 and early production 820 power steering worm, wormshaft housing, valve actuating sleeve and associated parts. Backlash between worm and steering gear is controlled by shims (54) and end play of the steering worm by shims (73).

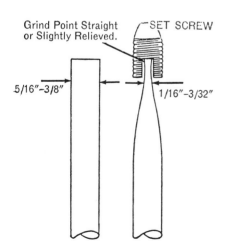

Grind Point Straight or Slightly Relieved.

SET SCREW

5/16"-3/8"   1/16"-3/32"

Fig. JD1510B — Dimensions for grinding a special screw driver for adjusting the early production valve actuating sleeve set screw.

48. Bushing	53. Plug	60. Pins	67. Worm cam
49. Dowel pin	54. Shims	61. Valve actuating sleeve	68. Bearing cup
50. Wormshaft housing	55. Woodruff key	62. Expansion plug	69. Bearing cone
51. Speed control opening button plug	56. Steering shaft	63. Nut	70. Steering worm
	57. "O" ring	64. Washer	73. Shims
52. Dowel pin	58. Roll pin	65. Spring	74. "O" ring
	59. Valve actuating sleeve set screw	66. Cam rod	75. Bearing housing

Clean and inspect all parts and renew any that are damaged and cannot be reconditioned. Renew the worm shaft if it is scored or worn at cam end, helix slot or worm. Burrs can be removed from the worm shaft keyway or the surface over which the actuating sleeve operates by using a fine stone. Check the valve actuating sleeve pins for damage, and bore of sleeve for being scored. Check fit of sleeve on worm shaft. If sleeve is tight on shaft or if bore is scored, remove the dowel pins and hone the sleeve to insure a free fit. Renew the pins if they are worn or damaged. Renew the actuating sleeve set screw if it has a loose fit in sleeve or if cone point is worn.

When reassembling, use a valve spring tester and measure the length of spring (65) at 70-80 lbs. pressure; then record the length.

Note: If spring will not exert 70-80 lbs. pressure at 2⅞-inches or more for a three inch free length spring or 1⅞-inches or more for a 2⅛-inch free length spring, obtain a new spring and measure its length at 70-80 lbs. pressure.

Then, install the spring on the cam rod and tighten the adjusting nut until spring is compressed to the recorded length which yielded the 70-80 lbs. pressure.

Reassemble the worm and actuating sleeve and adjust the end play as in paragraphs 24 and 25. Install the unit and adjust the backlash as in paragraph 26. Install the valve housing as outlined in paragraph 22.

STEERING WORM AND VALVE ACTUATING SLEEVE
Models 820 (After 8205349)-830
27B. **END PLAY ADJUSTMENTS.** Two end play adjustments are pro-

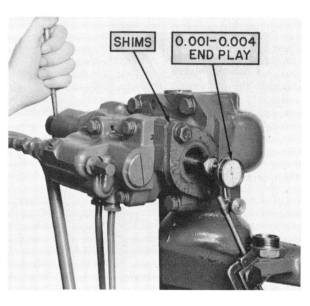

Fig. JD1512A — Using dial indicator to check the steering worm end play. This illustration applies to the 520, 620 and 720 tractors, but a similar procedure can be used on late production model 820 and model 830 tractors.

vided in the steering worm and valve actuating sleeve assembly. The number of shims between the worm housing and the front bearing housing should be varied to obtain a worm shaft end play of 0.001-0.004. Then, the worm helix width should be adjusted to provide the valve actuating sleeve with an end play of 0.0015-0.0025. To make the adjustments, proceed as follows: Remove hood, disconnect steering shaft and pull the steering shaft rearward and free from the valve actuating sleeve. On tractors equipped with the gasoline starting engine, remove the solenoid switch from starting motor. Note: Some mechanics prefer to remove the complete worm and housing assembly from tractor. On all models, remove the large plug (22—Fig. JD1512), install a cap screw in end of steering worm and mount a dial indicator in a manner

similar to that shown in Fig. JD1512A. Use a suitable drift in actuating sleeve roll pin hole, rotate worm assembly in both directions and note the amount of end play in the worm bearings as registered on the dial indicator. If the indicator reading is not 0.001-0.004, add or deduct shims (19—Fig. JD1512) until proper end play is obtained. It is important that end play be adjusted as close to 0.001 as possible.

With the worm shaft end play properly adjusted, mount a dial indicator in a manner similar to that shown in Fig. JD 1512B, reach into the large screw plug opening in front bearing housing with your fingers and while pressing on the worm to keep it from moving, use a suitable drift in actuating sleeve roll pin hole, move the valve actuating sleeve back and forth to determine the amount of end play between the hardened dowel and the helix. If the indicator reading is not 0.0015-0.0025, adjust the helix width by loosening the special lock screw (15—Fig. JD1512), which is located inside bushing (14), and using a ⅞-

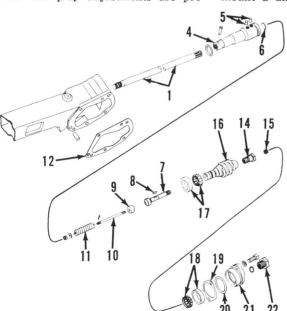

Fig. JD1512 — Exploded view of late production model 820 and model 830 steering worm, valve actuating sleeve and associated parts. Backlash between the worm and steering gear is controlled by shims (12).

1. Steering shaft
4. Actuating sleeve
5. Dowels
6. Welch plug
7. Worm shaft
8. Woodruff key
9. Cam
10. Spring rod
11. Cam spring
12. Adjusting shims
14. Adjusting bushing
15. Lock screw
16. Worm
17. Bearing cup and cone
18. Bearing cup and cone
19. Adjusting shims
20. "O" ring
21. Worm bearing housing
22. Screw plug

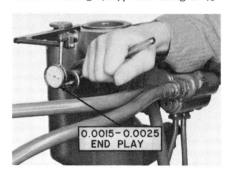

Fig. JD1512B—Checking end play of steering valve actuating sleeve. Specified end play is 0.0015-0.0025. This illustration applies to the 520, 620 and 720 tractors, but a similar procedure can be used on late production model 820 and model 830 tractors.

inch socket, turn the bushing **in** to decrease or **out** to increase the end play clearance. Adjust the clearance as close to 0.0015 as possible. When adjustment is complete, hold the bushing from turning and tighten the lock screw. Note: Never make the adjustment so tight that steering wheel will not return to neutral when released after being turned in either direction to operate the steering valve.

27C. **BACKLASH.** To check and/or adjust the backlash between the steering worm and gear, refer to paragraph 26.

27D. **R&R AND OVERHAUL.** Remove the steering worm, worm housing and valve actuating sleeve assembly as in paragraph 27 and proceed as follows: Remove the worm shaft front bearing housing (21—Fig. JD 1512) and carefully measure the distance the adjusting bushing protrudes from the worm as shown in Fig. JD1512C. Loosen the special lock screw, remove the adjusting bushing and withdraw the worm.

Extract the worm gear rear bearing cup and pull actuating sleeve assembly from housing. To remove the worm shaft from the actuating sleeve, use a pair of vise-grip pliers to turn the shaft about 90 degrees in a counter-clockwise direction so the cam and shaft keyways align.

Clean and inspect all parts and renew any that are damaged and cannot be reconditioned. Renew worm if it is worn or if helix end is scored. Renew shaft (7—Fig. JD1512) if it is scored at cam or helix locations. Small burrs can be removed from wearing surfaces by using a fine stone. Check the valve actuating sleeve dowels (5) for damage, and bore of sleeve for being scored. Check fit of parts by inserting them in the sleeve. If binding exists, remove the dowel pins and hone the sleeve bore to insure a free fit. Renew the dowel pins if they are worn or damaged.

When reassembling, use a valve spring tester and measure the length of spring (11) at 70-80 lbs. pressure. Note: If the spring will not exert 70-80 lbs. pressure at 1⅞-inches or more, obtain a new spring and measure its length at 70-80 lbs. pressure. Then install spring on spring rod and tighten the adjusting nut until spring is compressed to the same length which yielded a pressure of 70-80 lbs.

Assemble worm shaft (7) and cam (9) with keyways aligned. Then insert

Fig. JD1512C — Checking protrusion of adjusting bushing from steering worm. This illustration applies to the 520, 620 and 720 tractors, but a similar procedure can be used on late production model 820 and model 830 tractors.

the worm shaft assembly into the actuating sleeve so the keyways will slide under the dowels and when the cam is behind the front dowel, turn the worm shaft clockwise with a pair of vise-grip pliers until the cam seats. Install actuating sleeve and worm shaft assembly into housing and install bearing assembly (17). With Woodruff key properly positioned in worm shaft, slide worm into position.

Using a small screw driver as shown in Fig. JD1512D, move the worm shaft until it lacks about 1/32-inch of being flush with end of worm; then, carefully install the adjusting bushing and screw it into the worm until all actuating sleeve end play is removed (dowel tight in helix). Measure the bushing protrusion as shown in Fig.

JD1512C. If the measured dimension is not almost identical to that measured before the bushing was removed, unscrew the bushing and start over.

When proper bushing protrusion is obtained, install lock screw, front bearing housing and oil seal housing. Check the end play adjustments as in paragraph 27B. Install the worm housing assembly and adjust the gear backlash as in paragraph 26. Install the steering valve housing as in paragraph 22.

STEERING GEAR SHAFT, HOUSING AND CYLINDER

28. **R&R AND OVERHAUL.** It is possible to remove the worm shaft housing, gear housing and steering cylinder assembly as a unit but this

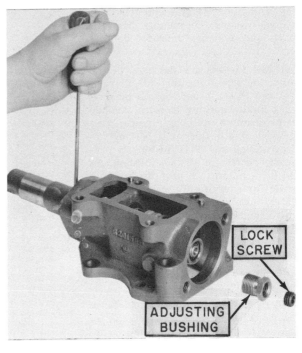

Fig. JD1512D — Positioning the worm shaft with respect to the worm prior to installing the adjusting bushing. This illustration applies to the 520, 620 and 720 tractors, but a similar procedure can be used on late production model 820 and model 830 tractors.

procedure is awkward and not always practical.

Remove the worm gear housing as in paragraph 27 and proceed as follows: Unclip the pressure hose from the gear housing and disconnect the suction tube from the reservoir. Unclip wires and lines from top of the power steering reservoir and starting engine timing gear cover. Disconnect the oil pressure gage line from the tractor main case and the distributor lead in wire from distributor. On tractors equipped with the gasoline starting engine, it is necessary to loosen or remove the electric starting motor. On all models, mark the steering gear arm and vertical steering shaft to aid in mating the correct splines in reassembly; then, disconnect the steering gear arm from the shaft. Remove bolts retaining the gear housing and steering cylinder assembly to the main case. Attach a suitable hoist and with all wires, rods and lines out of the way, lift the assembly from the tractor.

Note: Before disassembling the unit, mark the housing and steering gear shaft quill in relation to the cylinder to aid in assembly. Also, note the location of the two cap screws which have flat washers and packing washers.

Remove assembly cap screws and separate the cylinder from the quill and gear housing. Refer to Fig. JD 1513.

Remove the stop pins (91) and the stationary vane. Remove the two special cap screws (102) and remove the movable vane and bracket assembly. To remove the steering gear (81) and shaft (104), first remove the end play adjusting screw (77). Remove cotter pin, and nut (79) and pull the shaft out of housing. Thoroughly clean all parts and examine them for damage or wear.

If the threads are damaged in the adjusting screw adapter (78), drive it out and install a new one. If bushing (84) in the housing is worn, remove same using a suitable puller. Press new bushing in flush with bottom surface of housing. This bushing is pre-sized and if carefully installed should need no honing. Inspect the steering shaft for excessive wear, scored surfaces, worn splines or bends. If any of these conditions exist, renew the shaft. Inspect the gear (81) and renew same if damaged.

On models after 8200474, remove and renew the "O" ring packing (83) and backing washer (82).

NOTE: The backing washer is installed at the top of the groove. If not installed in this manner, the "O" ring packing will soon fail.

On models prior to 8200475, there is no seal between reservoir and cylinder. If leakage occurs at this point the housing should be renewed.

Inspect the vanes and seal contacting surfaces of quill, cylinder and housing. Renew any parts which are excessively worn, scored, cracked, chipped or broken. If the attaching pin for the movable steering vane has worn a groove in the quill, renew quill, pin and vane. The new type vane uses a headed pin to eliminate grooving of the quill. It is always a good practice to renew the vane packings whenever the unit is disassembled.

29. To reassemble, observe the following: On late models, lubricate "O" ring packing (83) in housing. Install the gear shaft (104), and gear (81) and tighten the retaining nut securely. When the headed vane pin is used, the vane must be installed on the bracket before the bracket is installed. Install vane bracket (100) and tighten special cap screws (102) to a torque of 208 Ft.-Lbs. Install vanes (89) and stop

Fig. JD1513 — Exploded view of power steering gear housing, cylinder, shaft and associated parts.

71. Cap nut
72. Washer
76. Jam nut
77. Vertical steering shaft end play adjusting screw
78. Adjusting screw adapter
79. Nut
80. Washer
81. Steering gear
82. Backing washer
83. "O" ring
84. Bushing
85. Cylinder gasket
86. Cylinder
87. Cylinder gasket
88. Vane pin
89. Steering vane
90. Steering vane packing
91. Stop pins
92. Steering gear shaft quill
93. Dowel pin
94. Packing washer
95. Flat washer
96. "O" ring
97. Backing washer
99. Woodruff key
100. Steering vane bracket
102. Steering vane bracket retaining cap screws
103. Vane pin
104. Vertical steering shaft
105. Bushing
106. Steering gear shaft tube

pins (91). Install a new cylinder gasket on the housing and install the cylinder, aligning the reference marks made before disassembly. Install a new cylinder gasket and assemble quill (92) to cylinder. Use reference marks for proper alignment. Install the assembly cap screws and nuts and tighten to a torque of 150 Ft.-Lbs.

NOTE: Make certain that the two cap screws with flat washers (95) and packing washers (94) are in the proper holes. The casting has a recess around the hole for the packing washers.

Install the end play adjusting screw (77), turn the screw snug against the end of the shaft, then back it up ⅛ turn and lock with jam nut. Before installing the assembled unit, examine the bushing (105) located in bottom of main case. Renew the pre-sized bushing if it is damaged or worn.

To install the unit, first make certain the "O" ring packing recess in the bottom of the steering gear shaft quill is clean. Lubricate new "O" ring and install it into recess. Install backing gasket (97), using grease to hold it

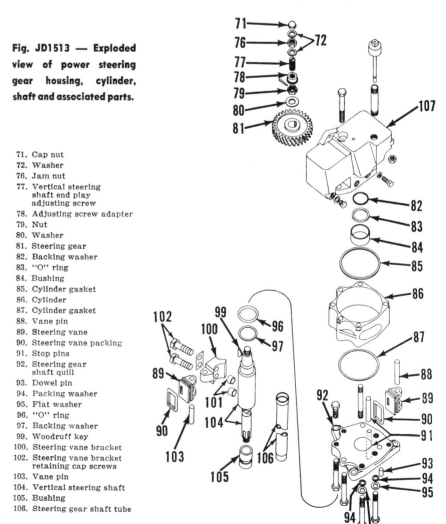

against "O" ring while installing unit on tractor. Install the power steering unit and tighten the attaching cap screws to a torque of 150 Ft.-Lbs. Install the worm shaft and housing assembly.

29A. **STEERING SHAFT TUBE — RENEW.** To renew the vertical steering shaft tube (106—Fig. JD1513), first remove the complete power steering gear unit from tractor. Drive bushing (105) out through bottom of main case and tube (106) out through top. Install new tube from above and using a piloted driver, bump the tube down until lower shoulder of tube bottoms in the main case. Install new, presized steering gear shaft bushing.

STARTING ENGINE, COMPONENTS AND ACCESSORIES

The 80 and 820 diesel tractors are equipped with a four cylinder, V-type, valve-in-head starting engine which has a bore of two inches, and a stroke of one and one-half inches.

John Deere 830 diesel tractors are available with either the gasoline starting engine which is covered in the following paragraphs or with the 24-volt electric starting motor which is covered in paragraph 128A.

R&R ENGINE ASSEMBLY

30. To remove the starting engine assembly, first drain the tractor cooling system, disconnect the battery ground strap and remove hood. Disconnect the starting engine throttle and fuel shut-off rods at couplings. Disconnect choke at bellcrank and wire from the automatic gasoline shut-off pressure switch. Remove the starting engine air cleaner, water intake pipe, water return pipe and exhaust pipe. Remove the diesel engine flywheel cover and the two cap screws retaining the flywheel cover back plate to the starting engine transmission. Unclip heat indicator tube and voltage regulator wires from timing gear housing and move same out of the way. Disconnect speed-hour meter cable from fan drive shaft rear bearing housing and pull cable to the rear, out of the way. Disconnect oil pressure gage line from the tractor main case and pull same as far to the side as practical. Disconnect wires from distributor and starting motor magnetic switch. Remove starting motor from starting engine; or remove the hydraulic pump. Disconnect the starting engine clutch rod. Remove the four cap screws which secure the starting engine to the main case and with a suitable hoist, remove the starting engine from the tractor.

When reinstalling the starting engine, reverse the removal procedure and be sure the oil drain hole gasket is in position in the diesel engine crankcase cover before lowering the starting engine into position. Tighten the engine attaching cap screws to a torque of 56 Ft.-Lbs.

CYLINDER HEADS

31. It is possible to remove either cylinder head from the starting engine without removing the starting engine from tractor as follows: Drain the tractor cooling system, disconnect the battery cable and remove the tractor hood. Disconnect the starting engine throttle and fuel shut-off rods at couplings. Disconnect choke at bellcrank, choke link from carburetor and governor spring from bracket. Unbolt and remove governor spring bracket from the starting engine manifold. Disconnect fuel line at automatic gasoline shut-off. Remove air cleaner, water by-pass pipe, water outlet pipe and exhaust pipe. Disconnect the throttle rod from the carburetor. Disconnect the oil line and wire from the automatic gasoline shut-off and unbolt shut-off from its bracket. Unbolt and remove the carburetor and gasoline shut-off assembly, then remove the starting engine manifold.

Remove the valve covers. Remove the machine screws (1—Fig. JD1516) from each end of the tappet lever shaft and also the Allen head set screw (4) which positions the shaft in the cylinder head. With valves closed, slide the tappet lever shaft out of the head and remove the tappet levers, spacer springs and push rods. Unbolt and remove the cylinder head from engine.

NOTE: Do not turn the engine crankshaft with cylinder heads off unless a 7/16-inch

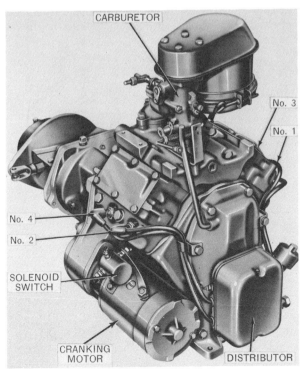

Fig. JD1515—Three quarter view of the gasoline starting engine. The V-type, valve-in-head, 4 cylinder engine has a 2-inch bore, a 1½-inch stroke and a rated speed of 4500 rpm.

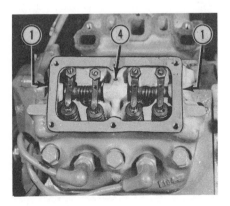

Fig. JD1516—The starting engine cylinder head with valve cover removed. Tappet lever shaft is retained by Allen head set screw (4).

cap screw and washer are installed as shown in Fig. JD1523 to hold the cylinder sleeves in position.

When reassembling, install head gasket on block with flange side up, then install head and tighten the cap screws finger tight. Install the manifold and tighten retaining cap screws securely; then tighten the cylinder head retaining cap screws in the sequence shown in Fig. JD1517 and to a torque of 67 Ft.-Lbs. Install the push rods, tappet levers and spacer springs. Slide the tappet lever shaft into position, making certain that the set screw hole aligns with the Allen head set screw (4—Fig. JD1518). Install a new packing washer (2) at each end of the tappet lever shaft and install and tighten machine screw (1) in each end of shaft.

VALVES AND SEATS

32. To remove the valves, remove the cylinder heads as outlined in paragraph 31, compress the valve springs and remove the split cone keepers.

Intake and exhaust valves are not interchangeable and seat directly in the cylinder head with a face angle of 44½ degrees and a seat angle in the cylinder head of 45 degrees. Seats should be concentric with guides within 0.005. Valve seat width is 0.052-0.072 for the intake, 0.030-0.050 for the exhaust. Seats can be narrowed using 20 and 70 degree stones. Valve stem diameter is 0.2445-0.2455 for both the intake and the exhaust.

Valve tappet gap is 0.008-0.010 for both the intake and the exhaust. Adjust the clearance for each valve when the other valve of the same cylinder is in the wide open position.

VALVE GUIDES

33. Intake and exhaust valve guides are interchangeable and can be driven from the cylinder head if renewal is required. When installing new guides, coat the outer surface of the guides with sealing compound and, using a cup shaped driver as shown in Fig. JD1519, drive the guides in until distance from end of guide to top of valve seat is $\frac{27}{32}$ inches.

Ream the guides after installation to provide a valve stem to guide diametral clearance of 0.0005-0.0025. Valve stem diameter is 0.2445-0.2455 for both the intake and the exhaust.

VALVE SPRINGS AND TAPPETS

34. Intake and exhaust valve springs are interchangeable and can be removed after removing the cylinder heads as outlined in paragraph 31. Springs should test 15.5-18.5 pounds when compressed to a height of $1\frac{1}{16}$ inches and 37-45 pounds at $\frac{13}{16}$-inch. Renew any spring which is rusted, discolored, or does not meet the foregoing pressure specifications.

35. The barrel type tappets (cam followers) ride directly in the cylinder block bores and are available in standard size only. Any tappet can be removed after removing the cylinder heads as outlined in paragraph 31.

VALVE TIMING

36. To check valve timing, first remove timing gear cover as outlined in paragraph 40. Valves are properly timed when "V" mark on the camshaft gear is in register with "V" mark on crankshaft gear as shown in Fig. JD1520.

VALVE TAPPET LEVERS

39. The tappet levers and shaft can

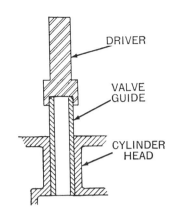

Fig. JD1519 — Starting engine valve guides should be installed with a cup shaped driver as shown.

be removed from either cylinder head without removing the starting engine from the tractor. Remove the tractor hood. Remove the valve covers and the machine screw (1—Fig. JD1518) from each end of the tappet lever shaft. Remove the Allen head set screw (4) which positions the tappet lever shaft in the cylinder head, and with the valves closed, slide the tappet lever shaft out of the head and remove the tappet levers and spacer springs. Intake and exhaust valve tappet levers are interchangeable.

Check all parts and renew any which are excessively worn.

Shaft diameter........0.310-0.311

Tappet lever bore.....0.3125-0.3145

Thoroughly clean the oil holes in the tappet levers and shaft before reassembly. Install the tappet levers and spacer springs and slide the tappet

Fig. JD1517—Sequence for tightening the starting engine cylinder head cap screws. Specified torque value is 67 Ft.-Lbs.

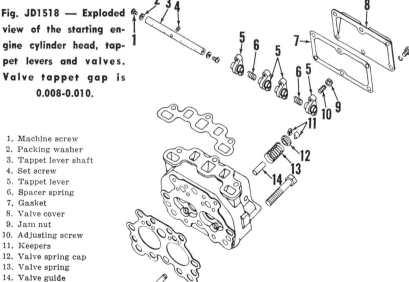

Fig. JD1518 — Exploded view of the starting engine cylinder head, tappet levers and valves. Valve tappet gap is 0.008-0.010.

1. Machine screw
2. Packing washer
3. Tappet lever shaft
4. Set screw
5. Tappet lever
6. Spacer spring
7. Gasket
8. Valve cover
9. Jam nut
10. Adjusting screw
11. Keepers
12. Valve spring cap
13. Valve spring
14. Valve guide
15. Valve

Fig. JD1520 — The starting engine valves are properly timed when "V" marks on camshaft gear and crankshaft gear are in register.

lever shaft into position, making certain that the set screw hole aligns with the Allen head set screw. Install and tighten the set screw. Install a new packing washer (2—Fig. JD1518) at each end of the tappet lever shaft and install and tighten the machine screw in each end of the shaft.

TIMING GEARS AND COVER

40. **TIMING GEAR COVER.** To remove the timing gear cover, first remove the tractor hood and on models so equipped, drain the power steering reservoir and remove the power steering return line. Drain the starting engine crankcase and disconnect the oil pressure gage line from the tractor main case. Unclip the heat gage tube and voltage regulator wires from the starting engine timing gear cover and remove the ignition distributor cover. Remove the four machine screws retaining the distributor to the timing gear cover and remove the distributor. Remove the two right starting engine retaining screws, loosen the two left screws and raise the engine slightly. Disconnect the governor linkage from throttle linkage, then unbolt and remove the timing gear cover. The crankshaft front oil seal can be renewed at this time and should be installed with lip facing flywheel end of engine.

When reassembling, check and adjust the ignition timing as outlined in paragraph 59.

41. **TIMING GEARS.** To remove the timing gears, first remove the timing gear cover as in preceding paragraph.

Fig. JD1521 — Using jack screws to pull the starting engine camshaft gear from the shaft. Refer to text.

To remove the camshaft gear, withdraw the governor ball race and balls and remove snap ring (27—Fig. JD 1520) and the ball retainer (26). Turn the camshaft gear until the two puller holes are directly over the two camshaft thrust plate retaining cap screws; then, using two ⅜-inch puller cap screws as shown in Fig. JD1521, pull the camshaft gear from shaft. The crankshaft gear can be pulled from shaft after removing its retaining snap ring.

When reassembling, mesh the "V" mark on the camshaft gear with the "V" mark on the crankshaft gear as shown in Fig. JD 1520.

CAMSHAFT AND BUSHINGS

42. To remove the camshaft, first remove the timing gear cover as outlined in paragraph 40 and the cylinder heads as outlined in paragraph 31. Withdraw the cam followers from the cylinder block bores. Withdraw the governor ball race and balls, remove snap ring (27 —Fig. JD1520) and remove the ball retainer (26). Turn the camshaft gear until the two puller holes are directly over the two camshaft thrust plate retaining cap screws; then, using two ⅜-inch puller screws as shown in Fig. JD1521, pull the camshaft gear from shaft. Unbolt and remove the camshaft thrust plate (24—Fig. JD1522) and withdraw the camshaft from the cylinder block.

Check the camshaft and bushings against the values which follow:

Camshaft Journal Diameter
Front 1.248-1.249
Center 1.060-1.061
Rear 0.748-0.749

Bushing Inside Diameter
Front 1.251-1.253
Center 1.063-1.065
Rear 0.751-0.753

Journal Diametral
Clearance 0.002-0.005
Camshaft End Play 0.003-0.009

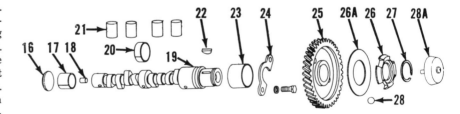

Fig. JD1522 — Exploded view of the starting engine camshaft, bearings and cam followers. Shaft end play is controlled by thrust plate (24).

16. Expansion plug	22. Woodruff key	26A. Thrust washer
17. Rear bushing	23. Front bushing	(830 and
18. Dowel pin	24. Thrust plate	late 820)
19. Camshaft	25. Camshaft gear	27. Snap ring
20. Center bushing	26. Governor ball	28. Governor flyball
21. Cam followers	retainer	28A. Governor ball race

If the camshaft end play is not as specified, renew the thrust plate (24).

To renew the camshaft bushings after camshaft is out, remove engine, unbolt clutch housing from cylinder block and remove clutch, flywheel and expansion plug behind the camshaft rear bushing. Drive out the old bushings and install the new ones using a piloted drift. Ream the bushings after installation, if necessary, to provide a journal diametral clearance of 0.002-0.005.

When installing the camshaft gear, mesh the valve timing marks as shown in Fig. JD1520. Retime the ignition as outlined in paragraph 59.

ROD AND PISTON UNITS

43. Connecting rod and piston units are removed from above as follows: Remove the starting engine as outlined in paragraph 30 and the cylinder heads as in paragraph 31. Install a $\frac{7}{16}$-inch cap screw and washer as shown in Fig. JD1523 to hold the cylinder sleeves in position when removing the rod and piston units. Drain the crankcase and remove the crankcase cover. Remove the water pump.

Pistons and connecting rods are not marked with cylinder numbers and they should be identified before removal. Bend the lock plate tabs away from connecting rod bearing cap screws, remove the connecting rod bearing caps and push the connecting rod and piston units out of cylinder sleeves.

44. When reassembling, install the rod and piston units in their respective cylinders so that depressions in top of pistons are toward top of engine as shown in Fig. JD1523. Install the connecting rod bearing caps so that raised projections on rod and cap are in register as shown at X in Fig. JD1524 and tighten the cap screws to a torque of 7 ft.-lbs.

PISTONS, RINGS AND SLEEVES

45. The aluminum alloy pistons have a skirt diameter of 1.996-1.997 and are equipped with two compression rings and one oil control ring. Pistons are available in standard size only and have a skirt clearance of 0.002-0.004 in the 1.999-2.000 diameter cylinder sleeves. Pistons and/or sleeves should be renewed if the skirt to sleeve diametral clearance exceeds 0.007.

Desired piston ring side clearance is 0.0015-0.0031 for the top compression ring, 0.0010-0.0026 for the second compression ring and the oil control ring. The oil ring can be installed with either side up. White dot on compression rings must face toward top of piston.

If the cylinder sleeves are worn, push them out of the cylinder block and remove any accumulation of cooling system sediment from engine water jacket and sleeve seating surfaces. Remove any foreign material or burrs which may prevent proper sleeve seating. Install a new "O" ring seal in cylinder block groove, lubricate the lower seating surface of the sleeve and push the sleeve into position.

PISTON PINS

46. The 0.4818-0.4822 diameter floating type piston pins are retained in the piston pin bosses by snap rings and are available in standard size only. Piston pins should have 0.0000-0.0008 diametral clearance in piston and 0.0002-0.0016 diametral clearance in the connecting rod. Piston can be installed either way on connecting rod, but rod and piston units must be installed as outlined in paragraph 44.

CONNECTING ROD

47. The aluminum alloy connecting rods ride directly on the 1.0615-1.0625 diameter crankshaft crankpins with a diametral clearance of 0.001-0.003. If clearance exceeds 0.005, renew the

worn part; do not file rod or cap. Connecting rod bearing bore is 1.0635-1.0645.

Connecting rod can be installed either way in piston, but rod and piston units must be installed as outlined in paragraph 44.

CRANKSHAFT AND MAIN BEARING BUSHINGS

48. The crankshaft is supported in two precision type main bearing bushings which can be renewed after removing the crankshaft. To remove the crankshaft, first remove the connecting rod and piston units as outlined in paragraph 43 and proceed as follows:

Remove the distributor cover, distributor and timing gear cover. Remove the gear retaining snap ring and pull the crankshaft gear from shaft. Unbolt clutch housing from cylinder block and remove the clutch and flywheel. Remove the crankshaft rear bearing housing, being careful not to lose the shims located between the bearing housing and crankcase and withdraw the crankshaft from the cylinder block as shown in Fig. JD 1525.

Check the crankshaft and main bearings against the values which follow:

Main Journal Diameter
Model 80 ....1.124-1.125
Model 820 prior 8203100..1.124-1.125
Model 820 after 8203099..1.499-1.500
Model 830 ....1.499-1.500

Bushing Inside Diameter
Model 80 ....1.127-1.128
Model 820 prior 8203100..1.127-1.128
Model 820 after 8203099..1.501-1.502
Model 830 ....1.501-1.502

Shaft Diametral Clearance in Bushings
Model 80 ....0.002-0.004
Model 820 prior 8203100..0.002-0.004
Model 820 after 8203099..0.001-0.003
Model 830 ....0.001-0.003

Fig. JD1523 — Cap screw and washer installed to hold the starting engine cylinder sleeves in position when cylinder heads are off.

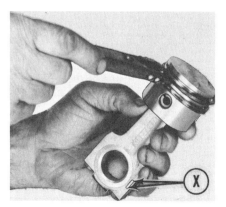

Fig. JD1524 — Installing rings on starting engine piston. When installing rod bearing caps, make certain that raised projections (X) are in register.

Fig. JD1525 — Removing crankshaft from starting engine cylinder block. Crankshaft end play is controlled by shims (37).

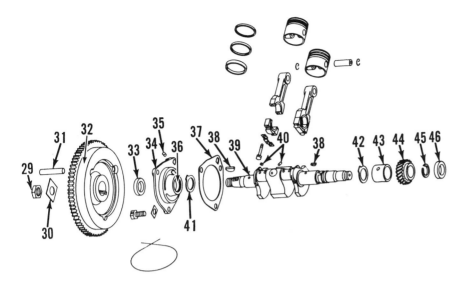

Fig. JD1526 — Exploded view of the starting engine pistons, rods, crankshaft and associated parts. Shaft end play of 0.005-0.008 is controlled by shims (37).

29. Nut	35. Dowel pin	41. Thrust washer
30. Lock plate	36. Rear bushing	42. Thrust washer
31. Clutch drive pin	37. Shims	43. Front bushing
32. Flywheel	38. Woodruff key	44. Crankshaft gear
33. Oil seal	39. Crankshaft	45. Snap ring
34. Main bushing housing	40. Pin	46. Oil seal

49. Main bearing bushings are available in standard size only. When installing new bushings, use a piloted driver and make sure the bushing oil holes are toward bottom of engine.

Install the crankshaft by reversing the removal procedure and be sure that tab on thrust washer (42—Fig. JD1526) fits into slot in cylinder block. The shaft rear oil seal (33) located in the bearing housing should be installed with lip facing inward. Install rear bearing housing and vary the number of shims (37) to obtain a shaft end play of 0.005-0.008. Securely tighten and safety wire the housing retaining cap screws.

Install the remaining parts by reversing the removal procedure. The crankshaft front oil seal (46) located in timing gear housing should be installed with lip facing the flywheel end of engine.

CRANKSHAFT REAR OIL SEAL

50. The crankshaft rear oil seal (33—Fig. JD1526) is located in the rear main bearing housing and can be renewed after removing the flywheel as outlined in paragraph 51. Seal should be installed with lips facing timing gear end of engine.

FLYWHEEL

51. To remove the flywheel, first remove the starting engine as outlined in paragraph 30, then unbolt and separate the clutch housing from the engine. Remove the clutch cover assembly and lined plate. Remove the flywheel retaining nut. To loosen flywheel from crankshaft, pry on flywheel from behind while bumping crankshaft with a soft hammer. To install starter ring gear, heat same to 550 deg. F. and install gear on flywheel so that beveled end of teeth will face toward starting motor pinion.

When installing the flywheel, tighten the retaining nut to a torque of 150 Ft.-Lbs. and bend tab of lock plate against flat of nut.

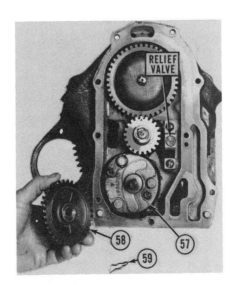

Fig. JD1527 — The oil pump driving gear can be removed after removing its retaining hairpin snap ring (59).

OIL PUMP AND RELIEF VALVE

52. **OIL PUMP.** To remove the oil pump, first remove the timing gear cover as outlined in paragraph 40. Remove the hairpin snap ring and withdraw the pump drive gear as shown in Fig. JD1527. Unbolt and withdraw pump from cylinder block, being careful not to lose or damage the shims located between oil pump and block.

Unbolt and remove the pump cover from the pump body and disassemble the remaining parts. The pump driving bevel gear (52—Fig. JD1528) can be pressed from shaft if renewal is required. New bevel gear should be installed flush with end of shaft. When reassembling, coat mating surfaces of pump body and cover with shellac and tighten the cover retaining cap screws finger tight. Shift the cover until pumping gears are free, then tighten the cap screws.

52A. Install pump by reversing the removal procedure and use the same number of shim gaskets (53) between pump body and cylinder block as were removed. Installing the original number of gaskets should provide a backlash of 0.004-0.006 between the water pump and oil pump bevel gears. If there is any reason to suspect that the bevel gear backlash is not as specified, remove the pump and apply a 0.003 thick piece of paper directly on the bevel gear teeth. Then install the pump and turn the pump drive gear, thereby passing the 0.003 piece of paper between the bevel gears. If the piece of paper shows a heavy impression but is not punctured, the backlash is within the specified limits.

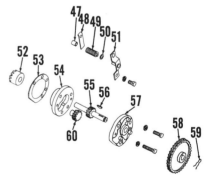

Fig. JD1528 — Exploded view of the starting engine oil pump and relief valve. Shims (53) control backlash between the oil pump and water pump bevel gears.

47. Relief valve seat	53. Shims
48. Leaf spring	54. Pump body
49. Relief valve spring	55. Driver gear
50. Shim washers	56. Woodruff key
51. Relief spring bracket	57. Pump cover
52. Bevel gear	58. Driving gear
	59. Snap ring
	60. Idler gear

53. **RELIEF VALVE.** The recommended starting engine oil pressure of 25-30 psi at high idle rpm is regulated by a spring loaded leaf type relief valve located within the timing gear cover. Refer to Fig. JD1529. Oil pressure can be checked on early models by removing the pipe plug, shown in Fig. JD1517, from side of cylinder block and installing a master gage. On later models, it is necessary to connect a gage in series with the oil line connected to the automatic fuel shut-off unit.

If the pressure is less than 20 psi, remove the timing gear cover as outlined in paragraph 40, and add adjusting washers (50—Fig. JD1528). The addition of one washer will increase the oil pressure approximately 5 psi.

The relief valve spring (49—Fig. JD1528) should have a free length of approximately $1\frac{1}{16}$-inches and should require 22-25 ounces to compress it to a height of $\frac{3}{4}$-inch. If the relief valve seat (47) is renewed, press it in until outer surface of seat is level with the relief valve bracket pads.

CARBURETOR

54. The starting engine on tractors prior to serial number 8203100 is equipped with a Zenith 0-11681, model TU3½ x 1C down draft type carburetor shown exploded in Fig. JD1530. After serial number 8203099, the starting engine is equipped with a Zenith D-12172 model TU3 x 1C carburetor. Distance from farthest face of float to machined surface of float bowl cover is $1\frac{3}{16}$-inches.

To adjust the carburetor, first start engine and let it run until warm. With speed control in the slow idle position, back the screw (74—Fig. JD1531) out until engine slows down or falters,

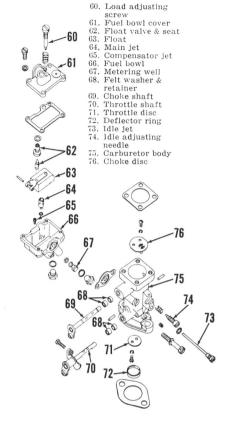

60. Load adjusting screw
61. Fuel bowl cover
62. Float valve & seat
63. Float
64. Main jet
65. Compensator jet
66. Fuel bowl
67. Metering well
68. Felt washer & retainer
69. Choke shaft
70. Throttle shaft
71. Throttle disc
72. Deflector ring
73. Idle jet
74. Idle adjusting needle
75. Carburetor body
76. Choke disc

Fig. JD1530 — Exploded view of Zenith carburetor used on the gasoline starting engine. The unit is provided with adjustments for both idling and power range.

then turn the screw in until engine runs smoothly.

With the starting engine motoring the diesel engine on full compression, turn the load adjusting screw (60) in

until engine slows down or falters and back the screw out until engine runs smoothly; then, back the screw out one or two more notches. Normal load screw adjustment is $\frac{5}{8}$-turn open.

GOVERNOR

55. The flyball type governor weight unit is mounted on the front of the camshaft as shown in Fig. JD1520 and can be overhauled after removing the timing gear cover as outlined in paragraph 40. When reassembling, make sure that slot in ball retainer (26) engages the Woodruff key in camshaft.

56. **ADJUSTMENT.** Before attempting to adjust the governed speed, free up all linkage and renew any parts causing lost motion. Disconnect the throttle rod (90—Fig. JD1531) from governor arm (80) and with throttle butterfly and governor arm in the wide open position, adjust the length of the throttle rod so it is short one full turn of the adjusting nut (N).

Reconnect the throttle rod, start engine and turn the throttle stop screw either way as required to obtain a slow idle speed of 4000 rpm. The high idle, no load speed of 5000 rpm is adjusted by loosening the two cap screws retaining the governor spring bracket (79) to manifold and moving it away from carburetor to increase speed or toward carburetor to decrease speed. With diesel engine at operating temperature, the starting engine should motor the diesel engine under full compression at 200 rpm.

Note: The starting engine speed can be checked at right end of crankshaft after removing the distributor cover.

Fig. JD1529 — Normal starting engine oil pressure of 25-30 psi at high idle rpm is regulated by leaf type spring (48). Pressure is increased by the addition of washers (50).

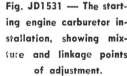

Fig. JD1531 — The starting engine carburetor installation, showing mixture and linkage points of adjustment.

60. Load adjusting screw
74. Idle adjusting needle
78. Governor spring
79. Spring bracket
80. Governor arm
90. Throttle rod
N. Jam nut

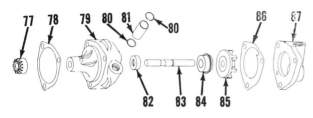

Fig. JD1532 — Exploded view of the starting engine water pump. Shims (78) control backlash between water pump and oil pump bevel gears.

77. Bevel gear	80. "O" ring	83. Pump shaft	85. Impeller
78. Shims	82. Oil seal	84. Pump seal	86. Gasket
79. Pump housing			87. Pump cover

COOLING SYSTEM

The starting engine cooling system is connected to the diesel engine cooling system. Water is circulated through the starting engine by an impeller type water pump mounted on the starting engine cylinder block.

57. R&R AND OVERHAUL WATER PUMP. To remove the water pump, drain the tractor cooling system and starting engine crankcase. Remove the water by-pass pipe. Remove the starting engine water inlet pipe by unbolting the water inlet casting from the right side of the diesel engine cylinder block and pulling the inlet pipe out of the water pump. Remove the diesel engine injection pump compartment cover. Remove the water pump retaining cap screws, turn the pump counter-clockwise to clear the short water pipe (81—Fig. JD1532) and withdraw the pump from cylinder block, being careful not to damage or lose the gasket shims (78) located between pump and block.

To disassemble the pump, press the pump shaft out of the bevel drive gear (77).

Install seal (82) with lip facing impeller. Press the bevel drive gear on shaft until hub of gear is flush with end of shaft. Impeller should be just flush with the machined surface of the pump body as shown in Fig. JD1533.

Install pump by reversing the removal procedure, use the original number of shims between water pump and cylinder block and renew all "O" rings. Backlash between the water pump and oil pump bevel gears should be a recommended 0.004-0.006. If there is any reason to suspect the backlash is not as specified, check the adjustment with paper as in paragraph 52A.

IGNITION SYSTEM

58. Wico distributor model B4027 is used. The unit is mounted on the timing gear cover and actuated by a cam which is retained to the end of the crankshaft by a cap screw. There is a coil, condenser and set of contacts for each two cylinders. The unit does not incorporate an automatic spark advance mechanism.

The gap for each set of breaker contacts is 0.020. Breaker contact spring tension is 18-26 oz. Firing order is 1 - 2 - 3 - 4. When viewed from flywheel end of engine, cylinders No. 1 and 3 are on left side (front of tractor) and cylinders No. 2 and 4 are on right hand side (rear of tractor). See Fig. JD1515.

59. TIMING. To time the distributor, remove the inspection cover from

clutch housing and using a screw driver or similar tool, turn the flywheel in a counter-clockwise direction (viewed from flywheel end) until number one piston is coming up on compression stroke and continue turning flywheel until "V" marks on flywheel and clutch housing are aligned as shown in Fig. JD1535. Loosen the four distributor mounting screws and rotate the distributor in a clockwise direction (viewed from distributor end of engine) as far as possible. Turn on the ignition switch and while holding the number one spark plug wire about ¼-inch from engine block, slowly rotate distributor in a counter-clockwise direction until a spark occurs at the end of the wire; then tighten the mounting screws securely.

STARTING MOTOR

60. The starting engine is equipped with a 6-volt, automotive type Delco-Remy No. 1107155 starting motor. Refer to paragraph 128 for test specifications.

CLUTCH

61. ADJUSTMENT. If the starting engine clutch slips when motoring the diesel engine under full compression, adjust the clutch and linkage as follows:

Remove hood, diesel engine flywheel cover, starting engine clutch adjusting hole cover and starting engine transmission inspection cover as shown in Fig. JD1537. Move the clutch operating lever to the engaged position and check to make certain that the transmission pinion (43) goes into full engagement with the diesel engine flywheel. If it does not, remove pin from yoke (52), turn the yoke to obtain full engagement, reinstall the pin and tighten the yoke jam nut.

With the clutch operating lever in the disengaged position, slowly move the clutch lever toward the engaged position, until the shifter bracket (47)

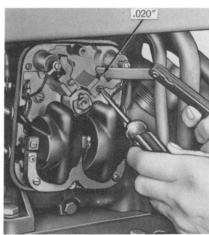

Fig. JD1533 — When the starting engine water pump is properly assembled, the impeller vanes should be just flush with machined surface of pump housing.

Fig. JD1534 — The starting engine distributor is equipped with two sets of contacts, two condensers and two coils. Contact gap is 0.020.

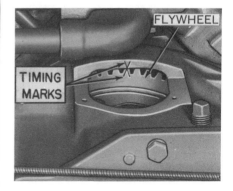

Fig. JD1535 — Starting engine ignition timing marks on flywheel and clutch housing.

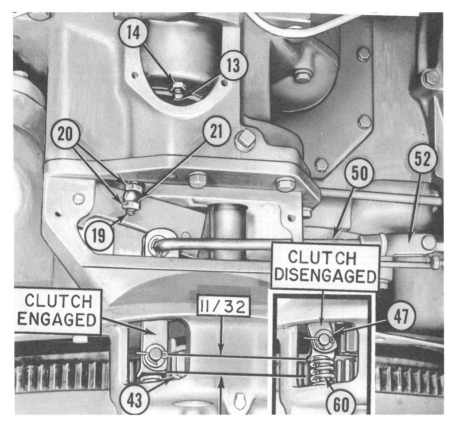

Fig. JD1539 — Starting engine clutch adjustment. Screw (13) should be adjusted to obtain 5/8-inch clearance between the bearing contact surface of each toggle and the rear face of the cover.

Fig. JD1537—The shifter bracket should compress the stop spring 11/32-inch as the clutch is engaged. If it does not, loosen jam nuts (20) and move swivel (21) in relation to shifter link (19) until the 11/32-inch travel is obtained. See text.

just contacts the stop spring (60). From this point, continue moving the clutch operating lever, and measure the distance the shifter bracket travels, until the clutch is fully engaged. If the measured distance is not $\frac{11}{32}$-inch as shown, loosen the jam nuts and turn each of the clutch adjusting screws (13) in, ½-turn at a time, until $\frac{11}{32}$-inch travel is obtained. If after this adjustment is made, the pinion (43) spins with the starting engine running and the clutch lever in the disengaged position, each of the clutch adjusting

screws (13) must be backed out an equal amount to release the clutch lined disc. Then in order to obtain the $\frac{11}{32}$-inch travel of the shifter bracket, shifter link (19) must be lengthened by loosening jam nuts (20) and changing the position of swivel (21) on the link. A combination of the two efforts will provide the $\frac{11}{32}$-inch travel and complete engagement and disengagement of the clutch.

64. **R&R AND OVERHAUL.** To remove the starting engine clutch, first remove the starting engine as outlined in paragraph 30 and proceed as follows: Remove the cap screws retaining the clutch housing to the engine and separate the two units.

The clutch release bearing (Fig. JD 1537A) can be removed after disconnecting the shifter link from the

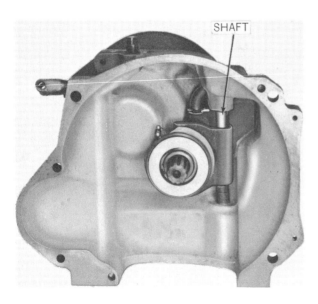

Fig. JD1537A — Front view of the starting engine clutch housing showing installation of release bearing.

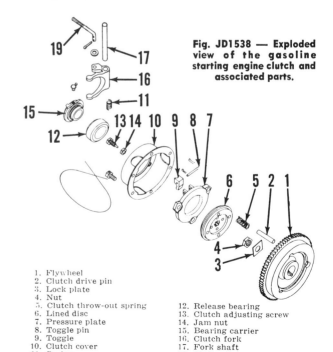

Fig. JD1538 — Exploded view of the gasoline starting engine clutch and associated parts.

1. Flywheel
2. Clutch drive pin
3. Lock plate
4. Nut
5. Clutch throw-out spring
6. Lined disc
7. Pressure plate
8. Toggle pin
9. Toggle
10. Clutch cover
11. Spring
12. Release bearing
13. Clutch adjusting screw
14. Jam nut
15. Bearing carrier
16. Clutch fork
17. Fork shaft
19. Shifter link

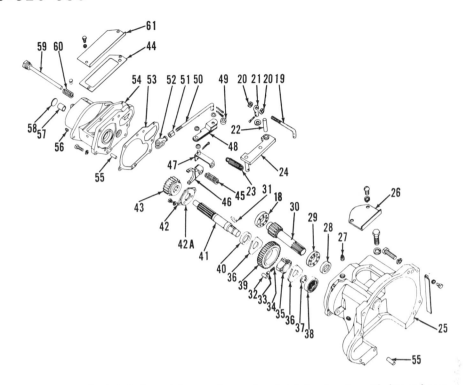

Fig. JD1540—Exploded view of the starting engine clutch housing, transmission and associated parts.

18. Clutch shaft outer bearing	33. Plunger	47. Shifter bracket
19. Shifter link	34. Spring	48. Shifter yoke
20. Jam nut	35. Over-running clutch hub	49. Shifter yoke roller
21. Shifter link swivel	36. Over-running clutch plate	50. Shifter rod
22. Follower retaining pin	37. Snap ring	51. Jam nut
23. Follower return spring	38. Bearing	52. Shifter rod yoke
24. Shifter follower	39. Transmission drive gear	53. Gasket
25. Clutch housing	40. Oil seal	54. Transmission housing
26. Inspection cover	41. Transmission drive shaft	55. Dowel pin
27. Pipe plug	42. Brake plate	56. Pipe plug
28. Oil seal	42A. Brake lining	57. Bushing
29. Clutch shaft inner bearing	43. Sliding pinion	58. Expansion plug
30. Clutch shaft	44. Gasket	59. Transmission shifter shaft
31. Woodruff key	45. Shifter spring	60. Stop spring
32. Clutch roller	46. Transmission shifter	61. Cover

shifter link swivel and pivoting the clutch fork arm enough to remove the bearing.

Unbolt and remove the clutch cover and lined disc from flywheel. The need and procedure for further disassembly is evident after an examination of Fig. JD1538.

When reassembling, install release springs (5) on the drive pins (2) and install the lined disc (6) with hub facing away from flywheel. Install pressure plate (7) and bolt cover assembly to flywheel. Be sure to safety wire the cover retaining cap screws. With the actuating toggles depressed until the pressure plate contacts the clutch lined disc, there should be a clearance of ⅝-inch between the release bearing contacting surface of each toggle and the rear surface of the clutch cover as shown in Fig. JD1539. If adjustment is not as specified, loosen the jam nut and turn each of the toggle adjusting screws (13) **in** to decrease or **out** to increase clearance.

Install the remaining parts by reversing the removal procedure.

TRANSMISSION

65. **R&R AND OVERHAUL.** To remove the starting engine transmission, it will be necessary to first remove the starting engine from the tractor as outlined in paragraph 30. Remove the cap screws retaining the clutch housing to the engine and separate the

units. The release bearing can now be removed after disconnecting the shifter link from the shifter link swivel and pivoting the clutch fork arm enough to remove the bearing. Drain the transmission. Remove the cap screws attaching the transmission housing to the clutch housing and bump the transmission housing off the locating dowels. Remove the clutch shaft and transmission shaft as shown in Fig. JD1540A. Using a suitable

Fig. JD1540A—When removing the starting engine clutch and transmission drive shafts it may be necessary to use a drift as shown to unseat the transmission shaft bearing.

puller (Fig. JD1541), remove bearing from transmission shaft. Remove snap ring (37—Fig. JD1542) and over-running clutch plate (36). Gear, hub and remaining clutch plate can now be removed from the shaft. Do not lose any of the rollers, plungers or springs. Using a pin punch, drive the follower retaining pin (22—Fig. JD1540) out of the housing. Unhook the follower spring (23) at the outer end of the housing. The follower (24) can then be lifted from the transmission housing. Remove the shifter shaft (59) and stop spring (60). Remove yoke (52) and jam nut (51) from shifter rod (50). Remove the cotter pin (Fig. JD1543) from the shifter rod then remove the shifter rod (50—Fig. JD1540), roller (49) and flat washer. The remainder of the linkage, the pinion and the transmission shifter can now be removed from the housing.

Inspect all parts and renew any that are scored, worn, cracked or in any other way damaged. Bushing (57—Fig. JD1540) is pre-sized and if carefully installed should need no final sizing. When installing bushing, be sure oil hole in bushing is aligned with the oil hole in the casting.

Use Figs. JD1538 and 1540 as a guide during reassembly and adjust the clutch and linkage as outlined in paragraph 61.

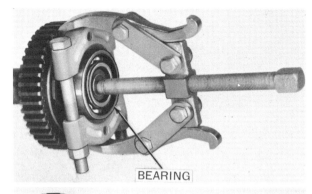

Fig. JD1541—Use a suitable puller to remove bearing from the transmission shaft.

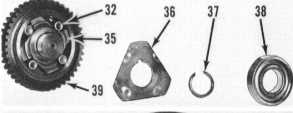

Fig. JD1542 — After removing bearing (38) and snap ring (37), the over-running clutch parts can be disassembled from the transmission drive shaft.

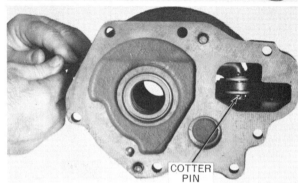

Fig. JD1543 — Remove cotter pin retaining shifter rod in shifter yoke, then remove shifter rod.

DIESEL ENGINE AND COMPONENTS

The diesel engine crankcase and tractor main case is an integral unit. A wall in the main case separates the diesel engine crankcase compartment from the transmission and differential compartment.

CYLINDER HEAD

The cylinder head on tractors prior to serial number 8203100 is fitted with hard drawn copper tubes to receive the diesel injection nozzles. Refer to paragraph 119 for removal and reinstallation of nozzle tubes. Model 820 tractors after serial number 820-3099 and 830 tractors are not equipped with renewable type nozzle sleeves. As shown at (N) in Fig. JD1545, the injection nozzle tips extend through small holes drilled into the cylinder head between the inlet and exhaust valves. A heat exchanger, which acts as an air inlet manifold for the diesel engine, is mounted on the cylinder head. Exhaust gasses from the starting engine flow into the heat exchanger and warm the initial incoming air for the diesel engine. Water outlet pipes and an exhaust manifold are also mounted on the head.

71. **REMOVE AND REINSTALL.** To remove cylinder head from tractor, proceed as follows: Remove hood, drain tractor cooling system and disconnect the battery ground strap. Unbolt the water inlet casting from bottom of head and remove the diesel engine exhaust manifold, water outlet pipes and air cleaner. On tractors equipped with a gasoline starting engine, remove the water vent pipe, starting engine exhaust pipe and heat exchanger. On tractors equipped with a 24-volt electric starting motor, re-

Fig. JD1545 — Rear view of the diesel engine cylinder head, showing injection nozzle tips (N) extending through small holes which are drilled between the inlet and exhaust valves. Before installing cylinder head, make certain that crankcase ventilating hole (V) and oil hole (O) are open and clean.

Fig. JD1546—Diesel engine cylinder head nut tightening sequence. Nuts should be tightened to a torque of 275 Ft.-Lbs. Cap screws in injection pump compartment must be tightened securely.

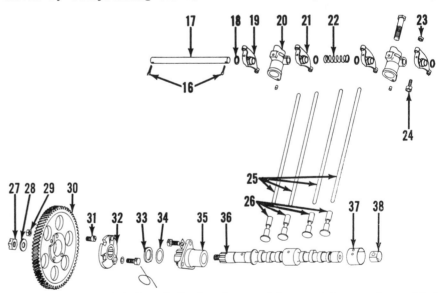

Fig. JD1547 — Exploded view of the diesel engine cylinder head. Renewable type injection nozzle sleeves (10) are only on models prior to serial number 8203100.

3. Split cone keepers	7. Coupling nut	12. Cylinder head
4. Valve spring cap	8. Machine screw	13. Lead washer
5. Valve spring	9. Washer	14. Gasket
6. Valve guide	11. "O" ring	15. Valve

move the air inlet manifold. On all models, remove the generator. Remove tappet lever cover, disconnect tappet lever oil line and remove the tappet levers assembly and push rods. Unbolt and remove the injection pump compartment cover. Remove both fuel lines which connect injection pumps to injection nozzles, and **immediately** cap the connectors on the pumps and nozzles to prevent the entrance of dirt. Protector caps are provided with each tractor for this purpose. Remove the head retaining stud nuts and the three cap screws located in the injection pump compartment. Slide head forward and remove the head from the tractor.

Before installing the cylinder head, clean carbon and all gasket surfaces, paying particular attention to the small holes through which the injection nozzle tips project. Check to make certain that the crankcase ventilating hole (V—Fig. JD1545), oil passage (O) and all oil lines are open and clean. Install head gasket with copper side toward head and use a sealing compound between gasket and cylinder block. Note: Lead washers (13—Fig. JD1547) are used under all cylinder head retaining nuts; however, spring steel lock washers are used on the three cap screws in the injection pump compartment.

Tighten the cylinder head retaining stud nuts in sequence shown in Fig. JD1546 and to a torque of 275 Ft.-Lbs. A torque wrench cannot be used on the three cap screws in the injection pump compartment, but they must be tightened securely.

If injection nozzles were removed from the cylinder head, the copper washers which are located between

the nozzles and the head must be renewed; or the old washers may be annealed by heating to a cherry red and quenching in water before installation. Adjust tappet gap to correct specifications listed in paragraph 72 and check diesel engine timing and tune-up as outlined in paragraph 77. Recheck the cylinder head nut tightening torque after engine is hot.

VALVES AND SEATS

72. Valve tappet gap (hot) should be set to 0.015 for the inlet and 0.020 for the exhaust. Valves may be removed after the cylinder head is removed by compressing the valve

springs and removing the split-cone type keepers.

Inlet and exhaust valves are not interchangeable, and seat directly in the cylinder head with a face angle of 44½ degrees and seat angle of 45 degrees. Desired seat width is 1/8-9/64-inch. Seats may be narrowed using 20 and 70 degree stones. Valve stem diameter is 0.496-0.497 for both inlet and exhaust.

VALVE GUIDES AND SPRINGS

73. The pre-sized inlet and exhaust valve guides are interchangeable and should be pressed into the cylinder head until the port end measures 2⅛

Fig. JD1548 — Exploded view of diesel engine camshaft, tappet levers and associated parts.

16. Cotter pin	24. Tappet lever adjusting screw	31. Eccentric screw
17. Tappet lever shaft	25. Push rods	32. Camshaft gear hub
18. Washer	26. Cam followers	33. Thrust washer
19. Exhaust tappet lever	27. Camshaft gear retaining nut	34. Shim washer (0.005)
20. Tappet lever shaft bracket	28. Washer	35. Camshaft left bearing
21. Inlet tappet lever	29. Lock nut	36. Camshaft
22. Spring	30. Camshaft gear	37. Center camshaft bushing
23. Jam nut		38. Right camshaft bushing

inches from the gasket surface of the cylinder head. The 0.496-0.497 diameter valve stems have a diametral clearance of 0.0041-0.0065 in the 0.5011-0.5025 inside diameter valve guides.

73A. Inlet and exhaust valve springs are interchangeable. Valve spring free length is approximately 4⅛ inches and should require 91-111 pounds to compress them to a height of 2¹¹⁄₁₆ inches. Renew any spring which is rusted, discolored or does not meet the foregoing pressure specifications. Springs can be renewed without R&R of head after removing valve tappet levers assembly.

VALVE TAPPET LEVERS

74. Valve tappet levers assembly can be removed from cylinder head after removing the hood, generator and tappet lever cover, and disconnecting the tappet lever oil line. Unbolt and remove the tappet lever assembly from the cylinder head. The need and procedure for further disassembly is evident after an examination of the unit and reference to Fig. JD1548. NOTE: Inlet and exhaust tappet levers are not interchangeable. The 0.8580-0.8590 diameter tappet lever shaft should have a diametral clearance of 0.001-0.007 in the tappet levers.

CAM FOLLOWERS

75. The mushroom type cam followers are supplied in standard size only and operate directly in machined bores cut into the tractor main case. The followers can be removed after removing the camshaft as outlined in paragraph 83.

Refer to paragraph 114A for information concerning the injection pump cam followers.

VALVE TIMING

76. Valves are properly timed when the timing gears are installed as in paragraph 82, and when the camshaft timing adjustment and decompression lever settings are as outlined in the following paragraphs.

GENERAL TIMING AND TUNE-UP

77. The following timing and tune-up adjustments should be checked after doing any major repair work or when the engine does not run properly. In order to check these adjustments, it is necessary to first remove hood, injection pump compartment cover, flywheel cover and tappet lever cover. With these parts removed, proceed as follows:

77A. **CAMSHAFT TIMING ADJUSTMENT.** The valve tappet gap must be adjusted to exactly 0.015 for the inlet and 0.020 for the exhaust. Turn the flywheel until exhaust valve of number one cylinder (left one) is fully closed. Check by looseness of push rod. Mount a dial indicator as shown in Fig. JD1549 so that the indicator contact button is preloaded and is resting on the number one exhaust valve spring cap, then zero the indicator face. NOTE: Make certain that the indicator body does not touch the tappet lever. Turn flywheel opposite to normal direction of rotation (clockwise) approximately ⅛ turn; then, turn flywheel in running direction (counter-clockwise) until "No.1 TD-C" mark on flywheel is in register with mark on flywheel cover back plate as shown. At this time, indicator should show that the exhaust valve

is open 0.025-0.030. If indicator reading is not as specified, remove timing gear housing cover as outlined in paragraph 78. Remove safety wire and loosen the three cap screws which secure the camshaft gear to the gear hub. Loosen the locknut and turn the eccentric (Fig. JD1550) until the dial indicator reading shows the exhaust valve to be open from 0.025 to 0.030. It is always a good practice to set the timing as close to 0.025 as possible. Tighten the locknut and recheck the indicator reading by repeating the above procedure. Tighten and safety wire the three camshaft gear retaining screws and stake the eccentric lock nut when adjustment is complete.

Check the injection pump timing as outlined in paragraph 111.

77B. **DECOMPRESSION LEVER SETTING.** Turn flywheel until number one cylinder completes its com-

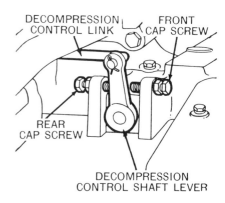

Fig. JD1551 — Adjust the amount of valve opening during decompression by rear cap screw and the position of the decompression lever in relation to the starting engine clutch lever by the front cap screw.

Fig. JD1549 — Dial indicator mounted for checking the diesel engine camshaft timing and/or decompression lever setting.

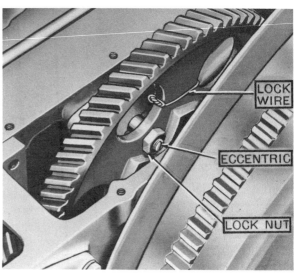

Fig. JD1550 — Diesel engine camshaft timing adjustment. The eccentric adjusting screw is accessible after removing the timing gear housing cover.

FAST IDLE ADJUSTING SCREW

STOP ADJUSTING SCREW

SPEED CONTROL ROD

Fig. JD1552—Disconnect the diesel engine speed control rod at rear.

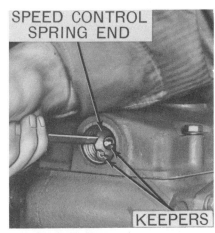

SPEED CONTROL SPRING END

KEEPERS

Fig. JD1552A—Removing the diesel engine speed control spring end keepers.

pression stroke (flywheel mark "No. 1 TDC" is in register or slightly past mark on flywheel cover back plate). Mount a dial indicator as shown in Fig. JD1549 so that indicator contact button is depressed (preloaded) to near the extreme inward position and contacting the exhaust valve spring cap. Zero the dial indicator face. NOTE: Make certain that indicator body does not touch tappet lever. Pull decompression lever. At this time, indicator should show that the decompression linkage has opened the exhaust valve 0.050-0.090. Make same check for number two cylinder. If indicator readings are not as specified, adjust at the rear cap screw (Fig. JD-1551).

Adjust decompression lever in relation to the starting engine clutch lever by turning the front cap screw until the decompression control lever cannot be locked in rear position. Then loosen the cap screw until the control lever can just be locked in rear position without forcing.

TIMING GEAR HOUSING AND COVER

78. TIMING GEAR HOUSING COVER. To remove timing gear housing cover, proceed as follows: Remove the gasoline starting engine as outlined in paragraph 30 or on tractors so equipped, the 24-volt electric starting motor. Drain hydraulic system, disconnect the hydraulic lines from the pump, and remove the pump from the timing gear housing cover. Disconnect the speed control rod (Fig. JD1552) at rear and decompression link (Fig. JD1551) at front. Remove plug from front of timing gear housing cover, then remove the cover retaining cap screws and slide the cover rearward. Using a screwdriver or similar tool as shown in Fig. JD1552A, compress the governor spring and re-

1. Speed control rod
14. Packing gland
15. Speed control rod packing
16. Speed control rod stop
18. Washer
19. Washer
20. Governor loading lever
21. Plug
22. Jam nut
23. Speed control spring end keepers
24. Speed control spring end (cap)
25. Speed control spring
26. Speed control spring fulcrum
27. Governor lever shaft
28. Governor lever stop adjusting screw
29. Governor lever
30. Injection pump control rod
31. & 33. "O" ring
32. "O" ring retainer
34. Injection pump control rod guide
35. Spring

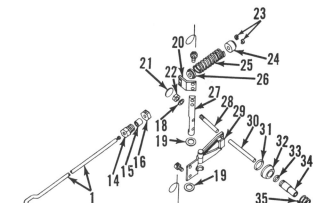

Fig. JD1552B — Exploded view of the diesel engine speed control linkage.

move the two half moon keepers (23—Fig. JD1552B), speed control spring end (24), speed control spring (25) and speed control spring fulcrum (26). NOTE: If any of these parts fall into the timing gear housing, it will be necessary to remove the timing gear housing to retrieve them. Pull the speed control rod rearward and out of the governor loading lever and remove the timing gear housing cover. When installing the speed control spring fulcrum, make certain that the raised portion of the fulcrum is vertical and faces inward.

79. TIMING GEAR HOUSING. To remove the timing gear housing, remove the timing gear housing cover as outlined in paragraph 78, and proceed as follows: Remove the flywheel as outlined in paragraph 97 and remove the flywheel cover back plate. Drain the timing gear housing, remove the timing gear housing retaining cap screws and lift the housing with governor unit from tractor. One of the housing retaining cap screws is located inside the housing. When reinstalling this cap screw, it is recom-

mended that shop towels be stuffed inside the housing to prevent the possibility of dropping the cap screw to the bottom of housing.

NOTE: When reinstalling the timing gear housing make sure parts shown in Fig. JD-1553 are arranged as shown.

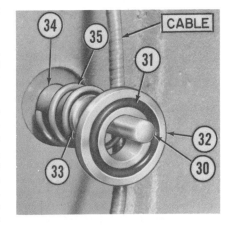

CABLE

Fig. JD1553 — When reinstalling the timing gear case, make certain that spring (35), packing retainer (32). "O" rings (31 & 33), speed hour meter cable and injection pump control rod (30) are installed properly.

Fig. JD1554 — Diesel engine timing gear train. Install camshaft gear and hub unit so that "V" mark on camshaft gear hub is in register with "V" mark on camshaft. Install crankshaft gear so that "V" mark on gear is in register with "V" mark on crankshaft and mesh "V" mark on crankshaft gear with "V" mark on camshaft gear.

TIMING GEARS

80. As shown in Fig. JD1554, the timing gear train consists of the camshaft gear, crankshaft gear, power shaft drive gear, the combination fan and diesel engine oil pump drive gear and two idler gears. The governor drive gear (not shown) which is driven by the camshaft gear is removed with the timing gear housing.

81. To remove the timing gears, first remove the timing gear housing as outlined in paragraph 79. Lift off both idler gears. NOTE: Thrust washers on oil pump and fan shaft idler gear are steel, while the power shaft idler gear is equipped with fiber thrust washers. Remove the oil slinger from the crankshaft gear. Remove the camshaft gear and hub unit by removing the hub retaining nut. Attach a suitable puller as shown in Fig. JD1555 and remove the crankshaft gear. Inspect the rubber packing which is located inside of crankshaft gear and renew same if it is damaged. Refer to paragraph 126 for information concerning the combination fan and oil pump drive gear and its idler and to paragraph 167 for information concerning the power shaft drive gear and its idler.

82. Install camshaft gear and hub unit so that "V" mark on camshaft

gear hub is in register with the "V" mark on camshaft, and tighten the hub retaining nut to a torque of 208 Ft.-Lbs. Install crankshaft gear so that "V" mark on gear is in register with "V" mark on crankshaft and "V" mark on crankshaft gear is in register with "V" mark on camshaft gear. Refer to Fig. JD1554. The remaining gears in the timing gear train may be meshed in any position, as timing is unnecessary. NOTE: Make certain that thrust washers on the idler gears are properly installed. Before installing the timing gear housing, remove the crankshaft oil slinger housing and the 0.005 thick oil slinger housing gasket.

Fig. JD1554A — Before tightening the slinger housing retaining screws, use 0.005 shim stock and center the housing about the crankshaft gear.

82A. Install the timing gear housing and start the crankshaft oil slinger on the crankshaft gear. To obtain the proper oil slinger position on the crankshaft, install and tighten the oil slinger housing, leaving out the 0.005 thick gasket; then remove the oil slinger housing and reinstall same, using the 0.005 thick gasket. Before tightening the slinger housing retaining screws, use 0.005 shim stock and center the housing about the crankshaft gear as shown in Fig. JD1554A. Install the remaining parts by reversing the removal procedure and check the diesel engine timing and tune-up as outlined in paragraph 77.

Fig. JD1555 — To remove the crankshaft gear from the diesel engine crankshaft, it is necessary to attach a suitable puller as shown.

CAMSHAFT

The diesel engine camshaft has six lobes, two of which actuate the injection pump cam followers. The camshaft is drilled for pressure lubrication and is carried in three bearings. The right hand and center bearings are pre-sized, steel-backed, bronze-lined bushings, the left hand bearing is pre-sized cast iron. The camshaft gear hub is positioned on the camshaft by splines, whereas the relative position of the camshaft gear with respect to the gear hub is adjustable by means of an eccentric adjusting screw. The fuel transfer pump which is attached to the right hand side of the main case is driven by a slotted plug which is screwed into the right end of the camshaft.

83. To remove the camshaft proceed as follows: Remove the tappet lever cover, tappet levers assembly and injection pump compartment cover. Remove the timing gear housing as outlined in paragraph 79. Remove the pressure lines connecting the injection pumps to the injection nozzles and **immediately** cap the four connectors. (Caps are provided with each tractor for this purpose.) Remove the injection pump rack plug and spring from the right side of the block then remove the four cap screws which secure pump and bracket assembly to the cylinder casting. Carefully pry pump and bracket assembly off dowel pins and lift assembly from the compartment. Cover the fuel inlet openings of the bracket with masking tape. Loosen the retaining cap screws (S— Fig. JD1556) and remove the injection pump cam followers as shown. Remove the fuel sediment bowl and fuel

transfer pump from right side of tractor main case. Remove the crankshaft oil slinger, remove the nut which retains the camshaft gear and hub assembly to the camshaft and using a suitable puller, pull the gear and hub assembly from the shaft. Remove the attaching cap screws and pull the camshaft, with left-hand bearing, from the crankcase. Right and center camshaft bearings can be renewed at this time. Bearings are pre-sized and do

Fig. JD1556 — Removing the diesel injection pump followers. Make certain followers are not bent during removal.

Fig. JD1557—The diesel engine camshaft end play is adjusted to 0.005-0.009 by varying number of 0.005 shims located between camshaft left bearing and thrust washer.

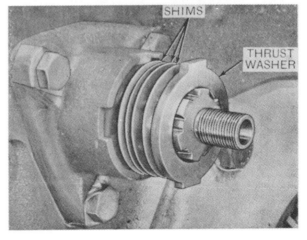

not normally require reaming if carefully installed.

Inspect drilled openings in camshaft and in main case at center camshaft bearing. Openings should be clean and free from obstructions. Check camshaft against values which follow:

Diam. of right journal 1.499-1.500
   Clearance in bearing 0.002-0.005
Diam. of center journal ..2.4975-2.4985
   Clearance in bearing 0.002-0.005
Diam. of left journal 1.4985-1.4995
   Clearance in bearing 0.002-0.004

Install camshaft by reversing the removal procedure and make certain that the fuel transfer pump is properly installed. Adjust the camshaft end play to 0.005-0.009 by varying the number of 0.005 thick shims (Fig. JD-1557) which are located between the camshaft left bearing and thrust washer. Refer to paragraphs 82 and 82A for correct installation of the camshaft gear and crankshaft oil slinger. Retime injection pumps as outlined in paragraph 111. Adjust tappet gap to the specified clearance and check diesel engine timing and tune-up as outlined in paragraph 77.

ROD AND PISTON UNITS

84. To remove connecting rod and piston units, proceed as follows: Remove the diesel engine cylinder head as outlined in paragraph 71 and the crankcase cover as in paragraph 99 or 99A. Disconnect connecting rods

from crankshaft and push piston and rod assemblies forward out of cylinder bores far enough to extract the piston pin retaining snap rings and piston pins as shown in Fig. JD1560. Connecting rods may now be removed from front after pistons are removed.

When reassembling, word "TOP" (stamped on front of piston) and numbers on connecting rods and caps must face toward top of engine. No. 1 cylinder is on left side of tractor. Tighten connecting rod bolts to a torque of 200-220 Ft.-Lbs.

PISTONS AND RINGS

85. Pistons and rings are available in standard size as well as 0.045 oversize. Each of the aluminum alloy pistons are tapered, cam ground and fitted with six rings. The top compression ring is chrome plated whereas the next three compression rings are unplated. The fifth is a ventilated oil ring and the sixth is a cast iron wiper ring.

Desired piston ring side clearance is as follows:

No. 1 (top)............0.004-0.006
Nos. 2, 3 and 4........0.0025-0.0045
Nos. 5 and 6..........0.0015-0.0035

Desired piston to cylinder diametral clearance is as follows:

Top of skirt
In horizontal plane....0.0208-0.0222
In vertical plane.......0.0138-0.0162

Bottom of skirt
In horizontal plane....0.0136-0.0150
(approximate)
In vertical plane.......0.0071-0.0095

Piston skirt clearance is best determined by measuring the cylinder bore and pistons.

Standard cylinder bore...6.1246-6.1260

Install piston rings as follows: The ventilated oil control ring can be in-

stalled with either side facing the closed end of piston but the non-ventilated wiper ring and all four compression rings must be installed with dot toward head of piston.

PISTON PINS AND BUSHINGS

87. The 2.3545-2.3550 diameter full floating type piston pins are retained in the piston bosses by snap rings and are available in standard size as well as oversizes of 0.003 marked yellow and 0.005 marked red. The piston pin end of the connecting rod is fitted with a renewable cast iron bushing which must be reamed after installation to provide a diametral clearance of 0.0010-0.0025 for the pin; or, with a light oil film, the pin should just slide through the bushing by its own weight when held with pin in a vertical position. Pin should have 0.000-0.0011 clearance (or a tight push fit) in piston.

CONNECTING RODS AND BEARINGS

88. The forged steel connecting rods are drilled for pressure lubrication to the cast iron piston pin bushings and are fitted with shimless, non-adjustable, slip-in, precision type bearing inserts which can be renewed after removing the crankcase cover as outlined in paragraph 99 or 99A. Connecting rod bearing inserts are steel-backed, copper-lead type and are available in standard size as well as undersizes of 0.002, 0.004, 0.020, 0.022, 0.030 and 0.032. Bearings have a diametrical clearance of 0.0015-0.0049 on the 3.9991-4.0005 diameter crankshaft crankpins.

89. When installing connecting rods and/or caps, make certain that numbers stamped on both rod and cap are in register and face toward top of

engine. Number one cylinder is located on left side of tractor and has the number "1" stamped on both the rod and the cap. Number two rod and cap have number "2" stamped on them. Tighten connecting rod nuts to a torque of 200-220 Ft.-Lbs.

CYLINDER BLOCK

90. To remove the cylinder block, first remove the rod and piston units as outlined in paragraph 84. Drain diesel engine crankcase and shut off fuel at tank. Disconnect fuel lines and remove fuel filters assembly from the right side of cylinder block. Remove the injection pump bleed line with union by pulling same outward from right side of cylinder block. Remove the injection pump rack plug and spring. Remove the four cap screws which secure the injection pumps and bracket assembly to cylinder casting. Carefully pry the assembly off dowel pins and lift assembly from the compartment. Cover the fuel inlet openings of the bracket with masking tape. Remove pump control rod and disconnect the heat indicator bulb. Remove the stud nuts retaining the block to the main case. NOTE: Four stud nuts are inside the injection pump compartment. Attach a suitable hoist and slide block forward on studs. NOTE: As the block is pulled forward remove the injection pump control rod packing retainer, injection pump control rod guide and spring as shown in Fig. JD1553. Lift the cylinder block from the tractor.

91. Before reinstalling block check to make certain that vent holes, oil passages and fuel passages are open and clean. Install the block by reversing the removal procedure and tighten the retaining nuts to a torque of 275

Fig. JD1560 — The piston and connecting rod cannot be removed as a unit unless the radiator has been removed.

Fig. JD1561 — View of the diesel engine cylinder block installation.

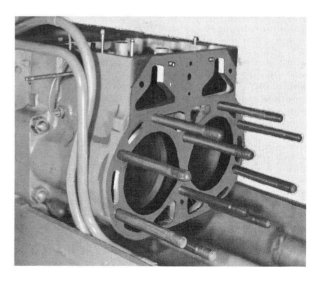

Ft.-Lbs. Bleed the fuel system as outlined in paragraph 108 and adjust the injection pump timing as outlined in paragraph 111. Adjust the tappet gap to correct specifications and diesel engine timing and tune-up as outlined in paragraph 77.

CRANKSHAFT, MAIN BEARINGS AND SEALS

The crankshaft is supported in the main case by three main bearings. The left and right hand bearing housings contain aluminum-alloy bushings whereas the center bearing is steel-backed, copper-lead-lined bearing insert. Left and right main bearing bushings are available as individual parts, or as a unit with the main bearing housings on a factory exchange basis. The exchange main bearing and housing units are available in standard size as well as 0.002 and 0.004 undersizes. When the left or right main bearing is not a factory exchanged unit the bearing must be sized after installation to provide a diametral clearance of 0.0055-0.0075 between it and the crankshaft. The right main bearing housing contains a double opposed oil seal which prevents mixing of transmission oil with the engine lubricating oil. This seal is furnished already installed in an exchange main bearing housing; or the seal is available as a separate part. The center main bearing inserts are available in standard size and 0.002, 0.004, 0.020, 0.022, 0.030 and 0.032 undersizes. Center main journal to bearing diametral clearance should be 0.0016-0.0050.

Fig. JD1561A — Right side of the main case showing the installation the first reduction gear.

92. **MAIN BEARINGS.** Although most repair jobs associated with the crankshaft will require the removal of all three main bearings there are infrequent instances where the failed or worn part is so located that repair can be accomplished safely by removing only one. In effecting such localized repairs, time will be saved by observing the following as a general guide.

Fig. JD1562 — Right side of the tractor main case with first reduction gear cover and first reduction gear removed.

93. RIGHT MAIN BEARING HOUSING AND SEAL. Remove the clutch and belt pulley assembly as outlined in paragraph 131. Drain transmission oil from the first reduction gear cover, disconnect hydraulic lines from the hydraulic pump and unclip the diesel engine fuel line from the first reduction gear cover. Disconnect fuel line from the filters assembly and raise the fuel and hydraulic lines as far as possible. Remove the cap screws attaching the first reduction gear cover to the main case and remove the first reduction gear cover. Remove castellated nut (Fig. JD 1561A) from sliding gear shaft and remove the first reduction gear. Refer to Fig. JD-1562 and remove the transmission oil return pipe. Unbolt and remove the right main bearing housing. Refer to Fig. JD1563. The two opposed oil seals (18) can be renewed at this time. Be sure to fill the space between the seals with gun grease to provide initial lubrication.

Reinstall main bearing housing, using a sleeve or shim stock to guide the opposed oil seals over the crankshaft. Make certain that gasket is properly installed and oil holes are open and clean. Tighten the bearing housing retaining cap screws to a torque of

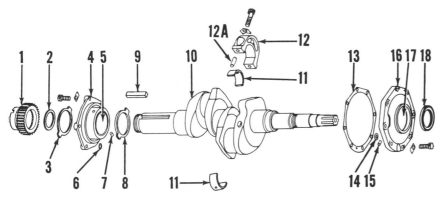

Fig. JD1563 — Exploded view of the diesel engine crankshaft and associated parts. Individual replacement main bearings (5 & 17) will require final sizing after installation. The hollow dowel (12A) is used on 830 tractors only.

1. Crankshaft gear
2. Packing
3. Thrust washer
4. Left main bearing housing
5. Left main bearing
6. Gasket
7. Dowel pin
8. Thrust washer
9. Key
10. Crankshaft
11. Center main bearing
12. Center main bearing cap
12A. Hollow dowel
13. Gasket
14. Gasket
15. Dowel pin
16. Right main bearing housing
17. Right main bearing
18. Oil seal

150 Ft.-Lbs. NOTE: Do not tighten the first reduction gear cover retaining cap screws until clutch and belt pulley assembly is permanently installed and rotated several revolutions to make certain that the pulley dust shield and clutch operating sleeve are not binding. When installing clutch drive disc align "V" marks as shown in Fig. JD1563A.

94. CENTER MAIN BEARING. To remove the center main bearing, first remove the cylinder block as outlined in paragraph 90. Disconnect oil line, remove center main bearing cap and renew bearing inserts in a conventional manner. When reinstalling, tighten cap screws to a torque of 150 Ft.-Lbs.

95. LEFT MAIN BEARING HOUSING. Remove the timing gear housing and cover as outlined in paragraphs 78 and 79. Remove the crankshaft oil slinger, camshaft gear and hub unit, both idler gears and crankshaft gear. Refer to Fig. JD1563B. Unbolt and remove the main bearing housing from the tractor main case. Renew thrust washers (3 and 8—Fig. JD1563) if they are scored or show excessive wear.

Reinstall left main bearing housing by reversing the removal procedure, and tighten the retaining cap screws to a torque of 150 Ft.-Lbs. Refer to paragraphs 82 and 82A for timing gear and oil slinger installation procedures.

96. CRANKSHAFT. To remove the crankshaft, it is necessary to remove all three main bearings as outlined in paragraphs 93, 94 and 95. Attach a rope hoist to crankshaft and carefully remove the crankshaft through the right side of tractor main case.

Crankpin diameter 3.9991-4.0005
  Bearing diametral
    clearance 0.0015-0.0049
Right and left main
  journals diameter . . . 4.498-4.499
  Bearing diametral
    clearance 0.0055-0.0075
Center main journal
  diameter 4.498-4.499
  Bearing diametral
    clearance 0.0016-0.0050
Pulley journal
  diameter 2.9961-2.9975
  Bearing diametral
    clearance 0.0015-0.0029
Rod bolt torque . . . . . 200-220 Ft.-Lbs.
Main bearing
  bolt torque . . . . 150 Ft.-Lbs.

Adjust crankshaft end play as outlined in paragraphs 97 and 98 and check diesel engine timing and tune-up as outlined in paragraph 77.

FLYWHEEL AND CRANKSHAFT END PLAY

97. To remove the diesel engine flywheel, proceed as follows: Remove flywheel cover, flywheel lock nut and loosen both flywheel retaining bolts. Swing flywheel in a hoist, attach a suitable puller and pull flywheel from crankshaft. Flywheel cover back plate can be removed at this time.

To install flywheel starter ring gear, heat same evenly to 550 degrees F. and place gear on flywheel so that the beveled ends of the teeth face toward the diesel engine crankcase. Install flywheel on crankshaft so that "V" mark on flywheel is in register with "V" mark on crankshaft.

98. The diesel engine crankshaft end play is determined by the position of the flywheel on the crankshaft. To ad-just crankshaft end play, proceed as follows: Drive flywheel on crankshaft and mount a dial indicator so that the indicator contact button is resting on the flywheel. Engage and disengage clutch with force several times and observe crankshaft end play as shown on the dial indicator. Continue driving flywheel on the crankshaft until the proper end play of 0.005-0.010 is obtained. Tighten both flywheel retaining bolts to a torque of 334 Ft.-Lbs. Install flywheel lock nut and secure same by peening a portion of the nut into one of the crankshaft keyways.

CRANKCASE COVER

99. On tractors equipped with the gasoline starting engine, the diesel engine crankcase cover can be removed from top of tractor main case after removing the starting engine as outlined in paragraph 30.

99A. On tractors equipped with a 24-volt starting motor, the diesel engine crankcase cover can be removed after first removing the hood, starting motor and bracket.

ENGINE OILING SYSTEM

100. **OIL PRESSURE — ADJUST.** The desired diesel engine oil pressure of 25 psi with engine at operating temperature and running at high idle speed is obtained by turning the oil pressure adjusting screw as shown in Fig. JD1564. Turn the adjusting screw **IN** to increase the oil pressure and **OUT** to decrease the oil pressure. To accurately check the oil pressure, connect a master pressure gage to the oil pressure gage line connection on top of main case.

Fig. JD1563A—When installing the diesel engine clutch drive disc, align "V" marks.

Fig. JD1563B — Left side of the main case with timing gear housing, camshaft gear and hub unit, both idler gears and crankshaft gear removed.

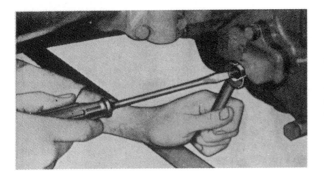

Fig. JD1564 — Diesel engine oil pressure of 25 psi is obtained by turning the oil pressure adjusting screw as shown. Turn the screw in to increase oil pressure.

101. **OIL PRESSURE REGULATOR — R&R AND OVERHAUL.** The oil pressure regulator assembly (items 26 through 33—Fig. JD1565) can be removed from the right side of the tractor main case after removing the retaining cap screws. The oil pressure relief coil spring (30) should have a free length of 1$\frac{25}{32}$-inches and should require 22-25 ounces to compress it to a height of ¾-inch.

102. **FILTER HEAD AND BY-PASS VALVE — R&R AND OVERHAUL.** The oil filter by-pass valve ball (38—Fig. JD1565) and related parts can be removed from oil filter head (39) after removing the crankcase cover as outlined in paragraph 99 or 99A. Remove by-pass retaining cap (35), spring (37) and ball (38). If the filter head is to be removed, complete the preceding steps and proceed as follows: Remove filter cap and element, disconnect the oil lines from the filter head and using a punch through the by-pass valve opening, drive the oil filter outlet (43) out of the filter head. Remove the filter head retaining cap screws from inside the filter body and withdraw the filter head through the crankcase opening.

Normal servicing of the filter head includes cleaning the oil passages and overhauling the by-pass valve assembly. If filter outlet was distorted during removal, be sure to thoroughly straighten same or install a new one.

103. **OIL FILTER BODY—RENEW.** Zero or low oil pressure can be caused by a distorted oil filter body (42—Fig. JD1565). Distortion is usually caused by over-tightening the filter bottom retaining nut. Nut (51) should be tightened only enough to eliminate oil leakage at this point. If leakage occurs, with normal tightening, renew gasket (48).

104. To renew the oil filter body, remove the filter head as outlined in paragraph 102 and proceed as follows: Place a cylindrical wooden block in filter body and jack up block, forcing filter body out of crankcase recess. Withdraw body through crankcase cover opening.

To install oil filter body, lubricate portion below the "bead" to prevent leaks and facilitate drawing body into recess in crankcase. Place filter head on the body to make certain that cap screw holes in body and head are aligned and that oil lines will align with filter head connections; then, remove filter head. To draw body into crankcase, use a long bolt and two steel plates as shown in Fig. JD1566. Tighten the nut until bead of body

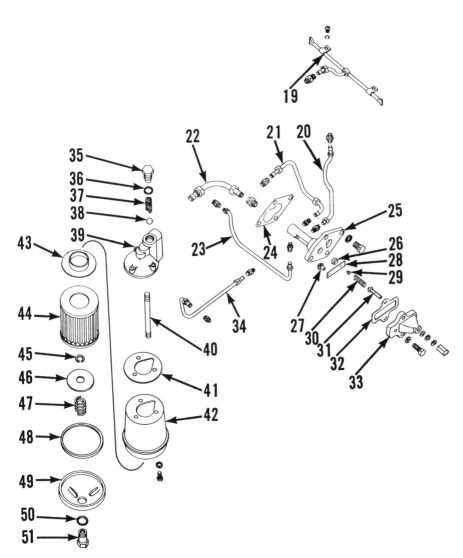

.Fig. JD1565 — Exploded view of the diesel engine oil filter, filter head, pressure regulator and associated parts.

19. Tappet lever oiler tube
20. Camshaft oil pipe
21. Center main bearing oil tube
22. Oil filter discharge tube
23. Left main bearing oil tube
24. Gasket
25. Oil manifold
26. Oil pressure relief valve bushing
27. Rivet
28. Oil pressure relief valve leaf spring

29. Oil pressure relief valve coil spring pilot
30. Oil pressure relief valve coil spring
31. Oil pressure adjusting screw
32. Gasket
33. Oil pressure relief valve housing
34. Right main bearing oiler tube
35. Oil filter by-pass valve cap
36. Gasket
37. Oil filter by-pass valve spring

38. Oil filter by-pass valve ball
39. Oil filter head
40. Stud
41. Gasket
42. Oil filter body
43. Oil filter outlet
44. Oil filter element
45. Snap ring
46. Oil filter bottom plate
47. Spring
48. Gasket
49. Oil filter cover
50. Copper washer

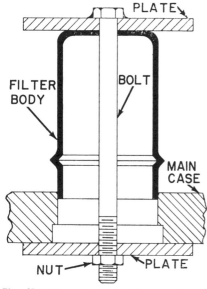

Fig. JD1566 — Suggested home-made tool for installing oil filter body. The long bolt can be welded to the upper plate.

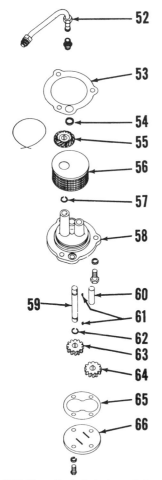

Fig. JD1567 — Exploded view of the diesel engine oil pump.

52. Oil pump discharge tube	59. Oil pump drive shaft
53. Shim gasket (0.005)	60. Oil pump idler gear shaft
54. Gasket	63. Oil pump drive gear
55. Oil pump drive bevel gear	64. Oil pump idler gear
56. Oil pump screen	65. Gasket
57 & 62. Snap rings	66. Oil pump cover
58. Oil pump body	

seats against crankcase. Install remainder of parts by reversing the removal procedure.

105. OIL PUMP—R&R AND OVERHAUL. To remove the diesel engine oil pump, first drain the crankcase, remove the cap screws retaining the pump to main case and remove pump. Note: Do not lose or damage the gaskets (53—Fig. JD1567) between the oil pump and main case as they are used to obtain proper backlash of the oil pump driving bevel gears. Make certain that oil passages in main case and pump are open and clean. Disassemble pump and check body, gears and shafts against the values listed below:

Drive shaft bore in pump
   body 0.627-0.629
Diameter of drive
   shaft 0.624-0.625
Diametral clearance between
   drive shaft and bore..0.002-0.005
Gear bore in pump
   body 2.090-2.092
Diameter of gear.......2.082-2.084
Radial clearance between
   teeth and body.......0.003-0.005
Idler gear shaft bore in
   pump body 0.627-0.629
Diameter of idler gear
   shaft 0.624-0.625
Diametral clearance
   between idler gear
   shaft and bore........0.002-0.005
Depth of body gear
   bore 0.618-0.622
Gear thickness 0.623-0.625
Thickness of gasket between
   body and cover.......0.018-0.022

When reinstalling the oil pump, it is important to consider the mesh position and backlash of the driving bevel gears. The gears should be meshed so that the heels are in register and backlash is from 0.004-0.006. The thickness of shims and gaskets (53) located between pump housing and main case control the backlash (and to some extent the mesh position) of the bevel gears. The act of setting the backlash of the bevel gears will generally serve to place them in the correct mesh position. Shims are 0.005 thick.

Install pump by reversing the removal procedure, use the same number and thickness of shims and gaskets (53) and apply a 0.003 thick piece of paper directly on the gear teeth. Tighten the retaining screws securely and turn the pump drive shaft, thereby passing the 0.003 thick piece of paper between the bevel gears. Remove the pump and check the piece of paper. If the paper shows a heavy impression but is not punctured, the backlash is within the specified limits.

If there is any reason to suspect that the gears are not meshed properly, refer to paragraph 126A.

DIESEL FUEL SYSTEM

As shown in Fig. JD1570, the diesel engine fuel system consists of four basic parts; the fuel transfer pump, fuel filters, injection pumps and injection nozzles. When servicing any unit associated with the fuel system, the maintenance of absolute cleanliness is of utmost importance.

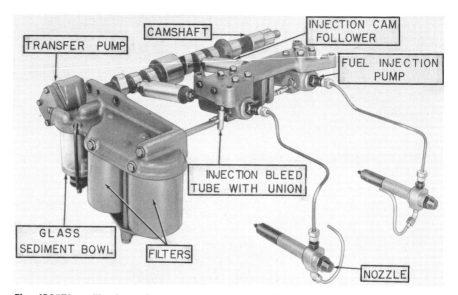

Fig. JD1570 — The fuel injection system consists of four basic parts; the fuel transfer pump, fuel filters, injection pumps and nozzles. Fuel leaves the transfer pump at 12-24 psi and leaves the nozzles at 2400-2600 psi.

DIESEL SYSTEM TROUBLE-SHOOTING CHART

	1. Engine Hard to Start	2. Lack of Power	3. Engine Surges (Idle Speeds)	4. Engine Smokes or Knocks	5. Excessive Fuel Consumption	6. Fuel in Pump Compartment	7. Sticking Pump Plunger
Lack of fuel	★						★
Water or dirt in fuel	★	★					★
Clogged fuel system	★	★					
Air in fuel and injection system	★	★					
Inferior fuel	★	★					
Cratered injection nozzle tips	★	★	★	★	★		
Clogged nozzle spray tip holes	★	★					
Sticky nozzle valves	★	★	★	★	★		
Faulty injection pump timing	★	★		★			
Worn or stuck pump plungers	★	★					
Improperly calibrated or balanced pumps	★	★			★		
Faulty transfer pump	★	★					
Engine speed too low		★					
Faulty or incorrectly adjusted governor			★	★			
Worn or sticky pump delivery valves			★	★			
Excessively worn pump rack and control sleeve			★				★
Dirty nozzle spray tip		★					
System leaks					★		
Loose pump delivery valve holder		★			★	★	
Damaged delivery valve holder gasket		★			★	★	
Loose or split fuel lines		★				★	
Leaking pump-bracket or bracket-cylinder gasket						★	
Delivery valve holder too tight							★
Bent racks or linkage			★				★

All troubles listed in columns 1, 2, 4 and 5 could be caused by derangement of parts other than the diesel pump and injector system.

107. QUICK CHECKS — UNITS ON TRACTOR. If diesel engine does not start or does not run properly, and the diesel fuel system is suspected as the source of trouble, refer to the Diesel System Trouble-Shooting Chart and locate points which require further checking. Many of the chart items are self-explanatory; however, if the difficulty points to the fuel filters, fuel transfer pump, injection nozzles and/or injection pumps, refer to the appropriate section which follows.

FILTERS AND BLEEDING

108. The fuel filtering system consists of a glass sediment bowl and strainer, and two stages of renewable element type filters. The fuel system should be purged of air whenever the filters have been removed or when the fuel lines have been disconnected.

The general procedure for testing the condition of the filters and bleeding the system is as follows:

Shut off the fuel, remove and thoroughly clean the glass sediment bowl which is located under the fuel tank. Reinstall the bowl loosely, turn on the fuel and allow it to flow until bowl is full, then tighten the bowl securely. Remove the relief valve plug (A—Fig. JD1571) or run engine at slow-idle speed during the bleeding process. Remove the first bleed plug (B), allow fuel to flow until free of air, then reinstall plug. Remove the second bleed plug (C), allow fuel to flow until free of air, then reinstall the plug. On early models so equipped, remove the third bleed plug (D), allow fuel to flow until free of air, then reinstall the plug. If fuel does not flow freely at a bleed point, it indicates a clogged filter.

TRANSFER PUMP

The gear type fuel transfer pump is located on the right side of the tractor main case and is driven by slotted plug (19—Fig. JD1574) which is screwed into the end of the diesel engine camshaft. A spring and ball type relief valve which is located in the fuel pump housing maintains a normal pres-

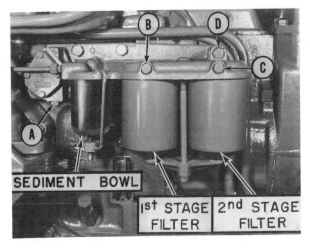

Fig. JD1571—Diesel fuel filters installation, showing bleed plugs (B, C & D) and relief valve plug (A).

sure of 12-24 psi. Seals (21) prevent fuel from entering the diesel engine crankcase and crankcase oil from entering the fuel pump.

109. TESTING. Remove first bleed plug and install a suitable pressure gage as shown in Fig. JD1573. Start diesel engine and observe pressure reading which should be 12-24 psi. If the pressure is not as specified, remove the relief valve plug, vary the number of washers (26—Fig. JD1574) as required and recheck the reading.

110. OVERHAUL. Removal and disassembly of pump is evident after an examination of the unit and reference to Fig. JD1574. Defective drive shaft seals (21) are indicated by fuel or crankcase oil leaking through the drain hole at bottom of pump body.

Failure of pump to deliver fuel at an adequate pressure may be caused by a faulty relief valve. Free length of relief valve spring is 1¼-inches and should require 5.5-7.5 ounces to compress it to a height of ⅞-inch. Check pump body, gears and shafts against the values which follow:

Drive shaft bore in
   pump housing 0.486-0.487
Drive shaft diameter.....0.484-0.485
Radial clearance between
   gear teeth and body...0.0015-0.003
Gear bore in pump
   body 1.113-1.115
Diameter of gears......1.109-1.110
Diameter of idler gear
   shaft 0.484-0.485
Idler gear shaft bore in
   pump housing 1.249-1.251
End clearance between
   gears and cover......0.003-0.004
Gear thickness 0.2805

When reassembling, use a thin coat of shellac on both sealing surfaces of pump body. Use a very small amount of No. 3 Permatex on O.D. of inner seal (21) and press the seal in until it bottoms. Install felt washer (21A). Outer seal should be installed flush with pump surface. Make certain that seal lips are opposed as shown in Fig. JD1574A.

NOTE: Early production tractors were not fitted with a felt washer between the seals. When working on these early models, obtain and install the felt washer (part no. R201-39R).

INJECTION PUMPS

The diesel injection pumps are located in a compartment in the diesel engine cylinder block. Fuel enters the injection pumps, from the filters, under a normal pressure of 12-24 psi and leaves the pumps under a pressure of 2400-2600 psi. The reciprocating pump plungers are actuated by cam followers from the diesel engine camshaft.

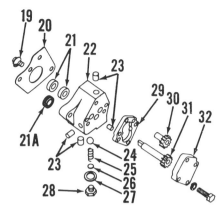

Fig. JD1574 — Exploded view of the diesel fuel transfer pump. The pump is driven by slotted plug (19) which screws into left end of the diesel engine camshaft. Be sure to install a felt washer (21A) on early models not so equipped.

19. Drive plug	26. Adjusting washer
20. Gasket	
21. Seals	27. Relief valve plug washer
21A. Felt washer	
22. Pump housing	28. Relief valve plug
23. Dowel pin	29. Pump body
24. Relief valve ball	30. Idler gear
25. Relief valve spring	31. Drive gear
	32. Cover

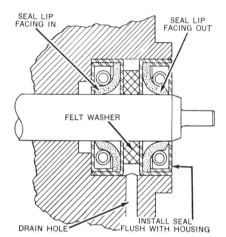

Fig. JD1574A — Partial sectional view of fuel transfer pump showing proper installation of the seals and felt washer.

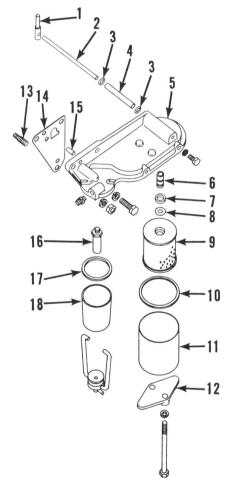

Fig. JD1572—Exploded view of the diesel fuel filtering system. The condition of filtering elements can be determined by bleeding the system (see text).

1. Injection bleed tube and union	9. Filter element
2. Bleed tube (long)	10. Case gasket
3. "O" ring	11. Filter case
4. Tube	12. Filter case clamp
5. Filter head	13. Stud
6. Filter element tube	14. Gasket
7. Felt washer	15. Dowel pin
8. "O" ring	16. Fuel strainer
	17. Gasket
	18. Fuel filter bowl

Fig. JD1573 — Checking fuel transfer pump pressure. Gage is installed in first bleed plug hole.

Injection pumps may be either Bendix-Scintilla or American-Bosch; and although both makes operate in the same manner and are similarly constructed, pumps of different makes should never be installed in any one tractor.

111. TIMING. Remove flywheel cover, tappet lever cover and injection pump compartment cover. Turn flywheel in running direction until the inlet valve for No. 1 (left hand) cylinder opens and closes and continue turning flywheel until the "No. 1 INJ" mark on flywheel is in register with mark on flywheel cover back plate, as shown in Fig. JD1575. At this time, the scribe mark on number 1 injection pump plunger follower should align with an index in the window of the pump body. On Bosch pumps, the index is a scribe line on the sides of the window as shown at right in Fig. JD 1576. On Scintilla pumps, the index is a sheet metal pointer as shown at the left. If index marks do not align, adjust the length of the injection pump push rod with screw (17—Fig. JD 1577), located just behind pump, until proper register is obtained. Turn flywheel ½-turn (180 degrees) and align "No. 2 INJ" mark on flywheel with mark on timing gear housing. Check and time number 2 injection pump in the same manner.

CAUTION: Make certain that scribe line does not disappear from window in No. 1 pump housing when flywheel is rotated ½-turn (180 degrees) from

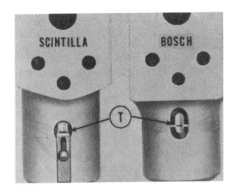

Fig. JD1576 — Either Scintilla or Bosch injection pumps are used. Notice differences in construction at timing windows (T).

injection timing mark. If scribe line does disappear, the pump is not properly timed and internal pump parts may be damaged.

112. TESTING. If diesel engine does not run properly, or not at all, and the quick checks outlined in paragraph 107 point to faulty injection pumps, the pumps must be renewed, cleaned and/or overhauled.

In order to obtain balanced fuel delivery to each cylinder, the pumps are synchronized by adjusting the connecting linkage between the pump racks. Since elaborate testing equipment (equipment which was especially designed for use in testing and adjusting the John Deere injection system) is required to make the adjustment, the linkage should NEVER be disturbed unless such equipment is available.

NOTE: It is recommended that any work which must be performed on the injection pumps be done by a John Deere factory approved service station. However, if adequate facilities are available for maintaining ABSOLUTE cleanliness, the pumps can be removed from the mounting bracket, disassembled, cleaned, and overhauled without disturbing the rack connecting linkage adjustment.

114. INJECTION PUMPS — R&R. To remove the injection pumps and mounting bracket unit, proceed as follows: Remove the heat exchanger outlet pipe, tappet lever cover and injection pump compartment cover. Remove fuel lines connecting injection pumps to injection nozzles and immediately cap the four connectors. Completely back off both injection pump push rod adjusting screws. Remove the rack spring plug from right side of cylinder block and extract the rack spring. Unbolt and remove the pump assembly from cylinder block.

CAUTION: Do not pry on the racks or any connecting linkage. If linkage is bent or distorted, it will be unsuitable for further use.

After assembly is removed from tractor, tape or cork the fuel inlet openings which are located in the underside of the pump mounting bracket. Note: Injection pump push rod units can be removed at this time.

CAUTION: In removing the injection pumps from mounting bracket, do not disturb the pump connecting linkage adjustment as in doing so, the pumps will be thrown out of synchronization.

Mark position of pumps on the mounting bracket. Pumps are not in-

Fig. JD1575 — "No. 1 INJ" mark in register with mark on flywheel cover back plate.

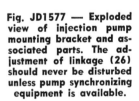

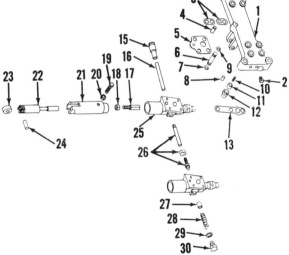

Fig. JD1577 — Exploded view of injection pump mounting bracket and associated parts. The adjustment of linkage (26) should never be disturbed unless pump synchronizing equipment is available.

1. Pump bracket
2. Plug
3. Gasket
4. Dowel pin
5. Gasket
6. Injection pump bracket oil inlet tube
7. "O" ring
8. Injection pump oil inlet tube
9. "O" ring
10. Injection pump fuel inlet filter spring
11. Fuel inlet filter
12. Fuel inlet tube
13. Gasket
15. Injection pump control rod guide
16. Injection pump control rod
17. Injection pump adjusting screw
18. Adjusting screw lock nut
19. Pump cam follower guide set screw

20. Lock nut
21. Follower guide
22. Cam follower
23. Roller
24. Roller pin
25. Injection pump

26. Rack connecting linkage
27. Rack spring guide
28. Rack spring
29. Washer
30. Injection pump rack spring plug

terchangeable from side to side and must be installed in their original position. Unbolt and remove pumps from bracket but don't disassemble unless test stand is available.

114A. After the injection pumps and mounting bracket unit is removed as outlined in the previous paragraph and the injection pump cam follower retaining cap screws (S—Fig. JD 1577A) are loosened, the followers can be extracted as shown in Fig. JD 1577A. Make certain followers are not bent or in any other way damaged during the removal.

INJECTION NOZZLES

Injection nozzles may be either Bendix-Scintilla or American-Bosch; and although both makes operate in the same manner and are similarly constructed, nozzles of different makes should never be installed in any one tractor.

115. **R&R AND TEST.** If the engine does not run properly, or not at all, and the quick checks outlined in paragraph 107 point to faulty nozzles, re-

move the injection pump compartment cover and tappet lever cover and proceed as follows: Disconnect fuel lines at nozzles and remove the nozzle retaining nuts and clamps. If a nozzle tester is available, check the nozzle as in paragraph 117. If a special tester is not available, mount nozzles as shown in Fig. JD1578.

WARNING: Fuel leaves the injection nozzle tips with sufficient force to penetrate the skin. When testing, keep your person clear of the nozzle spray.

Place diesel engine speed control lever in wide-open position, hold injection pump racks open, motor the diesel engine with the starting engine or starting motor and observe the nozzle spray pattern.

In a good nozzle, the fuel comes out in a finely-atomized mist. A lopsided spray pattern indicates that holes in nozzle tip are partially clogged, and a stream of fuel or a poorly-atomized spray indicates that nozzle tip holes are worn oversize.

NOTE: Never install a six orifice

nozzle tip on a nozzle which was made for a five orifice tip. If cleaning and/or nozzle tip renewal does not restore the unit and a special nozzle tester is not available for further checking, send the complete nozzle assembly to an official diesel service station for overhaul.

117. **NOZZLE TESTER.** A complete job of testing and adjusting the nozzle requires the use of a special nozzle tester. The nozzle should be tested for leakage, spray pattern and opening (or cracking) pressure. Operate the tester lever until oil flows and attach the nozzle and holder assembly.

Close the tester valve and apply a few quick strokes to the lever. If undue pressure is required to operate the lever, the nozzle valve is plugged and same should be serviced as in paragraph 118.

CHATTERING AND POPPING. When operating the tester handle at a steady rate of 25-35 strokes per minute, the needle valve should open and close with a distinct, audible chattering or popping sound. If the chattering or popping sound is not heard, the needle valve and seat are dirty, broken or excessively worn. Service the valves as in paragraph 118.

SPRAY PATTERN. Operate the tester handle and observe the spray pattern which should be a finely-atomized mist. A lopsided spray pattern indicates that holes in nozzle tip are partially clogged, and a stream of

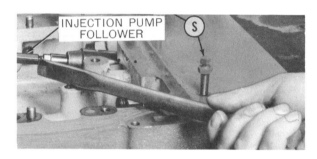

Fig. JD1577A — When removing the injection pump cam followers, make certain followers are not bent during the removal.

Fig. JD1578 — Diesel injection nozzle mounted in position for observing spray pattern.

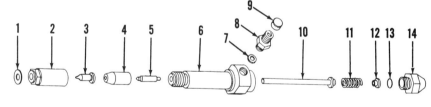

BENDIX SCINTILLA

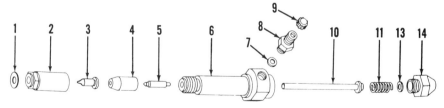

AMERICAN BOSCH

Fig. JD1579 — Exploded views of the injection nozzles. Pressure spring cap (14) should never be removed unless suitable equipment is available for checking the nozzle cracking pressure.

1. Copper washer	5. Needle valve	8. Inlet nipple	11. Pressure spring
2. Nozzle tip retaining nut	6. Nozzle holder body	9. Protector cap	12. Pressure spring seat
3. Nozzle tip	7. Gasket	10. Pressure spring pin	13. Pressure adjusting shims
4. Needle valve seat			14. Spring cap

fuel or a poorly-atomized spray indicates that nozzle tip holes are worn oversize. The nozzle tip should be cleaned or renewed. Note: Never install a six orifice nozzle tip on a nozzle which was designed for a five orifice tip.

OPENING PRESSURE. While operating the tester handle, observe the gage pressure at which the spray occurs. The gage pressure should be at least 2400 psi. If the pressure is not as specified, remove nozzle spring cap and add or deduct shims (13—Fig. JD 1579) until the opening pressure is 2400 psi. Note: If a new pressure spring has been installed in the nozzle holder, adjust the pressure to 2600 psi.

LEAKAGE. The nozzle valve should not leak at a pressure of 100-300 pounds less than the opening pressure of 2400-2600 psi. To check for leakage, actuate the tester handle slowly and as the gage needle approaches the specified opening pressure, observe the nozzle tip for drops of fuel. If the nozzle valve and seat are in good condition, the nozzle tip will be reasonably dry. Also, observe the nozzle leak-off pipe where a small amount of

leakage is normal. If the leakage through the leak-off pipe exceeds 1 cc. per minute or if the nozzle tip is wet, the parts should be reconditioned.

DRIBBLE. When operating the tester handle with quick strokes, the fuel injection should start and stop without producing any dribble at the spray tip. Dribble could be caused by the method of handling the tester. Therefore, make the check several times and wipe the tip after each check. A persistent dribble indicates worn or dirty parts.

118. **NOZZLES—R&R AND OVERHAUL.** To remove nozzles from tractor, remove the tappet lever cover and injection pump compartment cover. Remove fuel lines connecting injection pumps to injection nozzles and immediately cap the four connectors. Remove both nozzle retaining nuts and clamps and pull nozzles from cylinder head, taking care not to damage the fuel leak-off pipes.

Before disassembling, wash complete nozzle in clean diesel fuel and blow off with clean, dry, compressed air. Carbon accumulation from exterior of nozzle spray tip can be removed with a wire brush. Set up a pan with a clean cloth in the bottom and fill pan with clean diesel fuel. As parts are removed from nozzles, place same on cloth in pan, but do not allow parts to touch one another. Clamp nozzle in a soft jawed vise, and remove nozzle tip retaining nut (2—Fig. JD1579). Nozzle valve assembly (4 & 5) and spray tip (3) will be re-

moved with the nut. Push spray tip out of nut. If spray tip is frozen to the retaining nut, use a piece of steel tubing which will fit about halfway down on the tip and tap tip out of the nut as shown in Fig. JD1580.

CAUTION: Do not remove cap (14—Fig. JD1579) unless equipment is available for setting the nozzle cracking pressure. Be careful to keep each matching needle valve and seat together.

NOZZLE TIP. On models 80 and 820 prior to serial number 8203100, proceed as follows: Clean the five tip orifices with a 0.0098 broach (John Deere part No. AF 2291 R) mounted in a pin vise as shown in Fig. JD1581. Clean center passage in tip with a No. 50 drill. See Fig. JD1582. Test the size of the five tip orifices with a 0.012 piano wire. If the wire will **not** go in the orifices, the tip is all right. If wire **does** go in, orifices are excessively worn and nozzle tip should be renewed. Be sure to use a five orifice tip.

On models after serial number 8203099 proceed as follows: Clean the six tip orifices with a 0.0118 broach (John Deere part No. AR 20148 R) mounted in a pin vise, as shown in Fig. JD1581. Clean center passage in tip with a No. 50 drill as shown in Fig. JD1582. Test the size of the six tip orifices with a 0.014 piano wire. If the wire will **not** go in the orifices, the tip is all right. If wire **does** go in, orifices are excessively worn and nozzle tip should be renewed. Be sure to use a six orifice tip.

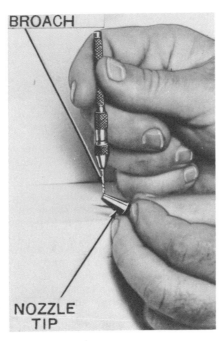

Fig. JD1581 — The orifices in the injection nozzle tip can be cleaned with a broach of the size indicated in text.

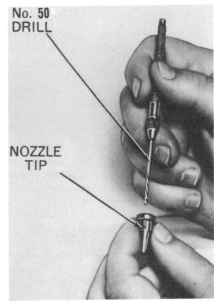

Fig. JD1582—Center passage of nozzle tips can be cleaned with a No. 50 drill. A wire brush will facilitate removal of carbon deposits from outside of tip.

Fig. JD1580—If injection nozzle tip is frozen in retaining nut, a steel tube and a hammer will facilitate removal.

NEEDLE VALVE AND SEAT. Inspect needle valve and seat with a magnifying glass. Renew valve and seat if seating surfaces are scored or if cylindrical surfaces are scratched or have a dull gray appearance. Both of these conditions are illustrated in Fig. JD1583. The needle valve and seat are available only as a matched unit. If valve and seat are in good condition, examine flat surface of valve seat where it contacts the nozzle holder. Discoloration and small pits can be removed by lapping with a coarse rouge until a smooth surface is obtained, then finishing with a polishing rouge — using the technique and equipment shown in Fig. JD1584.

Test fit of needle valve in its seat as follows: Hold seat in a vertical position and start needle valve in its seat. Valve should slide slowly into the seat under its own weight. Note: Dirt particles, too small to be seen by the naked eye, will restrict the valve action. If needle valve sticks, and it is known to be clean, free-up valve in seat, using a mixture of mutton tallow and clean diesel fuel and rotating the needle valve in its seat.

NOZZLE HOLDER BODY. Examine flat surface where it contacts the needle valve seat. Discoloration and small pits can be removed by lapping with a coarse rouge until a smooth surface is obtained, then finishing with a polishing rouge.

REASSEMBLE. Before reassembling any parts, rinse same in clean fuel. Reassemble parts by reversing the disassembly procedure, using Fig. JD 1579 as a general guide. Test nozzles as outlined in paragraph 117.

Reinstall nozzle in cylinder head by reversing the removal procedure, using a new gasket (1) between the nozzle and the head. If new copper gasket is not available, old gasket can be reclaimed by annealing, by heating same to a cherry red and quenching in cold water.

119. NOZZLE SLEEVE — RENEW. Prior to tractor serial number 8203100 the diesel engine cylinder head was equipped with renewable type sleeves to receive the injection nozzles. To remove the nozzle sleeves, a tool similar to the one shown in Fig. JD1585 can be used. Remove the injection nozzles as outlined in paragraph 118, then turn the tap into the sleeve before removing the sleeve retaining machine screw and washer. (This will keep the sleeve from turning in cylinder head.) After the tap is in place, remove the machine screw, install a length of pipe over the tap stud and using a washer

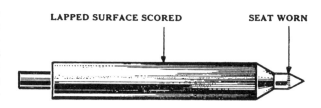

Fig. JD1583 — The injection nozzle needle valve and seat should be renewed if seating surfaces are worn or if cylindrical surfaces are scratched or have a dull gray appearance.

LAPPED SURFACE SCORED   SEAT WORN

Fig. JD1584 — When lapping flat surfaces, use the "figure eight" motion as shown.

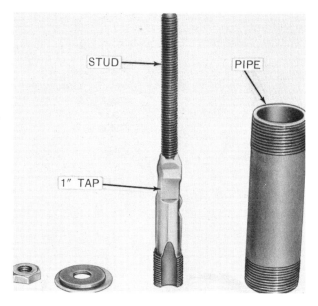

Fig. JD1585 — A tool for removing nozzle sleeves, similar to one shown, can be made by welding a stud to a 1-inch tap. See Fig. JD1586.

STUD   PIPE

1" TAP

and nut on the stud, pull the sleeve out of the head (Fig. JD1586). Make certain that all old material is removed from the head.

When installing the new sleeve, first make certain the opening is free from burrs and new "O" ring (11—Fig. JD 1547) is greased and properly installed. Coat sleeve with light film of Permatex, then slip the sleeve into the cylinder head and while rotating the sleeve slightly, press the sleeve in place. When the sleeve is properly aligned, use a suitable piloted arbor and drive the sleeve in until the sleeve bottoms in the head, then install the sleeve retaining screw and washer.

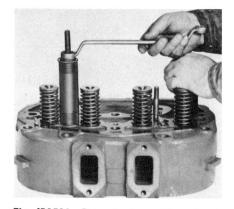

Fig. JD1586—Removing Deere 80 and early 820 injection nozzle sleeve with puller shown in Fig. JD1585.

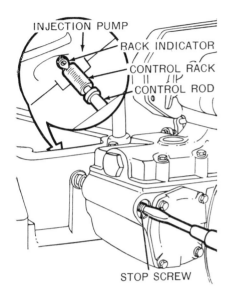

Fig. JD1587—With the speed control lever in fast idle position the rack reading should NEVER exceed 12½. If the engine has been operating satisfactorily and the setting is less than 12½, do not disturb the setting.

GOVERNOR

The gear driven mechanical fly-weight type governor is located within the timing gear housing. Action of the governor flyweights imparts motion to the governor lever which is linked to the injection pump control racks. The pump rack spring which is located between the right end of the right hand injection pump rack and the injection pump rack spring plug eliminates end play from the linkage and works **against** the governor flyweights by tending to move the injection pump racks toward the wide open position. Movement of the racks beyond the maximum permissible rack reading of "12½" is prevented by a stop screw located in the timing gear housing.

120. **ADJUSTMENTS.** Before starting the engine, remove the injection pump compartment cover and with the speed control lever in the fast idle (fully forward) position, observe the rack reading as shown in Fig. JD1587. If the rack reading is more than 12½, remove plug (21 — Fig. JD1588) and turn the stop screw (28) inward as required. Refer also to Fig. JD1587. Note: Under no circumstances should the rack reading ever be adjusted to more than 12½.

If, however, the engine has been operating satisfactorily and delivering rated power and the rack reading is found to be slightly less than "12½", do not alter the adjustment.

Remove the large hex. plug from front of timing gear housing cover and extract the Welch plug located in top of timing gear housing cover just in front of the "Powr-Trol" pump. Set the speed control lever halfway between the slow idle and stop position and adjust the coupling (Fig. JD1589) until distance from speed control rod spring end cap to machined surface of hex. plug opening is $\frac{11}{16}$-inch as shown in Fig. JD1590. Working through the Welch plug opening in top of timing gear housing cover, loosen the speed control rod stop set screw. Using a screw driver, push forward on the stop until the pump racks read "1" and securely tighten the stop set screw.

Start engine and run until operating temperature is reached. With the speed control lever in the slow idle position, turn the coupling (Fig. JD1589) to obtain an engine slow idle speed of 750 rpm. Now, with the engine running at 750 rpm, there should be $\frac{1}{16}$-inch clear-

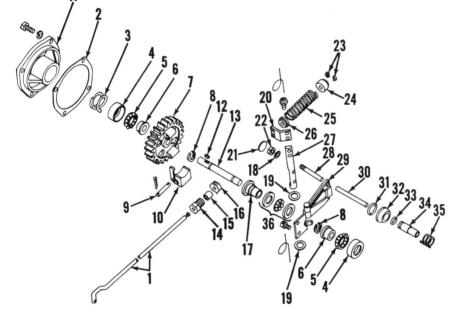

Fig. JD1588 — Exploded view of the diesel engine governor and associated parts.

1. Speed control rod
2. Gasket
3. Spring
4. Bearing cup
5. Bearing
6. Bearing cone
7. Governor drive gear
8. Snap ring
9. Governor weight pin
10. Governor weight
11. Governor bearing housing
12. Woodruff key
13. Governor shaft
14. Speed control rod packing gland
15. Packing
16. Speed control rod stop
17. Governor sleeve
18. Washer
19. Washer
20. Governor loading lever
21. Plug
22. Jam nut
23. Spring end keepers
24. Speed control spring end (cap)
25. Speed control spring
26. Speed control spring fulcrum
27. Governor lever shaft
28. Adjusting lever stop adjusting screw
29. Governor lever
30. Injection pump control rod
31. & 33. "O" ring
32. "O" ring retainer
34. Injection pump control rod guide
35. Spring
36. Thrust bearing

SLOW IDLE
ADJUSTMENT
COUPLING

Fig. JD1589—Turn the coupling until a slow idle speed of 750 rpm is obtained as checked at the belt pulley dust cover.

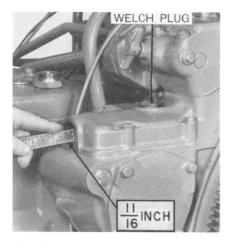

Fig. JD1590—With the speed control lever halfway between the slow idle and stop position the distance from the speed control spring end cap to the machined surface of the opening should be 11/16-inch. See text.

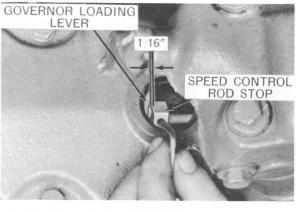

Fig. JD1591 — With the diesel engine running at slow idle speed, there should be 1/16-inch clearance between the speed control rod stop and the governor loading lever.

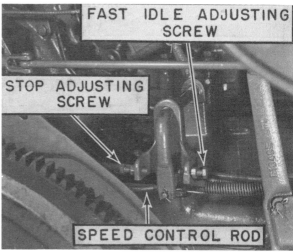

Fig. JD1592 — The fast idle speed should never be more than 1250 rpm as checked at the belt pulley dust cover.

ance between the speed control rod stop and the governor loading lever as shown in Fig. JD1591.

Move the speed control lever to the wide open (fully forward) position and turn the fast idle adjusting screw (Fig. JD1592) until the engine high idle (no load) speed is 1250 rpm when checked at the belt pulley dust cover.

Depress the speed control lever stop button and pull the speed control lever all the way back. If the engine does not stop, loosen the stop adjusting screw (Fig. JD1592) one or two turns and carefully pull the speed control lever back until the engine stops. Then, turn the stop screw in until it just contacts the stop arm of the control mechanism.

Failure of speed control lever to stay in forward position, when operating tractor, indicates that more tension is required on the speed control lever spring. Remove the speed control opening plug (14 — Fig. JD1593) and adjust tension of spring (16) by nut (15). Adjust nut until a recommended force of 10-15 pounds, at end of speed control lever, is required to move the lever.

122. **GOVERNOR — R & R AND OVERHAUL.** It is possible to remove the governor drive gear, weights and shaft assembly without removing the timing gear housing cover; but when reinstalling the unit, the inner bearing may be cocked and damaged. Therefore, the recommended procedure is as follows: Remove the timing gear housing cover as outlined in paragraph 78. Remove governor bearing housing (11—Fig. JD1588) and pull governor assembly from timing gear housing. The need and procedure for further disassembly will be evident from an

1. Speed control rod
2. Fast idle adjusting screw
3. Jam nut
4. Lock washer
5. Stop adjusting screw
6. Speed control shaft
7. Speed control arm
8. Nut
9. Taper bolt
10. Speed control lever end
11. Turnbuckle (coupling)
12. Jam nut
13. Speed control lever rod
14. Plug
15. Speed control lever tension nut
16. Spring
17. Washer
18. Speed control plate
19. & 28. Friction washer
20. Spacer
21. Speed control lever shaft
22. Dowel pin
23. Speed control stop pin
24. Spring
25. Roll pin
26. Speed control stop pin knob
27. Speed control lever
29. Woodruff key

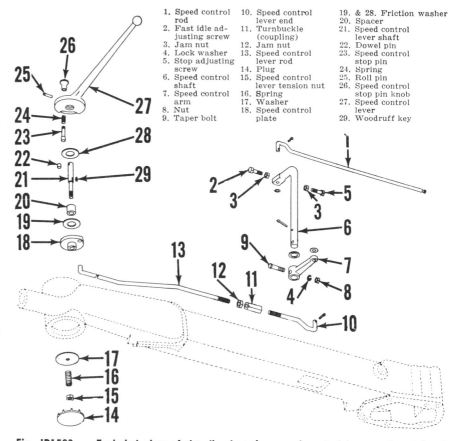

Fig. JD1593 — Exploded view of the diesel engine speed control lever and associated parts.

examination of the unit. Check to make certain that all governor parts and connecting linkage operate freely. Governor gear may be meshed with the camshaft gear in any position, as timing at this point is unnecessary. Speed control spring (25) has a free length of $3\frac{21}{32}$-inches and should require 39-53 pounds to compress it to a height of $2\frac{3}{4}$-inches. Check governor parts for wear and renew any which are questionable.

COOLING SYSTEM

RADIATOR

123. REMOVE AND REINSTALL. To remove the radiator assembly, first drain tractor cooling system and remove hood and grille. Remove diesel engine air cleaner and vent pipe. Remove radiator lower hose and both of the upper water pipes. Remove the water pump belt and the generator fan belt. Disconnect the shutter control rod and fan belt tensioner. Remove the radiator pivot pin retaining snap ring from rear of pivot pin and using a cap screw and washer as shown in Fig. JD1600, remove the pivot pin. Attach suitable hoist and remove the radiator and water pump assembly from tractor.

FAN

124. R&R AND DISASSEMBLE. To remove the fan assembly, remove hood. Disconnect fan belt from drive pulley and generator pulley. Disconnect fan belt tensioner from fan housing, remove the cap screws which retain the assembly to the radiator and lift assembly from tractor.

Disassembly procedure is obvious after an examination of the removed unit and reference to Fig. JD1601.

When reinstalling the fan assembly align the fan pulley with drive shaft pulley and tighten the retaining cap screws securely.

FAN DRIVE SHAFT UNIT AND FAN AND OIL PUMP DRIVE UNIT

The fan drive shaft assembly (A—Fig. JD-1602) is mounted in an opening at the front of the tractor main case and carries the fan drive pulley at its forward end. Tractors equipped with power steering have the power steering pump mounted on the forward end of this shaft. (Refer to paragraph 17.) At the opposite (rear) end, is a bevel gear that meshes with a similar type gear which is keyed to the inner end of a quill mounted gearshaft. A helical gear is keyed to the opposite or outer end of the gearshaft and is driven by a helical idler interposed between it and the crankshaft gear. This gearshaft is known as the fan and oil pump drive unit because the bevel gear on its inner end drives not only the fan drive shaft but also the diesel engine oil pump.

125. DRIVE SHAFT ASSEMBLY — R&R AND OVERHAUL. To remove the fan drive shaft assembly (A—Fig. JD1602), first remove the tractor hood.

Fig. JD1600 — Using a long cap screw and washer to remove the radiator pivot pin.

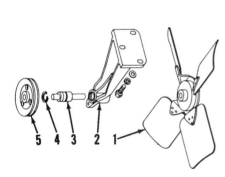

Fig. JD1601 — Exploded view of the fan assembly.

1. Fan
2. Fan shaft bearing housing
3. Bearing
4. Snap ring
5. Fan pulley

Fig. JD1602 — Partially exploded view of the fan and oil pump drive and associated parts, showing adjusting shims.

A. For complete exploded view of fan drive shaft assembly, refer to Fig. JD-1603.

B. For complete exploded view of combination fan and diesel engine oil pump drive assembly, refer to Fig. JD1604.

C. For complete exploded view of diesel engine oil pump, refer to Fig. JD-1567.

2. Fan drive bevel gear
5. Shims
35. Shims
36. Combination fan and oil pump drive bevel pinion
53. Shims
55. Oil pump drive bevel gear

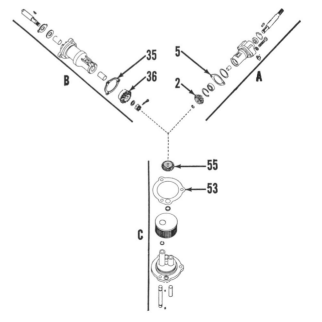

On tractors equipped with power steering remove the power steering pump as outlined in paragraph 17.

On tractors with manual steering remove the fan-generator belt and the water pump belt. Unbolt the front drive housing assembly from tractor front end support.

Pull fan drive shaft tube and coupling shaft forward from rear bearing housing and remove tube and shaft from tractor. Disconnect speed-hour meter cable, then unbolt and remove the rear drive assembly from the tractor main case. Be careful not to damage or lose the shims (5—Fig. JD1603).

Disassembly and overhaul procedures are obvious after an inspection of units and reference to Fig. JD1603.

When reinstalling the drive unit, it is important to consider the mesh position and backlash of the driving bevel gears. The gears should be meshed so that heels are in register and backlash is from 0.004-0.006. The thickness of shims (5) located between rear bearing housing and main case controls the backlash (and to some extent the mesh position) of the bevel gears. The act of setting the backlash of the bevel gears will generally serve to place them in the correct mesh position. Shims are 0.006 and 0.018 thick.

Install the drive assembly by reversing the removal procedure, use

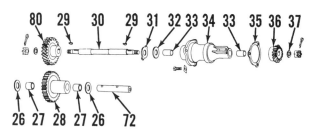

Fig. JD1604 — Exploded view of combination fan and diesel engine oil pump drive assembly. Idler gear (28) meshes with the crankshaft gear. Refer also to Fig. JD1554.

26. & 31. Thrust washer
27. Bushing
28. Idler gear
29. Woodruff key
30. Combination fan and diesel engine oil pump drive shaft
32. Shims
33. Bushing

34. Quill
35. Shims
36. Combination fan and diesel engine oil pump drive bevel pinion

37. Washer
72. Idler gear shaft
80. Combination fan and diesel engine oil pump drive gear

the same number and thickness of shims and gaskets (5) and apply a 0.0025 thick piece of paper directly on the gear teeth. Tighten the retaining screws securely and turn the drive shaft, thereby passing the 0.0025 thick piece of paper between the bevel gears. Remove the drive unit and check the piece of paper. If the paper shows a heavy impression but is not punctured, the backlash is within the specified limits.

If there is any reason to suspect that the gears are not meshed properly, refer to paragraph 126A.

126. FAN & OIL PUMP DRIVE UNIT — OVERHAUL. To remove the drive assembly (B—Fig. JD1602) first remove the timing gear housing and cover as outlined in paragraphs 78 and 79. Unlock and remove cap

screws retaining quill (34—Fig. JD 1604) to main case and withdraw the drive assembly but be careful not to damage or lose shims (35).

Disassembly of the combination fan and engine oil pump drive unit is evident after examination of the unit and reference to Fig. JD1604. Bushings (33) should be installed with a piloted driver. Press the bushings in until they are flush with inner edge of chamfer in edge of bore. If not distorted during installation, the bushings are pre-sized to provide a clearance of 0.0015-0.0039 for the 0.9990-1.0000 diameter drive shaft. When assembling, adjust drive shaft end play to 0.0025-0.0050 by adding or removing shims (32). Shims are 0.0025 thick.

When reinstalling the driving unit, it is important to consider the mesh position and backlash of the fan and oil pump driving bevel gears. The gears should be meshed so that heels are in register and backlash is from 0.004-0.006. The thickness of shims (35) located between the bearing quill and main case control the backlash (and to some extent the mesh position) of the bevel gears. The act of setting the backlash of the bevel gears will generally serve to place them in correct mesh position. Shims are 0.005 and 0.018 thick.

Install the assembled driving unit by reversing the removal procedure; use the same number and thickness of shims (35) and tighten the retaining screws securely. Mount a dial indicator as shown in Fig. JD1604A and check the bevel gear backlash. Since the diameter of the helical gear (80) is larger than the diameter of the bevel gear at inner end of drive shaft, the desired bevel gear backlash of 0.004-0.006 will be obtained if the dial indicator reads 0.006-0.008. If the backlash reading is not as specified, remove the driving unit and vary the number of shims (35—Fig. JD1604).

If there is any reason to suspect that the gears are not meshed properly, refer to paragraph 126A.

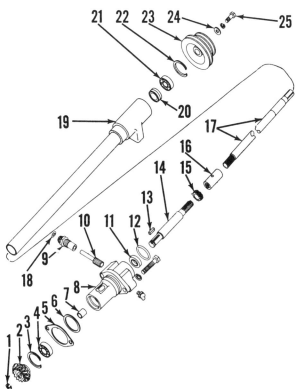

Fig. JD1603 — Exploded view of the fan drive shaft and associated parts. Items 17, 19, 20, 21, 22, 23, 24 and 25 are used on models with manual steering only. On tractors equipped with power steering, the power steering pump is mounted on the forward end of the fan drive shaft tube. Refer to Fig. JD1507.

1. Snap ring
2. Fan drive bevel gear
3. Snap ring
4. Bearing
5. Shims
6. "O" ring
7. Bushing
8. Rear bearing housing
9. Speed-hour meter cable connector
10. Speed-hour meter drive pinion
11. Oil seal (prior to serial number 8203100)
12. "O" ring
13. Key
14. Rear fan drive shaft
15. Speed-hour meter drive gear
16. Coupling
17. Front fan drive shaft
18. Oil seal (after serial number 8203099)
19. Front fan shaft bearing housing and tube
20. Oil seal (after serial number 8203099)
21. Bearing
22. Snap ring
23. Pulley
24. Washer
25. Cap screw

Fig. JD1604A — Dial indicator mounted for checking backlash between the fan and diesel engine oil pump drive bevel gears. Refer to text.

> 72. Idler gear shaft
> 80. Combination fan
>   and engine oil
>   pump drive gear
> 85. Dial indicator

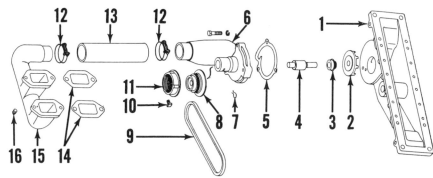

Fig. JD1605 — Exploded view of the diesel engine water pump and associated parts.

1. Radiator bottom tank	5. Gasket
2. Impeller	6. Water pump housing
3. Seal	7. Bearing retainer
4. Bearing	8. Pulley hub

9. Water pump belt	13. Hose
10. Set screw	14. Gasket
11. Pulley adjusting ring	15. Water inlet
12. Clamp	16. Pipe plug

To remove the driving idler gear (28—Fig. JD1604), remove oil slinger from crankshaft timing gear and withdraw the idler gear. Idler gear bushings (27) can be renewed if excessively worn. If idler gear shaft (72) is damaged or worn, it should be renewed. The shaft is tapped to accommodate a puller screw. When reassembling, be sure that thrust washers (26) are properly located. When installing the timing gear housing and crankshaft oil slinger, refer to paragraphs 79 and 82A.

126A. MESH POSITION OF BEVEL GEARS. Refer to Fig. JD1602. Bevel gears (2, 36 & 55) should be meshed so that heels are in register and backlash is from 0.004-0.006. To observe the mesh position, it is necessary to remove the crankcase cover as outlined in paragraph 99 or 99A. Desired mesh position and backlash of the bevel gears is obtained by using the proper combination of shims (5, 35 & 53) between the respective housings and the main case. Refer also to paragraphs 105, 125 and 126.

WATER PUMP

Coolant leakage at drain hole in pump housing usually indicates a leak in seal (3—Fig. JD1605).

127. R&R AND OVERHAUL. To remove the water pump, drain cooling system, disconnect water pump drive belt and remove the lower radiator hose (13). Unbolt water pump from radiator lower tank and remove pump from tractor.

To disassemble the pump, remove the bearing retainer (7) and press shaft and bearing assembly out of impeller and housing. Press shaft and bearing out of pulley. The need and procedure for further disassembly is evident.

When reassembling, press pulley on the shaft until edge of hole in pulley hub is flush with end of shaft. Coat O.D. of seal with a light coat of grease and press seal into housing until flange of seal is tight against housing. Install shaft assembly into housing, then install retainer (7) so that one of the prongs is over the lug on pump body. NOTE: If necessary to press the shaft assembly into the housing, do so by pressing only on outer race of the bearing. Lubricate wearing surface of oil seal with wheel bearing grease and press the impeller on the shaft.

NOTE: If any vane protrudes beyond the mounting face of pump housing, the protruding vane will strike the radiator lower tank when pump is installed.

Install pump on tractor by reversing the removal procedure.

ELECTRICAL SYSTEM

John Deere 830 diesel tractors are available with either a gasoline starting engine or a 24-volt electric starting motor. All tractors equipped with the gasoline starting engine have a 6-volt electrical system with positive ground polarity. Model 830 tractors equipped with the electric starter have a 24-volt split-load system using 12-volts for the lights and accessories and 24-volts for

cranking and charging. A wiring diagram of the 24-volt electrical system is shown in Fig. JD1606. A diagram of the 6-volt system is shown in Fig. JD1606A.

STARTING MOTOR
6-Volt System

128. On tractors equipped with the gasoline starting engine and a 6-volt electrical system, a Delco-Remy No.

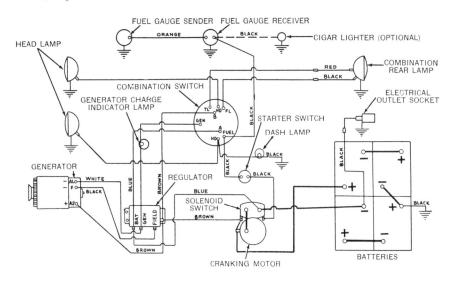

Fig. JD1606 — Wiring diagram for 830 diesel tractors equipped with the split load 24-12 volt electrical system. The four 6-volt batteries should be installed as shown.

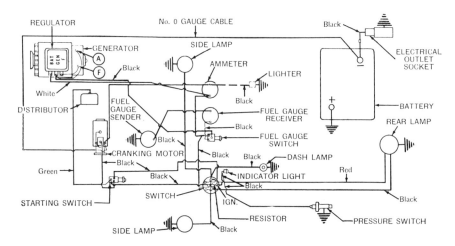

Fig. JD1606A — Wiring diagram for 80, 820 and 830 diesel tractors equipped with the gasoline starting engine and the 6-volt electrical system.

1107155, 6-volt electric starting motor is mounted on the starting engine clutch housing and is accessible after removing the tractor hood. Test specifications are as follows:

Brush spring tension (min.) . . . . 24 oz.

No load test

    Volts 5.65

    Amperes 70

    RPM 5500

Lock test

    Volts 3.25

    Amperes 550

    Torque (Ft.-Lbs.) 11

24-Volt System

128A. On 830 diesel tractors equipped with the 24-volt electrical system a Delco-Remy starter No. 111-3801 and solenoid No. 1119803, or starter No. 1113830 and solenoid No. 1119832 is mounted on a bracket located on the top of the tractor main case and is accessible after removing the tractor hood. If the starting motor solenoid switch is removed, or if for any other reason the starting motor pinion does not correctly engage the flywheel ring gear, adjust the solenoid

and pinion as in the following paragraph.

The starting motor solenoid and pinion can be adjusted with the starter and solenoid unit installed on the tractor; however, it is recommended that the unit be removed from the tractor and mounted in a vise or holding fixture. Connect one wire from a 6-volt battery to one of the solenoid control winding terminals, engage the solenoid manually and hold the solenoid in the engaged position by connecting another wire from the battery to the remaining solenoid control winding terminal as shown in Fig. JD1607. Hold the pinion (Fig. JD1608) toward the starting motor to remove any play in the linkage and measure the clearance between the forward edge of the pinion teeth and the inside edge of the motor drive housing. If the clearance is not 21/64-25/64-inch, disconnect the plunger stud from the starting motor shift lever and turn the stud in or out as required until the recommended clearance is obtained.

To renew the solenoid contact disc (Fig. JD1609), first remove the solenoid assembly from the starting motor. Remove the compression spring retaining pin, retainer and compression spring from one end of the solenoid and the terminal plate from the other end. Remove the cotter pin, the castellated nut and the contact disc. Install the new contact disc with the heavy rolled edge of the center bushing toward the plunger end. Install the castellated nut and turn same on the plunger until the distance from the contact disc to the edge of the housing is $1\frac{1}{32}$ inches (Fig. JD1609). Note: During this measurement it is important that the compression spring be removed. Reassembly of the solenoid is reverse of the disassembly procedure. Check and adjust the starting motor pinion setting as in the preceding paragraph.

Specification data for both starting motors and solenoids are as follows:

Voltage regulator

    Air gap 0.075

    Setting voltage range...... 6.8-7.4

    Adjust to 7.1

Fig. JD1607 — Jumper wires from a 6-volt battery to the two small posts on the solenoid will hold the solenoid in the engaged position for checking the pinion setting. Refer to text.

24-Volt System

128C. A Delco-Remy No. 1103021 24-volt generator is used on tractors equipped with the 24-volt electrical system. Generator output is controlled by a Delco-Remy No. 1119219, three unit voltage regulator. Specification data for both units are as follows:

Generator—1103021

    Brush spring tension........ 28 oz.

    Field draw

        Volts 24.0

        Amperes .... .75-.85

    Output (cold)

        Volts 28.5

        Amperes 10.0

        RPM 2500

Voltage regulator—1119219

    Ground polarity .... Negative

    Cut-out relay

        Air gap 0.017

        Point gap 0.032

        Closing voltage range...... 24-27

        Adjust to 25.5

    Voltage regulator

        Air gap 0.075

        Voltage setting range.... 27.5-29.5

        Adjust to 28.5

    Current regulator

        Air gap 0.075

        Regulator setting range.. 8.5-11.5

        Adjust to 10

Starting Motors—1113830 and 1113801

    Brush spring tension.... ...35 oz.

    No load test

        Volts 23.0

        Amperes 100

        RPM 8000

    Lock test

        Volts 3.0

        Amperes 500

        Torque (Ft.-Lbs.) 28

Solenoid—1119803

    At 80° F., consumption should be:

    Both windings

        Amperes 41.6-45.8

        Volts 24

    Hold-in winding

        Amperes 7.3-8.3

        Volts 24

Solenoid—1119832

    At 80° F., consumption should be:

    Both windings

        Amperes 46.1-52.3

        Volts 20

    Hold-in winding

        Amperes 6.1-6.8

        Volts 20

Field draw

    Volts 6.0

    Amperes 1.85-2.03

Output (cold)

    Volts 8.0

    Amperes 35.0

    RPM 2650

Voltage Regulator—1118786

    Ground polarity .... Positive

    Cut-out relay

        Air gap 0.020

        Point gap 0.020

        Closing voltage range..... 5.9-6.7

        Adjust to 6.3

GENERATOR AND REGULATOR
6-Volt System

128B. A Delco-Remy No. 1100027, 6-volt generator is used on tractors equipped with a gasoline starting engine. Generator output is controlled by a Delco-Remy No. 1118786, two unit regulator. Specification data for both units are as follows:

Generator—1100027

    Brush spring tension........ 28 oz.

Fig. JD1608 — With the solenoid engaged as in Fig. JD1607, dimension between front face of teeth and inside edge of motor housing should be 21/64-25/64-inch.

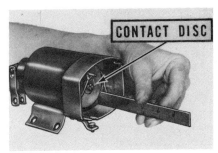

Fig. JD1609 — The solenoid contact disc should be 1 1/32 inches from the edge of housing when measured as described in text.

45

CLUTCH & BELT PULLEY

ADJUSTMENTS

All Models

129. **CLUTCH.** To adjust the clutch, first place clutch operating lever in the engaged position (lever fully forward). Remove the belt pulley dust cover and tighten each adjusting nut (Fig. JD1610) a little at a time and to the same tension. Check tightness of clutch after each adjustment by disengaging and re-engaging clutch. When the adjustment is correct, a distinct snap will occur when the clutch is engaged and 40-80 pounds pressure will be required at the end of the operating lever to lock the clutch in the engaged position with engine running at idle speed.

Fig. JD1611 — Clutch linkage and pulley brake installation.

11. Clutch operating rod
13. Yoke
14. Clutch fork shaft
27. Pulley brake adjusting screw
28. Jam nut
29. Pulley brake

Models 80-820

(Prior Serial No. 8203100)

129A. **PULLEY BRAKE.** Tractors before serial number 8203100 are equipped with a headed pulley brake operating pin (Fig. JD1613), which is shim adjusted to just prevent the clutch operating sleeve from striking the pulley gear when the clutch lever is moved to the disengaged position. Clashing of gears could result if the thickness of these shims is incorrect.

First adjust the tractor clutch as outlined in paragraph 129; then, loosen the pulley brake adjusting screw (27—Fig. JD1611). Vary num-

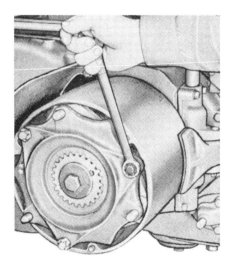

Fig. JD1610—To adjust the tractor clutch, tighten each adjusting nut a little at a time until a distinct snap is heard as clutch is engaged.

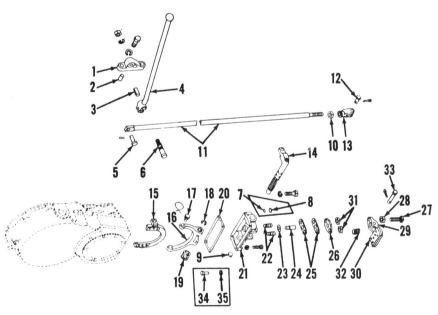

Fig. JD1612 — Exploded view of the clutch and pulley brake operating parts. Tractors prior to serial number 8203100 were equipped with a headed pulley brake operating pin (34) which must be adjusted by varying number of shim washers (35).

1. Clutch lever bracket
2. Dowel pin
3. Bushing
4. Clutch operating lever
5. Pin
6. Pivot bolt
7. "O" ring (prior to ser. No. 8203100)
8. Expansion plug (prior to ser. No. 8203100)
9. Expansion plug
10. Jam nut
11. Operating rod
12. & 33. Pivot pins
13. Yoke
14. Clutch fork shaft

15. Clutch collar
16. Clutch fork
17. Fork pivot
18. Snap ring
19. Spring
20. Gasket
21. Clutch fork bearing
22. Studs (after serial number 8203099)
23. "O" ring (after serial number 8203099)
24. Pulley brake operating pin (after serial number 8203099)
25. 0.005 and 0.062 shims after ser. No. 8203099)

26. Pulley brake operating pin stop (after ser. number 8203099)
27. Pulley brake adjusting screw
28. Jam nut
29. Pulley brake
30. Pulley brake lining
31. Nuts (after serial number 8203099)
32. Spring
34. Headed pulley brake operating pin (prior to serial number 8203100)
35. Shims (prior to serial number 8203100)

ber of shims (35—Fig. JD1612) under head of the operating pin until the pulley has the recommended free movement of $\frac{1}{16}$-$\frac{1}{8}$-inch on the crankshaft. Adjust the pulley brake with the adjusting screw as outlined in paragraph 129B.

129B. Adjustment of the pulley brake is made with the adjusting screw (27 —Fig. JD1611). The pulley brake should be adjusted so that when the clutch lever is moved slightly forward from the disengaged position, the pulley is free to turn.

NOTE: If the pulley brake is adjusted too tight, the brake will contact the pulley before the clutch is completely released. This will cause the pulley to become hot and melt the grease out of the pulley bearing.

Models 820 (After Serial No. 8203099)-830

129C. **PULLEY BRAKE.** Loosen nut (28—Fig. JD1611) and back out the adjusting screw (27) until the screw does not contact the pulley brake op-

erating pin when clutch lever is pulled back. Then, with the clutch lever in the released position, the belt pulley should rotate freely without any bind.

If binding exists, remove the pulley brake assembly and shim retainer (26—Fig. JD1612) and remove one adjusting shim (25) and recheck. When properly adjusted and with the clutch lever in the disengaged position, the pulley should rotate freely and have an end play of $\frac{1}{16}$-$\frac{1}{8}$-inch on the crankshaft. Refer to Fig. JD1614.

To adjust the pulley brake, pull forward on the brake shoe to insure that the operating pin is contacting the clutch fork. Then while holding the shoe firmly against the pulley, turn the adjusting screw (27—Fig. JD1611) in until it just contacts the operating pin. Turn the screw in $\frac{1}{4}$-turn more and tighten the jam nut (28).

CLUTCH
130. **REMOVE AND REINSTALL.** To remove clutch discs and facings,

proceed as follows: Remove belt pulley dust cover. Remove adjusting disc (38—Fig. JD1615) by removing the three adjusting nuts. Lift out the lined and unlined discs. Remove clutch drive disc retaining cap screw, attach a suitable puller as shown in Fig. JD1616 and remove clutch drive disc and inner facing.

130A. **OVERHAUL.** The three clutch release springs should test 42-58 pounds when compressed to a height of 2¼ inches. Renew any spring which is rusted, distorted, or does not meet the foregoing pressure specifications. Refer to Fig. JD1615. Renew facings that are glazed or oil soaked. If facing discs (43) are in good condition, new facings can be riveted to the discs. Renew sliding drive discs (42) if either face is scored or if teeth are excessively worn. Examine drive disc (46) for worn splines, excessively worn teeth or cracked faces. Small check-line cracks in the drive disc faces will not affect the clutch per-

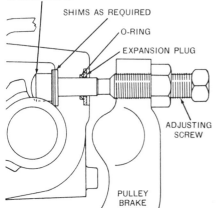

Fig. JD1613—On 80 & 820 tractors prior to serial number 8203100 the pulley brake operating pin must be adjusted by varying number of shims to just prevent the clutch operating sleeve from striking the pulley gear when the clutch operating lever is moved to the disengaged position.

Fig. JD1614 — On late 820 models and 830, the pulley brake operating pin adjusting shims are externally located in a manner similar to the 720 shown.

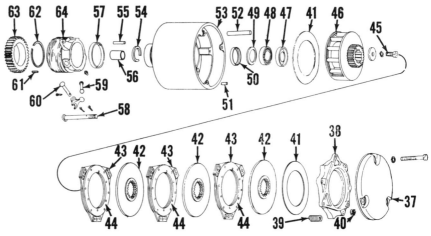

Fig. JD1615 — Exploded view of the clutch, belt pulley and associated parts. Bearing (48) must be packed with high temperature bearing grease.

37. Pulley dust cover	46. Clutch drive disc	55. Drive pin
38. Clutch adjusting disc	47. Bearing retainer	56. Bushing
39. Spring	48. Roller bearing	57. Grease retainer
40. Adjusting nut	49. Bearing washer	58. Clutch operating bolt
41. Clutch facing	50. Pulley oil retainer	59. Clutch dog toggle
42. Sliding drive disc	51. Dowel pin	60. Clutch dog pin
43. Clutch facing disc	52. Pin	61. Key
44. Facing	53. Belt pulley	62. Snap ring
45. Cap screw	54. Snap ring	63. Pulley gear

Fig. JD1616 — Using a suitable puller to remove clutch drive disc. Tapped holes are provided in the disc.

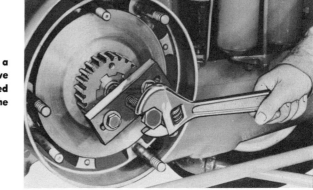

formance; however, the drive disc should be renewed if cracks are large and extend into the hub or if gear teeth and splines are excessively worn.

Reassemble and reinstall clutch, making certain that "V" mark on clutch drive disc is in register with "V" mark on crankshaft as shown in Fig. JD1617. Adjust clutch as outlined in paragraph 129. NOTE: Clutch operating sleeve and associated parts are accessible for overhaul after removing the belt pulley as outlined in the following paragraph.

BELT PULLEY

131. REMOVE AND REINSTALL. Disconnect clutch operating rod yoke (13—Fig. JD1611) from clutch fork, clutch fork shaft. Remove clutch discs and facings as outlined in paragraph 130. Remove pulley from tractor.

131A. OVERHAUL. Bearing (48—Fig. JD1615) and bushing (56) can be renewed at this time. Bushing (56) is pre-sized and if not distorted during installation, will require no final sizing. Bearing (48) must be packed with high temperature bearing grease before installation. The 2.9961-2.9975 diameter crankshaft has a clearance of 0.0015-0.0029 in the bushing (56). Belt pulley gear (63) can be removed from pulley by using a suitable puller. The need and procedure for further disassembly is evident after an examination of the unit. When reassembling heat gear (63) and press same on pulley (53) until hub of gear is against snap ring (62). Lubricate toggles (59) as follows:

Any one of the following special lubricants are recommended by John Deere for lubricating the clutch toggles:

    Calumet Viscous Lubricant, 10X, manufactured by Standard Oil Company of Indiana.

    No. 102 Cosmolube, manufactured by E. F. Houghton and Co., 303 West Lehigh Ave., Philadelphia 33, Pa.

On early production tractors so equipped, pack the recess in end of each clutch toggle with lubricant and install lubricated end of toggle into sockets in the clutch operating sleeve. Lubricate opposite end of each toggle with the same special lubricant.

On other models, lubricate end of toggle which goes into the clutch dog cup with one of the special lubricants. Place the operating sleeve in the engaged position (away from pulley gear) and force the special grease into the grease fitting until grease appears around toggle ends. This procedure fills the grease reservoir in the operating sleeve and provides toggle lubrication via small holes to the toggle sockets.

Install bearing washer (49) and drive the center section down to expand the washer against inner diameter of pulley hub. Pack bearing (48) with wheel bearing grease and assemble the remaining parts. When installing the pulley, center the dust shield so that it does not strike the operating sleeve when pulley is rotated. After the shield is centered, tighten the retaining screws to a torque of 15 Ft.-Lbs.

Check the clutch and pulley brake controls against the following specifications: Refer to Fig. JD1612.
Thickness of collar (15) ....0.393 max.
Pivot hole diameter
  in collar 0.499-0.501
Diameter of fork shaft
  upper bearing 1.0635-1.0665
Diameter of fork shaft
  upper journal 1.060-1.062
Diameter of fork shaft
  lower bearing 0.874-0.876
Diameter of fork shaft
  lower journal 0.807-0.809
Renew any parts which are worn and reassemble.

After clutch is installed, adjust the clutch and pulley brake.

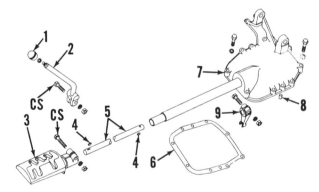

Fig. JD1620 — Exploded view of the transmission cover and associated parts.

1. Ball
2. Shift lever
3. Gear shift quadrant
4. Woodruff key
5. Shift lever shaft
6. Gasket
7. Transmission Cover
8. Dowel pin
9. Shifter arm

Fig. JD1617—When installing clutch drive disc, make certain that "V" mark on disc is in register with "V" mark on crankshaft as shown.

Fig. JD1621—Right hand side of the tractor main case with first reduction gear cover removed.

TRANSMISSION AND CONNECTIONS

The transmission shafts, differential assembly and diesel engine crankshaft are carried in the tractor main case. A wall in the main case separates the diesel engine crankcase compartment from the transmission and differential compartment. The sliding gear shaft drive gear (first reduction gear), fifth speed drive gear and the fifth speed sliding pinion are located in the reduction gear cover. Transmission oil is circulated throughout the transmission by an engine-driven lubrication pump.

To remove the transmission shifter mechanism, sliding gear shaft, countershaft and/or reverse gear shaft, it is necessary to remove the dash and transmission top cover as outlined in paragraphs 132 and 132A

R&R OF DASH AND TRANSMISSION TOP COVER

132. DASH. NOTE: The dash is the sheet metal panel which is located directly under the instrument panel.

To remove the dash, first remove the tractor hood. Disconnect the diesel engine speed control rod at top of transmission case cover. On tractors equipped with the gasoline starting engine, disconnect the starting engine controls at couplings. On all models, disconnect the shutter control rod from shutter control lever. Unbolt instrument panel and pull same upward enough to clear dash. Unbolt dash and pull rearward at top enough to detach brake pedal return springs from brackets on front face of dash. Remove dash from tractor.

132A. TRANSMISSION TOP COVER. To remove the transmission top cover, first remove the dash as outlined in paragraph 132. Remove the quadrant and shift lever retaining cap screws (CS—Fig. JD1620). Remove the shift lever (2) and quadrant (3). Remove the flywheel cover, flywheel and flywheel cover back plate. Disconnect the speed control lever rod from shaft and speed control rod from speed control arm. Disconnect the starting control and decompression control rods. Remove the transmission cover retaining cap screws, pry cover from its locating dowels and remove the transmission cover from tractor.

MAJOR OVERHAUL

Data on overhauling the various transmission components are outlined in the following paragraphs.

Fig. JD1622 — Exploded view of transmission shifter forks and shafts. Sixth speed shifter fork (19) is located in the first reduction gear case.

10. Sixth speed shifter
11A. Set screw
11B. Set screw
11C. Set screw
11D. Set screw
12. Sixth speed shifter shaft
13. Sixth speed shifter pawl retainer
14. Sixth speed pawl spring
15. Sixth speed shifter pawl
16. Shifter pawl
17. Adjusting screw cap screws
18. Shifter shaft adjusting screws
19. Sixth speed shifter fork
20. Pawl spring
21. Fourth and fifth speed shifter
22. Third speed stop screw
23. Shifter shaft lock plate
24. Lock plate
25. Second and third speed shifter shaft
26. Fourth and fifth speed shifter shaft
27. Fifth speed shifter stop cotter pin
28. Third speed shifter
29. Second speed shifter
30. First speed shifter
31. First speed shifter shaft
32. First speed shifter pawl
33. First speed pawl spring
34. First speed shifter pawl retainer
35. First speed shifter fork
36. Jam nut

133. PREPARATION. Remove dash and transmission cover as outlined in paragraphs 132 and 132A. Jack up rear portion of tractor, block in the raised position and remove both tire and rim units. Remove the clutch and belt pulley. Drain transmission lubricant from tractor main case and shut fuel off at tank. Disconnect fuel line at filters and hydraulic lines at pump. Unclip hydraulic lines and fuel line from the first reduction gear cover. Remove the cap screws attaching the first reduction gear cover to the main case, raise the hydraulic lines and the fuel line and remove cover from the right side of tractor main case. Remove first reduction gear retaining nut (Fig. JD1621) and pull gear from the sliding gear shaft. Remove the first speed gear cover from left side of main case.

134. SHIFTER SHAFTS & FORKS. Slide shifters along their respective shafts until pawls are out of detents and lock pawls in the raised position by inserting cotter pins or nails in holes provided.

Remove retainers (13 & 34—Fig. JD1622) and extract springs (14 & 33) and pawls (15 & 32). Remove the two shifter adjusting screw retaining cap

Fig. JD1623 — Adjusting screws (18), third speed stop screw (22) and first speed gears are exposed after removing the first speed gear cover from the left side of the main case.

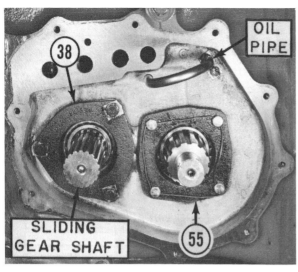

Fig. JD1624—Left side of main case showing the sliding gear shaft and countershaft bearing retainers.

speed shifter set screws are properly safety wired and fourth and fifth speed shifter stop cotter pin is properly installed before installing the transmission top cover. Also, be sure that plate (23—Fig. JD1622) is firmly seated against flats of shifter shafts. After gear shift lever and quadrant are installed, locate quadrant so that shift lever moves freely and easily through the gates in the neutral position.

135. SLIDING GEAR SHAFT. To remove the transmission sliding gear shaft (42—Fig. JD1625), first remove the shifter shafts and shifters as outlined in the previous paragraph and proceed as follows: Refer to Fig. JD1623 and remove the first speed gear retaining nut (52) and gear (53). Remove the transmission oil pipe (Fig. JD1624). Remove the sliding gear shaft bearing retainer (38). Bump sliding gear shaft toward left side of tractor until left hand bearing cone emerges from main case and using a suitable puller, as shown in Fig. JD1624A, remove the bearing. Remove the bearing retaining snap ring (48—Fig. JD1624B) and bump sliding gear shaft out right side of tractor main case and remove gears from above. Right bearing cone can be removed from shaft at this time.

Using Fig. JD1625 as a general guide, reinstall sliding gear shaft by reversing the removal procedure. Adjust sliding gear shaft end play to 0.001-0.004 by adding or removing shims (39) which are located under the bearing retainer. Steel shims are 0.006 thick.

136. COUNTERSHAFT. To remove the transmission countershaft (60—Fig. JD1625), remove the sliding gear shaft as outlined in paragraph 135 and proceed as follows: Remove spacer (54—Fig. JD1625). Remove the countershaft bearing retainer (55) from

screws (17), adjusting screws (18) and the third speed stop screw (22). Refer also to Fig. JD1623. Remove cotter pin (27—Fig. JD1622) from the fourth and fifth speed shifter shaft. Pull fourth and fifth speed shifter shaft (26) and the second and third speed shifter shaft (25) from left side of tractor main case. Remove the fourth and fifth speed shifter, second and reverse speed shifter and the third speed shifter from opening in top of tractor main case. Slide sixth speed shifter shaft to the right and remove the sixth speed sliding gear. Remove set screw (11A) from sixth speed shifter and withdraw the sixth speed shifter shaft (12) from the right side of main case. Remove the sixth speed shifter (10). Slide the first speed shifter shaft (31) to the left and remove the first speed sliding gear. Remove set screw (11D) from the first speed shifter, withdraw the shifter shaft from the left side of the main case and remove the shifter from above. Check shifter forks and yokes for being excessively worn. Examine the area around detents on shifter shafts for excessive wear. First and sixth speed shifter shaft pawl springs (14 & 33) should test 51-63 pounds when compressed to $1\frac{7}{8}$ inches. Their free length should be approximately $2\frac{15}{32}$ inches. Other pawl springs should test 47-53 pounds when compressed to $1\frac{1}{8}$ inches. Their free length should be approximately $1\frac{9}{16}$ inches.

Reinstall shifter shafts and shifters by reversing the removal procedure. Remove cotter pins from detents and install plugs (13 & 34). When all shifter shafts and forks are installed, align shifter fork gates as follows: Using gates on first and sixth speed shifters

(10 & 30) as a reference, turn adjusting screws (18) until all gates are aligned in neutral position. After shifters are aligned, shift the transmission into third gear and install third gear shifter stop cap screw (22). Turn cap screw in until the inner end just contacts shifter (28). Back screw up to allow $\frac{1}{32}$-inch between inner end of cap screw and shifter (28). Tighten jam nut securely. Refer to Fig. JD 1623. Make certain that first and sixth

Fig. JD1624A—Removing left hand bearing cone from sliding gear shaft.

Fig. JD1624B—Right side of main case showing the installation of the bearing retaining snap rings.

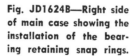

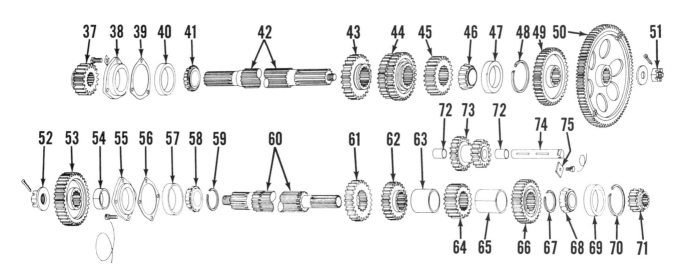

Fig. JD1625 — Exploded view of the transmission shafts and associated parts. Reverse gear bushings (72) are pre-sized.

37. First speed
 sliding pinion
38. Sliding gear
 shaft bearing
 retainer
39. Shims (0.006)
40. Bearing cup
41. Bearing cone
42. Sliding
 gear shaft

43. Third speed
 sliding pinion
44. Fourth and fifth
 speed sliding
 pinion
45. Second speed
 sliding pinion
46. Bearing cone
47. Bearing cup
48. Snap ring

49. Sixth speed
 drive gear
50. Sliding gear
 shaft drive gear
 (first reduction
 gear)
51. & 52. Nut
53. First speed gear
54. Spacer

55. Countershaft
 bearing
 retainer
56. Shims
 (0.006 & 0.018)
57. Bearing cup
58. Bearing cone
59. Snap ring
60. Countershaft

61. Third speed gear
62. Fourth speed
 gear
63. Spacer
64. Differential
 drive pinion
65. Spacer
66. Second speed
 gear
67. Snap ring

68. Bearing cone
69. Bearing cup
70. Snap ring
71. Sixth speed
 sliding pinion
72. Reverse gear
 bushing
73. Reverse gear
74. Reverse shaft
75. Lock plate

the left side of the main case and bump the countershaft toward left to remove bearing cup (57). Working through opening in left side of main case, remove snap ring (59). NOTE: It will be necessary to stretch or cut the snap ring (59) to get it out. Remove the snap ring (70) from right end of countershaft and bump the countershaft toward right to remove bearing cup (69). Cut snap ring (67) and remove same from the countershaft. Pull countershaft as far to the left as possible and with suitable puller (Fig. JD1626), remove bearing cone from left end of countershaft. The countershaft can then be removed

from right side of the main case while the countershaft gears and spacers are withdrawn from above.

Using Fig. JD1625 as a general guide, reinstall countershaft by reversing the removal procedure. Adjust countershaft end play to 0.001-0.004 by adding or removing shims (56) which are located under countershaft bearing retainer (55). Steel shims are 0.006 and 0.018 thick.

137. REVERSE GEAR AND SHAFT. To remove the shaft (74—Fig. JD1625) on which the reverse gear (73) rotates, first remove the counter shaft as out-

lined in paragraph 136 and proceed as follows: Remove lock plate (Fig. JD1627), pull shaft (74) out of main case and remove reverse gear from above. Reverse gear bushings (72—Fig. JD1625) can be renewed at this time; and if carefully installed, will require no final sizing. The 1.234-1.235 diameter reverse gear shaft has a clearance of 0.003-0.007 in the gear bushings.

Using Figs. JD1625 and JD1627 as a general guide, install reverse gear and shaft by reversing the removal procedure.

Fig. JD1626 — Removing countershaft left hand bearing cone.

Fig. JD1627—Right hand side of the tractor main case with sliding gear shaft and countershaft removed showing the lock plate which must be removed before removal of the reverse gear shaft.

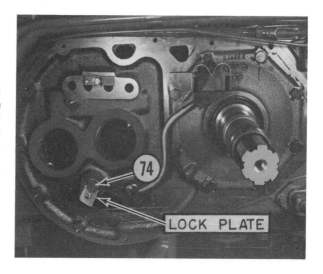

TRANSMISSION OIL PUMP

The transmission oil pump drive gear (12—Fig. JD1636) is engine driven by a worm type pinion on the power shaft (P.T.O.) bevel pinion. Refer to paragraph 164 for renewal of the worm gear. The pump draws oil through a screen in the bottom of the main case and delivers oil to the gears and bearings of the transmission by way of a perforated tube. The oil tube terminates in the reduction gear cover where it supplies lubricant to the enclosed gears.

138. The transmission oil pump does not deliver oil with sufficient pressure to make a gage check; however, pump operation can be checked be removing the pipe plug (Fig. JD1635) and observing the oil flow. Pump operation can be considered satisfactory when a steady stream of oil flows from the pipe plug hole. If the pump fails to produce a good volume of oil, it is possible that the "O" ring packing (5—Fig. JD1636) on the intake pipe (4) is leaking and must be renewed. Intake pipe (4) can be removed after removing the sump cover (1). Also remove the reduction gear cover drain plug and make certain that oil is being pumped into that area. If oil is not showing at this point, "O" ring seal (7) and/or (8) must be renewed. Discharge tube (6) can be removed after removing the transmission top cover as outlined in paragraph 132A. Tube (9) can be removed after removing the first speed gear cover from left side of main case.

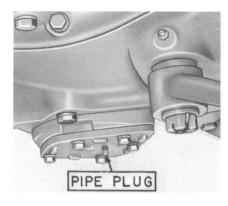

Fig. JD1635 — The transmission oil pump oil flow is checked by removing pipe plug. If the volume of oil delivered is sufficient, the pump is O.K.

Fig. JD1636 — Exploded view of the transmission oil pump and associated parts. Return tube (23) is located in the first reduction gear compartment.

1. Sump cover
2. Gasket
3. Screen
4. Pump intake tube
5. "O" ring
6. Pump discharge tube
7. "O" ring
8. "O" ring
9. Oil discharge tube
10. Gasket
11. Nut
12. Oil pump drive gear
13. Oil pump body
14. Woodruff key
15. Oil pump drive shaft
16. Pump gear
17. Snap ring
18. Idler gear
19. Idler gear shaft
20. Gasket
21. Oil pump cover
22. Pipe plug

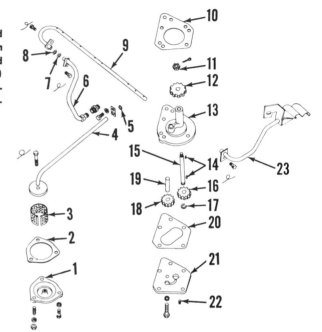

138A. **R&R AND OVERHAUL.** Drain transmission oil from the main case. Remove cap screws retaining oil pump unit to underside of main case and withdraw the pump as an assembly. Remove nut (11—Fig. JD1636) and using a suitable puller, remove drive gear (12) and key (14). Remove cover from pump body and disassemble the remaining parts.

Thoroughly inspect all parts and renew any which are damaged or worn. After pump is installed on tractor, check the oil flow as in paragraph 138.

DIFFERENTIAL, FINAL DRIVE AND REAR AXLE

Models 80, 820 and 830 diesel tractors are equipped with a three pinion type differential that is mounted on the spider of a spur (ring) gear which meshes with a driving pinion on the transmission countershaft. Pressed through the spider is the differential cross shaft which forms the journals for the differential side gears. The outer ends of the differential cross shaft carry taper roller bearings which support the differential unit. The remainder of the final drive includes the final drive (bull) gears, the wheel axle shafts and associated parts.

DIFFERENTIAL AND BULL PINION
139. **REMOVE AND REINSTALL.** To remove the differential, remove the rear axle housing as outlined in paragraph 144 and the brake assemblies as outlined in paragraph 152. Remove the transmission oil pump sump and intake pipe from bottom of transmission case. Remove the differential left hand bearing housing and withdraw the differential, right hand end first from transmission housing as shown in Fig. JD1641.

When reinstalling, use the same thickness of shims (3—Fig. JD1642) as were originally removed, mount a dial indicator so that contact button is resting on side of the ring gear and check the differential end play which should

be 0.001-0.004. If end play is not as specified, remove the left bearing quill and add or remove the required amount of shims.

140. **OVERHAUL.** With the differential removed as outlined in the preceding paragraph, proceed to disassemble the unit as follows: Remove snap rings (8—Fig. JD1642); or cap screws (5) and bearing retainers (7) from ends of shaft. Use a suitable puller and remove the bearing cones (9), thrust washers (10) and bevel gears (11). The three bevel pinions (15) can be removed after removing rivets (16) and pinion shafts (13).

Check the disassembled parts against the values which follow:

Thickness of washers (10)..0.225-0.230
I.D. of bull pinions (11)..2.6290-2.6310
O.D. of shaft (17 or 18)..2.6250-2.6260
I.D. of pinions (15).....1.2380-1.2400
O.D. of shafts (13).......1.2335-1.2350

NOTE: If shaft (17 or 18), differential spider or spur ring gear (14) is worn or damaged, press old shaft out. To install shaft, support the differential spider on a piece of pipe and press the new shaft in place.

When reassembling, reverse the disassembly procedure. Lubricate assembly before installation.

FINAL DRIVE GEARS

141. The final drive gears consist of two spur pinions (bull pinions) which are integral with the differential bevel

Fig. JD1644 — Removing the bull gear (27), using a long taper wedge (38).

gears (11—Fig. JD1642) and two bull gears (27—Fig. JD1645) which are splined to the inner ends of the wheel axle shafts.

142. **BULL PINION.** To remove a final drive bull pinion, it is necessary to remove the differential assembly as outlined in paragraph 139. Remove snap rings (8—Fig. JD1642) or cap screws (5) from ends of differential shaft. Attach a suitable puller and remove bearing cones (9), thrust washers (10) and gears (11) from the shaft.

143. **BULL GEARS.** To remove either bull gear, drain transmission oil from main case, remove the basic housing as outlined in paragraph 162 and proceed as follows: Block up rear portion of tractor and remove the respective wheel and tire unit. Loosen (but do not remove) adjusting nut (26—Fig. JD1644) and using a hammer and a long taper wedge as shown, force

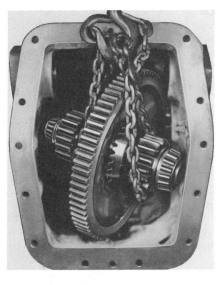

Fig. JD1641—Using a chain and hoist to remove differential assembly. Notice that unit is removed right end first.

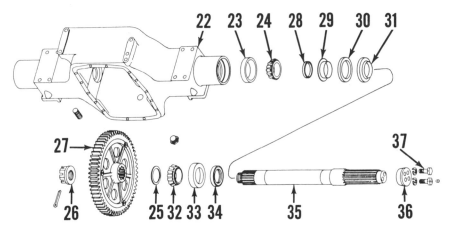

Fig. JD1645 — Exploded view of the rear axle housing, axle and associated parts.

22. Rear axle housing	25. Thrust washer (after serial number 8199999)	28. Spacer	32. Bearing cone
23. Bearing cup	26. Axle nut	29. Inner felt retainer	33. Bearing cup
24. Bearing cone	27. Final drive (bull) gear	30. Felt washer	34. Oil seal
		31. Outer felt retainer	35. Rear axle shaft
			36. Wheel retainer
			37. Cap screw

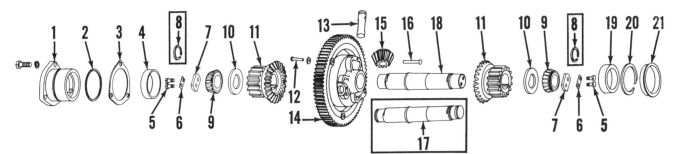

Fig. JD1642 — Exploded view of the differential. Tractors after serial number 8199999 are equipped with cap screws (5), lock plates (6) and bearing retainers (7) which hold the bearing cones (9) on the differential shaft (18). Early models use snap rings (8).

1. Differential left bearing housing	6. Lock plate (after serial number 8199999)	11. Differential bevel and spur pinion (bull pinion) gear	14. Differential ring gear and spider	18. Differential shaft (after serial number 8199999)
2. Packing	7. Bearing retainer (after serial number 8199999)	12. Rivet	15. Differential bevel pinion	19. Bearing cup
3. Shims (0.005)	8. Snap ring (early model)	13. Differential bevel pinion shaft	16. Rivet	20. Snap ring
4. Bearing cup	9. Bearing cone		17. Differential shaft (prior to serial number 8200000)	21. Differential right bearing cover
5. Cap screws (after serial number 8199999)	10. Thrust washer			

axle out of bull gear and remove the gear.

Install bull gear by reversing the removal procedure. Alternately tighten adjusting nut (26) and drive on outer end of the wheel axle shaft with a soft hammer to assure proper seating of the taper roller axle carrier bearings. Mount a dial indicator as shown in Fig. JD1646 and using a bar, shift the final drive gear. Tighten adjusting nut (26) until there is 0.001-0.004 end play as measured by dial indicator.

AXLE SHAFTS AND HOUSING

144. **R&R ASSEMBLY.** To remove the complete rear axle housing and associated parts as a unit, first remove tractor hood. On tractors equipped with the gasoline starting engine, disconnect starting engine controls at couplings. On all models, disconnect the shutter control rod from shutter control lever. Unbolt instrument panel and pull upward enough to clear dash. Unbolt dash and pull rearward enough at top to detach brake pedal return springs from brackets on back of dash, then remove dash from tractor. Remove cap screws retaining cowl to fenders and cowl to cowl side support. Disconnect starting engine fuel line and headlight wires. Move instrument panel back to clear cowl, raise cowl and unclip wires from rear underside of cowl, then remove cowl from tractor. Remove quadrant and shift lever retaining cap screws, then remove the shift lever and quadrant. Remove seat, steering wheel, battery and battery box. Disconnect starting engine clutch lever and the decompression lever operating rods at forward ends. Unbolt and remove levers assembly. Disconnect diesel engine clutch operating rod at forward end, engage power shaft clutch and remove powershaft clutch operating rod. Un-

bolt and remove diesel engine clutch and powershaft clutch levers assembly, disconnect "Powr-Trol" operating rods, remove breakaway couplings from brackets and remove remote control cylinder and mounting bracket. Remove brake pedals and disconnect hydraulic lines from "Powr-Trol". Unbolt and remove the platform assembly. Remove drawbar. Shut fuel off at tank and disconnect fuel line. Disconnect fuel gage wire from top of tank. Remove fuel tank hold down straps and lift tank from tractor. Block up under tractor main case so as to just remove weight from rear wheels. Remove cap screws which retain final drive housing to main case. Support rear axle housing and work housing rearward off dowel pins. NOTE: Make certain rear axle housing comes squarely away from main case to avoid bending or breaking the locating dowels.

For reinstallation, reverse removal procedure and tighten the axle housing to main case screws to a torque of 275 Ft.-Lbs.

145. **AXLE SHAFTS AND/OR BEARINGS OR INNER OIL SEAL RENEW.** Either axle shaft can be slipped out of axle housing after the corresponding bull gear has been removed as outlined in paragraph 143. Bearing cups (23 & 33—Fig. JD1645) and inner seal (34) can be removed from housing at this time.

146. **AXLE SHAFT OUTER FELT SEAL RENEW.** Outer seal can be removed from rear axle housing after removing the rear wheel. Using a punch or chisel, drive felt retainer (31—Fig. JD1645) from housing. Install new felt washer and outer retainer. Drive retainer in end of housing until it is seated against bottom of recess in housing.

BRAKES

150. **ADJUSTMENT.** Turn adjusting screw (23—Fig. JD1650) until wheel is locked, then back off to allow 2¾-3¼ inches of pedal travel before brake shoes contact drum.

151. **R&R SHOES.** Back off adjusting screw (23—Fig. JD1651) and remove brake drum retaining nut (28). Bump the brake shaft inward to free the drum. Pry shoes away from adjusting pins (22) and remove springs (19) and shoes (21) from brake housing. Install shoes by reversing removal procedure and adjust brakes as outlined in the preceding paragraph.

152. **R&R AND OVERHAUL BRAKE ASSEMBLY.** Remove seat assembly, battery and battery box. On tractors equipped with the gasoline starting engine, disconnect the starting engine clutch control lever at the forward end. On all models, disconnect the decompression and the diesel engine clutch control levers at their forward ends. Remove cap screws retaining lever brackets at front of platform; then, remove the powershaft operating rod. Remove brake pedals from levers and breakaway coupler carriers from brackets. Disconnect "Powr-Trol" control rods from valve housing and control lever assembly.

Remove platform attaching cap screws along sides and rear; then raise rear of platform. Remove the brake housing retaining stud nuts and withdraw brake assembly as a unit from tractor. Back off adjusting screw (23—

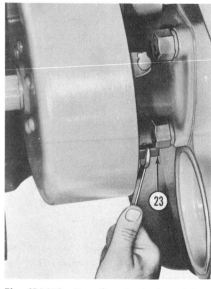

Fig. JD1650—To adjust the brakes, tighten the adjusting screw (23) until wheel is locked, then back off until 2¾-3¼ inches of pedal travel is obtained before brake shoes contact drum.

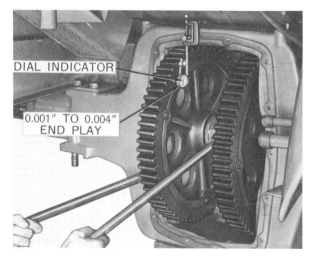

DIAL INDICATOR

0.001" TO 0.004" END PLAY

Fig. JD1646—Rear wheel axle shaft end play should be 0.001-0.004 when checked as shown.

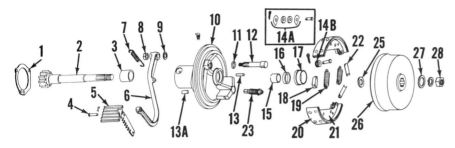

Fig. JD1651 — Exploded view of the brake assembly. Bushings (3 and 15) are pre-sized after serial number 8199999. (See text).

1. Gasket	12. Lever shaft	14B. Shoe roller	19. Spring
2. Brake shaft	13. Dowel pin stop	(after serial	20. Brake lining
3. Inner bushing	(prior to serial	number 8199999)	21. Brake shoe
6. Brake lever	number 8203100)	15. Outer bushing	22. Adjusting pins
7. Return spring	13A. Dowel pin	16. Oil seal	23. Adjusting screw
8. Nut	14A. Shoe roller	(after serial	25. Shim washers
9. Washer	assembly (prior	number 8199999)	26. Brake drum
10. Housing	to serial	17. Inner dust guard	27. Washer
11. "O" ring	number 8200000)	18. Cam	28. Nut

Fig. JD1651) and remove brake drum retaining nut. Bump brake shaft out of brake drum and remove shaft and drum. Pry shoes away from adjusting pins (22) and remove springs and shoes from housing. The need and procedure for further disassembly is evident after an examination of the unit.

Check the brake shafts and bushings against the following specifications:

Shaft diameter at
inner bushing 1.999-2.000
Shaft diameter at outer bushing
Model 80 . . . . 1.494-1.495
Model 820 and 830 . . . . . . 1.749-1.750
Shaft clearance in inner bushing
Model 80 . . . . 0.007-0.009
Model 820 and 830 . . . . 0.001 Min.
Shaft clearance in outer bushing
Model 80 . . . . 0.003-0.004
Model 820 and 830 . . . . 0.001 Min.

On model 80, the brake shaft bushings require final sizing after installation to obtain the specified clearance. On models 820 and 830, the bushings are pre-sized and will not require reaming if carefully installed. If bushings have oil holes, make certain they are aligned with oil holes in casting.

When reassembling, vary the number of washers (9) to obtain a pedal shaft end play of 0.004-0.044 and tighten the lever retaining nut to a torque of 63 Ft.-Lbs. Vary the number of shims (25) to obtain a brake shaft end play of 0.004-0.044 and tighten the drum retaining nut securely.

POWER SHAFT (PTO SYSTEM)

The power take-off system is independent of the diesel engine clutch and is in continuous operation when the engine is running and the power shaft clutch is engaged. The power shaft drive originates in the engine timing gear train where the power shaft drive gear is driven by an idler which meshes with the engine crankshaft gear. A view of the timing gear train is shown in Fig. JD1554. The bevel drive pinion, which is mounted on the same shaft as the power shaft drive gear, meshes with a similar bevel gear in bottom of main case. The power shaft clutch which is located at the aft end of the connecting power shaft is a multiple disc wet type. When the clutch is engaged, it drives a pinion which is in constant mesh with the internal gear of the output shaft assembly. An oil pump mounted on the front of the basic housing, provides a constant flow of oil for the power shaft clutch discs and other moving parts.

OUTPUT SHAFT

155. R&R AND OVERHAUL. The output shaft (13—Fig. JD1661) and bell housing (9) can be removed as an assembly from the basic housing after removing the tractor drawbar frame and the bell housing retaining cap screws.

Unbolt and remove oil seal housing (1). Remove snap ring (4) and bump the output shaft out of the bell housing. The need and procedure for further disassembly is evident. When reassembling, install seal (2) with lip facing toward front of tractor.

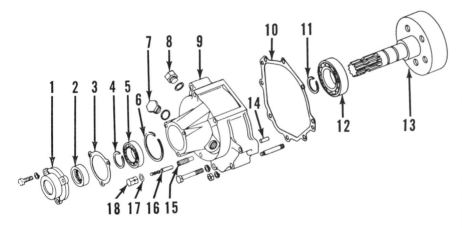

Fig. JD1661 — Exploded view of the power output (P.T.O.) shaft bell housing. Parts (15, 16, 17 & 18) are used after tractor serial number 8202143 and these plus (9) are contained in the power shaft conversion kit.

1. Oil seal housing	8. Plug	12. Bearing	15. Clutch adjust-
2. Oil seal	9. Power shaft	13. Power output	ment indicator
3. Gasket	housing	shaft and	guide
4. Snap ring	10. Gasket	internal gear	16. Clutch adjust-
5. Bearing	11. Snap ring	14. Dowel pin	ment indicator
6. Snap ring			17. Washer
7. Plug			18. Cap nut

PTO CLUTCH
Models 80-820 (Prior to 8202144)

NOTE: These tractors may have been converted to the later construction which is covered by the paragraphs beginning with number 159.

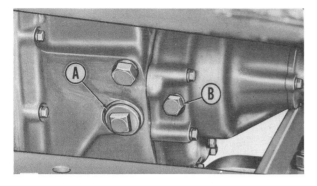

Fig. JD1662—Power shaft housing showing plugs which need to be removed to check and adjust the power shaft clutch. Plug "A" is where adjustment is made, plug "B" is where clutch adjustment is checked on tractors prior to serial number 8202144 without conversion kit.

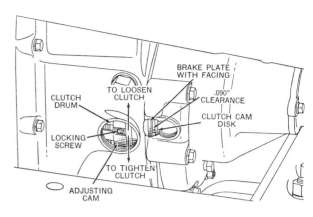

Fig. JD1663—Drawing showing power shaft clutch adjusting parts on early (non-converted) models.

156. **ADJUSTMENT**. Remove the small plug (B—Fig. JD1662) on left side of clutch housing, engage the clutch and check the distance the clutch cam disc has moved back from the brake plate as shown in Fig. JD-1663. The desired clearance is 0.090 and can be checked with a $\frac{3}{32}$-inch welding rod. If the adjustment is not as specified, remove the large plug (A—Fig. JD1662), engage the clutch and using a suitable wrench on the power output shaft, turn the shaft until the cam locking screw is accessible. Disengage clutch, turn the cam locking screw inward until its head clears slots in the adjusting cam.

To tighten the clutch, turn the adjusting cam in a counter-clockwise direction (viewed from rear) one notch, engage the clutch and check the clearance between the cam disc and brake plate. Continue this procedure until adjustment is correct when checked in two or three different positions to make certain that the adjustment is satisfactory in all positions. When adjustment is completed, turn the cam locking screw outward into one of the slots in the adjusting cam and reinstall plugs.

157. **REMOVE AND REINSTALL**. To remove the powershaft clutch, remove the output shaft and bell housing unit as outlined in paragraph 155, then disconnect and remove the clutch fork linkage. Remove bottom cover from basic housing and pull cotter pin from clutch fork and shaft. Pull shaft from fork and remove fork and shoe assembly from below. Remove cap screws (57—Fig. JD1664) and remove the complete power shaft clutch assembly. NOTE: Clutch drum cannot be removed at this time. If drum requires renewal, refer to paragraph 162A.

Reinstall clutch by reversing the removal procedure and adjust the unit as in paragraph 156.

Fig. JD1664 — Rear view of the basic housing with the PTO bell housing removed. Clutch cover (26) is retained to housing by cap screws (57).

Fig. JD1665 — Exploded view of the power shaft clutch assembly used prior to tractor serial number 8202144. If conversion kit has been installed refer to Fig. JD-1676.

21. Clutch shaft bearing retaining plate
22. P.T.O. clutch shaft
23. Bearing
24. Snap ring
25. Clutch cam disc
26. Clutch cover
27. Washer
28. Spring (nine used)
29. Spring (nine used)
30. Clutch cam
31. Ball (three used)
32. Brake plate shim
33. Brake plate
34. Clutch collar
35. Clutch adjusting cam assembly
36. Clutch plate
37. Spring (three used)
38. Steel drive disc
39. Facing disc
40. Washer
41. Snap ring
42. Cap screw
43. Cam and spring assembly
45. Locking screw
57. Cap screw

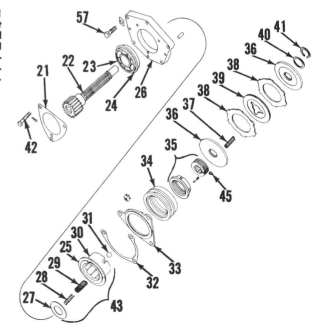

Fig. JD1667 — Disengaging the removed power shaft clutch with two screw drivers. Early clutch assembly is shown but method is the same for both types of clutches.

Fig. JD1670 — Press cam down until loading springs are compressed solid, release pressure until cam disc moves upward 5/32-inch, then determine the number of shims that will go freely between brake plate and clutch cover.

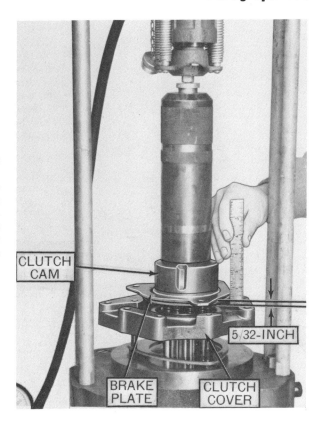

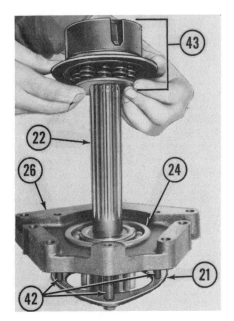

Fig. JD1668—Installing pto clutch cam and spring assembly on early models.

158. OVERHAUL. Disengage clutch using two screwdrivers as shown in Fig. JD1667. Remove snap ring (41—JD1665) and adjusting washers (40). Remove clutch plates (36 & 39), drive discs (38), release springs (37) and the adjusting cam assembly (35). Remove clutch collar (34) and balls (31). Remove cotter pins from cap screws (42) and remove brake plate retaining nuts. Loosen nuts evenly to avoid bending brake plate. Remove brake plate (33), shims (32) and clutch cam assembly (43).

The need and procedure for further disassembly is evident. Thoroughly inspect all parts and renew any which are damaged or worn. Clutch discs

must not be excessively worn or warped. Outer pressure springs (29) have a free length of approximately 2.00 inches and should require 83-93 pounds to compress them to a height of 1⅜ inches. Inner pressure springs (28) have a free length of approximately 1$\frac{25}{32}$ inches and should require 48-56 pounds to compress them to a height of 1 29/64 inches.

When reassembling, use Fig. JD1665 as a general guide and proceed as follows: Install clutch shaft (22) with bearing (23) into clutch cover (26). Place bearing retainer (21) and cap screws (42) in position and install clutch cam and spring assembly (43). Refer also to Fig. JD1668. Install brake plate (33—Fig. JD1665) with facing side toward clutch cover. Place assembly in press as shown in Fig. JD1670, press cam downward until pressure springs are compressed solid; then release pressure until cam disc moves upward $\frac{5}{32}$-inch. Hold the brake plate squarely and firmly down against cam disc and determine number of shims (32—Fig. JD1665) that will freely go between brake plate and cover. Install these shims and secure brake plate to clutch cover. Install clutch collar (34), balls (31) and adjusting cam assembly (35). NOTE: Clutch adjusting cam hub should be adjusted to 0-$\frac{1}{16}$ inch higher than clutch adjusting collar. Install clutch plate (36) with facing side toward end of shaft, install steel

Fig. JD1671—Install facing discs with one of the release spring openings adjacent to oil holes in clutch shaft.

drive disc (38), then a facing disc (39) with one of the release spring openings in the facing discs adjacent to oil holes in clutch shaft as shown in Fig. JD1671. Install remaining steel and facing discs. Install release springs (37—Fig. JD1665) and the other clutch plate (36) with facing side toward clutch disc. Push plate (36) down and install same number of adjusting washers (40) as were originally re-

Fig. JD1673 — Engaging removed P.T.O. clutch with pry bars.

Fig. JD1676 — Exploded view of the power shaft clutch assembly used after tractor serial number 820-2143. Model 80 and early 820 clutch assemblies are similar if conversion kit is installed.

21. Clutch shaft bearing retaining plate
22. P.T.O. clutch shaft
23. Bearing
24. Snap ring
25. Clutch cam disc
26. Clutch cover
27. Washer
28. Spring (nine used)
29. Spring (nine used)
30. Clutch cam
31. Ball (three used)
33. Brake plate
34. Clutch collar
35. Clutch adjusting cam assembly
36. Clutch plate
37. Spring (three used)
38. Steel drive disc
39. Facing disc
40. Washer
41. Snap ring
43. Cam and spring assembly

45. Locking screw
51. Brake plate pin (three used)
52. Snap ring (three used)
53. Washer

54. Adjusting washer
55. Clutch cam washer
56. Snap ring
57. Cap screw
58. Cap screw

moved. Install snap ring (41). Align discs with a spare clutch drum or straight edge; then engage clutch as shown in Fig. JD1673. Clearance between disc (25—Fig. JD1665) and brake plate facing (33) should be 0.090-0.100 at the closest point. If clearance is much less than 0.090 or more than 0.100, add or deduct adjusting washer (40). Small adjustments can be made as follows: Disengage clutch and turn adjusting cam lock screw (45) into hub to clear slot in adjusting cam. Turn adjusting cam to obtain correct clearance. Turn lock screw out after adjustment is complete. Align clutch discs, engage clutch and reinstall same.

Models 820 (After 8202143)-830

NOTE: The 80 and earlier 820 models may have been converted to this later construction; in which case, the subsequent paragraphs will apply.

159. **ADJUSTMENT.** Remove the cap nut, exposing the adjustment indicator rod (16—Fig. JD1675) and guide (15). Disengage the clutch, hold the adjustment indicator rod in and note the position of the rod with respect to end of rod guide (15). Engage

clutch and again note position of indicator rod (16) in relation to end of guide. If the clutch is properly adjusted, the distance the rod has moved will be equal to the length of one land and one groove, (0.090) on the rod as shown. If adjustment is not as specified, proceed as follows:

Remove the clutch adjusting hole plug from left side of clutch housing, and with the clutch lever in neutral (disengaged but not latched), turn the power (output) shaft until the adjusting cam locking screw is visible through opening in clutch housing. Refer to Fig. JD1663. Now, latch the pedal in the disengaged position and turn the locking screw in until head of same clears slot in adjusting cam. To tighten the clutch, turn the adjusting cam counter-clockwise (viewed from rear), one notch at a time and recheck the adjustment. Continue this procedure until adjustment is as specified, then turn the locking screw out-

ward into one of the notches of the adjusting cam. Reinstall plug and indicator rod cap nut.

If clutch lever strikes or comes too close to the dash, reposition the lever by means of the adjustable bracket.

If the power shaft fails to stop when clutch is disengaged and lever latched, adjust the power shaft brake as follows: Disconnect yoke at lower end of operating rod and lengthen the rod ½-turn at a time until proper adjustment is obtained. When lever is just unlatched, the shaft should be free to turn.

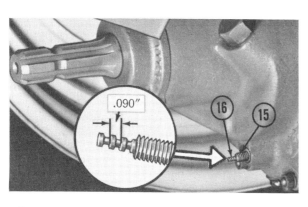

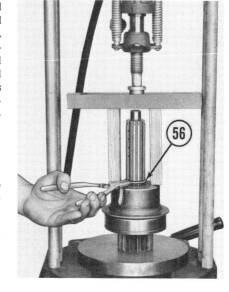

Fig. JD1675 — Late production 820 and 830 P.T.O. clutches and early converted 80 & 820 P.T.O. clutches are equipped with an indicator rod to check clutch adjustment. One land and one groove is equal to 0.090.

Fig. JD1677 — With shaft assembly in a press, snap ring (56) can be removed as shown.

Fig. JD1678 — Installing cam, springs and washer assembly on P.T.O. clutch shaft.

160. REMOVE AND REINSTALL. To remove the power shaft clutch, remove the output shaft and bell housing unit as outlined in paragraph 155, then disconnect and remove the clutch fork linkage. Remove bottom cover from basic housing and pull cotter pin from clutch fork and shaft. Pull shaft from fork and remove fork and shoe assembly from below. Remove cap screws (57—Fig. JD1664) and remove the complete power shaft clutch assembly. To remove clutch drum, refer to paragraph 162A. Reinstall clutch by reversing the removal procedure and adjust the unit as in paragraph 159.

161. OVERHAUL. Disengage clutch, remove snap ring (41—Fig. JD1676) and adjusting washers (40). Remove clutch plates (36 & 39), discs (38), release springs (37) and the adjusting cam assembly (35). Remove clutch collar (34) and balls (31). Remove retaining rings from brake plate pins and remove brake plate. Remove cap screws attaching bearing retainer to clutch cover and bump shaft and bearing out of cover enough to remove bearing snap ring (24) from the bearing. Bump bearing and shaft assembly out other side of cover. Place shaft assembly in press as shown in Fig. JD1677 and remove snap ring (56). Remove cam, springs assembly and adjusting washers from shaft

The need and procedure for further disassembly is evident. Thoroughly inspect all parts and renew any which are damaged or worn. Clutch discs must not be excessively worn or warped. Outer springs (29—Fig. JD1676) have a free length of approximately 2.00 inches and should require 83-93 pounds to compress them to a height of 1⅝ inches. Inner pressure springs (28) have a free length of approximately 1²⁵⁄₃₂ inches and should require 48-56 pounds to compress them to a height of 1 29/64 inches.

Use Fig. JD1676 as a general guide during reassembly and proceed as follows: Install cam, springs and washer assembly on shaft as shown in Fig. JD1678. Place assembly in press as shown in Fig. JD1680, compress springs solid then release pressure until the cam disc has moved up $\frac{5}{32}$-inch. Install as many adjusting washers (54-Fig. JD1676) under clutch cam washer (55) as possible and install snap ring (56). NOTE: Washer (55) is hardened and must be next to the snap ring. Install indicator rod and washer, drive clutch shaft assembly through cover and install snap ring (24). Drive assembly back into cover until bearing snap ring (24) seats in recess in cover casting. Install bearing retainer (21). Install brake plate (33) with facing toward clutch cam disc and secure retaining rings (52). Install clutch collar (34), balls (31) and adjusting cam assembly (35). NOTE: Clutch adjusting cam hub should be adjusted to 0-$\frac{1}{16}$ inch higher than clutch adjusting collar. Install clutch plate (36) with facing side toward end of shaft, steel drive disc (38), then a facing disc with one of the release spring openings adjacent to oil holes in clutch shaft. Install remaining steel and facing discs. Install release springs (37) and other clutch plate (36) with facing side toward clutch disc. Push plate (36) down and install same number of adjusting washers (40) as were removed at disassembly. Using a punch to hold the indicator rod with washer against the cam disc as shown in Fig. JD1681, measure distance between the washer and cover casting. Engage clutch as shown in Fig. JD1673 and again measure the distance between edge of the washer and casting. The difference between the two measurements should be 0.090-0.100 ($\frac{3}{32}$). If clearance is much less than 0.090 or more than 0.100, add or deduct adjusting washer (40—Fig. JD1676). Small adjustments can be made as follows: Disengage clutch and turn adjusting cam lock screw (45) into hub to clear slot in adjusting cam. Turn adjusting cam to obtain correct clearance. Turn lock screw out after adjustment is complete. Align clutch discs, with a spare clutch drum or straight edge, engage clutch and reinstall same.

$\frac{5}{32}$ INCH

Fig. JD1680—Compress springs solid, then release pressure until the cam disc has moved up 5/32-inch. Install as many adjusting washers under clutch cam washer as possible and install snap ring.

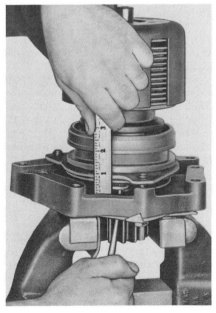

Fig. JD1681—Use a punch to hold the indicator rod with washer against the cam disc, then measure distance between the washer to the cover casting. Engage clutch as shown in Fig. JD1673, then measure again. Difference should be 0.090-0.100.

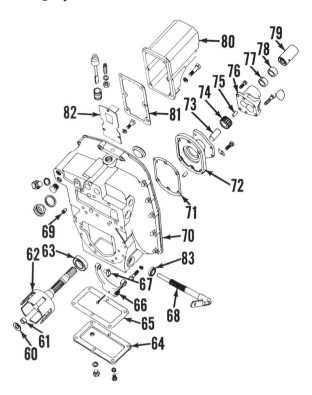

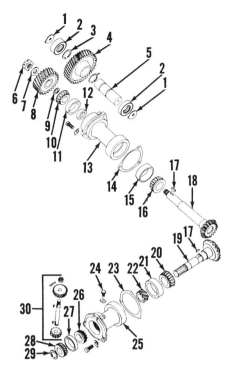

Fig. JD1685 — Exploded view of the basic housing and associated parts. Complete basic housing must be removed before P.T.O. clutch drum (62) can be removed, as the drive gear of the pump is part of the clutch drum shaft.

60. Thrust washer
61. Bushing
62. P.T.O. clutch drum
63. Bearing
64. Bottom cover
65. Gasket
66. Clutch fork
67. Clutch fork shoe
68. Clutch fork shaft
69. Thermal relief valve (only on model 80)
70. Basic housing
71. Gasket
72. Clutch drum shaft quill
73. Idler shaft
74. Power shaft pump idler gear
75. Dowel pin
76. Power shaft pump housing
77. Bushing
78. Oil seal
79. Coupling
80. Reservoir (Models 820 & 830)
81. Gasket (Models 820 & 830)
82. Reservoir baffle (Models 820 & 830)
83. Oil seal

Fig. JD1686—Exploded view of the power shaft, drive shaft and bevel pinion assemblies.

1. Thrust washer
2. Bearing
3. Snap ring
4. Power shaft idler gear
5. Power shaft idler gear shaft
6. Nut
7. Washer
8. Power shaft drive gear
9. Shim washer (0.060)
10. Bearing cone
11. & 15. Bearing cup
12. Oil seal
13. Drive shaft housing
14. Shims
16. Bearing cone
17. Woodruff key
18. Drive shaft
19. Driven shaft
20. Bearing cone
21. Bearing cup
22. Transmission oil pump drive pinion
23. Shims (0.005)
24. Lock screw
25. Housing
26. Bearing adjusting collar
27. Bearing cup
28. Bearing cone
29. Snap ring
30. Transmission oil pump drive

BASIC HOUSING

162. REMOVE AND REINSTALL. Drain "Powr-Trol" housing and power shaft clutch housing. Remove seat, battery, battery box and drawbar. Disconnect starting engine clutch lever, decompression lever and diesel engine clutch operating rods at forward ends. Engage power shaft clutch and remove the operating rod. Unbolt and remove the lever assemblies from the tractor. Disconnect "Powr-Trol" operating rods, breakaway couplings and hoses from housing. Remove cap screws retaining platform at rear and sides. Raise rear of platform enough to clear housing. Support housing assembly with a jack, remove attaching cap screws and roll housing from tractor.

CLUTCH DRUM AND OIL PUMP

162A. To remove the pto clutch drum and oil pump, first remove the basic housing as outlined in paragraph 162. Remove the clutch assembly as in paragraph 157 or 160. Remove cap screws retaining pump housing to drum shaft quill and remove the pump housing and idler gear. Bump the clutch drum and shaft out of basic housing. The drum shaft is supported in an anti-friction bearing (63—Fig. JD1685) which should be renewed if its condition is questionable. The front end of the clutch drum

shaft is supported by bushing (77). Bushing is pre-sized and will not require reaming if carefully installed. Renew the pre-sized bushing (61) if it is worn.

Pump specifications are as follows:
I.D. of pump housing idler
 gear bore 2.121-2.123
O.D. of idler gear 2.115-2.117
Idler gear diametral
 clearance in housing 0.004-0.008
I.D. of idler gear bushing .. 1.002-1.003
O.D. of idler gear shaft .. 0.9994-1.0000
Diametral clearance between
 shaft and bushing 0.0020-0.0036
Depth of idler gear bore .. 1.000-1.005
Thickness of idler gear ... 0.997-0.998
Idler gear end clearance ... 0.002-0.008
 When reassembling, use a thin coat of shellac on mating surfaces of pump housing and drum shaft quill

CONNECTING POWER SHAFT AND COUPLINGS

163. REMOVE AND REINSTALL. The connecting power shaft connects the rear of the bevel driven gear shaft to the forward end of the clutch drum shaft via splined coupling at both ends. The shaft and couplings can be removed after removing the basic housing as outlined in paragraph 162.

BEVEL DRIVEN GEAR ASSEMBLY

164. R&R AND OVERHAUL. To remove the bevel driven gear assembly,

first remove the differential as outlined in paragraph 139 and the transmission oil pump as outlined in paragraph 138A. Working through the rear of the tractor main case, remove the three cap screws retaining the assembly to the main case and remove the bevel gear assembly. Make certain shims (23—Fig. JD1686) are not lost or damaged.

Disassembly procedure is evident after removing snap ring (29) and pressing shaft and bearings out of housing. Transmission oil pump drive pinion (worm) (22) can be removed at this time. Use Fig. JD1686 as a general guide during reassembly and observe the following: Make certain that bearing cone (20) is firmly seated on shaft (19) and that bearing cup (21) is seated in housing. Install snap ring (29) and bump the shaft to be sure snap ring is tight against cone (28). Measure the shaft end play which should be 0.001-0.004. If end play is not as specified, remove adjusting collar lock screw (24) and turn adjusting collar (26) to obtain the proper end play. Install lock screw when adjustment is complete.

Reinstall assembly in main case, using the same number and thickness of shims (23) as were originally removed.

NOTE: If the bevel gear shaft (19) and/or housing are renewed, it may be necessary to adjust the mesh position and backlash of the power shaft bevel gears. Refer to paragraph 169.

DRIVE SHAFT AND GEAR

167. **R&R AND OVERHAUL.** In order to remove the power shaft drive shaft and drive gear unit it is necessary to first remove the diesel engine flywheel as outlined in paragraph 97 and the timing gear housing and cover as outlined in paragraphs 78 and 79. Remove the oil slinger from crankshaft gear. Remove the power shaft idler gear; then unbolt and remove drive gear assembly. Save shim pack (14—Fig. JD1686) for reinstallation.

Procedure for disassembly of the removed unit is evident after an examination of unit and reference to Fig. JD1686. Oil seal (12) prevents engine lubricating oil from entering the transmission compartment and should be renewed whenever the drive shaft unit is disassembled. Use Fig. JD1686 as a general guide during reassembly, and vary number of shim washers (9) to provide the recommended shaft end play of 0.001-0.005. Tighten nut (6) to the torque of 208 Ft.-Lbs. When reinstalling the unit, use the original shim pack, but make certain there is at least one paper shim in the pack.

NOTE: If the power shaft drive shaft and bevel gear and/or housing are renewed, it may be necessary to adjust the mesh position and backlash of the power shaft bevel gears. Refer to paragraph 169.

After timing gear housing is installed, position the crankshaft oil slinger as outlined in paragraph 82A.

BEVEL GEARS

169. **ADJUST.** Bevel gears (18 and 19—Fig. JD1686) should be meshed so that heels are in register and backlash is 0.006-0.008 as measured at bevel gears. The desired mesh position and backlash of the bevel gears is obtained by using the proper combination of shims (14 & 23) between respective housings and main case. There must be a paper shim in pack (14).

"POWR-TROL" (HYDRAULIC LIFT)

170. The hydraulic power lift system is composed of three basic units: The pump unit which is engine driven by the camshaft gear; the valve unit which is mounted on the rear face of the basic housing; and the remote control cylinder. Tractors equipped with a single remote cylinder valve housing will accommodate either a single acting or a double acting remote cylinder; whereas, tractors equipped with a dual remote cylinder valve housing will accommodate only double acting cylinders.

For operation of a single acting cylinder, remove the cap nut and slotted by-pass screw as shown in Fig. JD1690, turn the by-pass screw into the cap nut as far as it will go, then reinstall the screw and cap nut in the valve housing.

For operation of a double acting cylinder, remove the cap nut, turn the slotted by-pass screw into the valve housing until it seats, then reinstall and tighten the cap nut.

NOTE: The maintenance of absolute cleanliness of all parts is of utmost importance in the operation and servicing of the hydraulic system. Of equal importance is the avoidance of nicks or burrs on any of the working parts.

Fig. JD1690 — Single remote cylinder valve housings can be adjusted to accept a single-action cylinder. See text.

LUBRICATION AND BLEEDING

171. It is recommended that the "Powr-Trol" working fluid be changed twice a year. After the system is completely drained, refill the reservoir, operate the system several times to bleed out any trapped air and refill the reservoir to the full mark on dip stick.

NOTE: If other than John Deere cylinders are used, always retract the piston before checking the fluid level.

Capacity of 820 and 830 system is 12 U. S. quarts plus 1¾ quarts for each remote cylinder. Capacity of the 80 system is 10 U. S. quarts plus 1¾ quarts for each remote cylinder.

For temperatures above 90 degrees F., use SAE 30 single viscosity oil or SAE 10W-30 multi-viscosity oil; for 32 degrees F. to 90 degreees F., use SAE 20-20W single viscosity oil or SAE 10W-30 multi-viscosity oil; for temperatures below 32 degrees F., use SAE 10W single viscosity oil or SAE 5W-20 multi-viscosity oil.

TROUBLE-SHOOTING

172. The following paragraphs should facilitate locating troubles in the hydraulic system.

172A. CYLINDER WILL NOT LIFT LOAD OR WILL NOT FUNCTION WHEN NOT LOADED. Could be caused by:

1. Air in remote cylinder: Refer to paragraph 171.
2. Pump disengaged.
3. Pump scored or drive shaft sheared: Refer to paragraph 177.
4. Remote cylinder overloaded.
5. Relief valve pressure too low: Refer to paragraph 173 or 174.
6. Oil line or oil line packing failed.
7. Porous valve housing: Refer to paragraph 179 or 181.
8. Failed or missing control shaft Woodruff key: Refer to paragraph 179 or 181.
9. Missing check valve inner ball on return side: Refer to paragraph 179 or 181.
10. Single valve housing set for single-acting cylinder operation or hoses crossed.
11. Defective oil line coupler or coupler not completely engaged.
12. Missing thermal relief valve: Refer to paragraph 181.
13. Faulty gasket between valve housing and basic housing.

172B. REMOTE CYLINDER WILL NOT LOWER. Could be caused by:

1. Air in remote cylinder: Refer to paragraph 171.
2. Check valve inner ball missing: Refer to paragraph 179 or 181.
3. Failed or missing control shaft Woodruff key: Refer to paragraph 179 or 181.
4. Defective oil line coupler or coupler not completely engaged.

172C. CONTROL LEVER WILL NOT RETURN TO NEUTRAL. Could be caused by:

1. Failed control valve centering spring: Refer to paragraph 179 or 181.
2. Secondary relief valve balls missing: Refer to paragraph 181

3. Porous valve housing between relief valve and detent passages: Refer to paragraph 179 or 181.
4. Failed or disengaged control valve spring retaining snap ring: Refer to paragraph 179 or 181.
5. Control valve sticking: Refer to paragraph 179 or 181.
6. Detent sticking: Refer to paragraph 179 or 181.
7. Single valve housing set for single-acting cylinder.
8. By-pass screw not seated in housing.

172D. CONTROL LEVER WILL NOT LATCH IN FAST OPERATING POSITION. Could be caused by:

1. Overloaded remote cylinder.
2. Relief valve pressure too low: Refer to paragraph 173 or 174.
3. Detent stuck, failed detent or failed detent spring: Refer to paragraph 179 or 181.
4. Defective oil line coupler or coupler not completely engaged.

172E. REMOTE CYLINDER SETTLED UNDER LOAD. Could be caused by:

1. Dirt or metal particles under check valve: Refer to paragraph 179 or 181.
2. Oil line adapter packing failed: Refer to paragraph 179 or 181.
3. By-pass screw open while using double-acting cylinder.
4. Outer check valve missing: Refer to paragraph 179 or 181.
5. Thermal relief valve leaking: Refer to paragraph 181.
6. Pressure relief valve leaking: Refer to paragraph 173 or 174.
7. Remote cylinder rod seal leaking: Refer to paragraph 182.
8. Remote cylinder piston ring failure: Refer to paragraph 182.
9. Remote cylinder casting failure: Refer to paragraph 182.

10. Check valve outer ball leaking or missing: Refer to paragraph 179 or 181.

172F. OIL OVERHEATING. Could be caused by:

1. Control lever being held in engaged position after remote cylinder reaches end of stroke.
2. Low oil supply.
3. Relief valve pressure too high: Refer to paragraph 173 or 174.

172G. PUMP NOISY. Could be caused by:

1. Oil viscosity too high: Refer to paragraph 171.
2. Low oil supply.
3. Air leak in oil return line.
4. Pump drive shaft oil seal leaking: Refer to paragraph 177.

OPERATING PRESSURE AND RELIEF VALVE

(Single Remote Cylinder)

173. To check and adjust the relief valve opening pressure, mount a pressure gage of sufficient capacity and shut off valve as shown in Fig. JD 1691. With the shut off valve open, engage the hydraulic pump and start engine. Move the "Powr-Trol" operating lever to the fast extend position, allowing the working fluid to circulate through the tube and shut off valve and back to the valve housing. Close the valve slowly, and note the pressure reading as the control lever returns to neutral. If pressure is not within the 1230-1300 psi limits, raise rear of platform as outlined in paragraph 175, remove plug as shown in Fig. JD1692 and add or deduct washers as required. Each washer represents approximately 40 psi.

Fig. JD1691 — Checking operating pressure on "Powr-Trol" system. The shut-off valve must be located between the pressure gage and the valve housing on the return line.

Fig. JD1692 — Location of relief valve adjusting washers in single remote cylinder valve housings.

NOTE: If correct specified pressure cannot be obtained by adding washers, look for a failed or badly worn pump unit. Refer to paragraph 176.

NOTE: If correct specified pressure cannot be obtained by adding washers, look for a failed or badly worn pump unit. Refer to paragraph 176.

(Dual Remote Cylinder)

174. The housing contains two circuits with one relief valve for each circuit. To check and adjust a relief valve opening pressure for the No. 1 circuit, mount a pressure gage of sufficient capacity and shut off valve as shown in Fig. JD1691. With the shut off valve open, engage the hydraulic pump and start engine. Move the "Powr-Trol" operating lever to the fast extend position, allowing the working fluid to circulate through the tube and shut off valve and back to the valve housing. Close the valve slowly, and note the pressure reading as the control lever returns to neutral. The No. 1 circuit should have an opening pressure of 1430-1500 psi. If pressure is not as specified, raise rear of platform as outlined in paragraph 175, remove No. 1 cover (Fig. JD1692A) and add or deduct washers as required. Each washer represents approximately 40 psi.

To check the No. 2 circuit relief valve opening pressure of 1230-1300 psi, follow the same procedure as for the No. 1 circuit except:

A. Connect the shut off valve and gage test fixture to openings (O—Fig. JD1692A).

B. Make certain that the No. 1 circuit control lever is in neutral during the test.

C. Relief valve adjusting washers are accessible after removing the No. 2 cover. Refer also to Fig. JD1693.

RAISE PLATFORM

175. Remove seat, battery and battery box. Disconnect starting engine clutch lever, decompression lever and diesel engine clutch operating rods at forward ends. Engage the power shaft clutch and remove the operating rod. Unbolt and remove the lever assemblies from the tractor. Disconnect "Powr-Trol" operating rods, breakaway couplings and hoses from housing. Remove cap screws retaining platform at rear and sides, then raise rear of platform to clear.

PUMP

176. **REMOVE AND REINSTALL.** To remove the gear type hydraulic pump, first unbolt oil tube retainer and pull tubes from pump. Unbolt and remove pump from timing gear housing cover. Make certain that gaskets between pump and timing gear housing cover are not lost or damaged as they control backlash of the pump drive gear and camshaft gear.

Install pump in reverse of the removal procedure, installing same number of gaskets (18—Fig. JD1695) as were removed. Tighten the pump retaining bolts to a torque of 35 Ft.-Lbs. If the drive gears and/or housing (6) were renewed, remove expansion plug (Fig. JD1694) and check the backlash between the pump drive gear and the camshaft gear. Recommended backlash of 0.002-0.006 can be obtained by adding or deducting gaskets (18—Fig. JD1695).

Fig. JD1692A—Dual remote cylinder valve housing installation.

Fig. JD1693—Relief valve adjusting washers in dual remote cylinder valve housing are accessible after removing the cover as shown. The No. 1 circuit has a relief valve opening pressure of 1430-1500 psi, while the No. 2 circuit opening pressure is 1230-1300 psi.

Fig. JD1694 — "Powr-Trol" (hydraulic) pump installed showing the expansion plug removed to check backlash of gears.

177. OVERHAUL. With the pump unit removed, drive dowel pin (33B—Fig. JD1696) out and remove cap screws. Separate pump cover (17—Fig. JD1695), body (16) and housing (6), then remove pump gears (13 & (14) and Woodruff keys (10) from shafts (11 & 12). Remove idler shaft (12) and drive shaft (11). Note: Pumps on tractors prior to serial 8001879 must have snap ring (24) removed before drive shaft can be removed. Driving gear (26), shifter collar (22), spring (21) and washer (20) can now be removed. Remove cap screw (1) after which shifter yoke (29) and spring (32) can be removed.

Check all parts against the values which follow and renew any which are excessively worn.

ID of pump body (16)...2.452-2.454
OD of pump gears
  (13 & 14) 2.4485-2.4495
Radial clearance between
  pump body and
  gears 0.00125-0.00275
Thickness of pump body
  (16) 1.0625-1.0635
Thickness of pump gears
  (13 & 14) 1.0594-1.0600
End clearance between
  pump cover and
  gears 0.0025-0.0041
ID of pump gear (14)
  bore 1.0005-1.0011
OD of idler shaft (12)..0.9994-1.0000
ID of idler shaft
  bushings (7A) 1.0025-1.0035
ID of pump gear (13)
  bore 1.0005-1.0011
OD of drive shaft (11),
  pump gear end.....0.9994-1.0000
ID of drive shaft
  bushings (7B) 1.0025-1.0035
OD of drive shaft (11),
  drive gear end.....0.6865-0.6875
ID of drive shaft bushings
  (27 & 28) 0.6895-0.6905

Bushings (7A, 7B, 27 & 28) are pre-sized and if carefully installed will need no final sizing. Bushings (7A & 7B) should be pressed in housing and cover flush with the inner edge of the bore chamfer. When renewing oil seal (19), press the new seal in until it bottoms in its bore, then pack the space between the lips of the seal with gun grease.

Pre-lubricate all parts, then using Fig. JD1695 as a general guide, reassemble the pump. Make certain no parts are damaged prior to or during reassembly. Thrust washer (23), used on pumps after tractor serial number 8001879, should be installed with the shoulder toward the shifter collar and aligned with the flat section of splines on shaft (11). Tighten the cap screws evenly to 32 ft.-lbs. of torque, then make certain that the pump turns freely with no binding.

SINGLE REMOTE CYLINDER
VALVE HOUSING

178. REMOVE AND REINSTALL. To remove the valve housing, raise rear of platform as outlined in paragraph 175; drain oil from hydraulic system, then unbolt and remove the valve housing.

Reinstallation is reverse of removal procedure; tighten the small cap screws and stud nuts to a torque of

Fig. JD1696 — "Powr-Trol" (hydraulic) pump unit removed.

56 Ft.-Lbs. Tighten the large cap screw to a torque of 150 Ft.-Lbs.

179. OVERHAUL. To overhaul the removed unit, refer to Fig. JD1698 and proceed as follows: Remove the oil line adaptor, by-pass screw (27) and the detent assembly (29, 30, 31 & 32). Remove the relief valve assembly (34, 35, 36, 37, 38 & 39) making certain none of the adjusting washers (37) are lost or damaged. Remove the two check valve retainers (25), then remove both check valve assemblies (21, 22, 23, 24 & 26) making certain all valve parts are removed and none are lost. Remove cover (1), then loosen jam nut (4) and set screw (3). Control

Fig. JD1695 — Exploded view of "Powr-Trol" hydraulic) pump. Gaskets (18) are used to provide the recommended backlash of 0.002-0.006 between pump drive gear (26) and camshaft gear. Washer (25) and snap ring (24) are used only on pumps prior to tractor serial No. 8001879.

1. Shifter yoke
   retaining cap screw
2. Gasket
3. Expansion plug
4. Roll pin
5. Shifter handle
6. Pump housing
7A. Idler shaft bushing
7B. Drive shaft bushing
8. Dowel pin
9. Thrust washer
10. Woodruff key
11. Pump drive shaft
12. Idler shaft
13. Pump gear
14. Pump gear
15. Packing
16. Pump body
17. Pump cover
18. Gasket
19. Oil seal
20. Washer
21. Spring
22. Shifter collar

23. Thrust washer
    (after serial
    number 8001878)
24. Snap ring
    (prior to serial
    number 8001879)
25. Washer
    (prior to serial
    number 8001879)

26. Pump drive gear
27. Bushing
28. Bushing
29. Shifter yoke
30. "O" ring
31. Shifter shaft
32. Spring
33A. Dowel pin
33B. Dowel pin

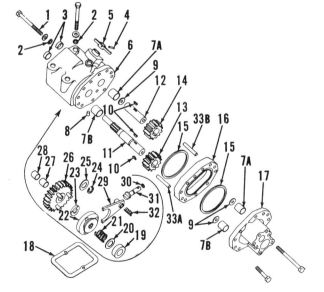

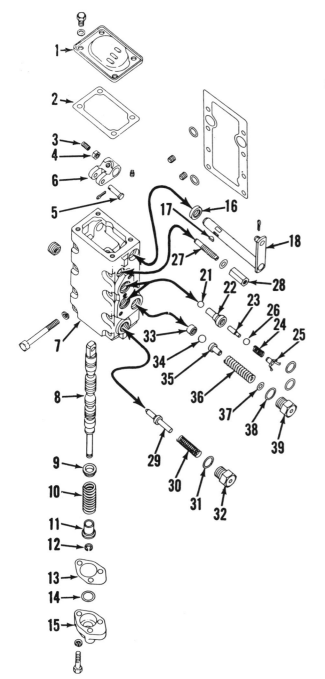

Fig. JD1698 — Exploded view of single remote cylinder valve housing.

1. Cover
2. Gasket
3. Set screw
4. Jam nut
5. Pin
6. Control valve arm
7. Valve housing
8. Control valve
9. Upper valve spring retainer
10. Control valve spring
11. Lower valve spring retainer
12. Snap ring
13. Gasket
14. Washers
15. Control valve cover
16. Oil seal
17. Woodruff key
18. Control shaft
21. Check valve inner ball (two used)
22. Outer check valve (two used)
23. Check valve metering shaft (two used)
24. Spring (two used)
25. Retainer (two used)
26. Check valve outer ball (two used)
27. By-pass screw
28. Cap nut
29. Control valve detent
30. Detent spring
31. Washer
32. Plug
33. Relief valve seat
34. Relief valve ball
35. Relief valve spring guide
36. Relief valve spring
37. Adjusting washer
38. Washer
39. Plug

shaft (18) can then be moved far enough to remove Woodruff key (17), after which shaft can be removed from valve housing. Unbolt and remove control valve cover (15) making certain washers (14) are not lost or damaged. Move control valve arm to position shown in Fig. JD1702 and remove pin. The control valve assembly can now be removed from housing.

Inspect for cracks, porous conditions, excessive wear, missing or failed parts, dirt or metal particles and valves and seats which do not seat properly.

The outer check valves (22) can be lapped, using fine lapping compound. If a valve seat (22 & 23) or steel ball (21, 26 & 34) is renewed, the ball can be seated in the valve seat by tapping it lightly against the seat. If oil seal (16) is renewed, install with lip facing inward.

All parts should be lubricated before reassembly. Reinstall control valve assembly, installing pin (5) through control valve and control valve arm. Slide control shaft (18) through oil seal (16) and position Woodruff key, then move the control shaft into the control arm and secure with set screw (3) and jam nut (4). Install control valve cover (15) using the same number of washers (14) as were removed. Check free end play of control valve with dial indicator as shown in Fig. JD1703 and add or deduct washers (14—Fig. JD1698) until the recommended free end play of 0.010 is obtained without compressing the neutralizing spring, then install cover (1). Install both check valve assemblies (21, 22, 23, 24 & 26) and be sure spring retainers (25) are secure. Install detent assembly (29, 30, 31 & 32) and by-pass screw (27). Install relief valve assembly (33, 34, 35, 36, 37, 38 & 39) using same number of adjusting washers (37) as were removed. After unit is reinstalled on tractor, check relief valve opening pressure as outlined in paragraph 173.

Fig. JD1702 — Single valve housing showing position of control valve arm when removing pin.

Fig. JD1703 — Single valve housing control valve end play should be adjusted to 0.010 by adding or deducting washers (14—Fig. JD1698).

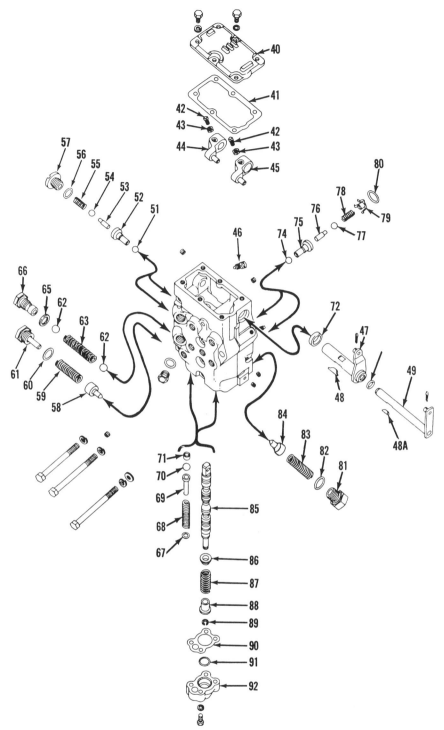

40. Cover
41. Gasket
42. Set screw
43. Jam nut
44. No. 2 circuit
control valve
arm
45. No.1 circuit
control valve
arm
46. Thermal relief
valve
47. No. 1 circuit
control shaft
48. Woodruff key
49. No. 2 circuit
control shaft
51. No. 2 circuit
check valve
inner ball
(two used)
52. No. 2 circuit
outer check
valve
(two used)
53. No. 2 circuit
check valve
metering shaft
(two used)
54. No. 2 circuit
check valve
outer ball
(two used)
55. Spring
(two used)
56. "O" ring
(two used)
57. Plug
(two used)
58. No. 2 circuit
control valve
detent
59. No. 2 circuit
detent spring
60. Washer
61. Plug
62. Secondary relief
valve ball
63. Secondary relief
valve spring
65. Gasket
66. Secondary relief
valve seat

67. Primary relief
valve adjusting
washers
68. Primary relief
valve spring
(two used)
69. Primary relief
valve spring guide
(two used)
70. Primary relief
valve ball
(two used)
71. Primary relief
valve seat
(two used)
72. Oil seal
74. No. 1 circuit check
valve inner ball
(two used)
75. No. 1 circuit outer
check valve
(two used)
76. No. 1 circuit check
valve metering shaft
(two used)
77. No. 1 circuit check
valve outer ball
(two used)
78. No. 1 circuit check
valve spring
(two used)
79. No. 1 circuit check
valve retainer
(two used)
80. "O" ring (two used)
81. Plug
82. Washer
83. No. 1 circuit
detent spring
84. No. 1 circuit control
valve detent
85. Control valve
(two used)
86. Upper valve spring
retainer (two used)
87. Control valve spring
(two used)
88. Lower valve spring
retainer (two used)
89. Snap ring (two used)
90. Gasket (two used)
91. Washer (two used)
92. Control valve cover
(two used)

secondary relief valves assembly (62, 63 & 65). NOTE: Make certain none of the parts from any of these assemblies are mixed, lost or damaged. Remove cover (40), loosen nuts (43) and set screws (42), then slide control arm (47) enough to remove key (48). Withdraw both control shafts and control arms. Remove Woodruff key (48A) and separate the control arms. Unbolt and remove both control valve covers (92), then withdraw both primary relief valve assemblies (67, 68, 69 & 70). NOTE: Washers (67 & 91) must not be lost, damaged or mixed with washers from the other circuit. Withdraw control valves from bottom of housing.

Inspect for cracks, porous conditions, excessive wear, missing or failed parts, dirt or metal particles and valves which do not seat properly.

The seat on outer check valves (52 & 75) can be lapped, using fine lapping compound. If ball seating sur-

DUAL REMOTE CYLINDER VALVE HOUSING

180. **REMOVE AND REINSTALL.** To remove the valve housing, raise rear of platform as outlined in paragraph 175; drain oil from hydraulic system, then unbolt and remove the valve housing.

Reinstall in reverse of removal procedure, tightening the small cap screws and stud nuts to a torque of 56 Ft.-Lbs. Tighten the large cap screw to a torque of 150 Ft.-Lbs.

181. **OVERHAUL.** To overhaul the removed unit, refer to Fig. JD1705 and proceed as follows: Remove retainers (79) and withdraw both of the No. 1 circuit check valve assemblies (74, 75, 76, 77 & 78). Unscrew plugs (57) and extract both of the No. 2 circuit check valve assemblies (51, 52, 53, 54, 55 & 56). Remove plug (81) and the No. 1 circuit detent assembly (82, 83 & 84). Remove plug (61) and the No. 2 circuit detent assembly (58, 59 & 60). Unscrew plug (66) and remove the

faces on parts (52, 71 & 75) or if steel balls (51, 54, 70, 74 & 77) are renewed, the ball can be seated by tapping it lightly against its seat with a brass drift and hammer. If oil seal (72) is renewed, install with lip facing inward. Both thermal relief valves (46) should be renewed if their condition is questionable.

All parts should be lubricated with "Powr-Trol" oil before reassembly. Reinstall both control valve assemblies (85, 86, 87, 88 & 89) and both primary relief valves (68, 69 & 70). Install the same number of washers (67 & 91) and install control valve covers (92). NOTE: Make certain none of the washers (67 and/or 91) are lost, damaged or mixed with those of the other circuit. Reinstall control shafts and control arms (44 & 45), then tighten set screws (42) and jam nuts (43). Check free end play of each control valve with a dial indicator in a manner similar to that shown in Fig. JD-1703 and add or deduct washers (91—Fig. JD1705) until the recommended free end play of 0.010 is obtained without compressing the neutralizing springs, then install cover (40). Reinstall all valve and detent assemblies in the sequence shown in Fig. JD1705. After unit is reinstalled on tractor, check each relief valve opening pressure as outlined in paragraph 174.

REMOTE CONTROL CYLINDER

182. DISASSEMBLY AND OVERHAUL. Remove oil lines and end cap (14—Fig. JD1716 and JD1717). Remove the piston retaining nut taking care not to damage any parts. Withdraw piston rod and yoke from cylinder, then remove piston rod stop (2) and adjusting rods (3). Seal retainer (4) and "V" seal assembly used on remote cylinder serial number 4RC-55637 or piston rod wiper (15), washer (16) and "O" ring (17) used on later cylinders can be removed.

Inspect face of end cap for nicks or burrs, adjusting rods for being bent and piston rod for burrs, scratches and/or being bent. Bent adjusting rods will be O.K., providing they can be thoroughly straightened. Small burrs and/or scratches can be removed from the piston rod by using a fine stone; however, the piston rod should be renewed if it is bent. Renew any other questionable parts.

On early remote cylinders (Fig. JD-1716) lubricate the piston rod and reassemble the cylinder, leaving out the "U" cup packings (10). Attach a spring scale as shown in Fig. JD-

1718 and check the pull required to move the lubricated piston rod through the "V" seal assembly. Add or deduct shims (5) until a pull of approximately 4 pounds is required. When adjustment is as specified, install the "U" cup packings (10) and tighten the piston retaining nut. Using a new gasket, install the end cap and tighten

the cap screws to a torque of 125 Ft.-Lbs.

On remote cylinders after cylinder serial number 4RC55636 (Fig. JD1717), lubricate all parts and using Fig. JD-1717 as a general guide during reassembly, tighten piston retaining nut and end cap retaining cap screws securely.

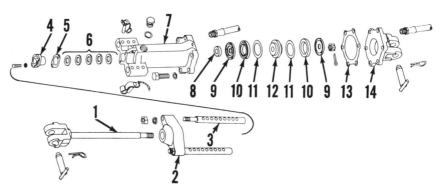

Fig. JD1716 — Exploded view of John Deere remote control cylinder prior to cylinder serial number 4RC55637.

1. Piston rod and yoke	4. Seal retainer	7. Cylinder	11. Paper washer
2. Piston rod stop	5. Shims	8. Spacer	12. Piston
3. Adjusting rod	6. "V" seal assembly	9. Retainer	13. Gasket
		10. "U" cup packing	14. End cap

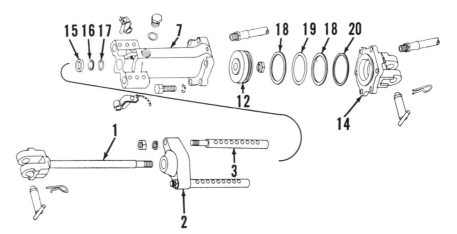

Fig. JD1717—Exploded view of John Deere remote cylinder after cylinder serial number 4RC55636.

1. Piston rod and yoke	3. Adjusting rod	14. End cap	17. & 19. "O" ring
2. Piston rod stop	7. Cylinder	15. Piston rod wiper	18. Paper washer
	12. Piston	16. Washer	20. Packing

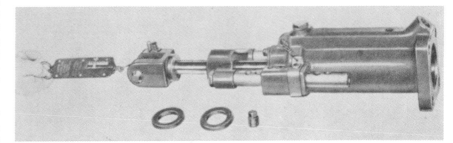

Fig. JD1718 — Checking pull required to move lubricated piston rod through "V" seal assembly on remote cylinders prior to cylinder serial number 4RC55637.

NOTES

Technical Information

Technical information is available from John Deere. Some of this information is available in electronic as well as printed form. Order from your John Deere dealer or call **1-800-522-7448**. Please have available the model number, serial number, and name of the product.

Available information includes:

- PARTS CATALOGS list service parts available for your machine with exploded view illustrations to help you identify the correct parts. It is also useful in assembling and disassembling.
- OPERATOR'S MANUALS providing safety, operating, maintenance, and service information. These manuals and safety signs on your machine may also be available in other languages.
- OPERATOR'S VIDEO TAPES showing highlights of safety, operating, maintenance, and service information. These tapes may be available in multiple languages and formats.

- TECHNICAL MANUALS outlining service information for your machine. Included are specifications, illustrated assembly and disassembly procedures, hydraulic oil flow diagrams, and wiring diagrams. Some products have separate manuals for repair and diagnostic information. Some components, such as engines, are available in separate component technical manuals
- FUNDAMENTAL MANUALS detailing basic information regardless of manufacturer:

  - Agricultural Primer series covers technology in farming and ranching, featuring subjects like computers, the Internet, and precision farming.
  - Farm Business Management series examines "real-world" problems and offers practical solutions in the areas of marketing, financing, equipment selection, and compliance.
  - Fundamentals of Services manuals show you how to repair and maintain off-road equipment.
  - Fundamentals of Machine Operation manuals explain machine capacities and adjustments, how to improve machine performance, and how to eliminate unnecessary field operations.